Metamaterials in Topological Acoustics

As an equivalent counterpart of topological research on photonics and condensed matter physics, acoustic metamaterials create an opportunity to explore the topological behaviors in phononics and physics of programmable acoustics. This book introduces the topological behavior of acoustics through the novel design of metamaterials. It provides valuable insight into acoustic metamaterials, from multidisciplinary fundamentals to cutting-edge research.

- Serves as a single resource on acoustic metamaterials
- Covers the fundamentals of classical mechanics, quantum mechanics, and state-of-the-art condensed matter physics principles so that topological acoustics can be easily understood by engineers
- Introduces topological behaviors with acoustics and elastic waves through quantum analogue Hall effects, quantum spin Hall effects, and quantum valley Hall effects and their applications
- Explains the pros and cons of different design methods and gives guidelines for selecting specific designs of acoustic metamaterials with specific topological behaviors
- Includes MATLAB® code for numerical analysis of band structures

This book is written for graduate students, researchers, scientists, and professionals across materials, mechanical, civil, and aerospace engineering, and those who want to enhance their understanding and commence research in metamaterials.

Metamaterials in Topological Acoustics

Sourav Banerjee

CRC Press
Taylor & Francis Group
Boca Raton London New York

CRC Press is an imprint of the
Taylor & Francis Group, an **informa** business

Designed cover image: Shutterstock

MATLAB® is a trademark of The MathWorks, Inc. and is used with permission. The MathWorks does not warrant the accuracy of the text or exercises in this book. This book's use or discussion of MATLAB® software or related products does not constitute endorsement or sponsorship by The MathWorks of a particular pedagogical approach or particular use of the MATLAB® software.

First edition published 2024
by CRC Press
2385 Executive Center Drive, Suite 320, Boca Raton, FL 33431

and by CRC Press
4 Park Square, Milton Park, Abingdon, Oxon, OX14 4RN

CRC Press is an imprint of Taylor & Francis Group, LLC

© 2024 Sourav Banerjee

Reasonable efforts have been made to publish reliable data and information, but the author and publisher cannot assume responsibility for the validity of all materials or the consequences of their use. The authors and publishers have attempted to trace the copyright holders of all material reproduced in this publication and apologize to copyright holders if permission to publish in this form has not been obtained. If any copyright material has not been acknowledged please write and let us know so we may rectify in any future reprint.

Except as permitted under U.S. copyright law, no part of this book may be reprinted, reproduced, transmitted, or utilized in any form by any electronic, mechanical, or other means, now known or hereafter invented, including photocopying, microfilming, and recording, or in any information storage or retrieval system, without written permission from the publishers.

For permission to photocopy or use material electronically from this work, access www.copyright.com or contact the Copyright Clearance Center, Inc. (CCC), 222 Rosewood Drive, Danvers, MA 01923, 978-750-8400. For works that are not available on CCC please contact mpkbookspermissions@tandf.co.uk

Trademark notice: Product or corporate names may be trademarks or registered trademarks and are used only for identification and explanation without intent to infringe.

ISBN: 978-1-032-12083-6 (hbk)
ISBN: 978-1-032-12690-6 (pbk)
ISBN: 978-1-003-22575-1 (ebk)

DOI: 10.1201/9781003225751

Typeset in Times
by KnowledgeWorks Global Ltd.

I would like to dedicate this book to
my daughter
Ms. Aishani Banerjee
and to my wife
Dr. Ritubarna Banerjee.

Contents

Preface.. xv
About the Author xix

Chapter 1   Acoustic Metamaterials.. 1

      1.1   Introduction 1
      1.2   Metamaterials.. 1
      1.3   Periodic Media and Phononic Crystals 2
      1.4   Connecting Metamaterial and Phononic Crystals................ 3
      1.5   Topological Phenomena... 4
      1.6   Future Direction 10
      1.7   Summary 13
      References 13

Chapter 2   Classical Mechanics and the Physics of Continua 15

      2.1   Introduction: History of Classical Mechanics...................... 15
      2.2   Fundamental Concept of Classical Mechanics..................... 16
      2.3   Governing Equation from Classical Mechanics.................... 18
      2.4   Fundamentals of Continuum Mechanics.............................. 20
            2.4.1   Lagrangian Coordinate or Material Coordinate
                  System 21
            2.4.2   Eulerian Coordinate or Spatial Coordinate
                  System 22
      2.5   Motion of a Deformable Body.. 24
            2.5.1   Material Derivatives... 24
            2.5.2   Path Lines and Streamlines...................................... 25
      2.6   Deformation and Strain in a Deformable Body.................... 25
            2.6.1   Cauchy's and Green's Deformation Tensors.............. 27
            2.6.2   Description of Strain in a Deformable Body.............. 27
            2.6.3   Strain in Terms of Displacement.............................. 27
      2.7   Mass, Momentum, and Energy... 27
            2.7.1   Mass of a Body.. 27
            2.7.2   Momentum of a Deformable Body............................ 29
            2.7.3   Angular Momentum of a Deformable Body 30
            2.7.4   Kinetic Energy Stored in a Deformable Body 31
      2.8   Fundamental Axiom of Continuum Mechanics 31
            2.8.1   Axiom 1: Principle of Conservation of Mass............ 31
            2.8.2   Axiom 2: Principle of Balance of Momentum 32
            2.8.3   Axiom 3: Principle of Balance of Angular
                   Momentum 32
            2.8.4   Axiom 4: Principle of Conservation of Energy......... 32

vii

viii Contents

2.9 Internal Stress State in a Deformable Body 33
2.10 External and Internal Load on a Deformable Body 35
2.11 Elastodynamic Equation or the Fundamental Wave
Equation..37
2.12 Energy Concept of Continua and its Relation to the
Classical Mechanics 38
    2.12.1 Conservation of Local Energy from Global
Energy Rule..39
    2.12.2 Conservation of Mechanical Energy (Kinetic,
Internal, and Potential Energy) 40
    2.12.3 Internal Energy and Strain Energy 41
    2.12.4 Energy Flux Density and Poynting Vector...............42
2.13 Lagrangian Formulation of Wave Equation 43
2.14 Legendre Transform and Hamiltonian Formulation of
Wave Equation..44
    2.14.1 Graphical Representation of Legendre Transform.......45
    2.14.2 Hamiltonian Form with Legendre Transform...........46
    2.14.3 Hamilton's Equation of Motion...............................47
2.15 Constitutive Law of Continua...48
    2.15.1 Materials with One Plane of Symmetry:
Monoclinic Materials 51
    2.15.2 Materials with Two Planes of Symmetry:
Orthotropic Materials...52
    2.15.3 Materials with Three Planes of Symmetry
and One Plane of Isotropy: Transversely
Isotropic Materials 52
    2.15.4 Materials with Three Planes and Three Axes of
Symmetry: Isotropic Materials 53
2.16 Appendix 54
    2.16.1 Important Equations in a Cartesian Coordinate
System 54
    2.16.2 Important Equations in a Cylindrical Coordinate
System 56
        2.16.2.1 Transformation to a Cylindrical
Coordinate System....................................56
        2.16.2.2 Gradient Operator in a Cylindrical
Coordinate System....................................57
        2.16.2.3 Strain-Displacement Relation in a
Cylindrical Coordinate System 58
        2.16.2.4 Governing Differential Equations of
Motion in a Cylindrical Coordinate
System..58
    2.16.3 Important Equations in a Spherical Coordinate
System 59
        2.16.3.1 Gradient Operator in a Spherical
Coordinate System....................................59

Contents ix

| | | 2.16.3.2 | Strain-Displacement Relation in a Spherical Coordinate System | 60 |

2.16.3.2 Strain-Displacement Relation in a
Spherical Coordinate System 60
2.16.3.3 Governing Differential Equations of
Motion in a Spherical Coordinate System 60
2.17 Summary 61
References 61

Chapter 3 Acoustics and Elastic Wave Propagation in Fluids
and Anisotropic Solids 63

3.1 Wave Propagation in Fluid Media 63
3.1.1 Governing Differential Equation with Pressure
in Fluid 63
3.1.2 Pressure Velocity Relation 64
3.1.3 Pressure and Velocity Potential in Fluid media 65
3.1.4 Wave Potential in Fluid 67
3.2 Wave Propagation in Solid Media 68
3.2.1 Navier's Equation of Motion 68
3.2.2 Homogeneous Isotropic and Anisotropic
Materials 69
3.2.2.1 Solving Navier's Equation of
Motion in Homogeneous Isotropic
Materials 70
3.2.2.2 Solving Navier's Equation of Motion
in Homogeneous Anisotropic
Materials 82
3.2.2.3 Understanding Wave Modes with
Normal and Anomalous Polarity 89
3.2.2.4 Exploring Abnormal Polarities 98
3.2.3 Nonhomogeneous Isotropic and Anisotropic
Materials 109
3.2.3.1 Wave Equations for Nonhomogeneous
Isotropic Media 109
3.2.3.2 Wave Equations for Nonhomogeneous
Anisotropic Media 114
3.2.3.3 Solution of Wave Equations for
Nonhomogeneous Media 117
3.3 Appendix: Understanding Nabla Hamiltonian
Operations 117
3.3.1 Nabla Hamiltonian 117
3.3.2 Strain Matrix Using Nabla Hamiltonian Form 119
3.3.3 Laplacian Using Nabla Hamiltonian 121
3.3.4 Nabla Operation on Stress Matrix and Spatially
Varying Material Constants 122
3.4 Summary 125
References 125

x   Contents

Chapter 4   Electromagnetic Wave Propagation 127

      4.1   Field and Field Theories 127
      4.2   Electric and Magnetic Fields 129
            4.2.1   Conservative and Nonconservative Fields 130
            4.2.2   Coulomb's Law and Gauss's Law 130
            4.2.3   Ampere's Law and Ampere-Maxwell Equation 131
            4.2.4   Faraday's Induction Law 134
      4.3   Maxwell's Electromagnetic Wave Equation 135
            4.3.1   Solution of Electromagnetic Wave Equations 135
      4.4   Comparison of Electromagnetic and Elastic Acoustic
            Wave Equations 138
      4.5   Appendix 140
            4.5.1   Divergence Theorem 140
            4.5.2   Stokes' Theorem 141
      4.6   Summary 142
      References 142

Chapter 5   Quantum Mechanics for Engineers 143

      5.1   Particle Waves and the Schrödinger Equation 143
            5.1.1   Quantized Energy 143
            5.1.2   Relativistic Particles 143
            5.1.3   Wave Particle Duality 144
                  5.1.3.1   Momentum of a Relativistic Particle 144
                  5.1.3.2   Relativistic Energy in Terms of
                              Momentum 145
            5.1.4   Wave Function 145
             5.1.5   The Schrödinger Equation 147
            5.1.6   Time-Independent Schrödinger Equation 149
            5.1.7   Time-Dependent Schrödinger Equation 154
            5.1.8   Schrödinger Equation for a Particle in an
                      Electromagnetic Field 156
      5.2   Quantum Operators 156
            5.2.1   Hamiltonian Operator 157
            5.2.2   Ladder Operators and Properties 158
            5.2.3   Angular Momentum Operators 160
                  5.2.3.1   Orbital Angular Momentum (OAM) 163
                  5.2.3.2   Spin Angular Momentum (SAM) 165
                  5.2.3.3   Total Angular Momentum (TAM) 166
                  5.2.3.4   Square Orbital Angular Momentum 167
                  5.2.3.5   Square Spin Angular Momentum 168
            5.2.4   Observable Operators 168
                  5.2.4.1   Heisenberg Uncertainty Principle 169
                  5.2.4.2   Uncertainty Principle for Angular
                              Momentum 171

Contents

xi

        5.2.4.3   Convention to Express the Angular Momentum 173

    5.2.5   Pauli's Matrix 175

5.3   Solution of Schrödinger Equation in Periodic Potential 178

    5.3.1   Bloch Solution 180

    5.3.2   Reduced-Order Solution and Brillouin Zone 185

    5.3.3   Finding Energy Bands Using the Perturbation Method 187

        5.3.3.1   Modified Hamiltonian in Periodic Media ... 187

        5.3.3.2   k.p Perturbation Method 187

5.4   Relativistic Particles with Spin Zero and Spin Half 195

    5.4.1   Klein-Gordon Equation for Spin 0 Relativistic Particles 195

    5.4.2   Energy Square Root Parity 196

    5.4.3   Dirac Equation for Spin Half Particles 197

    5.4.4   Hamiltonian for Spin ½ Fermions and Spinors 199

5.5   Dirac Cones and Dirac-Like Cones 202

    5.5.1   Dirac Cones 202

    5.5.2   Dirac-Like Cones 204

5.6   Introduction to Topology and the Geometric Phase 204

    5.6.1   What is Topology? 205

    5.6.2   What Is the Geometric Phase? 207

    5.6.3   How Are They Connected? 208

    5.6.4   The Berry Phase 210

        5.6.4.1   The Berry Phase in Bloch Media 211

        5.6.4.2   The Chern Number at Degenerated Band Structure 214

    5.6.5   The Zak Phase 215

5.7   Understanding Symmetry and Invariance 217

    5.7.1   Geometric Symmetries 218

    5.7.2   Time-Reversal Symmetries or T-symmetry 220

5.8   Connecting Symmetry Breaking and Geometric Phase 223

5.9   Quantum Hall Effects 224

5.10 Summary 225

References 225

Chapter 6   Waves in Periodic Media: Quantum Analogous Application of Acoustics and Elastic Waves 229

6.1   Periodic Media for Acoustic and Elastic Waves 229

    6.1.1   Introduction 229

    6.1.2   Periodicity and Symmetry 229

    6.1.3   Brillouin Zones in Periodic Media 231

6.2   Acoustic Waves in Periodic Media 233

    6.2.1   Governing Differential Equations 233

    6.2.2   Bloch Solution for Acoustic Waves 235

		6.2.2.1	One-Dimension Waves in a Continuous Periodic Chain	235

6.2.2.1 One-Dimension Waves in a Continuous Periodic Chain 235

6.2.2.2 Plane Waves with Out-of-Plane Polarity 239

6.2.2.3 Plane Waves with In-Plane Polarity 243

6.2.2.4 Computer Code in MATLAB to Find Wave Dispersion 248

6.2.2.5 Dispersion Curves 258

6.3 Bloch Wave Vectors for Other Lattice Structures 259

6.3.1 Generalized Bloch Wave Solution 261

6.3.1.1 Fast-Plane Wave Expansion Method 261

6.3.1.2 Finite Element Simulation Method 270

6.4 Features of Wave Dispersion $(\omega - k)$ 270

6.4.1 Phonons 272

6.4.2 Equifrequency Contours 272

6.4.3 Band Degeneracies 273

6.4.4 Deaf Bands 275

6.4.5 Dirac Cones at K Point 276

6.4.6 Dirac-Like Cones 277

6.4.7 Weyl Point 278

6.4.8 Spawning Rings at Exceptional Points 279

6.4.9 Double Dirac Cones and Spinors 281

6.4.10 Topological Charge and Invariant 283

6.5 Examples: Counterintuitive Non-Topological Wave Phenomena 284

6.5.1 Acoustic Transparency, Beam Splitting, Negative Refraction, and Super Lensing 284

6.5.1.1 Butterfly Crystal Dispersion 285

6.5.1.2 Wave Bifurcation 288

6.5.1.3 Negative Refraction and Wave Focusing 289

6.5.1.4 Superlensing: Beyond the Diffraction Limit 291

6.5.2 Orthogonal Wave Transport at Dirac-like Cone 292

6.5.3 Acoustic Computing at Dirac-like Cone 297

6.6 Active Breaking of Time-Reversal Symmetry 300

6.6.1 Topological Band Gaps 300

6.6.1.1 Spatio-Temporal Modulation of Material Coefficients 301

6.6.1.2 Governing Differential Equations 302

6.6.1.3 Dispersion Bands with Directional Bandgaps 303

6.6.1.4 How Directional Band Gaps Are Topological 304

6.6.1.5 Manifold with Parallel Transport of the Wave Function 306

Contents   xiii

6.7 Quantum Analogous Elastic Waves 307
   6.7.1 Hamiltonian and Ladder Operation for Elastic Waves 308
     6.7.1.1 Elastic Hamiltonian 308
     6.7.1.2 Super Symmetry (SUSY) Ladder Operations 310
   6.7.2 Klein-Gordon Equation and Dirac Equation 314
     6.7.2.1 Elastic Klein-Gordon Equation 315
     6.7.2.2 Elastic Dirac Equation 315
     6.7.2.3 Pseudospin State of Elastic Wave Modes 318
     6.7.2.4 Spring-mass System for Topological Elastic Waves 321
   6.7.3 Intrinsic Spin States of Elastic Wave 325
     6.7.3.1 Elastic Spin Angular Momentum 325
     6.7.3.2 Elastic Spin Operators 328
     6.7.3.3 Topological Behavior with Wave Vortex 330
     6.7.3.4 Power Flow with Spin Angular Momentum 331
     6.7.3.5 Elastic and Acoustic Spin Mediated Skyrmion 331
6.8 Appendix 332
   6.8.1 Circular Phononic Crystal in a Host Matrix 332
   6.8.2 Square Phononic Crystal in a Host Matrix 335
   6.8.3 Parallel Transport of a Vector 338
     6.8.3.1 Christoffel Symbol and Geodesic 339
     6.8.3.2 Parallel Transport along a Curved Path 341
     6.8.3.3 Example: Parallel Transport 343
     6.8.3.4 Notes on Parallel Transport for Wave Vectors 344
   6.8.4 Computer Code to Explore Elastic Spin 344
6.9 Summary 348
References 349

Chapter 7 Topological Acoustics in Metamaterials 353

7.1 Topology 353
   7.1.1 Topology in Crystals 354
   7.1.2 Topology in Phononics 359
     7.1.2.1 Breaking $\mathcal{T}$-symmetry: QHE 359
     7.1.2.2 Without Breaking $\mathcal{T}$-symmetry: QSHE 361
     7.1.2.3 Without Breaking $\mathcal{T}$-symmetry: QVHE 366
7.2 Topological Phononics and Quantum Trio: A Gateway of Quantum Transportation 369

		7.2.1	Berry Phase and Band Topology in Phononics	372
		7.2.2	Phononic Topological States	372
			7.2.2.1 Topological Insulators in Acoustics	375
			7.2.2.2 Acoustic Spinning and Topological Edge State in Acoustics	375
	7.3	Emergence of Topological Black Holes		375
		7.3.1	Modal Analysis at Dirac Frequency	378
		7.3.2	Tuning of Geometric Parameters for TBHs	382
		7.3.3	Real-time Tunable Metamaterial for TBHs	383
	7.4	Polarization Anomaly: Demonstration of Acoustic Spin		388
		7.4.1	Formation of Black Hole-Like Sink	388
		7.4.2	Counterinteractive Spin and OAM in TBH	389
			7.4.2.1 Mathematical Treatment to Find Polarity Anomaly	391
			7.4.2.2 A Possible Link to the Acoustic Skyrmions	394
	7.5	Summary		394
	References			395

Index 401

Preface

Admiring the intricate artwork of historical structures, staring at never-ending geometric patterns while getting lost, I always wondered if they had any greater purpose – if those aesthetically pleasing ideas that simply came from within may be manifestation of harmony over chaos. It will be hard to get an answer. But 21st-century physics may give an opportunity to partially quench my childhood curiosity.

The obsession of human civilizations with geometric patterns is undeniable. Periodic geometric patterns and their repetition throughout history created numerous intricate artworks. Visualizing and rejuvenating these periodic structures from a different perspective, while understanding the physics of acoustics, is the subject of this book. Although these patterns were in plain sight, they were never realized as a tool for discovering new physics before the 21st century. Physics and tools that are required to explore unknown phenomena, such as their interaction with electromagnetic, acoustic, and elastic waves, were only discovered in the 20th century. We started to call these patterned materials 'metamaterials'. Materials and structures that may not exist naturally but were created to explore functionalities of beyond are called metamaterials. With a deeper look into nature, at a scale that is smaller than our eye can see, we found that our nature has created these metamaterials in different spaces, in fact for a purpose. Biomimetics has given us endless opportunities to study such articulated metamaterials. Here is this book, however, only the acoustic physics of these metamaterials will be discussed.

Metamaterial has become a serious topic of research since 1990s. First it was realized theoretically. An artificially designed metamaterial for electrodynamic application was proposed in 1990. A material that can create double negative material properties was realized. It took another 10 years to experimentally demonstrate the functionality of that metamaterial in 2000. Further researchers from multiple disciplines argued that if metamaterials are designed for electrodynamics, then why can they not be designed for elastodynamics or acoustics? All are waves within their respective physics! Acoustic metamaterials are the result of that curiosity. Materials with negative mass density and elastic modulus were first to be realized. Further, beyond these properties, numerous other properties also came into reality that help us recognize that the topological phenomena exist in acoustic metamaterials. This book presents the journey of the past 20 years of exploration, evolution of ideas, and the story of unravelling the physics of acoustics in this new class of materials. Topological understanding of acoustics and elastodynamics in metamaterials came at a much later date, only after 2010.

Topological phenomena are quantum related. Quantum mechanics, its operators, and their unique mathematical treatments were used for explaining the topological behavior. At first this came very naturally to the physicist who was working with the electromagnetic waves. Physicists are fundamentally trained with the understanding of quantum mechanics. Further physicists and material scientists also worked with acoustic and elastic waves to explain their topological behavior, adapting the understanding from quantum mechanics that they naturally have. With close observation

xv

of our society, it could be realized that the application of acoustic waves and elastic waves for practical use is mostly dominated by engineers. Engineers on the other hand are naturally trained with mechanics of solids, mechanics of fluids, thermodynamics, manufacturing, and many other interesting transdisciplinary skills. But not quantum mechanics!

In my perspective, this is soon going to change. Application of metamaterials and their implementation will require engineers to design them in the most innovative and unique ways possible. Engineers have already come up with metamaterials for energy harvesting, metamaterials for nondestructive evaluation, metamaterials for vibration isolation and impact mitigation, etc., to name a few. These applications were possible even without any active knowledge of topological physics and understanding of geometric phases. Knowing the topological behavior of metamaterials under different exposure to the different types of waves will soon be necessary for engineers. Being engineers, when we first started researching this subject, I and my students stumbled on a road block that I felt was like climbing Mt. Everest. It was quantum mechanics. At first, we could not find a single book on topological acoustics that we needed as a guide for us, a guide for an engineer, not for a physicist. There are many books on quantum mechanics and even quantum mechanics for engineers; however, unfortunately those did not resonate with my mechanical engineering background and perspectives. We needed multiple books to comprehend a single small mathematical equation, because none were targeted to the topological acoustic behavior to be specific. It was a vast ocean with no direction. So much to comprehend, and we had no idea how to optimize our effort quickly.

It's known as a criticism or praise that an engineer's mind may work totally differently from others. They are trained to be problem solvers. To solve problems, they need tools. Optimizing learning and utilizing the tools as quickly as possible in the most effective way is a skillset that engineers practice almost every day. This fundamental barrier as an engineer and the vast ocean that my students and I swam, made me realize that we need a guidebook on topological acoustics for the new students and new researchers who will potentially come to contribute to this field. This book is a testimony of our journey to find the most optimized and effective ways to quickly contribute to the field while learning the fundamentals of quantum mechanics, its mathematical tools, and its relation to topological behaviors. I feel that the book is very timely because since 2016 there has been an explosion in the number of publications on this subject, mostly by physicists, only a few by engineers. I found that the growth originated from the physicists but has propagated to the field of engineering, and engineers are propagating it further but have stumbled on the road block as we did. More and more engineering researchers have found metamaterials useful for numerous novel applications that otherwise were not possible before. Knowing the physics and designing new materials for acoustic insulation, acoustic and elastic wave guiding for underwater applications, acoustic computing, acoustic delay lines, acoustic topological insulators, conductors, etc., are a few areas to which topological acoustics will contribute if properly utilized.

To expedite the research activities productively, I feel this book will help better prepare engineering students and researchers. Those who are new to this field will find this book immensely valuable, providing them with all the necessary

Preface

xvii

fundamental tools with just the right amount of knowledge needed. Instead of swimming in the vast ocean again, they can easily leverage the effort already spent by other researchers and can use the fruits for their own contribution as quickly as possible.

I started my research on elastic waves in periodic media in 2003. My PhD dissertation was on guided waves in sinusoidally corrugated periodic media. There we found negative and zero group velocity of waves being a unique characteristic of periodic system. Beyond band gaps, my interest was to find alternate quantum analogous behavior in acoustics and with elastic waves, which by then researchers had started to demonstrate for electrodynamics in electromagnetic metamaterial. I have made a deep dive into this subject since 2012, when I joined the mechanical engineering department at the University of South Carolina as an assistant professor. Immediately I realized that, to an engineer, physics is obscured. This was partially due to my lack of training in quantum mechanics. While working on various designs of acoustic metamaterials (e.g., split ring materials, metamaterials for energy harvesting, nondestructive evaluation) and diligently exploring new acoustic behaviors, since 2019 we have started to demonstrate multiplexing behavior of Dirac-like cones and their modulation based on deaf bands in periodic metamaterial. In addition to well-demonstrated quantum trio effects, we found counter behavior that shows in topological insulators. We found a topological sink. Having been published in more than 30 publications on waves in periodic metamaterials and having long-standing research experience since 2003, I feel I am well positioned to write this book and present how quickly one can achieve the knowledge needed to contribute to this field.

I started writing this book in December 2021 and finished in February 2023. I am thankful to my family, who wholeheartedly supported me during this time. When I was almost submerged in this book and pulled myself away from many household and social activities, my family always understood me. Specifically I would like to thank and share my love with my daughter Ms. Aishani Banerjee, with whom I could have played and painted more. I am thankful to my former students Dr. Riaz Ahmed, Dr. Hossain Ahmed, Dr. Fariha Mir, and especially Dr. Mustahseen M. Indaleeb, who took the roller coaster ride with me during the time of the exploration of this physics. This book is the testimony of their contributions. I am also thankful to my current student, Mr. Debdyuti Mandal, for his administrative support whenever it was necessary. This book cites many books and research articles written by multiple researchers; but they are cited selectively due to the restricted page limit. This is not an ignorance of the vast research that equally contributed to the field of metamaterials for topological acoustics. With due respect to all the authors and contributors, let's have a new engineering journey with the new physics of topological waves.

Sourav Banerjee
Irmo, SC

About the Author

Dr. Sourav Banerjee, a fellow of American Society of Mechanical Engineers (ASME) is a professor in the Department of Mechanical Engineering Department at the University of South Carolina (USC), Columbia, South Carolina, USA. He is currently serving as a director of the Integrated Material Assessment and Predictive Simulation Laboratory (i-MAPS) at USC since 2012. Before joining USC Dr. Banerjee served as a senior research scientist and then the director of Product Development in Acellent Technologies Inc. during 2008 and 2011. Dr. Banerjee's research is focused on Ultrasonic and Acoustics. Dr. Banerjee has published more than ~190 research articles out of which more than ~89 articles are in peer reviewed international Journals. He has authored two books and three book chapters on diverse topics of wave propagation. Dr. Banerjee has given research and academic lectures at multiple countries including, Canada, Germany, Italy, China, India, Russia, Malaysia to name a few. He works with many industries and government agencies around the world. He is the recipient of structural health monitoring (SHM) Person of the year award in 2019. Same year he also received Breakthrough Star award, awarded by the Office of Vice President of Research of USC.

Dr. Banerjee received Ph.D. in Engineering Mechanics (major) and Applied Mathematics (minor) from University of Arizona, Tucson, USA, in 2005. He received M.Tech in Structural Engineering from Indian Institute of Technology (IIT), Bombay, India. In 2002. He is a former DAAD scholar and worked in the Institut für Statik und Dynamik der Luft- und Raumfahrtkonstruktionen (ISD), at the University of Stuttgart, Germany during 2001-2002 for his master's project which later won the best master thesis award given by the Indian Society of Technical Education (ISTE). Dr. Banerjee received his bachelor's in civil engineering (B.E. Civil) from the Indian Institute of Engineering Science and Technology (IIEST), Shibpur, formerly known as Shibpur BE College, India.

Dr. Banerjee is currently serving as an executive committee member for the ASME NDPD division, committee member for the SPIE Smart Structure and NDE conference etc. He served as a track and symposium organizer on the topic related to NDE at the ASME IMECE conference during 2016- 2021. He serves on the editorial board of *Scientific Reports* published by Nature Publishing Group, *Ultrasonics*, by Elsevier, *International Aeronautics Journal, and International Journal of Aeronautics and Aerospace Engineering*. Dr. Banerjee also an associate editor of the ASME Journal of Nondestructive Evaluation, Diagnostics and Prognostics of Engineering Systems.

1 Acoustic Metamaterials

1.1 INTRODUCTION

In this chapter, the understanding of metamaterials is presented in a historical outline. The research field is significantly vast, and thus not all papers are cited. With due respect to all significant contributors and researchers, this book cites relevant articles pertaining to the discussion in respective chapters.

1.2 METAMATERIALS

The word 'meta' is now more popular than ever due to social media activities. 'Meta' is a word originated from Greek which means 'beyond'. Thus, a metamaterial is the material that by definition does not exist in the world and its features are beyond the capability of current materials. A metamaterial is artificially made material by researchers. While comprising various internal components and exploring various geometric patterns, counter-intuitive physics or phenomena were explored by the researchers in metamaterials. Generally speaking, metamaterials developed in acoustics are periodic in nature.

Wave propagation in periodic media has been studied since the time of Lord Rayleigh [1] and Maxwell [2] (during the late 19th century), with elastodynamics and electrodynamics, respectively. Most of the work was theoretical and was proposed in such a way that regular patterned structured could be handled mathematically. Further, through mathematical notion and considering application to quantum mechanics, respectively, Gaston Floquet [3] in 1983 and Felix Bloch [4] in 1929 transformed the understanding of waves in periodic media. However, not until the second half of the 20th century were the counterintuitive wave phenomena in periodic media discovered. It was much criticized at that time but exploded into significantly high-impact research field afterwards. In 1967, a Soviet scientist, Veselago, [5] proposed the possibility of negative permeability and negative permittivity of a material that could result in a unique electromagnetic wave propagation phenomena. It took 23 years to prove the practicality of such materials when artificially designed material for elastodynamics was proposed in 1990. The 'metamaterial' name was coined, but it took another ten years to experimentally demonstrate [6] the unique wave behavior. This started the explosion of research with metamaterials with the physics of electrodynamics. This intrigued the physicists and the acoustic and elastodynamic researcher community to discover new physics that are analogous to quantum behavior but present in acoustic pressure waves and elastic stress waves. As it is a vast field to be concise, only acoustic metamaterials and related topics are discussed in this chapter.

DOI: 10.1201/9781003225751-1

1.3 PERIODIC MEDIA AND PHONONIC CRYSTALS

Periodic acoustic media can be one dimensional (1D), two dimensional (2D), or three dimensional (3D) in nature (Fig. 1.1). Periodicities are created artificially along the respective directions in a metamaterial. Periodic media consist of periodically arranged geometric configurations of more than one material type. Hence, the media must have a host medium that creates a unit cell called 'matrix'. There could be one or more constituent materials with material properties different from the constituent matrix (creating acoustic impedance mismatch) but housed inside the unit cell. Together they create a new unit cell with host and constituent materials. Once the unit cell is created, the unit cell could be repeated along any specific dimension to create a metamaterial with geometrical periodicity. The unit cell length in 1D, 2D, or 3D are called the lattice parameters in the respective directions. In this regard, phononic crystals must be discussed.

Phononic crystals are also periodic. A specific material type is repeated in a host matrix, ensuring the acoustic impedance mismatch to achieve a specific behavior of acoustic or elastic wave propagation. They are also called band gap materials. Before the 21st century, the artificial design of phononic crystals was conceptualized in such a way that a specific wavelength of the incident wave could interact with the material. The interaction was mostly in the range of their lattice parameters. The lattice parameters were mostly kept within half of the wavelength of the acoustic and elastodynamic waves to ensure Bragg scattering [7], which was originally found through electromagnetic wave interaction in periodic crystals. Hence, the phononic crystals and related research at the later stage of 20th century were mostly linked to the wavelengths that strongly interact with the crystal orientations. Beyond the concept of discrete phononic crystals that initially simulated the material crystals, many elastic continuums and piezoelectric materials were created with periodic gratings for various wave applications and nondestructive evaluation applications during the second half of the 20th century. Many such detailed discussions and materials can be found in Ref. [8]. Further, during the first decade of the 21st century, continuum elastic materials were artificially created with periodic gratings to demonstrate the Bragg phenomenon [9–11] for guided Rayleigh-Lamb waves. Hence, the Bragg phenomena was not limited to the phononic crystals. A situation with negative group

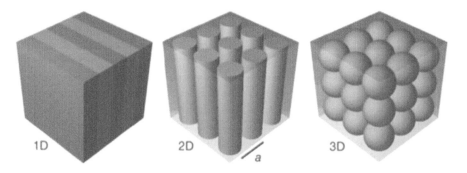

FIGURE 1.1 Periodic structures.

velocity of guided elastic wave was also demonstrated [12]. Still the wavelength of interaction was limited by the grating frequency in the material.

1.4 CONNECTING METAMATERIAL AND PHONONIC CRYSTALS

In phononic crystals, wave interactions were limited by the periodicity of the crystals. Simultaneously came a new concept. What if the wave with larger wavelengths can also interact with the material with lower order periodicity? This became the next generation question to answer by the researchers of 21st century. In 2000, Liu et al. proposed in *Science* journal a new acoustic material that demonstrated negative density at the resonance frequencies [13, 14]. At local resonance it was shown that the transmissibility of acoustic wave declined significantly. That means that the group velocity of the waves was zero at the local resonance frequency. A negative effective mass density was discovered.

Permeability and permittivity are the two primary material properties that govern electrodynamics. Similarly, elastodynamics is governed by the mass density and the elastic modulus of the material. It has been known since 1967 that permeability and permittivity can both be negative simultaneously and/or individually. In electrodynamics these properties would be the density and elastic modulus. In acoustics, these properties would be the density and bulk modulus. But for elastodynamic can both these properties be negative? Figure 1.2 shows four quadrants of the material property axes that are further to be explored by the acoustic and elastodynamic researchers. Negative mass density was proved in 2000 and thus the second quadrant was reached. Creating negative mass density was easier than creating negative bulk or elastic modulus. While the community was looking for such materials, auxetic materials had been already discovered in 1987 [15], which presents the negative effective poisons ratio of a structure. It was recognized that negative elastic modulus must be possible. Starting with Helmholtz resonators [16], a quest for metamaterial with negative effective modulus began in 2005. Later researchers found various innovative cases of metamaterials with simultaneous effective negative bulk modulus and

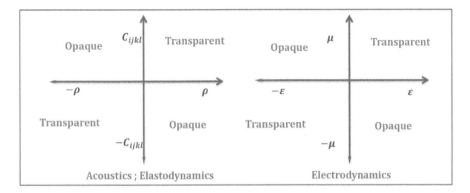

FIGURE 1.2 Four quadrants of material properties in acoustics, elastodynamics, and electrodynamics.

negative density [17]. A split ring metamaterial-like property in electrodynamics was demonstrated for elastodynamics to generate negative material properties [18] such as negative Poisson ratio in continuum.

In all these cases, effective medium theory was used. This means that the metamaterial proposed had inner inclusions and their dimensions are much smaller than the wavelength of the elastodynamic wave of interest. That also means that the dimensions of the internal geometries of the metamaterials are independent of the wavelength. This is certainly different from the phononic crystals discussed above. Metamaterials discussed above needed a homogenization technique that is frequently used in continuum mechanics. The metamaterials and their effective medium properties involved local resonance of the internal inclusions and their architecture.

Thus, the research community never mixed the term 'phononic crystals' with 'metamaterials' as if they were separate entity. However, as discussed later in this chapter, many new phenomena with quantum analogous effects were demonstrated through phononic crystals with much larger wavelength than their lattice parameters. Hence, according to the author's opinion, with new research their individual consideration should not be discrete but rather mutually inclusive. A metamaterial could be constructed using phononic crystals, and different sets of phononic crystals could create periodic metamaterials. Thus, in this book metamaterials and phononic crystals are used synonymously. Together in this book the term 'phononic' will be used as an analogue to the photonics, which include phonon propagation in both phononic crystals and metamaterials. Some sample metamaterials are shown in Fig. 1.3.

When the negative effective modulus and the negative effective densities were found in the metamaterials, many unique wave phenomena were also discovered utilizing those properties. Some phenomena are acoustic negative refraction, acoustic superlensing capability, acoustic wave bifurcation, wave tunneling, acoustic cloaking, acoustic Anderson localization, and acoustic Talbot effect, to name a few. Some of these features with respective references are discussed in Chapter 6. The research did not stop there. As electromagnetic waves in photonics progressed very fast with much financing from funding agencies, phononics had to catch up. The field of photonics research discovered numerous quantum effects by 2015, exploring the band structures of unique orientation of artificially created metamaterials. Similar investigations started with phononics while investigating the physics of acoustics and elastodynamics. Out of all different quantum phenomena topological behavior prevails.

1.5 TOPOLOGICAL PHENOMENA

In this section, many new terminologies are used and summarized in historical outline. Respective references related to these terminologies and topics are vast and not explicitly cited in this section but can be found in Chapters 2 through 7 as they appear while discussing their respective physics.

Topological phenomena in the 21st century are predominantly described through a quantum mechanics approach. Many topological phenomena in materials are in fact related to quantum phenomena. Hence, all such phenomena in acoustics and elastodynamics must also be described through their quantum counterparts. However, finding and explaining the topological behaviors of acoustic and elastic waves are

Acoustic Metamaterials

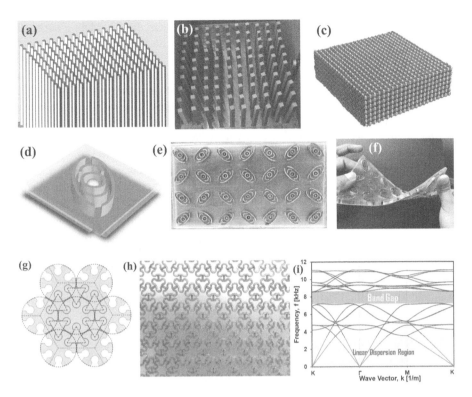

FIGURE 1.3 Phononic crystals and metamaterials: a) circular columns in fluid matrix, b) broken symmetry rotated square phononic crystals in air matrix, c) 3D spherical magnetic phononic crystals in fluid, d) a model for acoustic split ring metamaterial with soft and hard material constituent, named butterfly metamaterial, e) a metamaterial made with steel inclusion in transparent epoxy matrix, f) flexibility of the butterfly metamaterial (the wave physics in this metamaterials are discussed in Chapter 6), g) a model showing simultaneously triangular and hexagonal periodic metamaterial, h) the metamaterial made with polymethyl methacrylate (PMMA) with epoxy lining, and i) the wave dispersion or, synonymously, the band diagram in frequency wave number domain, showing band gap. More details on band diagrams can be found in Chapters 5 and 6.

challenging, because fundamentally the acoustic and elastic waves do not have the most important analogous quantum degree of freedom, which is spin.

With the discovery of a topological insulator owing to the quantum Hall effect (QHE), photonics researchers subsequently explored the quantum spin Hall effect (QSHE) and quantum valley Hall effect (QVHE) in the hexagonal structure of periodic metamaterials. Together, these three quantum effects are called the quantum trio effect. While the quantum trio behaviors were taking shape, the phononic community had already explored the hexagonal periodic metamaterials for the counterintuitive wave phenomena mentioned in the previous section. Hence, at a much later stage, after 2012 a surprising new look to the phononic band structures (an example is shown in Fig. 1.3i and is further referred to in Chapters 5 and 6) reveals similar

6 Metamaterials in Topological Acoustics

features that mediated the photonic quantum trio effects. Immediately a surge of research activities was seen that populated the field with a race to find topological behavior in acoustics and elastodynamics.

QHE was discovered in 1980 when an external magnetic field was applied to a thin layer of silicon-based metal-oxide-semiconductor. In these specific materials, it was found that the traditional Hall voltage is quantized. Due to a very thin layer of the material, electrons had mobility only on one plane and out of plane mobility of the electron was restricted. It is well known from classical electrodynamics that when a magnetic field is applied to a two-dimensional electron layer, the electrons observe cyclotron orbits around the magnetic field lines. However, when this same problem is analyzed quantum mechanically, the cyclotron orbits are quantized and only occupies the Landau levels. The Schrodinger equation was solved by introducing the magnetic field to explain this phenomenon which is presented in Chapters 4 and 5. It was immediately realized that one thing is fundamentally different in phononics. There is no equivalent magnetic field for acoustic or elastic waves.

Based on QHE, theoretically the concept of topological insulator was proposed in 2005. Experimentally it was demonstrated in 2007 in a material made of HgTe quantum sandwiched between CdTe. It was found that the only boundary of the material was conductive, but no current flowed through the bulk. Later a 3D topological insulator was discovered in 2007. Subsequently, many other metamaterials with bulk-boundary distinction were discovered and designed. It was found that in a trivial state of such materials, the conduction band and the valance band have a small gap and can be artificially modulated to access a nontrivial state. This fundamental logic governed the design of photonic metamaterials. As photons do not have an inherent electric charge, to access QHE an artificial magnetic field was required. They were created using experimental arrangements such as optical resonators and phase modulators. This was still possible for electrodynamics.

In 2006, it was found that the primary purpose of the external magnetic field was to break the time reversal symmetry of the system. So, the QHE was primarily the result of breaking the time reversal symmetry. The importance of topological invariants was also observed in QHE and governed the design of different metamaterials as topological insulators. Simultaneously, the concept of a geometric phase also appeared. The concept of the Zak phase for 1D periodic media and the Berry phase for 2D and 3D periodic media was brought by the researchers to connect the understanding through topological charges. Chern numbers were used to explain the topological charge of a metamedia, and the concept of geometric phase and its link to the topological behavior was explained. It is confirmed that an active breaking of time reversal symmetry was required. Time reversal breaking of symmetry was causing the geometric phase and thus causing the topological behavior. Discussion of their respective physics and their connections are presented in Chapters 5 and 6.

That means that to access topological behavior in acoustic metamaterials, the symmetry must be broken too. Unlike photonics, where electric fields and magnetic fields interact with each other and it is possible to create artificial magnetic field for photonics, no such field exists for acoustics or elastodynamics. It was a road blocker. When the magnetic field is applied to photonic metamaterial, a spin state is activated around the magnetic field lines and electrons orbiting at the Landau levels. Spinning

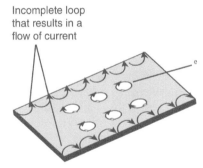

FIGURE 1.4 The state of electrons in a topological quantum Hall insulator. (The picture is adapted from Ref. [19].)

cyclotronic electrons around the magnetic field lines are created by introducing the magnetic field. This causes an incomplete rotation of the electrons at the boundary (Fig. 1.4). The half spin around the boundary of the metamaterial causes the current to flow through the material when the internal material is insulated. Current flowed through the boundary, irrespective of the geometry and orientation and with or without defects. A backscattering immune wave propagation was achieved.

The question was asked, Is that something that could be done for acoustics? Then came an ingenious idea. What if, instead of creating the artificial magnetic field, the artificial state that is induced by the magnetic field is created. Is it possible to create the field around the phononic crystals in such a way that it can simulate the state of the spinning electrons? Researchers demonstrated that it was possible by inducing artificial flow of fluid around the phononic crystal in the unit cell of the metamaterial. Interestingly, the acoustic topological insulator emerged (Fig. 1.5). The phenomenon was named quantum anomalous Hall effect (QAHE) as it is not exactly the Hall effect but replicates the Hall phenomenon.

Still, something was lacking with the topological acoustics. The artificial induction of fluid flow around the phononic crystals required additional energy to the system, which was not wholeheartedly accepted by the community. Hence, to access topological phenomena the acoustic community needed a state of metamaterial in which no additional energy is required to create an artificial external field. Fortunately, researchers in photonics surprisingly saw an additional opportunity to achieve bulk-boundary distinction and a type of topological behavior where the active breaking the time reversal symmetry is not required. So no artificial magnetic field was required either. That is exactly what the acoustic community was waiting for. No need for the magnetic field to access topological behavior. Here the topological phenomena were achieved by breaking the space inversion symmetry. Photonic researchers, while investigating graphene, found that two bands intersect at the K point of the Brillouin zone at certain energy level, creating a mirror image at the Brillouin boundary. And it was found that the bands are actually linear at the close proximity of that energy level. Two straight bands form a Dirac cone at the K (K′) point. (Refer to Chapter 6.)

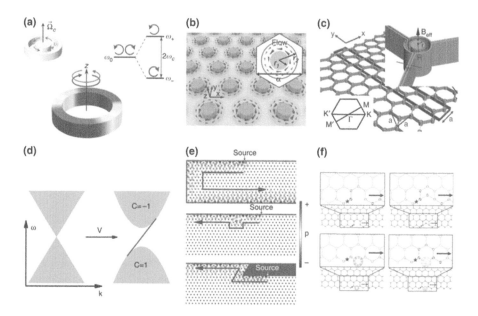

FIGURE 1.5 The quantum Hall effect in phononic crystals: a) the airflow-induced acoustic nonreciprocity, b) and c) two acoustic quantum Hall lattices incorporated with circulating airflow, d) illustration of the band gap opening induced by the airflow, associated with a one-way edge state that spans the bulk gap region, and e) and f) the robust edge state propagation against various defects. (The figures are adapted from Refs. [20–23].)

It was pointed out that for QHE, the system must acquire a geometric phase to access the topological behavior, and that could be achieved by active breaking the time reversal symmetry. As the time reversal symmetry is not broken in this case for the Dirac cone at the K point, does it still acquire the geometric phase? It was found that locally it did, and two states – trivial and non-trivial – existed. When these two states of the metamaterial were placed side by side, creating a domain wall, the wave always propagated through the domain wall. A backscattering immune topological wave was achieved, and the wave followed the geometry of the domain wall. This phenomenon was named quantum valley Hall effect (QVHE). This phenomenon, with examples, is discussed in Chapter 7. A similar QVHE was also discovered in acoustics by exploring hexagonal periodic metamaterials simulating graphene-like structure. If hexagonal or triangular geometric patterns are explored, almost certainly the Dirac cones are observed at the Brillouin boundary. Here Dirac cones are formed, degenerating two linear bands. Dirac cones and degeneracies are discussed in Chapter 6.

While exploring band structures of different geometric patterns, researchers found unique three- or four-band degeneracies at the center of the Brillouin zone. These degeneracies were accidental. Specific geometric configuration or parameters were the reason for these degeneracies. Such degeneracies are discussed in Chapters 5 to 7. The four bands became almost linear near the energy level where the degeneracy was detected. It is not like an absolute Dirac cone with two bands but

Acoustic Metamaterials

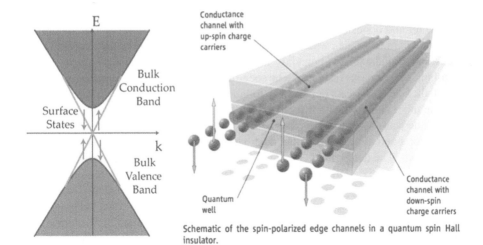

Schematic of the spin-polarized edge channels in a quantum spin Hall insulator.

FIGURE 1.6 Schematics of quantum spin Hall topological insulator.

demonstrates a Dirac-like phenomena. The point was then called a Dirac-like cone, which is further discussed in Chapter 6. Dirac-like cones can have three- or four-band degeneracies. At the degenerated energy level, depending on the number of bands, the physics of wave propagation was found to be different. Based on quantum analogous behavior, these bands at Dirac-like cones are spin-activated bands, which are discussed in Chapters 6 and 7. Generally speaking, there are no actual spins in acoustics. Thus, they are called pseudospin. However, in elastodynamics, intrinsic spin may exist for elastic waves discussed in Chapter 6. Pseudospin states are used in acoustics to explain similar physics found in quantum spin Hall insulators (Fig. 1.6), where actual spin states dominate the topological behavior. There are two situations in acoustics:

- Four-band degeneracies with pseudospin ½ – fermion-like behavior
- Three-band degeneracies with pseudospin 1 – boson-like behavior

When four bands are degenerated, two top bands and two bottom bands are classified into two electron states synonymous to P and d orbitals. Surprisingly it was found that at a specific geometric configuration (i.e., internal geometric parameters, dimensions, rotations, etc., of phononic crystals) the Dirac-like cone is formed as an accidental degeneracy. But beyond this parameter identity, the space is inverted. It means that if the metamaterial is analyzed along the parameter space by increasing or decreasing the parameter (positive increment of the parameter), two bottom bands and two top bands switch their energy levels beyond the Dirac-like cone parameter. These bands are idealized as spin ½ bands. In Fig. 1.6 it can be seen as a quantum state in photonics, where the spin up and spin down states are counter propagating in the material at the boundary while the bulk is a quantum well. A similar situation with p and d orbital bands and their fermion-like behavior with spin ½ particle is explained in Chapters 6 and 7 for phononics. This situation is called quantum spin

Hall effect (QSHE). Metamaterial made of two slabs, one side carrying negative parameter and opposite side carrying positive parameter (with respect to the Dirac-like cone parameter, i.e., where the Dirac-like cone is formed) creates a topological transition boundary. This boundary is capable of carrying the wave only while the two-bulk side remains insulated. The specific design of the domain wall between the two-bulk side will create opportunity to guide the wave as desired.

One more situation with three-band degeneracy could be found to be interesting for unique wave behavior. This degeneracy is also accidental. There are three bands, named top band, bottom band, and deaf band, which is sandwiched between the top and bottom. The deaf band is an antisymmetric modal band. At a specific geometrical parameter (i.e., internal geometric parameters, dimensions, rotations, etc., of phononic crystals), the Dirac-like cone is formed accidentally, called the Dirac cone parameter. Here the top and the bottom bands form the Dirac cone, keeping their local linear dispersion intact but trapping the deaf band in between. It was found that the deaf band mostly controls the wave phenomena at this state. Zero index material was also described at this state with orthogonal wave transport, acoustic cloaking, acoustic pseudo diffusion, wave bifurcation, etc. (Refer to Chapter 6.) If the parameter space is modulated like it was done for QSHE, it was found that the energy state (i.e., frequency of the deaf bands) always remain almost unaltered. Along the negative parameter space (with regard to the Dirac-like cone parameter) the bottom band is degenerated with the deaf band, and along the positive parameter space the top band is degenerated with the deaf band. This situation could be reversed depending on the energy state of the Dirac-like cone. Simultaneous occurrence of two Dirac-like cone states with reverse properties were also presented in the literature. Owing to the curvature of the top and bottom bands beyond the triple degeneracy governed by the deaf band a unique state was realized where the wave never leaves the metamaterial but rather spins and remains trapped inside the material block. In this case, the boundary of the material remains insulated but the bulk is acoustically conductive. This phenomenon was found to be topologically protected. This stable spin-mediated state is very closely related to the skyrmion-like phenomenon found in magnetic metamaterials. Irrespective of size and shape of the metamaterial block, the wave is always trapped inside bulk. It was named a topological black hole (TBH). This was due to the pseudospin-1 and boson-like behavior of the deaf band in acoustics. Similar behavior could be achieved in elastodynamics exploiting their intrinsic spin state owing to the abnormal polarity of the displacement fields. Discussion on this behavior is presented in Chapter 7. Figure 1.7 shows some wave states related to this phenomenon.

1.6   FUTURE DIRECTION

Fermion with ½ spin and boson-like spin 1 particles are synonymously found in acoustics through their modal behavior. However, spin is not something immediately realized in acoustics and elastodynamics. A particle subjected to a transversely polarized shear wave could be visualized as two self-rotating particles with opposite polarity (i.e., opposite spin) on a plane perpendicular to the propagating directions. A particle subjected to a longitudinal wave could be visualized as two spin states of the

Acoustic Metamaterials

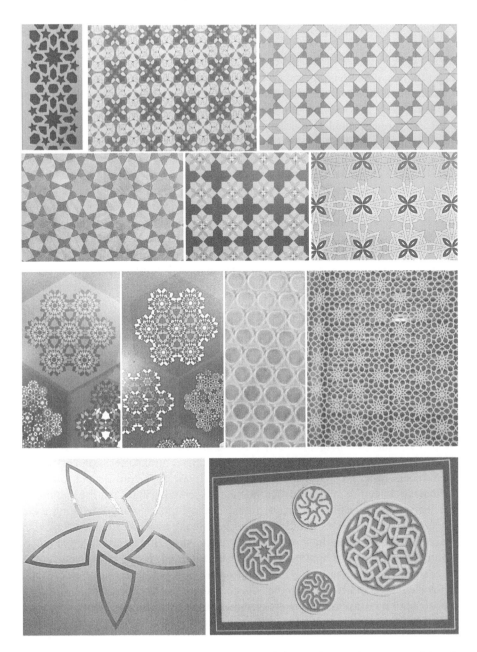

FIGURE 1.7 Mathematical patterns that may result in novel acoustic metamaterial realizing their respective band structures: a) square periodic architecture, b) hexagonal periodic architecture, and c) chiral architecture. These arts and structures are displayed around the Sheikh Zayed Grand Mosque in Abu Dhabi, UAE.

particle rotating with opposite polarity in the same plane as the propagation direction. These quantum analogous spin states are discussed in Chapter 6. Additionally, when two transversely polarized or longitudinally polarized waves propagating in two different directions are superposed, an intrinsic spin state emerges. Orbital angular momentum and spin angular momentum densities are used to express them mathematically. However, as of now, beyond few hypothetical thought experiments presented through simulation, it is not known how and when these states could be accessed/exploited. These novel wave behavior could be experimentally achieved exploiting the abnormal polarity which is beyond what are presented in this book. This could be an open research field in metamaterials for the next ten years.

The band structures discussed in Chapters 5 and 6 are the signature of any periodic metamaterial for all possible behavior/characteristic that they may manifest. However, it is almost impossible to investigate all unique features that a band structure may present. The Dirac cones at the Brillouin boundary and at the center of Brillouin zone were overlooked in 20th century and recently came to the limelight. Similarly, it is possible some other unique feature is still overlooked at this date.

Metamaterials were found to have many characteristics owing to their periodicity. Historically, different cultures have imagined different periodic patterns for architectural purposes. Mathematical possibilities of different architectures were explored and designed throughout history. Starting from the Sumerian civilization to the Islamic era, mathematical discoveries created many such beautiful periodic patterns in religious and civil structures like in mosques and buildings. Periodic architecture can be found in various objects and structures in the Indus valley civilization. Mathematical treatment with geometry created novel Hindu architectures. Those periodic patterns and architectures were considered holy and can be found in Hindu temples and in kings' palaces. Figure 1.7 shows some unique geometric patterns that are mathematically possible and can be seen around the Sheikh Zayed Grand Mosque in Abu Dhabi, UAE, as architectural patterns and artworks. Researchers have not investigated many of the vast opportunities that different periodic architecture made of different materials could bring. Creating and investigating periodic architecture in the future could discover many intriguing wave phenomena such as are presented in this book. So far very specific situations with specific band structures have been investigated. Analyzing many different metamaterials with tailored applications could be achieved. Many phenomena could be accidental, depending on geometrical or material parameters like QSHE and topological black hole. But with the advent of artificial intelligence (AI) surrogate models could tell us what geometric configuration and material constituent have to be used for certain specific tailored applications. Of course, through machine learning (ML) algorithms AI models must be trained, and many such simulations would be required. But with increasing computing capabilities with graphical processing units (GPUs) multithread simulations are going to be the reality for finding metamaterials. An acoustic metamaterial genome-like project could lead to finding novel materials that could be simultaneously stable and would present unique but required wave behavior, predictably. The quantum trio effect should be the phenomena to start with. Different new material constituents, and numerous different material configurations in novel architectural patterns could present new

Acoustic Metamaterials **13**

opportunities for exploring topological acoustics, and the research community has only just begun to realize the underlying physics. The field is vast, and a deep dive would be necessary. With this motivation, this book presents a seedling activity to train the students and prepare them with fully guided knowledge required for the field of topological acoustics. A much bigger research activity by the future generation is awaiting.

1.7 SUMMARY

The chapter presents the historical development of the concept of topological acoustics in metamaterials. While discussing the necessary topics on topological acoustics, the chapters that discuss physics in detail are referred to in the text.

REFERENCES

1. Rayleigh, J.W.S., *The Theory of Sound*. 2nd ed. 1945, London: Dover Publications.
2. Forbes, N., Mahon, B., *Faraday, Maxwell, and the Electromagnetic Field: How Two Men Revolutionized Physics*. 2014, Amherst, New York: Prometheus Books.
3. Floquet, G., *Sur les équations différentielles linéaires à coefficients périodiques*. Annales Scientifiques de l'École Normale Supérieure, 1883, **12**: p. 47–88.
4. Bloch, F., *Über die quantenmechanik der elektronen in kristallgittern*. Zeitschrift für Physik, 1929, **52**: p. 555–600.
5. Veselago, V.G., *The electrodynamics of substances with simultaneously negative values of ε and μ*. Soviet Physics Uspekhi, 1967, **10**: p. 509–514.
6. Smith, D.R., Padilla, W.J., Vier, D.C., Nemat-Naser, S.C., Schultz, S., *Composite medium with simultaneously negative permeability and permittivity*. Physical Review Letters, 2000, **84**(18): p. 4184–4187.
7. Bragg, W.H., Bragg, W.L., *The reflexion of x-rays by crystals*. Proceedings of the Royal Society of London. Series A, 1913, **88**(605): p. 428–438.
8. Auld, B.A., *Acoustic Fields and Waves in Solids*. 1973, Malabar, Florida: Krieger Publishing Company, Inc.
9. Banerjee, S., Kundu, T., *Symmetric and anti-symmetric Rayleigh–Lamb modes in sinusoidally corrugated waveguides: An analytical approach*. International Journal of Solids and Structures, 2006, **43**(21): p. 6551–6567.
10. Banerjee, S., Kundu, T., *Scattering of Ultrasonic Waves by Internal Anomalies in Plates Immersed in a Fluid*, in *Health Monitoring and Smart Nondestructive Evaluation of Structural and Biological Systems V*. 2006. Bellingham, WA: SPIE.
11. Kundu, T., Banerjee, S., Jata, K.V., *An experimental investigation of guided wave propagation in corrugated plates showing stop bands and pass bands*. The Journal of the Acoustical Society of America, 2006, **120**(3): p. 1217–1226.
12. Banerjee, S., Kundu, T., *Elastic wave propagation in sinusoidally corrugated waveguides*. The Journal of the Acoustical Society of America, 2006, **119**(4): p. 2006–2017.
13. Liu, Z., Zhang, X., Mao, Y., Yang, Z., Chan, C.T., Sheng, P., *Locally resonant sonic materials*. Science, 2000, **289**(5485): p. 1734–1736.
14. Liu, Z., Chan, C.T., Sheng, P., *Analytic model of phononic crystals with local resonances*. Physical Review B, 2005, **71**: p. 014103-1 to 014103-8.
15. Lakes, R.S., *Foam structures with a negative Poisson's ratio*. Science, 1987, **235**(4792): p. 1038–1040.
16. Fang, N., Xi, D., Xu, J., Ambati, M., Srituravanich, W., Sun, C., Zhang, X., *Ultrasonic metamaterials with negative modulus*. Nature Materials, 2006, **5**: p. 452–456.

17. Liu, X.N., Hu, G.K., Huang, G.L., Sun, C.T., *An elastic metamaterial with simultaneously negative mass density and bulk modulus.* Applied Physics Letters, 2011, **98**: p. 251907.
18. Ahmed, R.U., Banerjee, S., *Wave propagation in metamaterial using multiscale resonators by creating local anisotropy.* International Journal of Modern Engineering, 2013, **13**(2): p. 51.
19. Lindner, N.H., Refael, G., Galitski, V., *Floquet topological insulator in semiconductor quantum wells.* Nature Physics, 2011, **7**: p. 490–495.
20. Zhang, X., et al., *Topological Sound*, 2018, **1**(1): p. 1–13.
21. Fleury, R., et al., *Sound Isolation and Giant Linear Nonreciprocity in a Compact Acoustic Circulator*, 2014, **343**(6170): p. 516–519.
22. Ni, X., et al., *Topologically Protected One-Way Edge Mode in Networks of Acoustic Resonators with Circulating Air Flow*, 2015, **17**(5): p. 053016.
23. Yang, Z., et al., *Topological Acoustics*, 2015, **114**(11): p. 114301.

2 Classical Mechanics and the Physics of Continua

2.1 INTRODUCTION: HISTORY OF CLASSICAL MECHANICS

Understanding the wave propagation in various media requires the understanding of motion and dynamics. To understand the motion of a particle that is later considered as energy or wave in a generalized form, we must gauge the understanding of classical mechanics. However, before we jump into the descriptions, definitions, and mathematical treatments of classical mechanics that took more than 2000 years to take its current shape, it would be wise to appreciate the contribution of the legendary philosopher, mathematicians, and scientists. Numerous philosophers, starting from the Greeks during 600 BC to 200 AD, considerably contributed to the understanding of matter and motion of a body. During these 800 years of modern evolution of science, chronologically, Democritus, Aristotle, Archimedes, Heron of Alexandria, and Ptolemy contributed many key concepts of classical mechanics [1]. 'What is lost in *Velocity* is gained in *Force*' by Aristotle at about 300 BC is the first seedling concept of virtual work principle that is yet to mature during the 'age of enlightenment'. The concept of infinitesimal quantity, which was yet to mature with Newton in 1687, was first conceptualized by Archimedes while calculating the value of π at about 200 BC. While describing the principle of reflection of light, Heron of Alexandria about 50 AD mentioned that light takes the shortest path, which could be the earliest depiction of the principle of least time. Many other contributions from Aristotle to Ptolemy (400 BC – 200 AD) could have been more valuable but went into oblivion due to the adoption of the earth-centered model. While in the next 1000 years European would go into the dark ages, bogged down with religious dogmatisms, classical mechanics had to wait till the 15th century for real scientific breakthroughs. Classical mechanics in its infancy needed much more support from mathematics, which also did not mature by the end of 400 AD. However, during the western dark age (400 AD – 1400 AD), the eastern world thrived with science, mathematics, business, and intercontinental communication [2]. A fundamental difference between the East and the West was the understanding of the universe by virtue of their religious teaching. (The East presumed a sun-centric universe, opposed to the earth-centric universe in the West.) Specifically, India, Persia, and other Arabic countries in the Middle East flourished with their ingenious mathematical ideas on arithmetic, astronomy, trigonometry, algebra, infinite series, and abstract form of infinitesimal qualities to give a formal birth to calculus [2–5]. The classical golden age of mathematics in the East started with Aryabhata right after the fall of the Roman Empire in the West in about 400 AD. Aryabhata was followed by Varahamihir, Bhaskara I, Shridharacharya, Bhaskaracharya, and Madhava to name a few in India during 400 AD to the 13th century. They significantly contributed to the concept of zero, algebra, development of series expressions, and even to the concept of infinitesimal qualities that

DOI: 10.1201/9781003225751-2

15

presented the derivates of the trigonometric functions [6]. Many of these were not known until the middle of 19th century by the Western world. Moreover, during the 11th and 12th centuries AD, Ibn Sina from Persia and Abu'l-Barakat from Iraq individually contributed to the few key concepts of classical mechanics. For example, the concept of inertia and the projectile motion was vaguely presented, without much mathematical treatment [7, 8]. Together the Greek age of philosophers and the golden age of mathematics in the East paved the way for the contemporary classical mechanics that took its final form by the end of 18th century. When the West was recovering from the dark age it moved from the earth-centered universe to a heliocentric cosmology. With Nicolaus Copernicus, Leonardo da Vinci, Johannes Kepler, and Galileo Galilei, living through brutal resistance during 1400 AD – 1600 AD, classical mechanics found its way out of the dark. Then came the *age of enlightenment*. With a few key building block concepts by Descartes, Fermat, and Pascal, ultimately Sir Isaac Newton contributed his master stroke to mature classical mechanics to its next level [9]. Western formulation of calculus, simultaneously presented by Gottfried Leibniz and Sir Isaac Newton, further paved the way for Leonhard Euler, Joseph-Louis Lagrange, William Rowan Hamilton, and Carl Jacobi to mature the classical mechanics the way we learn it today. Primarily two books that are hundred years apart, one by Newton in 1687 and another by Lagrange in 1788 [9, 10], are the pioneers. During the post-Newton era, variational methods, the concept of virtual work that was first conceptualized by Aristotle, and the principle of minimum action ultimately found their mathematical framework through a few brilliant strokes by Johann Bernoulli, Leonhard Euler, Jean le Rond d'Alembert, and Joseph-Louis Lagrange. After Lagrange, the next master stroke came from William Hamilton, with the development of Hamiltonian mechanics formalism [1], which later proudly adopted in quantum mechanics. It made a distinct approach compared to the Lagrangian mechanics. A few other contributors, such as Carl Friedrich Gauss, Simeon Poisson, Carl Jacobi, James Clerk Maxwell, Ludwig Boltzmann, and Henri Poincaré, helped mature the field of classical mechanics further during the 18th and 19th centuries. In this book we will present the concept of classical mechanics and quantum mechanics in relation to the fundamental understanding of wave propagation and wave characteristics in metamaterials.

2.2 FUNDAMENTAL CONCEPT OF CLASSICAL MECHANICS

The study of wave propagation in any medium, metamaterial or not, requires the understanding of elastodynamics. It means that wave propagation in an elastic system is a dynamic behavior, and its material points are continuously evolving over time. According to classical mechanics [1] a continuously evolving system can be described by Fig. 2.1. Let's assume that a system is at state $\mathbb{S}_I$ at time t_I and has evolved to a state $\mathbb{S}_{II}$ after time t_{II}. Next, assume the system took a path $q(t)$ to evolve from the state $\mathbb{S}_I$ to $\mathbb{S}_{II}$ shown in Fig. 2.1. Then a classical question was asked during the early development of classical mechanics. Is it necessary for the system to take this particular path $q(t)$? Or was any other path possible? This can be answered by perturbing the path $q(t)$ slightly by δq. According to the principle of least action

Classical Mechanics and the Physics of Continua

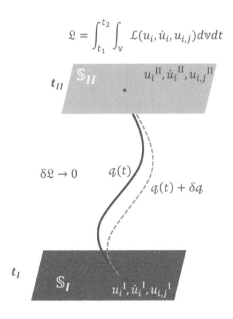

FIGURE 2.1 A continuously evolving system: Evolution of the system from state $\mathbb{S}_1$ to $\mathbb{S}_2$ through a most optimized path $q(t)$.

in classical mechanics, it is said that action ($\mathfrak{L}$), which is the spatio-temporal integration of a scalar function called Lagrangian ($\mathcal{L}$) defined for a system, should be minimum during an evolution.

In plain words, a system that tries to minimize the work by performing minimum action during the evolution is very lazy. This small universal statement is true for any physical system, biological system, and our universe, which is continuously evolving. The Lagrangian scalar function of a system is nothing but the difference between the kinetic energy and the potential energy of the system. Based on the *principle of least action*, if the path is slightly perturbed, the perturbed action ($\delta\mathfrak{L}$) must be equal to zero. A state $\mathbb{S}_I$ or $\mathbb{S}_{II}$ can be defined by few state variables appropriate for the physics we intend to investigate. For example, in a deformable body to understand wave propagation, the displacements u_i, velocities $\dot{u}_i$, and spatial derivatives $u_{i,j}$ can be used as state variables. Please note that u_i are displacements in all possible directions in a specific co-ordinate system, written using index notation. Understanding of index notation relevant to wave propagation is assumed in this book but can be referred to elsewhere [11]. For example, following the index notation in a Cartesian coordinate system $u_i \equiv [u_1, u_2, u_3]$, where i takes values 1, 2, and 3. This means that between the states $\mathbb{S}_I$ and $\mathbb{S}_{II}$, all the state variables have evolved from one set of values $(u_i^I, \dot{u}_i^I, u_{i,j}^I)$ to another set of values $(u_i^{II}, \dot{u}_i^{II}, u_{i,j}^{II})$. Irrespective of the changes in the state variables, through a path $q(t)$ or slightly a deviated path $q(t) \pm \delta q$, the variation of the action $\delta\mathfrak{L}$ must vanish. This principle is called the Hamilton's principle in classical mechanics [1, 12]. Mathematically the variation of action can be written as

$$\delta\mathfrak{L} = \delta\int_{t_I}^{t_{II}}\int_V \mathcal{L}\left(u_i, \dot{u}_i, u_{i,j}\right)dvdt = \delta\int_{t_I}^{t_{II}}\int_V \mathcal{L}\left(u_i, \dot{u}_i, u_{i,j}\right)dx_jdt = 0 \qquad (2.1)$$

where dv is a small elemental volume of the body to perform the integral over the entire volume v. Using the principle of least action ($\delta\mathfrak{L} \to 0$), the governing equation of the system could be derived.

2.3 GOVERNING EQUATION FROM CLASSICAL MECHANICS

The governing equation could be derived using an integral approach of classical mechanics. The most prominent function that is used in classical mechanics is called the Lagrangian ($\mathcal{L}$) of a system or simply the Lagrangian density. In this book on studying wave propagation in any media, our system is a material body that is deformable in space and time. The Lagrangian of a system is defined as a function of the generalized coordinates of the system and their respective temporal and spatial derivatives. If the generalized coordinates of system, for example, displacements are designated as u_i in a multidimensional system, the Lagrangian can be defined as

$$\mathcal{L} = \mathcal{L}\left(u_i, \dot{u}_i, u_{i,j}\right) \qquad (2.2)$$

where, $\dot{u}_i$ are the velocities (the temporal derivatives of displacements) and $u_{i,j}$ are the displacement gradients (the spatial derivates of displacements). The Lagrangian density of a deformable body can be defined as the difference of the kinetic energy density and the strain energy density.

When a wave propagates in a material, wave energy is transported from one point to the other. The action is a temporal and spatial action. That means that at two different time points, say at t_I and t_{II}, the system is in different states and the values of the arguments of the Lagrangian would be different. Action is a scalar function defined as a spatial and temporal integral of the Lagrangian density, and can be written as

$$\mathfrak{L} = \int_{t_I}^{t_{II}}\int_v \mathcal{L}\left(u_i, \dot{u}_i, u_{i,j}\right)dvdt = \int_{t_I}^{t_{II}}\int_v \mathcal{L}\left(u_i, \dot{u}_i, u_{i,j}\right)dx_jdt \qquad (2.3)$$

According to the classical mechanics theorem [1, 12], a system evolves from t_I and t_{II} in such a way that the scalar function defined in Eq. 2.3 is minimized. This principle is called the principle of least action or the Hamilton's principle [12]. As the Lagrangian is a function of u_i, $\dot{u}_i$ and $u_{i,j}$, in generalized coordinate system it is related to the displacements during wave propagation. Perturbation of these generalized coordinates will cause the change in the action during the wave propagation. Thus, a small variation (δ) to these generalized coordinate systems (here displacements), while satisfying all the boundary conditions, will cause a small variation (δ) in action $\mathfrak{L}$ as follows.

$$\delta\mathfrak{L} = \mathfrak{L}\left(u_i + \delta u_i, \dot{u}_i + \delta\dot{u}_i, u_{i,j} + \delta u_{i,j}\right) - \mathfrak{L}\left(u_i, \dot{u}_i, u_{i,j}\right) \qquad (2.4)$$

Classical Mechanics and the Physics of Continua 19

According to the Hamilton's principle, $\delta\mathcal{L} \to 0$ to satisfy the principle of least action in classical mechanics. That means that the system could have evolved during the wave propagation only in one way, by minimizing the action. Perturbation of the generalized coordinates will not perturb the action and will remain the same, making $\delta\mathcal{L} \to 0$. Next by expanding the first term in Eq. 2.4, using a Taylor's series, the $\delta\mathcal{L}$ can be written as

$$\delta\mathcal{L} = \mathcal{L}\left(u_i, \dot{u}_i, u_{i,j}\right) + \frac{\partial\mathcal{L}}{\partial u_i}\delta u_i + \frac{\partial\mathcal{L}}{\partial \dot{u}_i}\delta\dot{u}_i + \frac{\partial\mathcal{L}}{\partial u_{i,j}}\delta u_{i,j} - \mathcal{L}\left(u_i, \dot{u}_i, u_{i,j}\right)$$

(2.5)

$$\delta\mathcal{L} = \frac{\partial\mathcal{L}}{\partial u_i}\delta u_i + \frac{\partial\mathcal{L}}{\partial \dot{u}_i}\delta\dot{u}_i + \frac{\partial\mathcal{L}}{\partial u_{i,j}}\delta u_{i,j}$$

After substituting the Eq. 2.5 into the equation of action in Eq. 2.3, the variation of action can be written as

$$\delta\mathcal{L} = \int_{t_I}^{t_{II}} \int_v \left[\frac{\partial\mathcal{L}}{\partial u_i}\delta u_i + \frac{\partial\mathcal{L}}{\partial \dot{u}_i}\delta\dot{u}_i + \frac{\partial\mathcal{L}}{\partial u_{i,j}}\delta u_{i,j} \right] dx_j dt$$

(2.6)

Now, by interchanging the conventional derivatives and variational symbol, one can write

$$\delta\mathcal{L} = \int_{t_I}^{t_{II}} \int_v \left[\frac{\partial\mathcal{L}}{\partial u_i}\delta u_i + \frac{\partial\mathcal{L}}{\partial \dot{u}_i}\frac{d(\delta u_i)}{dt} + \frac{\partial\mathcal{L}}{\partial u_{i,j}}\frac{d(\delta u_i)}{dx_j} \right] dx_j dt$$

(2.7)

After performing the integral by parts of the second and third terms with respect to time and space, respectively in Eq. 2.7, can be shown as

$$\int_{t_I}^{t_{II}} \frac{\partial\mathcal{L}}{\partial \dot{u}_i}\frac{d(\delta u_i)}{dt}dt = \frac{\partial\mathcal{L}}{\partial u_i}\delta u_i \Big|_{t_I}^{t_{II}} - \int_{t_I}^{t_{II}} \frac{d}{dt}\left(\frac{\partial\mathcal{L}}{\partial \dot{u}_i}\right)\delta u_i dt$$

(2.8)

Please note, dx_j is absent because the above integral is over time only.

$$\int_{v_I}^{v_{II}} \frac{\partial\mathcal{L}}{\partial u_{i,j}}\frac{d(\delta u_i)}{dx_j}dx_j = \frac{\partial\mathcal{L}}{\partial u_{i,j}}\delta u_i \Big|_{v_I}^{v_{II}} - \int_{v_I}^{v_{II}} \frac{d}{dx_j}\left(\frac{\partial\mathcal{L}}{\partial u_{i,j}}\right)\delta u_i dx_j$$

(2.9)

Please note, dt is absent because the above integral is over space only. Substituting Eqs. 2.8 and 2.9 into Eq. 2.7, the variation of action modifies to

$$\delta\mathcal{L} = \int_{t_I}^{t_{II}} \int_v \left[\frac{\partial\mathcal{L}}{\partial u_i} - \frac{d}{dt}\left(\frac{\partial\mathcal{L}}{\partial \dot{u}_i}\right) - \frac{d}{dx_j}\left(\frac{\partial\mathcal{L}}{\partial u_{i,j}}\right) \right] dx_j \delta u_i dt$$

(2.10)

According to the Hamilton's principle, $\delta\mathcal{L} \to 0$, and hence

$$\frac{\partial \mathcal{L}}{\partial u_i} - \frac{d}{dt}\left(\frac{\partial \mathcal{L}}{\partial \dot{u}_i}\right) - \frac{d}{dx_j}\left(\frac{\partial \mathcal{L}}{\partial u_{i,j}}\right) = 0 \qquad (2.11)$$

The equation in Eq. 2.11 is the fundamental governing differential equation of the system that is evolving over space and time presented using the index notation. This equation is called the Euler-Lagrange equation [1]. The Euler-Lagrange equation may contain multiple derivative terms beyond the three terms presented in Eq. 2.11, depending on the number of axes in a coordinate system considered.

2.4 FUNDAMENTALS OF CONTINUUM MECHANICS

Conventionally, discussion on continuum mechanics starts with the concept of stress and strain. However, instead of starting our discussion from stress and strains, we will start our introduction to continuum mechanics [13–17] with the description of the coordinate systems. Appropriate use of the state variables in appropriate coordinate system is the most important concept required in continuum mechanics.

Let's assume a continuous object consists of numerous material particles interconnected by imaginary vectors shown in Fig. 2.2. A sample material point (P_1) is assumed to occupy a region in space that consists of the material volume ϑ_1. Let's assume the vector pointing to the material point P_1 can be written as $\mathbf{P}_1 = X_{11}\widehat{\mathbf{E}}_1 + X_{21}\widehat{\mathbf{E}}_2 + X_{31}\widehat{\mathbf{E}}_3$. Similarly, another material point (P_2) is assumed to occupy a region in space which consists of the material volume ϑ_2. Let's assume the vector pointing to the material point P_2 can be written as $\mathbf{P}_2 = X_{12}\widehat{\mathbf{E}}_1 + X_{22}\widehat{\mathbf{E}}_2 + X_{32}\widehat{\mathbf{E}}_3$. If the material body in Fig. 2.2

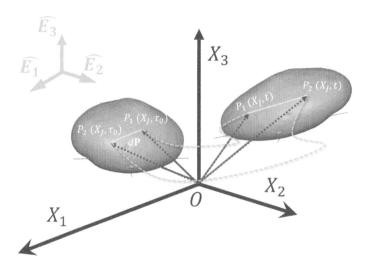

FIGURE 2.2 Deformation of a deformable body: Deformed position of two particles and a vector connecting them in Cartesian coordinate system is compared with their initial position and the vector before deformation.

Classical Mechanics and the Physics of Continua

has N numbers of material points, then the system will have N numbers of vectors that point to all the material points. We can call any one of them by $\mathbf{P}_N$. But the $\mathbf{P}_N$ here is not a static vector, because the material occupies the space is subjected to deform during wave propagation. The external boundary forces and the internal stresses in a material body are generated due to pure mechanical or coupled thermomechanical, electromechanical, mechanochemical, or biomechanical actions. Thus, an arbitrary point P_N can move in space relative to the other points in the material space, and as a result the components of the vector $\mathbf{P}_N$ change. Sometimes, these motions are small. Both the deformed and the undeformed states can be referred to by the same parent coordinate system fixed at the origin O (Fig. 2.2). In other words, it's like watching the event of the motion standing at a specific point in space. It means that the coordinate system, its axes, its origin, and the bases are static and are frozen in space and time. This also means that the observer (or the coordinate system) is fixed in space and time. This is applicable when the motion or deformation is small.

When the motion is large, the coordinate must move with the material points. It's like the observer is moving during the motion to track the particles. It is assumed that when the motion is complete, the observer comes to its final stop. Then it is obvious that the observer sees the final state of the material from its current location. But now, to describe the current state, can the observer refer to the current position of the points with respect to the original state from where both the object and the observer started in the first place? That means, in that description, we have to consider both the deformation of the body and the motion of the observer. Hence, the word 'both' here carries a significant mathematical meaning that many tend to confuse. In such description an additional velocity term appears that resembles the motion of the coordinate system itself. The first case, where the coordinate system or the observer is static, is called the Lagrangian coordinate or the material coordinate system. The second case, where the observer is moving with the motion, is called the Eulerian coordinate or the spatial coordinate system. In first case, we will designate the basis vectors of a material coordinate system with a capital $\hat{\mathbf{E}}$ and the points in space with capital X_1, X_2 and X_3 or simply the X_j. In second case, we will designate the basis vectors in a spatial coordinate system with a small $\hat{\mathbf{e}}$ and the points in space with small x_1, x_2 and x_3 or simply the x_j.

2.4.1 Lagrangian Coordinate or Material Coordinate System

Let's assume a material body at time $t = \tau_0$ in Fig. 2.2. If the material body has N numbers of material points, then the position vectors of N material particles can be described by their respective position using X_1, X_2 and X_3 variables in a fixed $\hat{\mathbf{E}}$ coordinate system, which has a fixed origin at O. These N position vectors can be written as

$$\mathbf{X}_N(X_J,t) = X_{IN}(t)\widehat{\mathbf{E}_I} \qquad (2.12)$$

where, I and J are the index indicated with capital letters used specifically for Lagrangian description. Here, I and J takes values 1, 2, and 3. Here, I is the dummy index and N is the free index. Next, extending the idea, an infinitesimal vector $d\mathbf{P}$ (see Fig. 2.2) in the material body can be similarly expressed as $d\mathbf{P} = dX_I \widehat{\mathbf{E}}_\mathbf{I}$. We can write the infinitesimal length of an element in the material body described in Lagrangian coordinate system [16] as

$$dL^2 = d\mathbf{P}.d\mathbf{P} = dX_I \widehat{\mathbf{E}}_\mathbf{I}.dX_J \widehat{\mathbf{E}}_\mathbf{J} = dX_I.dX_J \widehat{\mathbf{E}}_\mathbf{I}.\widehat{\mathbf{E}}_\mathbf{J} = dX_I.dX_I \quad (2.13)$$

where, $\widehat{\mathbf{E}}_\mathbf{I}.\widehat{\mathbf{E}}_\mathbf{J} = \delta_{IJ}$ is Kronecker delta.

Similarly, any tensor in a material coordinate system will be a function of the variables, X_1, X_2, and X_3 and time t. For example, stress – a second-order (2) tensor at any point in the material body that has ($2^3 = 9$) nine elements in a three-dimensional (3) coordinate system – is a function of X_1, X_2, and X_3, and time t and can be expressed as $\sigma_{IJ}(X_j,t)$.

2.4.2 Eulerian Coordinate or Spatial Coordinate System

Starting at time $t = \tau_0$, say a material body has deformed and moved to a new position after $t = t$. The material particles can now be described by a new coordinate system $\hat{\mathbf{e}}$ as shown in Fig. 2.3. The spatial coordinates of the particles can be described by x_1, x_2, and x_3 or simply the x_j. It is obvious that the new coordinates of any particle in the new coordinate system $\hat{\mathbf{e}}$ are inherently related to the original position of the particle in the old coordinate system $\widehat{\mathbf{E}}$. Hence, we can say

$$x_k = x_k(X_J,t) \quad (2.14)$$

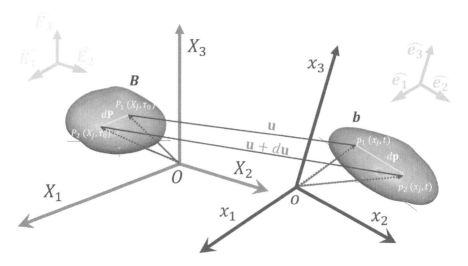

FIGURE 2.3 Schematic presentation of the original state of the material in the original coordinate system and deformed state in a new coordinate system that moves with the deformable body. Each point in the body is traced before and after deformation and shows the displacement of the respective points. Change in length of a vector in undeformed and deformed state with respect to their local coordinate system.

Classical Mechanics and the Physics of Continua 23

However, it is important to know the relation between the $\hat{\mathbf{E}}$ and $\hat{\mathbf{e}}$ coordinate system at all the instances during the motion. Please note that the $\hat{\mathbf{E}}$ and $\hat{\mathbf{e}}$ vectors are the contravariant vectors, and their relation gives us the Jacobian matrix. The Jacobian matrix that establishes the relation between the bases of $\hat{\mathbf{E}}$ and $\hat{\mathbf{e}}$ vectors transforming $\hat{\mathbf{E}}$ to $\hat{\mathbf{e}}$ can be written as

$$\mathbf{J} = jacobian = \left| \frac{\partial x_i}{\partial X_I} \right| = \begin{vmatrix} \dfrac{\partial x_1}{\partial X_1} & \dfrac{\partial x_1}{\partial X_2} & \dfrac{\partial x_1}{\partial X_3} \\[2mm] \dfrac{\partial x_2}{\partial X_1} & \dfrac{\partial x_2}{\partial X_2} & \dfrac{\partial x_2}{\partial X_3} \\[2mm] \dfrac{\partial x_3}{\partial X_1} & \dfrac{\partial x_3}{\partial X_2} & \dfrac{\partial x_3}{\partial X_3} \end{vmatrix} \tag{2.15}$$

Let's now consider Fig. 2.3 in detail. The material points P_1 and P_2 in body B are located at points p_1 and p_2 in body b, respectively after the deformation. Body B was located in $\hat{\mathbf{E}}$ coordinate system, and body b is now located in the new $\hat{\mathbf{e}}$ coordinate system. However, there is a possibility of the inverse motion in space, and points p_1 and p_2 in body b at time t can be traced back to points P_1 and P_2 in body B. Hence, it is feasible that the original coordinates of points in $\hat{\mathbf{E}}$, can be represented by the new coordinate system $\hat{\mathbf{e}}$ as follows.

$$X_K = X_K\left(x_j, t\right) \tag{2.16}$$

With this definition, we must state our assumption about a continuous body (B or b), which is fundamental to the continuum mechanics called axiom of continuity. It is assumed that the equations above for particle mapping are single valued, but they must have the continuous partial derivatives with respect to the arguments. There are a total of six equations in Eqs. 2.14 and 2.16 combined, and it is assumed that each member of this set of equations is a unique inverse of the other in a neighborhood of any material point, such as P_1 or P_2. This axiom of continuity restricts any region of positive volume in the material body to collapse to a zero volume or inflate to an infinite volume during the deformation. Because matter is *indestructible,* it also assumed that the matter is *impenetrable* and ensures that an infinitesimal material volume will never penetrate the other neighboring material volume. Hence, we can say that the material particles are infinitesimal deformable volumes that follow the above axiom to make a continuous material body.

Similarly, we can write the infinitesimal length of the element between the points p_1 and p_2 in the body b at time t described in Eulerian coordinate system as

$$dl^2 = d\mathbf{p}.d\mathbf{p} = dx_i\widehat{e_i}.dx_j\widehat{\mathbf{e_j}} = dx_i.dx_j\widehat{e_i}.\widehat{e_j} = dx_i.dx_i,$$

where $\widehat{e_i}.\widehat{e_j} = \delta_{ij}$ is Kronecker delta.

With this understating, next we will discuss the deformation of the material body in brief.

2.5 MOTION OF A DEFORMABLE BODY

The motion of a continuous medium can be seen as a movie of the deformation over time. In order to describe the motion of a body we must know the following definitions of the material derivatives, path lines, and streamlines. Path lines and stream lines frequently occur in fluid mechanics but carry some significance in wave propagation and thus are discussed in this chapter.

2.5.1 MATERIAL DERIVATIVES

Material derivatives are considered only with respect to time. Let's assume an arbitrary tensor $\mathbf{T}$. Now, to find the material time rate of change of $\mathbf{T}$, when the material axis (i.e., the $\hat{\mathbf{E}}$ coordinate system) is fixed for the original material points (keeping $\mathbf{X}$ constant), we can write the expression for a vector as

$$\frac{d\mathbf{T}}{dt} = \frac{\partial \mathbf{T}}{\partial t}\bigg|_{\mathbf{X}}; \quad \mathbf{T} = \mathbf{T}(\mathbf{X},t) = T_K(X_J,t)\hat{\mathbf{E}}_K; \quad \frac{d\mathbf{T}}{dt} = \frac{\partial T_K}{\partial t}\hat{\mathbf{E}}_K \qquad (2.17)$$

where, $\hat{\mathbf{E}}_K$ are the constant unit vectors and the tensor is defined as the material coordinate or the Lagrangian coordinate system as discussed in section 2.4.1. Next if the vector is a spatial function as described in section 2.4.2 (i.e., Eulerian coordinate system), then the material derivatives can be written as

$$\mathbf{T} = \mathbf{T}(x_j,t) = T_k(x_j,t)\widehat{\mathbf{e}_k}; \quad \frac{d\mathbf{T}}{dt} = \left(\frac{\partial T_k}{\partial t}\bigg|_{\mathbf{X}} + \frac{\partial T_k}{\partial x_j}\frac{\partial x_j}{\partial t}\right)\hat{\mathbf{e}}_k \qquad (2.18)$$

Here, the tensor is the function of the spatial coordinate system $\mathbf{x}$ with unit vectors $\hat{\mathbf{e}}$, and the spatial coordinates $\mathbf{x}$ are a function of material coordinates $\mathbf{X}$ with unit vector $\hat{\mathbf{E}}$ (refer to Eq. 2.14). Hence, we can write the material derivatives in a general form as follows:

$$\frac{d\mathbf{T}}{dt} = \dot{\mathbf{T}} = \frac{DT_k}{Dt}\hat{\mathbf{e}}_k; \quad where, \quad \frac{DT_k}{Dt} = \frac{\partial T_k}{\partial t} + T_{k,j}\frac{\partial x_j}{\partial t} \qquad (2.19)$$

where, the generalized material derivative $\dfrac{DT_k}{Dt}$ consists of two terms namely the summation of the nonstationary rate and the convective time rate, respectively. If THE Tensor is defined in the Lagrangian system, the convective time rate would be zero; or if the tensor is defined in the Eulerian coordinate system, both the terms will survive. With this understanding we can write the expression for acceleration (a_k) as rate a of change of velocity (v_k) in the Eulerian coordinate system as follows:

$$\frac{Dv_k}{Dt} = a_k(x,t) = \frac{\partial v_k}{\partial t} + v_{k,l}v_l \qquad (2.20)$$

With this similar understanding, material derivatives of a few common terms that are frequently used in continuum mechanics can be found in Ref [14], and expressions using similar notation presented in this book can be found in Ref [11]

Classical Mechanics and the Physics of Continua 25

2.5.2 Path Lines and Streamlines

Path lines and streamlines in a moving body are essentially related to the velocity field of the material particles as they appear in the material derivatives. Now imagine that all the material particles that are in motion in a continuous body are to be traced (each particle) with reference to a fixed coordinate origin. It's like following a bunch of racing cars after they started from the start line, with respect to the start line itself, which is known as $\mathbf{X}$. As the material particles are followed, the local coordinate changes continuously in the Eulerian coordinate system. The trace of these particles with respect to $\mathbf{X}$ as time passes is called the path line. We can write the equation of a path line as follows:

$$x_k = x_k\left(X_J,t\right)\text{when }\mathbf{X}\text{ is fixed} \qquad (2.21)$$

The path lines can also be achieved as an integral curve of $dx_k = v_k dt$ that passes through $\mathbf{X}$ at $t = 0$ [16]. That is joining the dots along which the particles have moved, visualized through a long exposure time.

Alternatively, stream lines are the curve that represents the tangent to the velocity field of the material particles at an instant of time. Stream lines could be achieved from the integral curves to $v_k = Cdx_k$, where, C is any arbitrary constant. This also means that along a streamline, the ratio of the incremental distance traveled by a material particle (at any instant along any direction) and the velocity of the particle along that same direction is always constant and can be presented as

$$\frac{dx_1}{v_1} = \frac{dx_2}{v_2} = \frac{dx_3}{v_3} = \frac{1}{C} \qquad (2.22)$$

The mathematical treatment above can be explained using the following example. Imagine a photograph of a waterfall. If we take the photograph of a beautiful waterfall using our smartphone, we will capture only one instant of the flow of the water. This picture from the smartphone shows the location of the water particles at the instant when the capture button was pressed. If we join the lines with similar velocity, we will get the stream line. However, if a professional photographer takes a picture of the waterfall with an expensive lens with a long exposure time, we will see a beautiful picture of the waterfall. In this picture we see the trace of the particles over time and see the white path lines that are nicely aligned. So, the smart phone picture is a Lagrangian device, but the expensive camera with long exposure time is a Eulerian device, just to understand the difference.

2.6 DEFORMATION AND STRAIN IN A DEFORMABLE BODY

In this section, we will refer to Figs. 2.3, 2.4a, and 2.4b to briefly understand the deformation of a deformable body. It is mentioned in previous sections that any arbitrary points P_1 and P_2 in the original body have moved to the new points p_1 and p_2, respectively, in the body b after deformation. Point P_1 is displaced by $\mathbf{u}$, where $\mathbf{u}$ is the displacement vector and has three components. Similarly, point P_2 at a $d\mathbf{P}$

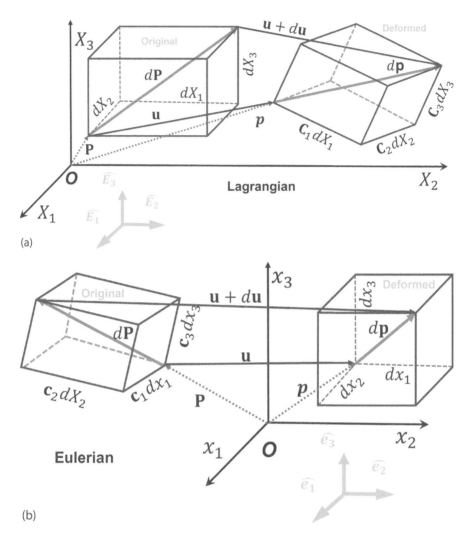

FIGURE 2.4 a) Lagrangian coordinate system showing how the deformation is measured in reference coordinate system. b) Eulerian coordinate system showing how the deformation is measured in deformed coordinate system.

distance from P_1 is displaced by $\mathbf{u} + d\mathbf{u}$. The new location of points p_1 and p_2 are at $d\mathbf{p}$ distance. Please refer to Fig. 2.4, which is a more elaborate description of the material element inside the body. In Lagrangian system (Fig. 2.4a) the deformation is observed with respect to a fixed coordinate system from where the deformation was started. However, in the Eulerian system (Fig. 2.4b) the material points are described with respect to its current position and are traced back to the original state from where the deformation was started. Please note that in Fig. 2.4 the $d\mathbf{P}$ vector is deformed to a $d\mathbf{p}$ vector, with the displacement of two opposite corners of the parallelepiped being $\mathbf{u}$ and $\mathbf{u} + d\mathbf{u}$, respectively. Original components of the $d\mathbf{P}$ vector

Classical Mechanics and the Physics of Continua

before deformation were dX_K. However, after deformation, all sides of the parallelepiped were changed due to the deformation and can be expressed by multiplying the deformation vectors to the respective components. In the following section, the original and deformed states are mathematically explained in both the Lagrangian and Eulerian systems in a tabular form. To understand the left column, please refer to Fig. 2.4a, and for the right column, please refer to Fig. 2.4b.

2.6.1 Cauchy's and Green's Deformation Tensors

Please refer to the definition of an arc length of an element line in a deformable body described above. Table 2.1 presents the step-by-step derivation of the deformation tensors in both coordinate systems.

2.6.2 Description of Strain in a Deformable Body

Please refer to Fig. 2.3, where the new square of the arc length of the element between the two material points p_1 and p_2 is dl^2. This new element length has been modified from the original length dL^2 between the two original material points P_1 and P_2. Hence the change in length can be measured in two coordinate system differently and are derived in Table 2.2.

2.6.3 Strain in Terms of Displacement

Please refer to Figs. 2.3 and 2.4, where the displacement of particles P_1 and P_2 from the original state to the new state located at p_1 and p_2 are $\mathbf{u}$ and $\mathbf{u} + d\mathbf{u}$, respectively. Next the resultant vectors [14] can be written as

$$d\mathbf{P} = d\mathbf{p} - d\mathbf{u} \ \ or \ \ d\mathbf{p} = d\mathbf{P} + d\mathbf{u} \tag{2.23}$$

Using the expression in T1.5, we can write the following equation that connects the deformation tensor and the displacement vectors (Table 2.1).

2.7 MASS, MOMENTUM, AND ENERGY

Instead of discussing stresses in a material body right after discussing strains, we will discuss a few axioms of continuum mechanics that will be used in the stress hypothesis. In order to describe the axioms of continuum mechanics mathematically, it is necessary to define the following terms in a mathematical language.

2.7.1 Mass of a Body

Mass is considered continuous in a material body and expressed using the local density of the body $\rho(\mathbf{x})$ per unit volume. The total mass of the body can be expressed as

$$M = \int_V \rho(\mathbf{x}) \, dv \tag{2.24}$$

TABLE 2.1

Cauchy's and Green's Deformation Tensors [14]

	Lagrangian *Description of the deformed state*	Eulerian *Description of the original state*
T1.1	$d\mathbf{p} = dx_k \hat{\mathbf{e}}_k$	$d\mathbf{P} = dX_K \hat{\mathbf{E}}_K$
T1.2	$dl^2 = d\mathbf{p}.d\mathbf{p}$ $= dx_k dx_l \hat{\mathbf{e}}_k.\hat{\mathbf{e}}_l$ $= \delta_{kl} dx_k dx_l$ $dl^2 = dx_k dx_k$	$dL^2 = d\mathbf{P}.d\mathbf{P}$ $= dX_K dX_L \hat{\mathbf{E}}_K.\hat{\mathbf{E}}_L$ $= \delta_{KL} dX_K dX_L$ $dL^2 = dX_K dX_K$
T1.3	$dx_k = \dfrac{\partial x_k}{\partial X_K} dX_K$	$dX_K = \dfrac{\partial X_K}{\partial x_k} dx_k$
T1.4	$dl^2 = \dfrac{\partial x_k}{\partial X_K} \dfrac{\partial x_k}{\partial X_K} dX_K dX_K$	$dL^2 = \dfrac{\partial X_K}{\partial x_k} \dfrac{\partial X_K}{\partial x_k} dx_k dx_k$
T1.5	$d\mathbf{p} = \dfrac{\partial x_k}{\partial X_K} dX_K \hat{\mathbf{e}}_k$ $= \left(\dfrac{\partial x_k}{\partial X_K} \hat{\mathbf{e}}_k \right) dX_K$ $d\mathbf{p} = \mathbf{C}_K dX_K$	$d\mathbf{P} = \dfrac{\partial X_K}{\partial x_k} dx_k \hat{\mathbf{E}}_K$ $= \left(\dfrac{\partial X_K}{\partial x_k} \hat{\mathbf{E}}_K \right) dx_k$ $d\mathbf{P} = \mathbf{c}_k dx_k$
T1.6	$dl^2 = d\mathbf{p}.\,d\mathbf{p}$ $= \mathbf{C}_K dX_K.\mathbf{C}_L dX_L$ $= \left(\mathbf{C}_K.\mathbf{C}_L \right) dX_K dX_L$	$dL^2 = d\mathbf{P}.d\mathbf{P}$ $= \mathbf{c}_k dx_k.\mathbf{c}_l dx_l$ $= \left(\mathbf{c}_k.\mathbf{c}_l \right) dx_k dx_l$
T1.7	$\mathbf{C}_K.\mathbf{C}_L = \dfrac{\partial x_k}{\partial X_K} \cdot \dfrac{\partial x_l}{\partial X_L} \delta_{kl}$ $= \dfrac{\partial x_k}{\partial X_K} \cdot \dfrac{\partial x_k}{\partial X_L}$ $\mathbf{C}_K.\mathbf{C}_L = C_{KL}(\mathbf{X},t)$ $C_{KL}(\mathbf{X},t)$ **is Green's deformation tensor**	$\mathbf{c}_k.\mathbf{c}_l = \dfrac{\partial X_K}{\partial x_k} \cdot \dfrac{\partial X_L}{\partial x_l} \delta_{KL}$ $= \dfrac{\partial X_K}{\partial x_k} \cdot \dfrac{\partial X_K}{\partial x_l}$ $\mathbf{c}_k.\mathbf{c}_l = c_{kl}(\mathbf{x},t)$ $c_{kl}(\mathbf{x},t)$ **is Cauchy's deformation tensor**
T1.8	Inverse $\mathbf{C}_K^{-1} = \dfrac{\partial X_K}{\partial x_k} \hat{\mathbf{e}}_K$ $\mathbf{C}_K^{-1}.\mathbf{C}_L^{-1} = \dfrac{\partial X_K}{\partial x_k} \cdot \dfrac{\partial X_L}{\partial x_l} \delta_{kl}$ $C_{KL}^{-1} = B_{KL}(\mathbf{X},t)$ $C_{KL}^{-1} C_{LM} = \delta_{KM}$ $C_{KL}^{-1} = B_{KL}$ *is the Piola deformation tensor* **Right Cauchy-Green tensor**	Inverse $\mathbf{c}_k^{-1} = \dfrac{\partial x_k}{\partial X_K} \hat{\mathbf{E}}_K$ $\mathbf{c}_k^{-1}.\mathbf{c}_l^{-1} = b_{kl}(\mathbf{x},t)$ $c_{kl}^{-1} = \dfrac{\partial x_k}{\partial X_K} \cdot \dfrac{\partial x_l}{\partial X_L} \delta_{KL}$ $c_{kl}^{-1} c_{lm} = \delta_{km}$ $c_{kl}^{-1} = b_{kl}$ *is the Finger deformation tensor* **Left Cauchy-Green tensor**

Note: Please note that the bold parameters are the vectors and the tensors are expressed with their indices.

Classical Mechanics and the Physics of Continua

TABLE 2.2

Lagrangian and Eulerian Strain Tensor and Strain Rate [13–15]

	Lagrangian (Fig. 2.4a)	Eulerian (Fig. 2.4b)
T2.1	Change in length.	Change in length.
	$\begin{aligned} dl^2 - dL^2 &= dx_k dx_k - dX_k dX_k \\ &= C_{KL}(\mathbf{X},t) dX_k dX_L - dX_k dX_k \\ &= dX_K dX_L\left(C_{KL}(\mathbf{X},t) - \delta_{KL}\right) \end{aligned}$	$\begin{aligned} dl^2 - dL^2 &= dx_k dx_k - dX_k dX_k \\ &= dx_k dx_k - c_{kl}(\mathbf{x},t) dx_k dx_l \\ &= dx_k dx_l\left(\delta_{kl} - c_{kl}(\mathbf{x},t)\right) \end{aligned}$
T2.2	*Lagrangian strain tensor*	*Eulerian strain tensor*
	$2\mathbb{E}_{KL} = C_{KL}(\mathbf{X},t) - \delta_{KL}$	$2e_{kl} = \delta_{kl} - c_{kl}(\mathbf{x},t)$
	With respect to undeformed state Fig. 2.4a	With respect to deformed state Fig. 2.4b
T2.3	*Lagangian to Eulerian strain transformation*	*Eulerian to Lagrangian strain transformation*
	$e_{kl} = \mathbb{E}_{KL} X_{K,k} X_{L,l}$	$\mathbb{E}_{KL} = e_{kl} x_{k,K} x_{l,L}$
T2.4	*Lagrangian strain Rate Tensor*	*Eulerian strain rate tensor*
	$2\dot{\mathbb{E}}_{KL} = \dot{C}_{KL}(\mathbf{X},t)$	$2\dot{e}_{kl} = -\dot{c}_{kl}(\mathbf{x},t)$
	$\dot{C}_{KL}(\mathbf{X},t) = \dfrac{D}{Dt}\left(\dfrac{\partial x_k}{\partial X_K} \cdot \dfrac{\partial x_k}{\partial X_L}\right)$	$\dot{c}_{kl}(\mathbf{x},t) = \dfrac{D}{Dt}\left(\dfrac{\partial X_K}{\partial x_k} \cdot \dfrac{\partial X_L}{\partial x_l}\right)$
	using Eq. 3.12 we get	using T2.3 and Eq. 3.13 we get
	$\dot{\mathbb{E}}_{KL} = \dfrac{1}{2}\left(v_{k,l} + v_{l,k}\right)\dfrac{\partial x_k}{\partial X_K} \cdot \dfrac{\partial x_l}{\partial X_L}$	$\dot{e}_{kl} = \dfrac{D}{Dt}\left(\mathbb{E}_{KL} X_{K,k} X_{L,l}\right)$
	Or $\dot{\mathbb{E}}_{KL} = d_{kl} x_{k,K} x_{l,L}$	as $\dot{\mathbb{E}}_{KL} = d_{kl} x_{k,K} x_{l,L}$
	Where, $d_{kl} = \dfrac{1}{2}\left(v_{k,l} + v_{l,k}\right)$	$\dot{e}_{kl} = d_{kl} - e_{pk} v_{p,l} - e_{pl} v_{p,k}$
	$\dot{C}_{KL}(\mathbf{X},t) = 2d_{kl} x_{k,K} x_{l,L}$	$\dot{c}_{kl}(\mathbf{x},t) = -c_{pk} v_{p,l} - c_{pl} v_{p,k}$
	Initial condition when the medium is not strained.	Initial condition when the medium is not strained.
	$\dot{C}_{KL}(\mathbf{X},0) = 2d_{kl}\delta_{kK}\delta_{lL}$	$\dot{c}_{kl}(\mathbf{x},0) = -2d_{kl}$

However, the total mass with discrete lumped masses distributed in a continuous body will give the following equation for the total mass as

$$M = \int_V \rho dv + \sum_\alpha M_\alpha \tag{2.25}$$

where, M_α is the lumped mass at the α-th point in the body.

2.7.2 Momentum of a Deformable Body

The momentum of a discrete particle is defined as the product of the mass of the particle and the velocity of the particle in a discrete sense. However, in a continuous medium, the density of the body can be different at different points. Similarly,

30 Metamaterials in Topological Acoustics

TABLE 2.3
Strain-Displacement Relation

	Lagrangian	Eulerian
T3.1	$$\mathbf{C}_K = \frac{\partial \mathbf{p}}{\partial X_K} = \frac{\partial \mathbf{P}}{\partial X_K} + \frac{\partial \mathbf{u}}{\partial X_K}$$	$$\mathbf{c}_k = \frac{\partial \mathbf{P}}{\partial x_k} = \frac{\partial \mathbf{p}}{\partial x_k} - \frac{\partial \mathbf{u}}{\partial x_k}$$

Lagrangian

T3.1

$$\mathbf{C}_K = \frac{\partial \mathbf{p}}{\partial X_K} = \frac{\partial \mathbf{P}}{\partial X_K} + \frac{\partial \mathbf{u}}{\partial X_K}$$

Displacement vector

$$d\mathbf{u} = dU_M \hat{\mathbf{E}}_M$$

M is an arbitrary index takes values, 1, 2, and 3

T3.2

$$\mathbf{C}_K = \hat{\mathbf{E}}_K + U_{M,K}\hat{\mathbf{E}}_M$$

T3.3 **Green's deformation tensor**

$$C_{KL} = \mathbf{C}_K \cdot \mathbf{C}_L$$
$$= \left(\hat{\mathbf{E}}_K + U_{M,K}\hat{\mathbf{E}}_M\right) \cdot \left(\hat{\mathbf{E}}_L + U_{N,L}\hat{\mathbf{E}}_N\right)$$
$$= \delta_{KL} + U_{L,K} + U_{K,L} + \delta_{NM}U_{M,K}U_{N,L}$$

T3.4 Substituting the expression for deformation tensor in the **Lagrangian strain tensor** expression in T2.2, we get

$$2\mathbb{E}_{KL} = U_{L,K} + U_{K,L} + \delta_{NM}U_{M,K}U_{N,L}$$

T3.5 With the study of small deformation, the nonlinear terms can be approximated to zero.

Lagrangian strain tensor

$$\mathbb{E}_{KL} = \frac{1}{2}\left(U_{K,L} + U_{L,K}\right)$$

Eulerian

$$\mathbf{c}_k = \frac{\partial \mathbf{P}}{\partial x_k} = \frac{\partial \mathbf{p}}{\partial x_k} - \frac{\partial \mathbf{u}}{\partial x_k}$$

Displacement vector

$$d\mathbf{u} = du_m \hat{\mathbf{e}}_m$$

m is an arbitrary index takes values, 1, 2, and 3

$$\mathbf{c}_k = \hat{\mathbf{e}}_k - u_{m,k}\hat{\mathbf{e}}_m$$

Cauchy's deformation tensor

$$c_{kl} = \mathbf{c}_k \cdot \mathbf{c}_l$$
$$= \left(\hat{\mathbf{e}}_k - u_{m,k}\hat{\mathbf{e}}_m\right) \cdot \left(\hat{\mathbf{e}}_l - u_{n,l}\hat{\mathbf{e}}_n\right)$$
$$= \delta_{kl} - u_{l,k} - u_{k,l} + \delta_{nm}u_{m,k}u_{n,l}$$

Substituting the expression for deformation tensor in the **Eulerian strain tensor** expression in T2.2, we get

$$2e_{kl} = u_{l,k} + u_{k,l} - \delta_{nm}u_{m,k}u_{n,l}$$

With the study of small deformation, the nonlinear terms can be approximated to zero.

Eulerian strain tensor

$$e_{kl} = \frac{1}{2}\left(u_{l,k} + u_{k,l}\right)$$

---

the velocities can also be different at different material particles. Hence, the volume integral of momentum $\mathbf{P}$ can be expressed as

$$\mathbf{P} = \int_V \rho(\mathbf{x})\,\vec{\mathbf{v}}\,dv \text{ or } P_i = \int_V \rho v_i dv \tag{2.26}$$

where $\vec{\mathbf{v}} = \dot{\mathbf{x}}$

2.7.3 Angular Momentum of a Deformable Body

The moment of momentum is called the angular moment, which is a cross product between the momentum of a particle and the distance of the particle from the axis about which the moment is taken. Hence, in a continuous medium the total angular momentum can be expressed by the volume integral of the local angular momentum and can be expressed as follows:

$$\mathbf{H} = \int_V \vec{\mathbf{p}} \times \rho(\mathbf{x})\,\vec{\mathbf{v}}\,dv \text{ or } H_i = \int_V \rho(x_m) \in_{ijk} x_j v_k\,dv \tag{2.27}$$

Classical Mechanics and the Physics of Continua

where $\vec{\mathbf{p}}$ is the respective distance of the material particles in a continuous body from a specific point about which the total angular momentum is to be calculated. Hence, the vector $\vec{p}$ inherently depends on the position vectors of the particles. Here, the fixed point was considered point O in the Fig. 2.3.

2.7.4 KINETIC ENERGY STORED IN A DEFORMABLE BODY

Kinetic energy of a material particle is defined as the half of the product of the mass of the particle and the square of the velocity of the particle. In a continuous body, specifically in metamaterials, as density and the velocity of the material particles are changing across the body, the total kinetic energy can be written as the volume integral of the local kinetic energy. The volume integral of the local kinetic energy stored in a deformable body can be expressed as

$$K = \frac{1}{2} \int \rho(\mathbf{x}) \left(\vec{\mathbf{v}}.\vec{\mathbf{v}} \right) dv \text{ or } K = \frac{1}{2}\int_v \rho v_k v_k \, dv \tag{2.28}$$

where, k is the index notation taking values 1, 2, and 3 and v_k is the particle velocity along the k-th direction.

2.8 FUNDAMENTAL AXIOM OF CONTINUUM MECHANICS

Axioms are statements that are considered universally true and should be used as fundamental assumptions that are regarded as being established, accepted, or self-evidently true. We may find the following axioms reasonable for our further discussion, which will provide key relationships without which the mechanics of continuum are impossible.

2.8.1 AXIOM 1: PRINCIPLE OF CONSERVATION OF MASS

As mass cannot be created or destroyed, it is reasonable to assume that the total mass in a continuous body is unchanged during the motion. If the statement is valid for a small region around a material particle, it can be said that the mass is conserved locally. But if the mass is locally conserved, then it is obvious that the total mass of a continuous body is also conserved globally. To ensure this condition, we can say that the mass before and after deformation is equal. Alternatively, the rate of change of mass of a continuous body is equal to zero during the motion, which gives the following equations.

$$\int_v \rho(\mathbf{x}) \, dv = \int_v \rho_0(\mathbf{X}) dv \ or \ \frac{D}{Dt}\int_v \rho dv = \int_v \frac{D}{Dt}(\rho dv) = 0 \tag{2.29}$$

where the local density of the body is expressed as $\rho_0(\mathbf{X})$ and $\rho(\mathbf{x})$, before and after deformation, respectively. The above integral can be expressed both in material and spatial coordinate system as describe in Table 2.4.

TABLE 2.4

Conservation of Mass in Lagrangian and Eulerian System

	Lagrangian (Fig. 2.4a)	Eulerian (Fig. 2.4b)
T4.1	$\int_{\mathbb{V}} (\rho_0 - \rho \mathbf{J})\, d\mathbb{V} = 0$	$\int_{\mathbb{V}} (\rho - \rho_0 \mathbf{J}^{-1})\, dv = 0$

2.8.2 Axiom 2: Principle of Balance of Momentum

It is fundamentally like Newton's second law, which says that the rate of change of velocity of a particle is directly proportional to the force applied to the particle along the direction of the motion. However, in this axiom it is said that the rate of change of momentum ($\boldsymbol{P}$) of a continuous body as an integral of the local momentum in a particular direction is equal to the resultant force ($\boldsymbol{F}$) acting on the body along the same direction.

$$\frac{d\boldsymbol{P}}{dt} = \boldsymbol{F}, \qquad \frac{D}{Dt} \int_{\mathbb{V}} \rho v_k \, dv = F_k \tag{2.30}$$

2.8.3 Axiom 3: Principle of Balance of Angular Momentum

It is established that a material body under deformation and motion should be under balance, when the total forces in all three directions and the moments with respect to all three axes are conserved. That means that the internal forces and moments in the body are equal to the applied forces and moments on the body. The balances of forces are explained in Eq. 2.30, but the balance of moments can be explained by the following equations. The rate of change of integral of the total local moment of momentum ($\boldsymbol{H}$) or the angular momentum should be equal to the total applied global moment ($\boldsymbol{M}$) as written in Eq. 2.31:

$$\frac{d\boldsymbol{H}}{dt} = \boldsymbol{M} \ or \ \frac{D}{Dt} \int_{v} \rho \epsilon_{klm} x_l v_m dv = M_k \tag{2.31}$$

where ϵ_{klm} is the permutation symbol, also called the Levi-Civita antisymmetric tensor.

2.8.4 Axiom 4: Principle of Conservation of Energy

During and after the deformation, it is necessary that the total energy of the system should be conserved. This is because the energy cannot be destroyed or created during an enclosed process but could be transformed from one form to another. Hence, the rate of change of addition of the total kinetic energy and the total internal energy is equal to the total work done by the external forces plus other external or internal energy (U_α) that enters or leave the system, respectively. Based on this axiom, the equation for continuum mechanics can be written as follows

$$\frac{D}{Dt}(\mathcal{K} + \mathcal{E}) = \mathcal{W} + \sum_\alpha U_\alpha \tag{2.32}$$

Classical Mechanics and the Physics of Continua   33

where $\mathcal{K}$ and E are the total kinetic energy and total internal energy, respectively. $\mathcal{W}$ is the work done on the system by the external forces per unit time. The other external energy that can enter or leave the system per unit time can be classified as heat energy, electrical energy, and chemical energy. According to this axiom, the energies are additive. If the internal energy density per unit mass is defined as ε, then the total internal energy can be written as follows.

$$\mathcal{E} = \int_v \rho\varepsilon \, dv \tag{2.33}$$

The internal energy in a thermodynamic sense is a state function that is independent of any process. *In continuum mechanics, the total internal energy is the summation of the total internal strain energy, the total internal heat energy and the total dissipative stress power, which is equal to the product of dissipative stress and the symmetric part of the velocity gradient during motion.*

2.9 INTERNAL STRESS STATE IN A DEFORMABLE BODY

A deformable body subjected to external and internal forces under motion must satisfy the global laws of motion. It also must follow the principle of least action that is described in section 2.2 and 2.3, thus following the Euler-Lagrange equation in Eq. 2.11. During deformation or motion such as happens during wave propagation in a medium, the internal forces are the result of action and reaction between two neighboring particles. These internal forces act along a line joining two particles, following Newton's third law. Hence, the resultant force between the pair of particles is zero but the internal stresses are nonzero. Augustin-Louis Cauchy [14] first explained the internal stress state in a deformable body based on a hypothesis called stress hypothesis. The orthogonal surfaces cutting through the line joining any two neighboring particles interact with each other through traction forces to keep two neighboring elements in balance. In Fig. 2.5, a 3D arbitrary object is considered, and an arbitrary white particle block is identified to be interrogated. The particle block identified might have multiple particles inside, and each of them can be further imagined to be surrounded by six imaginary planes, making a 3D array of imaginary cubes enclosing each particle. Taking out an arbitrary cube from that system, the stress state is explained. The traction ($\mathbf{T}^{(\hat{e}_j)}$ or T_j in Fig. 2.5) acting on the surfaces along the line joining the neighboring particles must have three components in a three-dimensional Cartesian coordinate system. The components on each plane have the normal stresses perpendicular to the plane and the shear stresses along the surface of the planes as shown in the Fig. 2.5. In a three-dimensional coordinate system, there are six such planes through which the neighboring particles are interacting with each other. Hence, there are six traction forces, out of which only three are unique to keep the element (the imaginary cube enclosing a single particle) in equilibrium. In other words, all planes have one resultant direction of stress but have three components with one normal stress and two shear stresses, according to Cauchy as shown in Fig. 2.5. $\mathbf{T}^{(\hat{e}_j)}$ or T_j or the three values $T_1, T_2,$ and $T_3,$ are sufficient to describe the stress state at

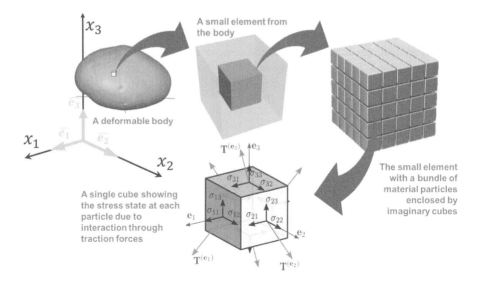

FIGURE 2.5 A 3D arbitrary object, its smallest part enclosing a material particle and a cube on which stress state is expressed.

a point. However, each traction has three components; thus nine values of stresses are unique to a point. These nine values are described in a two-dimensional tensor form as σ_{ij}, where i and j take values 1, 2, and 3. It is now obvious that the traction and stresses at a point are related to each other in terms of the normal direction of the six planes that are imagined surrounding the material particle. If the normal direction of each plane is expressed with respect to their direction cosine with n_1, n_2, and n_3, then the traction and the stresses are related using the following equation

$$T_i = \sigma_{ij} n_j \tag{2.34}$$

The stress tensor σ_{ij} can be written in a matrix form as follows:

$$\sum = \sigma_{ij} = \begin{bmatrix} \sigma_{11} & \sigma_{12} & \sigma_{13} \\ \sigma_{21} & \sigma_{22} & \sigma_{23} \\ \sigma_{31} & \sigma_{32} & \sigma_{33} \end{bmatrix} \tag{2.35}$$

The normal stresses in Eq. 2.35 occupy the diagonal elements, and the off-diagonal elements are the shear stresses. The matrix is symmetric because the shear stresses are in balance and the local concentrated moment at the point is assumed to be zero. Hence, there are six independent stress components in a two-dimensional tensor.

Sometimes, it is necessary to know the stress state on another plane, particularly to know the principal direction of the stress state in which the normal stresses are maximum but no shear stresses are present. A second-order tensor could be

Classical Mechanics and the Physics of Continua

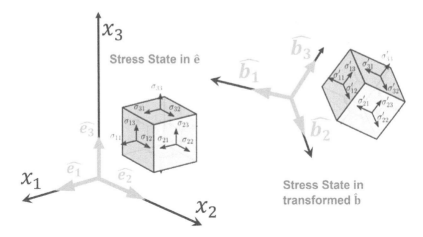

FIGURE 2.6 Stress state of a 3D stress element in an undeformed and deformed coordinate system, the deformed stress element shows the transformed stress states.

transformed from a coordinate system with basis vector $\hat{e}$ to a different coordinate system with basis vector $\hat{b}$. Here we write the stress state on a coordinate system with basis vector $\hat{b}$ as shown in Fig. 2.6. The stress state (σ'_{ij}) in the coordinate system with basis vector $\hat{b}$ can be written as

$$\sigma'_{ij} = \Psi_{im} \Psi_{nj} \sigma_{mn} \tag{2.36}$$

where Ψ_{im} and Ψ_{nj} are the second-order tensor called transformation matrix that holds the relation between the $\hat{e}$ and $\hat{b}$ coordinate systems.

2.10 EXTERNAL AND INTERNAL LOAD ON A DEFORMABLE BODY

A deformable body is subjected to deformation when the body is exposed to loads. These loads could be external and internal as shown in Fig. 2.7 on the surface Γ and inside the material body Ω, respectively. They can be categorized into three types:

a. Extrinsic body loads are the forces and couples that act externally on the body. Generally, these forces act through the center of mass of the body and the load density per unit mass is assumed under this category. Gravity and electrostatics forces are generally considered as extrinsic body forces. Let **f** be the body force per unit mass.

b. Extrinsic surface loads are the forces and couples that act externally on the surface of the body. Extrinsic surface forces generally arise from the interaction between two bodies (solid or fluid) through surface contacts. Surface force per unit area is called the surface traction, and the couple per unit area is called the surface couple. Hydrostatic forces, surface traction by electrostatic field, and piezoelectric effects are considered in this category.

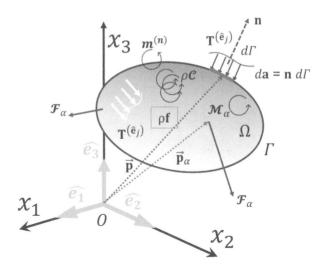

FIGURE 2.7 A deformable body with external and internal loads on the surface Γ and in Ω.

Boundary condition in continuum mechanics is greatly influenced by these surface tractions and surface couples. Let $\mathbf{T}^{(n)}$ be the traction force per unit surface area on the surface with direction normal $(\mathbf{n})$.

c. Intrinsic loads arise from mutual interaction between the neighboring particles as shown in Fig. 2.5. Two particles interact with their mutual traction forces. These forces (as traction on the surfaces of the cubes, Figs. 2.5 and 2.6) act along the lines connecting the points but act in opposite directions, causing the resultant internal force to be equal to zero. Traction force on any surface will have three components in a Cartesian coordinate system. The three force components per unit area will give three stresses on any surface shown in Figs. 2.5 and 2.6. Assuming concentrated force ($\mathcal{F}_\alpha$) and concentrated couple ($\mathcal{M}_\alpha$) acting on the body at the points $\vec{p}_\alpha$ (Fig. 2.7), we can write the total external force F and the total moment M as follows [14]:

$$\mathbf{F} = \oint_a \mathbf{T}^{(\hat{e}_j)} \cdot d\mathbf{a} + \int_v \rho\, \mathbf{f}\, dv + \sum_\alpha \mathcal{F}_\alpha \tag{2.37}$$

$$\mathbf{M} = \oint_a \left(m^{(n)} + \vec{p} \times \mathbf{T}^{(\hat{e}_j)} \right) d\mathbf{a} + \int_v \rho(\mathcal{C} + \vec{p} \times \mathbf{f}) dv$$
$$+ \sum_\alpha (\mathcal{M}_\alpha + \vec{p}_\alpha \times \mathcal{F}_\alpha) \tag{2.38}$$

where $\mathcal{C}$ is the couple per unit mass, $m^{(n)}$ is the surface couple per unit area on the surface with direction normal (n), and $\vec{p}$ is the vector that signifies the directional distance between the traction and the point about which the moment is calculated in the body (Fig. 2.7).

Classical Mechanics and the Physics of Continua

2.11 ELASTODYNAMIC EQUATION OR THE FUNDAMENTAL WAVE EQUATION

Applying axioms 2 and 3 presented in sections 2.8.2 and 2.8.3, respectively the Eqs. 2.30 and 2.31 can be revised, where Eqs. 2.37 and 2.38 are used further as follows.

$$\frac{D}{Dt}\int_v \rho \mathbf{v}\, dv = \oint_a \mathbf{T}^{(\hat{e}_j)}.d\mathbf{a} + \int_v \rho\, \mathbf{f}\, dv + \sum_\alpha \mathcal{F}_\alpha \tag{2.39}$$

$$\frac{D}{Dt}\int_v \rho\left(\vec{\mathbf{p}}\times\mathbf{v}\right)dv = \oint_a \left(\boldsymbol{m}^{(n)} + \vec{\mathbf{p}}\times\mathbf{T}^{(\hat{e}_j)}\right)d\mathbf{a} + \int_v \rho\left(\mathcal{C}+\vec{\mathbf{p}}\times\mathbf{f}\right)dv$$
$$+ \sum_\alpha \left(\mathcal{M}_\alpha + \vec{\mathbf{p}}_\alpha \times \mathcal{F}_\alpha\right) \tag{2.40}$$

These are the equations for global balance of momenta of the entire body subjected to the external loads and moments. In case of a nonpolar material body, we can ignore the terms $\mathcal{C}$ and $\boldsymbol{m}^{(n)}$. Additionally, if the material body is not subjected to any concentrated forces or the concentrated local couples at a discrete point, we can ignore the end terms in Eqs. 2.39 and 2.40. Performing the material derivatives and applying the result from axiom 1 (Eq. 2.29), which says $\dfrac{D}{Dt}\int_V \rho dv = 0$, we get.

$$\int_v \rho \frac{d\mathbf{v}}{dt}\, d\mathbb{V} = \oint_a \mathbf{T}^{(\hat{e}_j)}.d\mathbf{a} + \int_v \rho\, \mathbf{f}\, dv \tag{2.41}$$

$$\int_v \rho\left(\vec{\mathbf{p}}\times\frac{d\mathbf{v}}{dt}\right)dv = \oint_a \left(\vec{\mathbf{p}}\times\mathbf{T}^{(\hat{e}_j)}\right)d\mathbf{a} + \int_v \rho\left(\vec{\mathbf{p}}\times\mathbf{f}\right)dv \tag{2.42}$$

The above equations can be written in index notations as follows:

$$\int_v \rho \frac{dv_k}{dt}\, dv = \oint_a \sigma_{kj}\, n_j\, d\Gamma + \int_v \rho\, f_k\, dv \tag{2.43}$$

$$\int_v \rho \in_{klm} x_l \frac{dv_m}{dt}\, dv = \oint_a \in_{klm} x_l \sigma_{mj} n_j\, d\Gamma + \int_v \rho \in_{klm} x_l f_m\, dv \tag{2.44}$$

where $d\Gamma$ is the elementary area on the surface of the body with direction cosine n_j. Now, applying divergence theorem, we could transfer the surface integral to the volume integral and the above equations modify to a local balance equation. Ignoring the volume integrals in all the terms, imposing the condition at each particle point, we get the governing differential equation.

$$\sigma_{ij,j} + \rho f_i = \rho \ddot{u}_i \tag{2.45}$$

38 Metamaterials in Topological Acoustics

and

$$\sigma_{ij} = \sigma_{ji} \qquad (2.46)$$

The above two equations are the two fundamental equations of continuum mechanics, one of which describes the elastodynamic behavior in the material (Eq. 2.45) and the other one says that in absence of local moment at any point in the material, the local stress matrix is always symmetric. These two equations are fundamental to the wave propagation in any media.

2.12 ENERGY CONCEPT OF CONTINUA AND ITS RELATION TO THE CLASSICAL MECHANICS

Connecting classical mechanics formulation with continuum mechanics formulation seems obvious but not easy. There are many articles that present the coherent understanding of both, to present continuum mechanics in the light of Hamiltonian mechanics [18–20]. Before we make the connection through the derivation of wave equation or the governing differential equation, it is necessary we understand the concept of energy in continua.

Wave propagation in a continuum must account for the conservation of energy in the body. During wave propagation in solid or fluid media, the body must satisfy the equation of equilibrium that leads to Eqs. 2.45 and 2.46. In addition to these equations, global conservation of energy must be satisfied during the propagation of the wave. As the particles in the body move with certain velocity during the wave propagation, the material body will be subjected to continuously changing kinetic energy ($\mathcal{K}$). In addition to the kinetic energy, the internal energy ($\mathcal{E}$) of body is also continuously changing. The internal energy comes from the change in the strain energy due to the deformation of the body. Dissipation of the energy is due to the material's damping. It is obvious that the wave propagation in a body can initiate only by probing external energy to the body. The work ($\mathcal{W}$) done by the external forces should contribute to the changes of the kinetic and internal energies. Considering this fact, we can write the general thermodynamic equation of the body imposing the principle of conservation of global energy, which says that the *time rate of change of the sum of the kinetic energy and the internal energy is equal to the work done on the body* and all the other energies supplied to the body such as heat energy, electrical energy, electromagnetic or chemical energy designated as U_α, when α signifies the identification of different non mechanical external energies. Hence, the conservation of global energy will be expressed as

$$\dot{\mathcal{K}} + \dot{\mathcal{E}} = \mathcal{W} + \sum_\alpha U_\alpha \qquad (2.47)$$

Ignoring all other energies except the heat energy ($\mathcal{Q}$) Eq. 2.47 is as follows:

$$\dot{\mathcal{K}} + \dot{\mathcal{E}} = \mathcal{W} + \mathcal{Q} \qquad (2.48)$$

Classical Mechanics and the Physics of Continua

2.12.1 Conservation of Local Energy from Global Energy Rule

Now it is time to expand the global energy equation in terms of the energies at local points. At any point, if the velocity of the particle with density ρ is $\mathbf{v} = v_j \hat{e}_j$ and displacement is $\mathbf{u} = u_j \hat{e}_j$ due to the wave propagation, then the global kinetic energy can be written as the volume integral of the local kinetic energy as follows:

$$\mathcal{K} = \frac{1}{2} \int_v \rho v_j v_j \, dv \quad v_j = \dot{u}_j \tag{2.49}$$

or

$$\mathcal{K} = \frac{1}{2} \int_v \rho (\dot{u}_j)^2 \, dv \tag{2.50}$$

Similarly, repeating Eq. 2.33, the global internal energy can be expressed as volume integral of local internal energy density per unit mass.

$$\mathcal{E} = \int_v \rho \, \varepsilon \, dv; \ \varepsilon = internal \ energy \ density \tag{2.51}$$

In Fig. 2.7 the exterior surface of the body is subjected to traction forces. The surface integral of the time rate of displacement (i.e., the velocity of a particle on the surface multiplied with the traction force) will be the measure of the work done on the body due to the surface loads. Similarly, the volume integral of the body force multiplied with the particle velocity will measure the work done on the body by the body forces. Hence, the total work done on the material body by the external and internal forces is summarized as follows.

$$\mathcal{W} = \oint_a \sigma_{ji} \dot{u}_i n_j \, d\Gamma + \int_v \rho \, f_i \dot{u}_i \, dv \tag{2.52}$$

If the material body is subjected to the thermal loads on the surface and the body, assuming the heat vector $\mathfrak{q}$ acting on unit surface area and the distributed heat source $\mathfrak{h}$ per unit mass, the heat energy $\mathcal{Q}$ can be written as

$$\mathcal{Q} = \oint_a \mathfrak{q} \cdot \mathbf{n} \, d\Gamma + \int_v \rho \, \mathfrak{h} \, dv \tag{2.53}$$

Applying the material derivatives (like Eq. 2.20) of kinetic energy density ($\dot{\mathcal{K}}$) and internal energy density ($\dot{\mathcal{E}}$) in Eulerian frame and applying the divergence theorem to transfer the surface integral to volume integral, we can write the energy terms in the Eq. 2.48 more elaborately:

$$\frac{D}{Dt} \left(\frac{1}{2} \int_v \rho \, (\dot{u}_k)^2 \, dv \right) + \frac{D}{Dt} \left(\int_v \rho \varepsilon dv \right) = \int_v \left(\sigma_{ji} \dot{u}_i \right)_{,j} dv$$

$$+ \int_v \rho \left(f_i \dot{u}_i \right) dv + \int_v \left(\mathfrak{q}_{k,k} + \rho \mathfrak{h} \right) dv \tag{2.54}$$

The material derivative of *kinetic energy* and *internal energy*, is

$$\dot{\mathcal{K}} = \frac{1}{2} \int \left[(\rho \; d\mathrm{v}) \frac{D}{Dt}(\dot{u}_k . \dot{u}_k) + \dot{u}_k \dot{u}_k \frac{D}{Dt}(\rho d\mathrm{v}) \right]$$

$$= \int_{\mathrm{v}} (\rho \dot{u}_k \; \ddot{u}_k + \frac{1}{2} \rho \dot{u}_k \dot{u}_k \dot{u}_{l,l} + \frac{1}{2} \dot{\rho} \dot{u}_k \dot{u}_k) \; d\mathrm{v} \tag{2.55}$$

$$\dot{\mathcal{E}} = \frac{D}{Dt} \int_{\mathrm{v}} \rho \varepsilon d\mathrm{v} = \int_{\mathrm{v}} (\rho d\mathrm{v}) \dot{\varepsilon} + \varepsilon \frac{D}{Dt}(\rho d\mathrm{v}) = \int_{\mathrm{v}} (\rho \dot{\varepsilon} + \varepsilon \dot{\rho} + \rho \varepsilon \dot{u}_{k,k}) d\mathrm{v} \tag{2.56}$$

Ignoring the volume integral, using the equation of conservation of mass and the conservation of linear momentum in Eqs. 2.29 and 2.30, the global conservation of energy in Eq. 2.54 will result a simpler form of an equation [14] for the *local conservation of energy*, which reads

$$\rho \dot{\varepsilon} = \sigma_{ji} \dot{u}_{i,j} + q_{k,k} + \rho h \tag{2.57}$$

2.12.2 CONSERVATION OF MECHANICAL ENERGY (KINETIC, INTERNAL, AND POTENTIAL ENERGY)

The body force vector (f_i) acting on a local particle can be described as a function of the position vector of the local particle. This is because a force field in a deformable body is developed due to an arbitrary potential $\mathbb{U}$ and can be expressed as the spatial derivatives of that potential function. For example, electrical and magnetic force fields contributed to the body force are the derivatives of the electrical and magnetic potential functions, respectively. Similarly, we can write, $f_i = -\mathbb{U}_{,i}$ and the work done in Eq. 2.52 can be revised to

$$\mathcal{W} = \oint_{\scriptscriptstyle\triangle} \sigma_{ji} v_i n_j \; d\Gamma - \int_{\mathrm{v}} \rho \; \mathbb{U}_{,i} v_i \; d\mathrm{v}$$

$$= \left[\oint_{\scriptscriptstyle\triangle} \sigma_{ji} v_i n_j \; d\Gamma - \dot{\mathcal{U}} + \int_{\mathrm{v}} \mathbb{U} \frac{D}{Dt} \left(\int_{\mathrm{v}} \rho d\mathrm{v} \right) \right] \tag{2.58}$$

where, $\mathcal{U} = \int_{\mathrm{v}} \rho \mathbb{U} d\mathrm{v}$ is the potential density function or the potential energy. Substituting $\mathcal{W}$ (Eq. 2.58) in Eq. 2.48, we can write the expression for the summation of kinetic energy, internal energy and potential energy as

$$\dot{\mathcal{K}} + \dot{\mathcal{E}} + \dot{\mathcal{U}} = \oint_{\scriptscriptstyle\triangle} \sigma_{ji} v_i n_j \; d\Gamma + \mathcal{Q} \tag{2.59}$$

where axiom 1 in continuum mechanics is enforced. Axiom 1 specifies that during a deformation process the total potential carried by the locally created or annihilated mass is equal to zero.

Classical Mechanics and the Physics of Continua

Now, if the deformable body during wave propagation is totally insulated, and the energy of the surface traction is zero, then we get the conservation of mechanical energy. When the local mass is conserved in an insulated body with vanishing surface integral of traction forces, the summation of kinetic energy, internal energy, and the potential energy is constant. Mathematically we can write

$$\mathcal{K} + \mathcal{E} + \mathcal{U} = constant \tag{2.60}$$

2.12.3 INTERNAL ENERGY AND STRAIN ENERGY

There has been a persistent misconception among students that concerns the internal energy and strain energy in a deformable body. On most occasions, they are considered the same, but that is not true. Here in this section, we explain the meaning of both. In the previous sections we introduced the potential function to explain the body force that gave us the conservation of mechanical energy in Eq. 2.60. However, a question may arise, Why only the body force? Why don't we also define the potential function for surface traction? The answer is yes, we can also define the potential for the surface traction $\sigma_{ji} = \sigma_{ij}$. However, the nature of surface traction is bit different from the body force. The body force is always recoverable. But the surface stresses can be divided into two parts, recoverable or reversible and non-recoverable or irreversible or dissipative. Please note that the potential functions can be defined only for the recoverable variables, such as body force. We define a potential function $\mathcal{T}$ to capture the recoverable part of the stress tensor multiplied with the spatial velocity gradient, i.e., $\sigma_{ji}\dot{u}_{i,j}$. Please recollect Table 2.3, where we showed that the strain is the gradient of the displacements. In general, strain energy is the product of stress, and the displacement gradient and strains are considered unknowns in the governing equations. The time rate of change of this strain energy is partially shown on the right side of the Eq. 2.59. Let's also recollect the volume integral of the Eq. 2.57 to describe the global balance of energy and rewrite the equation using the new functions we described in this section.

$$\int_v \rho\dot{e}dv = \int_v \rho\dot{T}dv + \int_v [\sigma]^D_{ji}\dot{u}_{i,j}dv + \int_v \left(q_{k,k} + \rho h\right)dv \tag{2.61.1}$$

where $[\sigma]^D_{ji}$ is the dissipative or irreversible part of the stress tensor [14]. When the conservation of mass is satisfied, we can express the above equation in the global form as follows.

$$\dot{\mathcal{E}} = \dot{\mathcal{T}} + \mathcal{D} + \mathcal{Q} \tag{2.61.2}$$

where $\mathcal{T} = \int_v \rho T dv$, is the total strain energy and $\mathcal{D} = \int_v [\sigma]^D_{ji} v_{i,j}dv$ is the total dissipative power. $\dot{\mathcal{T}}$ is the rate of change of strain energy. Hence, the total internal energy and the strain energy are not the same. However, in case of two scenarios described below they are equal, (i.e., internal energy is equal to the strain energy) if and only if

- a deformable body is totally insulated and the dissipated stress tensor is equal to zero, $\mathcal{E} = \mathcal{T}$, if $\mathcal{D} = \mathcal{Q} = 0$

or

- in a deformable body, the total dissipated energy is totally converted into heat energy, $\mathcal{E} = \mathcal{T}$ if, $\mathcal{D} + \mathcal{Q} = 0$

Hence, the above results (internal energy is equal to the strain energy) are true only under the condition of the *adiabatic process*. It implies that no heat source may be present in the body and no heat is lost or gained to and from the environment. Hence, in a thermodynamically admissible elastic solid, recollecting Eq. 2.57, without the loss of generality we can rewrite the equation of internal energy density as follows using the expressions in Table 2.3.

$$\rho\dot{\varepsilon} = \sigma_{kl}\dot{u}_{k,l} \quad \text{or} \quad \rho\varepsilon = \sigma_{kl}u_{k,l} \quad \text{or} \quad \rho\varepsilon = \sigma_{kl}e_{kl} \quad \text{or} \quad \rho\varepsilon = \sigma : e \text{ or} \qquad (2.62)$$

where different indices are used, but they still take values, 1, 2, and 3. As we just proved, under adiabatic conditions the internal energy is equivalent to the strain energy. We can express the internal energy density as a function of deformation gradient tensor or, more inclusively, Green's deformation tensor. Please note ':' is a symbol for matrix multiplication or double dot product.

2.12.4 Energy Flux Density and Poynting Vector

This topic is more relevant when the acoustic and elastic wave propagation is discussed in Chapter 3. In a thermodynamically admissible solid under dynamic deformation condition with small displacement, the energy flows in the body must satisfy the following condition. Followed by Eqs. 2.58 through 2.62 the following equation could be written.

$$\oint -\sigma_{ji}v_i n_j \, d\Gamma = \int_v v_i f_i \, dv - \int_v \left(\rho v_i \frac{\partial v_i}{\partial t} + \sigma_{kl}\dot{u}_{k,l} \right) dv \qquad (2.63.1)$$

Or in vector notation

$$\oint -\left(\sigma^* . v \right) . \hat{n} \, d\Gamma = \int_v -v \cdot f \, dv + \int_v \left(\rho v \cdot \frac{\partial v}{\partial t} + \sigma : \frac{\partial e}{\partial t} \right) dv \qquad (2.63.2)$$

This is derived from the complex Poynting theorem [14]. Here, the left-hand side represented the complex power flow through the surface along the direction $\hat{n}$ is the direction of the Poynting vector. The right-hand side represented the power supplied to the body and the rate of change of stored energy in the body. The real part of the

Classical Mechanics and the Physics of Continua   **43**

surface integral is the average power and imaginary part is the reactive power. Along the Poynting vector $\mathbf{P}_E$, the energy flux density could be written as

$$\mathbf{P}_E = -\frac{1}{2} Re\left[\sigma^* . \mathbf{v}\right] \tag{2.63.3}$$

$\mathbf{P}_E$ describes the directional energy flux density of elastic wave. Similarly in acoustics, the Poynting vector $(\mathbf{P}_A)$ could be written as

$$\mathbf{P}_A = -\frac{1}{2} Re\left[\mathbb{p}^* . \mathbf{v}\right] \tag{2.63.4}$$

where $\mathbb{p}$ is the acoustic pressure field.

2.13   LAGRANGIAN FORMULATION OF WAVE EQUATION

Please refer to the Euler-Lagrange equation in Eq. 2.11 that is obtained from the Hamiltonian principle of least action in classical mechanics. To find the governing differential equation of wave propagation in a deformable body, the Lagrangian $\mathcal{L}$ functional in Eq. 2.11 should be first expressed as a function of the displacement (u_i), velocity $(\dot{u}_i)$, and displacement gradient $(u_{i,j})$ functions. Now, defining the Lagrangian as the difference of the kinetic energy and the potential energy, which consists of strain energy and potential energy due to body force in a deformable continuum, we can express the Lagrangian as follows.

$$\mathcal{L} = \mathcal{K} - (\mathcal{E} - \mathcal{U}) \tag{2.64}$$

If the displacement in the body is defined as u_i, the kinetic energy retrieved from Eq. 2.50, and the strain energy retrieved from Eqs. 2.51 through 2.63, substituting the integral form of kinetic energy and strain energy in Eq. 2.64, the Lagrangian takes the form

$$\mathcal{L} = \int_v \frac{1}{2} \rho (\dot{u}_i)^2 \, dv - \int_v \sigma_{ij} e_{ij} dv + \int_v \rho \mathbb{U} dv \tag{2.65}$$

Please note that the potential energy has a positive contribution to $\mathcal{L}$ (i.e., because the potential function is described with a negative sign in $f_i = -\mathbb{U}_{,i}$ in our assumption). Further, substituting $\mathcal{L}$ in Eq. 2.11 can be rewritten in Eq. 2.66, when the first term $\frac{\partial \mathcal{L}}{\partial u_i} = 0$.

$$-\frac{d}{dt}\left(\frac{\partial \mathcal{L}}{\partial \dot{u}_i}\right) - \frac{d}{dx_j}\left(\frac{\partial \mathcal{L}}{\partial u_{i,j}}\right) = 0 \tag{2.66}$$

Substituting $\mathcal{L}$ from Eq. 2.65 in Eq. 2.66, the Hamilton principle from classical mechanics results in

$$-\frac{1}{2}\frac{d}{dt}\left(\frac{\partial\rho(\dot{u}_i)^2}{\partial\dot{u}_i}\right)+\frac{d}{dx_j}\left(\frac{\partial\sigma_{ij}e_{ij}}{\partial u_{i,j}}\right)-\frac{d}{dx_j}\left(\rho\mathbb{U}\right)=0 \tag{2.67}$$

Using the expression of strain from Table 2.3 $e_{ij}=\frac{1}{2}\left(u_{i,j}+u_{j,i}\right)$, we can further simplify.

$$-\rho\ddot{u}_i+\frac{d}{dx_j}\left(\sigma_{ij}\frac{1}{2}\frac{\partial\left(u_{i,j}+u_{j,i}\right)}{\partial u_{i,j}}\right)-\frac{d}{dx_j}\left(\rho\mathbb{U}\right)=0 \tag{2.68}$$

$$-\rho\ddot{u}_i+\frac{d}{dx_j}\left(\sigma_{ij}\frac{1}{2}\left[1+1\right]\right)-\frac{d}{dx_j}\left(\rho\mathbb{U}\right)=0 \tag{2.69}$$

$$-\rho\ddot{u}_i+\left(\sigma_{ij,j}\right)-\rho\mathbb{U}_{,i}=0 \tag{2.70}$$

$$\sigma_{ij,j}+\rho f_i=\rho\ddot{u}_i \tag{2.71}$$

Because $f_i=-\mathbb{U}_{,i}$, equation Eq. 2.71 is the same as Eq. 2.45. The wave equation without body force will be

$$\sigma_{ij,j}=\rho\ddot{u}_i \text{ or } \nabla.\sigma=\rho\ddot{\mathbf{u}} \tag{2.72}$$

Here, ∇ is a nabla Hamiltonian operator or simple a gradient operator.

2.14 LEGENDRE TRANSFORM AND HAMILTONIAN FORMULATION OF WAVE EQUATION

At the beginning of the 19th century, Adrien-Marie Legendre proposed to transform a function of a set of variables to another function of its conjugate set of variables. When we say conjugate, that is nothing but the gradient of the original function. The Legendre transform is the key to present the Hamiltonian mechanics in a new form. The mechanics would be the same; however. Hamiltonian formulation is an alternative representation of the classical mechanics, which is immensely valuable in the formulation of quantum mechanics. The key concept of the Legendre transform is to present a function using two conjugate pair of variables, here displacement (u_i) and velocity ($\dot{u}_i$ or v_i), into another form of a function evaluating the intercept of the gradient of the function to the function axis. It seems complicated; however, pictographically it is very intuitive. In a nutshell, the Legendre transform alters the Lagrangian, which is the difference of kinetic energy and potential energy, to Hamiltonian, which is the addition of kinetic energy and potentials energy. This concept will be used in Chapter 5 while deriving the Schrodinger equation.

2.14.1 GRAPHICAL REPRESENTATION OF LEGENDRE TRANSFORM

Let's refer to Fig. 2.8, where any arbitrary concave function $\mathcal{L}$ is shown, which is a function of u_i and v_i. Assume u_i are the passive variables and v_i are the active variables and $\mathcal{L}$ is plotted against the v_i variables. Let's consider an arbitrary point on the v_i axis at v_i^0; distance is C. A vertical line passing through the point C intercepts the function $\mathcal{L}$ at A. If we compute the slope of the function $\mathcal{L}$ at point A, we get $\tan\varphi = (\partial \mathcal{L}/\partial v_i)$ @ v_i^0. It can be seen that the slope line starting from point A can be extended all the way to the $\mathcal{L}$ axis and can intercept at point B. Two orthogonal lines passing through points A and B, respectively, form a triangle with the vertices D. The height of the triangle AD is composed of AC+CD. In Fig. 2.8 the intercept I can be calculated subtracting AC from AD. The length AC in Fig. 2.8 is $a = \mathcal{L}(u_i, v_i^0)$. The total length of AD can be found from a simple trigonometry rule and can be found by multiplying v_i^0 with the slope $\tan\varphi$. Hence, $AD = v_i^0 \tan\varphi = a + I = \mathcal{L}(u_i, v_i^0) + I$. Using these terms, the intercept I can be written as

$$I = v_i^0 \tan\varphi - \mathcal{L}(u_i, v_i^0) = v_i^0 \frac{\partial \mathcal{L}}{\partial v_i} - \mathcal{L}(u_i, v_i^0) \tag{2.73}$$

Defining a new variable that is conjugate to v_i as p_i, the above equation will take the form

$$I = v_i^0 p_i^0 - \mathcal{L}(u_i, v_i^0) \tag{2.74}$$

The intercept I is the fundamental definition of the Legendre transform of the function $\mathcal{L}$.

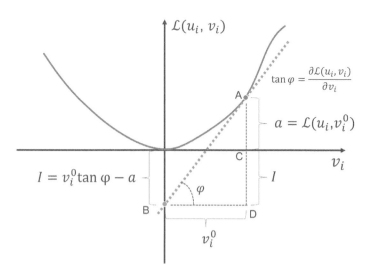

FIGURE 2.8 Derivation of Lagrangian with displacement and velocity (time derivative of displacement).

2.14.2 Hamiltonian Form with Legendre Transform

If the function above happens to be our Lagrangian $\mathcal{L}(u_i,\dot{u}_i) \equiv \mathcal{L}(u_i,v_i)$, then the Legendre transform of the Lagrangian can be written as

$$H(u_i, p_i) = p_i \dot{u}_i - \mathcal{L}(u_i, \dot{u}_i) \tag{2.75}$$

where volume integral over the entire domain is neglected to impose the condition at every point in the solid. This new function is called Hamiltonian of the system where index i takes values 1, 2, and 3 and standard index sum rule implies. The p_i is momentum conjugate variable to velocity. In the above equation we have considered only the velocity terms in the Lagrangian but ignored the spatial gradient of displacements. If we consider the Lagrangian as a function of displacement (u_i) and displacement gradients ($u_{i,j}$) and make the displacement gradients ($u_{i,j}$) as active variables, then we can reevaluate the Fig. 2.8 where the horizontal axis is modified to $u_{i,j}$ in Fig. 2.9. With a similar description above the intercept function as Legendre transform can be written as follows

$$I = u_{i,j}^0 \tan\varphi - \mathcal{L}(u_i, u_{i,j}^0) = u_{i,j}^0 \frac{\partial \mathcal{L}}{\partial u_{i,j}} - \mathcal{L}(u_i, u_{i,j}^0) \tag{2.76}$$

Defining a new variable that is conjugate to $u_{i,j}$, assuming $\dfrac{\partial \mathcal{L}}{\partial u_{i,j}}$ at $u_{i,j}^0$ equal to ϕ_{ij}^0, the above equation will take the form

$$I = u_{i,j}^0 \phi_{ij}^0 - \mathcal{L}(u_i, u_{i,j}^0) \tag{2.77}$$

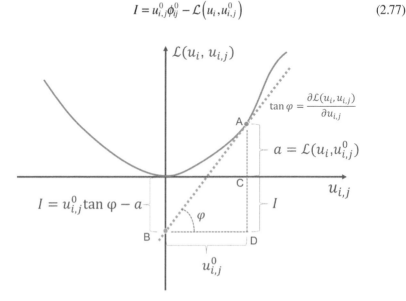

FIGURE 2.9 Derivation of Lagrangian with displacement and displacement gradient (spatial derivative of displacement).

Classical Mechanics and the Physics of Continua

Like Eq. 2.75, for an elastic body without the time dependent displacement or velocity, the Hamiltonian with stresses can be written as follows.

$$H\left(u_i,\phi_{ij}\right) = \phi_{ij}u_{i,j} - \mathcal{L}\left(u_i,u_{i,j}\right) \tag{2.78}$$

2.14.3 HAMILTON'S EQUATION OF MOTION

Considering the Hamiltonian in Eq. 2.75 further, we derive the equation of motion. Taking the derivative of the Hamiltonian (H), which is a function of u_i, p_i and time t as a contravariant vector can be written as

$$dH = \frac{\partial H}{\partial u_i}du_i + \frac{\partial H}{\partial p_i}dp_i + \frac{\partial H}{\partial t}dt \tag{2.79}$$

Similarly, taking the derivative of Lagrangian ($\mathcal{L}$), which is a function of u_i, $\dot{u}_i$ and time t as a contravariant vector can be written as

$$d\mathcal{L} = \frac{\partial \mathcal{L}}{\partial u_i}du_i + \frac{\partial \mathcal{L}}{\partial \dot{u}_i}d\dot{u}_i + \frac{\partial \mathcal{L}}{\partial t}dt \tag{2.80}$$

where the sum over index i is implied according to index notation. Further, taking the derivative of the Eq. 2.75, we get

$$dH\left(u_i,p_i,t\right) = p_i d\dot{u}_i + \dot{u}_i dp_i - d\mathcal{L}\left(u_i,\dot{u}_i,t\right) \tag{2.81}$$

Substituting Eq. 2.80 in Eq. 2.81, we get

$$dH\left(u_i,p_i,t\right) = p_i d\dot{u}_i + \dot{u}_i dp_i - \frac{\partial \mathcal{L}}{\partial u_i}du_i - \frac{\partial \mathcal{L}}{\partial \dot{u}_i}d\dot{u}_i - \frac{\partial \mathcal{L}}{\partial t}dt \tag{2.82.1}$$

Please note that $p_i = \partial \mathcal{L} / \partial \dot{u}_i$ is a conjugate variable to the $\dot{u}_i$, previously introduced in section 2.14.1 in Eqs. 2.73 and 2.74. Hence, substituting p_i and cancelling the term $p_i d\dot{u}_i$ the final form will be

$$dH\left(u_i,p_i,t\right) = -\frac{\partial \mathcal{L}}{\partial u_i}du_i + \dot{u}_i dp_i - \frac{\partial \mathcal{L}}{\partial t}dt \tag{2.82.2}$$

Comparing Eqs. 2.82 and 2.79, we can say

$$\frac{\partial H}{\partial u_i} = -\frac{\partial \mathcal{L}}{\partial u_i}; \quad \frac{\partial H}{\partial p_i} = \dot{u}_i; \quad \frac{\partial H}{\partial t} = -\frac{\partial \mathcal{L}}{\partial t} \tag{2.83}$$

Alternatively,

$$\frac{\partial \mathcal{L}}{\partial u_i} = \frac{d}{dt}\frac{\partial \mathcal{L}}{\partial \dot{u}_i} = \frac{d}{dt}p_i = \dot{p}_i \tag{2.84}$$

And hence, the Hamiltonian equation of motion can be written as

$$\frac{\partial H}{\partial u_i} = -\dot{p}_i \tag{2.85.1}$$

$$\frac{\partial H}{\partial p_i} = \dot{u}_i \tag{2.85.2}$$

Similarly, using the Hamiltonian written in Eq. 2.78, where displacement and displacement gradient contributed to the Lagrangian, will result a set of Hamiltonian equation as follows.

$$\frac{\partial H}{\partial u_i} = -\frac{\partial \phi_{ij}}{\partial x_j} \tag{2.86.1}$$

$$\frac{\partial H}{\partial \phi_{ij}} = \frac{\partial u_i}{\partial x_j} \tag{2.86.2}$$

Please note that in the above expression the displacement gradient or the strain is the active variable and displacement is the passive variable for deriving the Hamiltonian form through Legendre transform. The ϕ_{ij} is equivalent to the stress σ_{ij} function in solids.

2.15   CONSTITUTIVE LAW OF CONTINUA

At any point in a continuum, six stresses and three displacements are the unknowns and must be found from the balance equations to completely describe the wave field. Axioms of continuum mechanics says that in addition to globally satisfying boundary conditions, it is necessary for the local mass to be conserved. Also, the linear and angular momentums must also be conserved at the local points in the body. These four conditions gave the following equations.

$$\left. \begin{array}{c} \dfrac{\partial \rho}{\partial t} + \left(\rho v_i \right)_{,i} = 0 \\[2mm] \sigma_{ij,j} + \rho f_i = \rho \ddot{u}_i \\[2mm] \sigma_{ij} = \sigma_{ji} \end{array} \right\} \tag{2.87}$$

They are a total of four independent equations with six stresses and three displacements. For given displacements, the strains at a point can be calculated and will be automatically compatible. However, if the strains are unknown, not just any displacements satisfy the equations. Strains are the function of displacement gradients and must obey certain rules. Thus, we need to bring six new strain variables. The strain displacement equation is reiterated here from Table 2.3.

$$e_{kl} = \frac{1}{2}\left(u_{k,l} + u_{l,k}\right) \tag{2.88}$$

Classical Mechanics and the Physics of Continua

This will increase the number of unknowns and will give us ten equations but sixteen unknowns. Further, a compatibility condition will help make additional six equations that will bind the deformation of the system in a geometrically admissible form. Although not derived, the compatibility equation is given below.

$$e_{ji,lm} + e_{lm,ji} - e_{jm,il} - e_{il,jm} = 0 \qquad (2.89)$$

Indices take values 1, 2, and 3 without the loss of generality. In the case of displacement formulation, the displacements will automatically make the strains compatible, and Eq. 2.89 will be automatically satisfied. However, in stress formulation, the strain displacement relation in Eq. 2.88 cannot be exploited, but the compatibility condition in Eq. 2.89 must be imposed. In either case we have ten equations and sixteen unknowns. There is always a deficit of six equations. These six equations are supplied by the constitutive law of thermodynamically admissible elastic materials where stress and strains are related as follows.

$$\sigma_{kl} = \frac{\partial \Sigma(e_{kl})}{\partial e_{kl}} \qquad (2.90)$$

The work function $\Sigma(e_{kl})$ in Eq. 2.90 can be expressed as a polynomial of the strain components in a general nonlinear elastic solid as follows.

$$\Sigma(\mathbf{x}) = \Sigma_o(\mathbf{x}) + \Sigma_{kl}(\mathbf{x})e_{kl} + \frac{1}{2}\Sigma_{klij}(\mathbf{x})e_{kl}e_{ij} + \ldots \qquad (2.91)$$

where Σ_o, Σ_{kl}, and Σ_{klij} are the functions of position vector $\mathbf{x}$ in an inhomogeneous solid and are constants in homogeneous solids. Metamaterials are, in general, heterogeneous materials and thus will keep the functions in Eq. 2.91 as a function of position $(\mathbf{x})$. In linear material we could retain up to the quadratic terms. Applying Eq. 2.90 or 2.91 and with many mathematical and logical steps we get the following constitutive law.

$$\sigma_{ij} = \mathbb{C}_{ijkl}(\mathbf{x})e_{kl} \qquad (2.92)$$

where $\mathbb{C}_{ijkl}(\mathbf{x}) = \frac{1}{2}\left(\Sigma_{klij}(\mathbf{x}) + \Sigma_{ijkl}(\mathbf{x})\right)$ is the constitutive matrix (a fourth-order tensor) and to be experimentally determined. However, $\mathbb{C}_{ijkl}(\mathbf{x})$ is subjected to the restrictions as follows.

$$\mathbb{C}_{ijkl} = \mathbb{C}_{jikl} = \mathbb{C}_{jilk} \ ; \ \mathbb{C}_{ijkl} = \mathbb{C}_{klij} \qquad (2.93)$$

For the sake of the brevity, based on earlier discussion, we could further write a polynomial expansion of stress in general nonlinear materials as follows.

$$\sigma_{ij} = \mathbb{C}_{ij}(\mathbf{x}) + \mathbb{C}_{ijkl}(\mathbf{x})e_{kl} + \mathbb{C}_{ijklmn}(\mathbf{x})e_{kl}e_{mn} \qquad (2.94)$$

Please note that σ_{ij} is a second-order tensor with nine elements. Similarly, the strain tensor (Eq. 2.88) will also result in six independent strain elements in the strain matrix. Instead of writing them in a matrix, it is convenient to write their relation in a vector form. Hence, the stress matrix will become a vector of six stress elements, and so does the strain matrix. These six stresses and six strains are related by the constitutive law, resulted fourth-order constitutive tensor $\mathbb{C}_{ijkl}(\mathbf{x})$. However, it needs only 36 elements to be in the constitutive matrix to relate six stresses to six strains. Using first three equalities in Eq. 3.76, we could expand the stress-strain relation through constitutive matrix as follows.

$$
\begin{Bmatrix} \sigma_{11} \\ \sigma_{22} \\ \sigma_{33} \\ \sigma_{23} \\ \sigma_{31} \\ \sigma_{12} \end{Bmatrix} = \begin{bmatrix} \mathbb{C}_{1111} & \mathbb{C}_{1122} & \mathbb{C}_{1133} & \mathbb{C}_{1123} & \mathbb{C}_{1131} & \mathbb{C}_{1112} \\ \mathbb{C}_{2211} & \mathbb{C}_{2222} & \mathbb{C}_{2233} & \mathbb{C}_{2223} & \mathbb{C}_{2231} & \mathbb{C}_{2212} \\ \mathbb{C}_{3311} & \mathbb{C}_{3322} & \mathbb{C}_{3333} & \mathbb{C}_{3323} & \mathbb{C}_{3331} & \mathbb{C}_{3312} \\ \mathbb{C}_{2311} & \mathbb{C}_{2322} & \mathbb{C}_{2333} & \mathbb{C}_{2323} & \mathbb{C}_{2331} & \mathbb{C}_{2312} \\ \mathbb{C}_{3111} & \mathbb{C}_{3122} & \mathbb{C}_{3133} & \mathbb{C}_{3123} & \mathbb{C}_{3131} & \mathbb{C}_{3112} \\ \mathbb{C}_{1211} & \mathbb{C}_{1222} & \mathbb{C}_{1233} & \mathbb{C}_{1223} & \mathbb{C}_{1231} & \mathbb{C}_{1212} \end{bmatrix} \begin{Bmatrix} e_{11} \\ e_{22} \\ e_{33} \\ 2e_{23} \\ 2e_{31} \\ 2e_{12} \end{Bmatrix} \quad (2.95)
$$

Please note that the position dependency on $(\mathbf{x})$ in Eq. 2.95 and the following constitutive equations are dropped without the loss of generality. Further, using the fourth equality in Eq. 2.93 obtained from consistent strain energy density function or the work function, we have fewer independent coefficients. Because of this thermodynamically admissible state of linear elastic material there are 21 independent constitutive coefficients in the $\mathbb{C}_{ijkl}(\mathbf{x})$ matrix. The matrix in Eq. 2.95 should be a symmetric matrix. Hence, we can redefine the Eq. 2.95 in the following form.

$$
\begin{Bmatrix} \sigma_{11} \\ \sigma_{22} \\ \sigma_{33} \\ \sigma_{23} \\ \sigma_{31} \\ \sigma_{12} \end{Bmatrix} = \begin{bmatrix} \mathbb{C}_{11} & \mathbb{C}_{12} & \mathbb{C}_{13} & \mathbb{C}_{14} & \mathbb{C}_{15} & \mathbb{C}_{16} \\ \mathbb{C}_{21} & \mathbb{C}_{22} & \mathbb{C}_{23} & \mathbb{C}_{24} & \mathbb{C}_{25} & \mathbb{C}_{26} \\ \mathbb{C}_{31} & \mathbb{C}_{32} & \mathbb{C}_{33} & \mathbb{C}_{34} & \mathbb{C}_{35} & \mathbb{C}_{36} \\ \mathbb{C}_{41} & \mathbb{C}_{42} & \mathbb{C}_{43} & \mathbb{C}_{44} & \mathbb{C}_{45} & \mathbb{C}_{46} \\ \mathbb{C}_{51} & \mathbb{C}_{52} & \mathbb{C}_{53} & \mathbb{C}_{54} & \mathbb{C}_{55} & \mathbb{C}_{56} \\ \mathbb{C}_{61} & \mathbb{C}_{62} & \mathbb{C}_{63} & \mathbb{C}_{64} & \mathbb{C}_{65} & \mathbb{C}_{66} \end{bmatrix} \begin{Bmatrix} e_{11} \\ e_{22} \\ e_{33} \\ 2e_{23} \\ 2e_{31} \\ 2e_{12} \end{Bmatrix} \quad (2.96)
$$

in index notation, $\sigma_m = \mathbb{C}_{mn}(\mathbf{x}) e_n$.

How the indices of the fourth-order tensor $\mathbb{C}_{ijkl}(\mathbf{x})$ are mapped to the indices of a second-order tensor $\mathbb{C}_{mn}(\mathbf{x})$ or the constitutive matrix with 21 independent coefficients is described in Table 2.5.

Further, due to material symmetry, the number of independent constitutive coefficients is reduced. Without any symmetry a material with 21 independent coefficients is called triclinic material.

Classical Mechanics and the Physics of Continua

TABLE 2.5
Constitutive Mapping from Fourth-Order Tensor to a Second-Order Tensor

$\sigma_{ij} \xrightarrow{yields}$	σ_m	$e_{kl} \xrightarrow{yields}$	e_n	ij or ji		m or n
σ_{11}	σ_1	e_{11}	e_1	11	$\longrightarrow$	1
σ_{22}	σ_2	e_{22}	e_2	22	$\longrightarrow$	2
σ_{33}	σ_3	e_{33}	e_3	33	$\longrightarrow$	3
σ_{23}	σ_4	e_{23}	e_4	23, 32	$\longrightarrow$	4
σ_{31}	σ_5	e_{31}	e_5	31, 13	$\longrightarrow$	5
σ_{12}	σ_6	e_{12}	e_6	12, 21	$\longrightarrow$	6

2.15.1 Materials with One Plane of Symmetry: Monoclinic Materials

Referring to a Cartesian coordinate system (x_j), if the material is symmetric with respect to a plane described by an arbitrary axis x_1 (i.e., x_1 is normal to the $x_2 x_3$ plane), then stress and strain states must have mirror symmetry with respect to the x_1 plane of symmetry. This will cause only 13 independent constitutive coefficients to survive [16, 17] as follows.

$$
\begin{Bmatrix} \sigma_{11} \\ \sigma_{22} \\ \sigma_{33} \\ \sigma_{23} \\ \sigma_{31} \\ \sigma_{12} \end{Bmatrix} = \begin{bmatrix} \mathbb{C}_{11} & \mathbb{C}_{12} & \mathbb{C}_{13} & \mathbb{C}_{14} & 0 & 0 \\ & \mathbb{C}_{22} & \mathbb{C}_{23} & \mathbb{C}_{24} & 0 & 0 \\ & & \mathbb{C}_{33} & \mathbb{C}_{34} & 0 & 0 \\ & & & \mathbb{C}_{44} & 0 & 0 \\ & Sym & & & \mathbb{C}_{55} & \mathbb{C}_{56} \\ & & & & & \mathbb{C}_{66} \end{bmatrix} \begin{Bmatrix} e_{11} \\ e_{22} \\ e_{33} \\ 2e_{23} \\ 2e_{31} \\ 2e_{12} \end{Bmatrix}
\tag{2.97}
$$

If the plane of symmetry changes to x_2 plane, then the symmetric stress and strain states will result in the constitutive relation as follows in Eq. 2.98. Please note that Eqs. 2.97 and 2.98 both have the same number (13) of independent coefficients, but the zeros are placed at different locations in the matrix depending on the plane of symmetry.

$$
\begin{Bmatrix} \sigma_{11} \\ \sigma_{22} \\ \sigma_{33} \\ \sigma_{23} \\ \sigma_{31} \\ \sigma_{12} \end{Bmatrix} = \begin{bmatrix} \mathbb{C}_{11} & \mathbb{C}_{12} & \mathbb{C}_{13} & 0 & \mathbb{C}_{15} & 0 \\ & \mathbb{C}_{22} & \mathbb{C}_{23} & 0 & \mathbb{C}_{25} & 0 \\ & & \mathbb{C}_{33} & 0 & \mathbb{C}_{35} & 0 \\ & & & \mathbb{C}_{44} & \mathbb{C}_{45} & 0 \\ & Sym & & & \mathbb{C}_{55} & 0 \\ & & & & & \mathbb{C}_{66} \end{bmatrix} \begin{Bmatrix} e_{11} \\ e_{22} \\ e_{33} \\ 2e_{23} \\ 2e_{31} \\ 2e_{12} \end{Bmatrix}
\tag{2.98}
$$

2.15.2 Materials with Two Planes of Symmetry: Orthotropic Materials

Drawing a conclusion from the above discussion, if the material has both an x_1 plane and an x_2 plane of symmetry, then the constitutive equation must be inclusive of both Eqs. 2.97 and 2.98. When the material is symmetric about the two orthogonal planes, then it makes the material symmetric automatically about the third plane (i.e., symmetric about the x_3 plane). The constitutive matrix of such material will have only nine independent coefficients as follows.

$$\begin{Bmatrix} \sigma_{11} \\ \sigma_{22} \\ \sigma_{33} \\ \sigma_{23} \\ \sigma_{31} \\ \sigma_{12} \end{Bmatrix} = \begin{bmatrix} \mathbb{C}_{11} & \mathbb{C}_{12} & \mathbb{C}_{13} & 0 & 0 & 0 \\ & \mathbb{C}_{22} & \mathbb{C}_{23} & 0 & 0 & 0 \\ & & \mathbb{C}_{33} & 0 & 0 & 0 \\ & & & \mathbb{C}_{44} & 0 & 0 \\ & Sym & & & \mathbb{C}_{55} & 0 \\ & & & & & \mathbb{C}_{66} \end{bmatrix} \begin{Bmatrix} e_{11} \\ e_{22} \\ e_{33} \\ 2e_{23} \\ 2e_{31} \\ 2e_{12} \end{Bmatrix} \qquad (2.99)$$

These materials are called orthotropic or orthorhombic materials.

2.15.3 Materials with Three Planes of Symmetry and One Plane of Isotropy: Transversely Isotropic Materials

In addition to the three planes of symmetry described in the previous category, if the material has one additional plane of isotropy (i.e., the material behaves like an isotropic material about an axis), then the plane perpendicular to the axis or the transverse plane has infinite plane of symmetry. If the material is isotropic about the x_3 axis, then the constitutive matrix will have only six coefficients but five independent coefficients as follows.

$$\begin{Bmatrix} \sigma_{11} \\ \sigma_{22} \\ \sigma_{33} \\ \sigma_{23} \\ \sigma_{31} \\ \sigma_{12} \end{Bmatrix} = \begin{bmatrix} \mathbb{C}_{11} & \mathbb{C}_{12} & \mathbb{C}_{13} & 0 & 0 & 0 \\ & \mathbb{C}_{11} & \mathbb{C}_{13} & 0 & 0 & 0 \\ & & \mathbb{C}_{33} & 0 & 0 & 0 \\ & & & \mathbb{C}_{44} & 0 & 0 \\ & Sym & & & \mathbb{C}_{44} & 0 \\ & & & & & \dfrac{\mathbb{C}_{11} - \mathbb{C}_{12}}{2} \end{bmatrix} \begin{Bmatrix} e_{11} \\ e_{22} \\ e_{33} \\ 2e_{23} \\ 2e_{31} \\ 2e_{12} \end{Bmatrix} \qquad (2.100.1)$$

These materials are called transversely isotropic materials. They have same modulus along the x_1 and x_2 axis and the same shear modulus along x_3 axis, irrespective of the plane on which it acts.

Classical Mechanics and the Physics of Continua 53

Please note that in all above cases, if the primary axis is x_1, the plane of symmetry changes; then the nomenclature of the constitutive parameters in the constitutive matrix also changes. If the primary axis is x_1 the above equation Eq. 2.100.1 will read

$$
\begin{Bmatrix} \sigma_{11} \\ \sigma_{22} \\ \sigma_{33} \\ \sigma_{23} \\ \sigma_{31} \\ \sigma_{12} \end{Bmatrix} = \begin{bmatrix} \mathbb{C}_{11} & \mathbb{C}_{12} & \mathbb{C}_{13} & 0 & 0 & 0 \\ & \mathbb{C}_{22} & \mathbb{C}_{13} & 0 & 0 & 0 \\ & & \mathbb{C}_{22} & 0 & 0 & 0 \\ & & & \mathbb{C}_{44} & 0 & 0 \\ & Sym & & & \mathbb{C}_{44} & 0 \\ & & & & & \dfrac{\mathbb{C}_{11}-\mathbb{C}_{12}}{2} \end{bmatrix} \begin{Bmatrix} e_{11} \\ e_{22} \\ e_{33} \\ 2e_{23} \\ 2e_{31} \\ 2e_{12} \end{Bmatrix} \quad (2.100.2)
$$

According to linear algebra, any number of rows can be swapped. In some literature it is customary to use the constitute equation for σ_{12} before σ_{23}. With this swap, the above equation will read

$$
\begin{Bmatrix} \sigma_{11} \\ \sigma_{22} \\ \sigma_{33} \\ \sigma_{12} \\ \sigma_{23} \\ \sigma_{31} \end{Bmatrix} = \begin{bmatrix} \mathbb{C}_{11} & \mathbb{C}_{12} & \mathbb{C}_{13} & 0 & 0 & 0 \\ & \mathbb{C}_{22} & \mathbb{C}_{13} & 0 & 0 & 0 \\ & & \mathbb{C}_{22} & 0 & 0 & 0 \\ & & & \dfrac{\mathbb{C}_{11}-\mathbb{C}_{12}}{2} & 0 & 0 \\ & Sym & & & \mathbb{C}_{66} & 0 \\ & & & & & \mathbb{C}_{66} \end{bmatrix} \begin{Bmatrix} e_{11} \\ e_{22} \\ e_{33} \\ 2e_{12} \\ 2e_{23} \\ 2e_{31} \end{Bmatrix} \quad (2.100.3)
$$

Hence, before using the constitutive matrix it is necessary to verify the convention of the primary axis and the axes of symmetries used in writing that equation. Also, it is necessary to verify where the stress σ_{12} is placed. Is it placed in the fourth row or in the sixth row? If the σ_{12} is placed in the fourth row, then $\mathbb{C}_{44} = \dfrac{\mathbb{C}_{11}-\mathbb{C}_{12}}{2}$ and $\mathbb{C}_{55} = \mathbb{C}_{66}$. Either $\mathbb{C}_{55}$ or $\mathbb{C}_{66}$ could be known. In most cases it is customary to use $\mathbb{C}_{66}$ as written in Eq. 2.100.3.

2.15.4 Materials with Three Planes and Three Axes of Symmetry: Isotropic Materials

Drawing a conclusion from the above, adding one more axis of symmetry will cause one more plane of isotropy. When two orthogonal planes are isotropic, then the third plane must also be isotropic. We can write the constitutive matrix of an isotropic material as follows. In this case there are only two independent material constants in the constitutive equation.

$$
\begin{Bmatrix} \sigma_{11} \\ \sigma_{22} \\ \sigma_{33} \\ \sigma_{23} \\ \sigma_{31} \\ \sigma_{12} \end{Bmatrix} = \begin{bmatrix} \mathbb{C}_{11} & \mathbb{C}_{12} & \mathbb{C}_{12} & 0 & 0 & 0 \\ & \mathbb{C}_{11} & \mathbb{C}_{12} & 0 & 0 & 0 \\ & & \mathbb{C}_{11} & 0 & 0 & 0 \\ & & & \dfrac{\mathbb{C}_{11}-\mathbb{C}_{12}}{2} & 0 & 0 \\ & Sym & & & \dfrac{\mathbb{C}_{11}-\mathbb{C}_{12}}{2} & 0 \\ & & & & & \dfrac{\mathbb{C}_{11}-\mathbb{C}_{12}}{2} \end{bmatrix} \begin{Bmatrix} e_{11} \\ e_{22} \\ e_{33} \\ 2e_{23} \\ 2e_{31} \\ 2e_{12} \end{Bmatrix}
$$

$$\tag{2.101}$$

The above coefficients can be further expressed in terms of Lamé constants $\mathbb{L}$ and m, where

$$
\mathbb{C}_{11} = \mathbb{L} + 2\mathrm{m} \text{ and } \mathbb{C}_{12} = \mathbb{L}. \tag{2.102}
$$

Substituting Eq. 2.102, the Eq. 2.101 can be further written in index notation as follows.

$$
\sigma_{ij} = \mathbb{L}\delta_{ij}e_{kk} + 2\mathrm{m}e_{ij} \tag{2.103}
$$

Conversely, strains in terms of the stresses (Hook's Law) in isotropic material can be written as

$$
e_{ij} = \frac{\sigma_{ij}}{2\mathrm{m}} + \delta_{ij}\frac{\mathbb{L}\sigma_{kk}}{2\mathrm{m}(3\mathbb{L}+2\mathrm{m})} \tag{2.104}
$$

2.16 APPENDIX

2.16.1 Important Equations in a Cartesian Coordinate System

In this chapter, most of the equations are presented using index notations; however, a few equations will be extensively used in the following chapters. Hence, for reference they are expanded in this appendix. Primarily Eqs. 2.87, 2.88, and 2.89 are expanded in the following set of equations, respectively.

Governing Differential Equations of Motion

$$
\frac{\partial \sigma_{11}}{\partial x_1} + \frac{\partial \sigma_{12}}{\partial x_2} + \frac{\partial \sigma_{13}}{\partial x_3} + f_1 = \rho \frac{\partial^2 u_1}{\partial t^2} \tag{A.2.1.1}
$$

$$
\frac{\partial \sigma_{21}}{\partial x_1} + \frac{\partial \sigma_{22}}{\partial x_2} + \frac{\partial \sigma_{23}}{\partial x_3} + f_2 = \rho \frac{\partial^2 u_2}{\partial t^2} \tag{A.2.1.2}
$$

$$
\frac{\partial \sigma_{31}}{\partial x_1} + \frac{\partial \sigma_{32}}{\partial x_2} + \frac{\partial \sigma_{33}}{\partial x_3} + f_3 = \rho \frac{\partial^2 u_3}{\partial t^2} \tag{A.2.1.3}
$$

Classical Mechanics and the Physics of Continua

55

Strain-Displacement Relations

$$e_{11} = \left(u_{1,1}\right) \tag{A.2.2.1}$$

$$e_{22} = \left(u_{2,2}\right) \tag{A.2.2.2}$$

$$e_{33} = \left(u_{3,3}\right) \tag{A.2.2.3}$$

$$e_{12} = \frac{1}{2}\left(u_{1,2} + u_{2,1}\right) \tag{A.2.2.4}$$

$$e_{23} = \frac{1}{2}\left(u_{2,3} + u_{3,2}\right) \tag{A.2.2.5}$$

$$e_{31} = \frac{1}{2}\left(u_{3,1} + u_{1,3}\right) \tag{A.2.2.6}$$

Strain Compatibility Equations

$$\frac{\partial^2 e_{11}}{\partial x_2{}^2} + \frac{\partial^2 e_{22}}{\partial x_1{}^2} = 2\frac{\partial^2 e_{12}}{\partial x_1 x_2} \tag{A.2.3.1}$$

$$\frac{\partial^2 e_{22}}{\partial x_3{}^2} + \frac{\partial^2 e_{33}}{\partial x_2{}^2} = 2\frac{\partial^2 e_{23}}{\partial x_2 x_3} \tag{A.2.3.2}$$

$$\frac{\partial^2 e_{33}}{\partial x_1{}^2} + \frac{\partial^2 e_{11}}{\partial x_3{}^2} = 2\frac{\partial^2 e_{31}}{\partial x_3 x_1} \tag{A.2.3.3}$$

$$\frac{\partial^2 e_{11}}{\partial x_2 x_3} = \frac{\partial}{\partial x_1}\left(-\frac{\partial e_{23}}{\partial x_1} + \frac{\partial e_{31}}{\partial x_2} + \frac{\partial e_{12}}{\partial x_3}\right) \tag{A.2.3.4}$$

$$\frac{\partial^2 e_{22}}{\partial x_3 x_1} = \frac{\partial}{\partial x_2}\left(+\frac{\partial e_{23}}{\partial x_1} - \frac{\partial e_{31}}{\partial x_2} + \frac{\partial e_{12}}{\partial x_3}\right) \tag{A.2.3.5}$$

$$\frac{\partial^2 e_{33}}{\partial x_1 x_2} = \frac{\partial}{\partial x_3}\left(+\frac{\partial e_{23}}{\partial x_1} + \frac{\partial e_{31}}{\partial x_2} - \frac{\partial e_{12}}{\partial x_3}\right) \tag{A.2.3.6}$$

Stress-Strain ($\sigma_{ij} - e_{kl}$) Relation for Triclinic material

$$\begin{Bmatrix} \sigma_{11} \\ \sigma_{22} \\ \sigma_{33} \\ \sigma_{23} \\ \sigma_{31} \\ \sigma_{12} \end{Bmatrix} = \begin{bmatrix} \mathbb{C}_{11} & \mathbb{C}_{12} & \mathbb{C}_{13} & \mathbb{C}_{14} & \mathbb{C}_{15} & \mathbb{C}_{16} \\ \mathbb{C}_{21} & \mathbb{C}_{22} & \mathbb{C}_{23} & \mathbb{C}_{24} & \mathbb{C}_{25} & \mathbb{C}_{26} \\ \mathbb{C}_{31} & \mathbb{C}_{32} & \mathbb{C}_{33} & \mathbb{C}_{34} & \mathbb{C}_{35} & \mathbb{C}_{36} \\ \mathbb{C}_{41} & \mathbb{C}_{42} & \mathbb{C}_{43} & \mathbb{C}_{44} & \mathbb{C}_{45} & \mathbb{C}_{46} \\ \mathbb{C}_{51} & \mathbb{C}_{52} & \mathbb{C}_{53} & \mathbb{C}_{54} & \mathbb{C}_{55} & \mathbb{C}_{56} \\ \mathbb{C}_{61} & \mathbb{C}_{62} & \mathbb{C}_{63} & \mathbb{C}_{64} & \mathbb{C}_{65} & \mathbb{C}_{66} \end{bmatrix} \begin{Bmatrix} e_{11} \\ e_{22} \\ e_{33} \\ 2e_{23} \\ 2e_{31} \\ 2e_{12} \end{Bmatrix} \tag{A.2.4}$$

Stress-Strain ($\sigma_{ij} - e_{kl}$) Relation for isotropic material

$$\begin{Bmatrix} \sigma_{11} \\ \sigma_{22} \\ \sigma_{33} \\ \sigma_{23} \\ \sigma_{31} \\ \sigma_{12} \end{Bmatrix} = \begin{bmatrix} \mathbb{L}+2m & \mathbb{L} & \mathbb{L} & 0 & 0 & 0 \\ & \mathbb{L}+2m & \mathbb{L} & 0 & 0 & 0 \\ & & \mathbb{L}+2m & 0 & 0 & 0 \\ & & & m & 0 & 0 \\ & Sym & & & m & 0 \\ & & & & & m \end{bmatrix} \begin{Bmatrix} e_{11} \\ e_{22} \\ e_{33} \\ 2e_{23} \\ 2e_{31} \\ 2e_{12} \end{Bmatrix} \quad (A.2.5)$$

2.16.2 Important Equations in a Cylindrical Coordinate System

2.16.2.1 Transformation to a Cylindrical Coordinate System

Here, the polar coordinate system and cylindrical coordinate system are synonymous. To initiate the discussion, it is necessary to refer to Fig. A.2.1. First a right-hand coordinate system is shown at the corner of the figure to visualize the transformation. In a cylindrical coordinate system the x_3 axis remains the same. Hence, the basis vector along the x_3 axis remains the same, which is $\hat{\mathbf{e}}_3$. But other two basis vectors are one along the radial direction (r), which is $\hat{\mathbf{e}}_r$, and the next is along the angular direction (θ), which is $\hat{\mathbf{e}}_\theta$ as shown in Fig. A.2.1. The x_3 axis is rotated by an angle θ. Let's consider any arbitrary point P, which is now located at x_1, x_2, x_3. After coordinate transformation, the location of the point P does not change, but the representation of the point changes to, say, R, Θ, x_3. OA and OC in Fig. A.2.1 are x_1 and x_2, respectively. Similarly, OB and OD in Fig. A.2.1 are R and Θ, respectively. Based on geometry, OB can be seen as a summation of OF and FB. Based on these geometrical relations, following expressions can be written as

$$OA = CP = x_1 \; ; \; OC = AP = x_2 \; ; \; OB = OF + FB = x_1 \cos\theta + x_2 \sin\theta \quad (A.2.6)$$

Similarly,

$$OD = BP = GP - GB = GP - AF = x_2 \cos\theta - x_1 \sin\theta \quad (A.2.7)$$

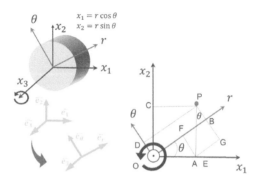

FIGURE A.2.1 Cylindrical coordinate system: Description of a particle in transformed coordinate system.

Classical Mechanics and the Physics of Continua

Since $OB = R$ and $OD = \Theta$, the relation between R, Θ, x_3 and x_1, x_2, x_3 can written as

$$\begin{Bmatrix} R \\ \Theta \\ x_3 \end{Bmatrix} = \begin{bmatrix} \cos\theta & \sin\theta & 0 \\ -\sin\theta & \cos\theta & 0 \\ 0 & 0 & 1 \end{bmatrix} \begin{Bmatrix} x_1 \\ x_2 \\ x_3 \end{Bmatrix} \tag{A.2.8}$$

Hence, the basis vectors for the respective coordinate systems are also related by

$$\begin{Bmatrix} \hat{\mathbf{e}}_r \\ \hat{\mathbf{e}}_\theta \\ \hat{\mathbf{e}}_3 \end{Bmatrix} = \begin{bmatrix} \cos\theta & \sin\theta & 0 \\ -\sin\theta & \cos\theta & 0 \\ 0 & 0 & 1 \end{bmatrix} \begin{Bmatrix} \hat{\mathbf{e}}_1 \\ \hat{\mathbf{e}}_2 \\ \hat{\mathbf{e}}_3 \end{Bmatrix} \tag{A.2.9}$$

The transformation matrix Ψ_{ij} for transforming the coordinates or tensors from Cartesian to cylindrical coordinate will be

$$\Psi = \begin{bmatrix} \cos\theta & \sin\theta & 0 \\ -\sin\theta & \cos\theta & 0 \\ 0 & 0 & 1 \end{bmatrix} \tag{A.2.10}$$

Using this rule, the displacement vector u_i in a Cartesian coordinate system can be transformed to polar coordinate as follows.

$$\begin{Bmatrix} u_r \\ u_\theta \\ u_3 \end{Bmatrix} = \begin{bmatrix} \cos\theta & \sin\theta & 0 \\ -\sin\theta & \cos\theta & 0 \\ 0 & 0 & 1 \end{bmatrix} \begin{Bmatrix} u_1 \\ u_2 \\ u_3 \end{Bmatrix} \tag{A.2.11}$$

2.16.2.2 Gradient Operator in a Cylindrical Coordinate System

The nabla Hamiltonian operator ∇ in a Cartesian coordinate system can be expressed by

$$\nabla = \left(\frac{\partial}{\partial x_1} \hat{\mathbf{e}}_1 + \frac{\partial}{\partial x_2} \hat{\mathbf{e}}_2 + \frac{\partial}{\partial x_3} \hat{\mathbf{e}}_3 \right) \tag{A.2.12}$$

which is similarly written in a cylindrical coordinate system as follows.

$$\nabla = \left(\frac{\partial}{\partial r} \hat{\mathbf{e}}_r + \frac{1}{r} \frac{\partial}{\partial \theta} \hat{\mathbf{e}}_\theta + \frac{\partial}{\partial x_3} \hat{\mathbf{e}}_3 \right) \tag{A.2.13}$$

- The gradient of a scalar field $\varphi(r, \theta, x_3, t)$ in cylindrical coordinate system can be written as

$$Grad\left(\varphi(r, \theta, x_3, t)\right) = \nabla\varphi = \left(\frac{\partial\varphi}{\partial r} \hat{\mathbf{e}}_r + \frac{1}{r} \frac{\partial\varphi}{\partial \theta} \hat{\mathbf{e}}_\theta + \frac{\partial\varphi}{\partial x_3} \hat{\mathbf{e}}_3 \right) \tag{A.2.14}$$

- The divergence of a vector field $\mathbf{V} = V_r(r,\theta,x_3,t)\hat{\mathbf{e}}_r + V_\theta(r,\theta,x_3,t)\hat{\mathbf{e}}_\theta + V_3(r,\theta,x_3,t)\hat{\mathbf{e}}_3$ in a cylindrical coordinate system can be written as

$$\nabla.\mathbf{V} = \left(\frac{\partial V_r}{\partial r} + \frac{1}{r}\frac{\partial V_\theta}{\partial \theta} + \frac{\partial V_3}{\partial x_3} \right) \tag{A.2.15}$$

- The curl of a vector field $\mathbf{V} = V_r(r,\theta,x_3,t)\hat{\mathbf{e}}_r + V_\theta(r,\theta,x_3,t)\hat{\mathbf{e}}_\theta + V_3(r,\theta,x_3,t)\hat{\mathbf{e}}_3$ in a cylindrical coordinate system can be written as

$$\nabla \times \mathbf{V} = \hat{\mathbf{e}}_r \left(\frac{1}{r}\frac{\partial V_3}{\partial \theta} - \frac{\partial V_\theta}{\partial x_3} \right) + \hat{\mathbf{e}}_\theta \left(-\frac{\partial V_3}{\partial r} + \frac{\partial V_r}{\partial x_3} \right) + \hat{\mathbf{e}}_3 \left(\frac{1}{r}\frac{\partial(rV_\theta)}{\partial r} - \frac{1}{r}\frac{\partial V_r}{\partial \theta} \right)$$

$$\tag{A.2.16}$$

2.16.2.3 Strain-Displacement Relation in a Cylindrical Coordinate System

Strain-displacement relations in a Cartesian coordinate system are written in Eqs. A.2.2.1 through A.2.2.6. The strain displacement relations in cylindrical coordinate system can be written as

$$e_{rr} = \left(\frac{\partial u_r}{\partial r} \right) \tag{A.2.17.1}$$

$$e_{\theta\theta} = \left(\frac{u_r}{r} + \frac{1}{r}\frac{\partial u_\theta}{\partial \theta} \right) \tag{A.2.17.2}$$

$$e_{33} = \left(\frac{\partial u_3}{\partial x_3} \right) \tag{A.2.17.3}$$

$$e_{r\theta} = \frac{1}{2}\left(\frac{1}{r}\frac{\partial u_r}{\partial \theta} - \frac{u_\theta}{r} + \frac{\partial u_\theta}{\partial r} \right) \tag{A.2.17.4}$$

$$e_{\theta 3} = \frac{1}{2}\left(\frac{1}{r}\frac{\partial u_3}{\partial \theta} + \frac{\partial u_\theta}{\partial x_3} \right) \tag{A.2.17.5}$$

$$e_{3r} = \frac{1}{2}\left(\frac{\partial u_r}{\partial x_3} + \frac{\partial u_3}{\partial r} \right) \tag{A.2.17.6}$$

2.16.2.4 Governing Differential Equations of Motion in a Cylindrical Coordinate System

The governing differential equation of motion in a Cartesian coordinate system is written in Eqs. A.2.1.1 through A.2.1.3. Similarly, the governing equation of motion in a cylindrical coordinate system can be written as

$$\frac{\partial \sigma_{rr}}{\partial r} + \frac{1}{r}\frac{\partial \sigma_{r\theta}}{\partial \theta} + \frac{\partial \sigma_{r3}}{\partial x_3} + \frac{1}{r}(\sigma_{rr} - \sigma_{\theta\theta}) + f_r = \rho\frac{\partial^2 u_r}{\partial t^2} \tag{A.2.18.1}$$

Classical Mechanics and the Physics of Continua

$$\frac{\partial \sigma_{\theta r}}{\partial r} + \frac{1}{r}\frac{\partial \sigma_{\theta\theta}}{\partial \theta} + \frac{\partial \sigma_{\theta 3}}{\partial x_3} + \frac{2}{r}(\sigma_{r\theta}) + f_\theta = \rho \frac{\partial^2 u_\theta}{\partial t^2} \qquad (A.2.18.2)$$

$$\frac{\partial \sigma_{3r}}{\partial r} + \frac{1}{r}\frac{\partial \sigma_{3\theta}}{\partial \theta} + \frac{\partial \sigma_{33}}{\partial x_3} + \frac{1}{r}(\sigma_{3r}) + f_3 = \rho \frac{\partial^2 u_3}{\partial t^2} \qquad (A.2.18.3)$$

2.16.3 Important Equations in a Spherical Coordinate System

In a spherical coordinate system the unit vectors are $\hat{\mathbf{e}}_r$, $\hat{\mathbf{e}}_\theta$, and $\hat{\mathbf{e}}_\phi$ as shown in Fig. A.2.2. The displacement field in a transformed coordinate system is

$$\mathbf{u} = u_r(r,\theta,\phi,t)\hat{\mathbf{e}}_r + u_\theta(r,\theta,\phi,t)\hat{\mathbf{e}}_\theta + u_\phi(r,\theta,\phi,t)\hat{\mathbf{e}}_\phi \qquad (A.2.19)$$

2.16.3.1 Gradient Operator in a Spherical Coordinate System

The nabla Hamiltonian operator ∇ in a cylindrical spherical coordinate system is

$$\nabla = \left(\frac{\partial}{\partial r}\hat{\mathbf{e}}_r + \frac{1}{r\sin\phi}\frac{\partial}{\partial \theta}\hat{\mathbf{e}}_\theta + \frac{1}{r}\frac{\partial}{\partial \phi}\hat{\mathbf{e}}_\phi \right) \qquad (A.2.20)$$

- The gradient of a scalar field $\varphi(r,\theta,\phi,t)$ in a spherical coordinate system can be written as

$$Grad\left(\varphi(r,\theta,x_3,t)\right) = \nabla\varphi = \left(\frac{\partial \varphi}{\partial r}\hat{\mathbf{e}}_r + \frac{1}{r\sin\phi}\frac{\partial \varphi}{\partial \theta}\hat{\mathbf{e}}_\theta + \frac{1}{r}\frac{\partial \varphi}{\partial \phi}\hat{\mathbf{e}}_\phi \right) \qquad (A.2.21)$$

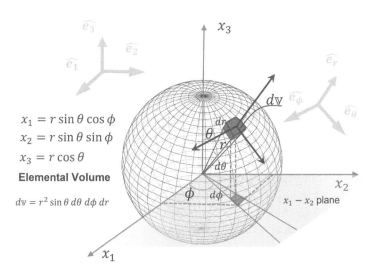

$x_1 = r\sin\theta\cos\phi$
$x_2 = r\sin\theta\sin\phi$
$x_3 = r\cos\theta$

Elemental Volume

$dv = r^2 \sin\theta\, d\theta\, d\phi\, dr$

FIGURE A.2.2 Spherical coordinate system: Description of a particle and a volume element in transformed coordinate system.

- The divergence of the following vector field $\mathbf{V} = V_r\left(r,\theta,\phi,t\right)\hat{\mathbf{e}}_r + V_\theta\left(r,\theta,\phi,t\right)\hat{\mathbf{e}}_\theta + V_\phi\left(r,\theta,\phi,t\right)\hat{\mathbf{e}}_\phi$ in a spherical coordinate system can be written as

$$\nabla.\mathbf{V} = \left(\frac{\partial V_r}{\partial r} + \frac{1}{r\sin\phi}\frac{\partial V_\theta}{\partial \theta} + \frac{1}{r}\frac{\partial V_\phi}{\partial \phi}\right) \tag{A.2.22}$$

- The curl of the following vector field $\mathbf{V} = V_r\left(r,\theta,\phi,t\right)\hat{\mathbf{e}}_r + V_\theta\left(r,\theta,\phi,t\right)\hat{\mathbf{e}}_\theta + V_\phi\left(r,\theta,\phi,t\right)\hat{\mathbf{e}}_\phi$ in spherical coordinate system can be written as

$$\nabla \times \mathbf{V} = \frac{1}{r^2\sin\phi}\left[\hat{\mathbf{e}}_r\left(\frac{\partial\left(rV_\phi\right)}{\partial\theta} - \frac{\partial\left(r\sin\phi V_\theta\right)}{\partial\phi}\right) + \hat{\mathbf{e}}_\theta r\sin\phi\left(-\frac{\partial\left(rV_\phi\right)}{\partial r} + \frac{\partial V_r}{\partial\phi}\right)\right.$$
$$\left. + \hat{\mathbf{e}}_\phi r\left(\frac{\partial\left(r\sin\phi V_\theta\right)}{\partial r} - \frac{\partial V_r}{\partial\theta}\right)\right] \tag{A.2.23}$$

2.16.3.2 Strain-Displacement Relation in a Spherical Coordinate System

The strain displacement relations in a spherical coordinate system can be written as

$$e_{rr} = \left(\frac{\partial u_r}{\partial r}\right) \tag{A.2.24.1}$$

$$e_{\theta\theta} = \left(\frac{u_r}{r} + \frac{1}{r\sin\phi}\frac{\partial u_\theta}{\partial\theta} + \frac{u_\phi}{r}\cot\phi\right) \tag{A.2.24.2}$$

$$e_{\phi\phi} = \left(\frac{u_r}{r} + \frac{1}{r}\frac{\partial u_\phi}{\partial\phi}\right) \tag{A.2.24.3}$$

$$e_{r\theta} = \frac{1}{2}\left(\frac{1}{r\sin\phi}\frac{\partial u_r}{\partial\theta} - \frac{u_\theta}{r} + \frac{\partial u_\theta}{\partial r}\right) \tag{A.2.24.4}$$

$$e_{\theta\phi} = \frac{1}{2}\left(\frac{1}{r\sin\phi}\frac{\partial u_\phi}{\partial\theta} + \frac{1}{r}\frac{\partial u_\theta}{\partial\phi} - \frac{u_\theta}{r}\cot\phi\right) \tag{A.2.24.5}$$

$$e_{\phi r} = \frac{1}{2}\left(\frac{1}{r}\frac{\partial u_r}{\partial\phi} + \frac{\partial u_\phi}{\partial r} - \frac{u_\phi}{r}\right) \tag{A.2.24.6}$$

2.16.3.3 Governing Differential Equations of Motion in a Spherical Coordinate System

The governing differential equation of motion in a Cartesian coordinate system is written in Eqs. A.2.1.1 through A.2.1.3. Similarly, the governing equation of motion in a spherical coordinate system can be written as

$$\frac{\partial\sigma_{rr}}{\partial r} + \frac{1}{r\sin\phi}\frac{\partial\sigma_{r\theta}}{\partial\theta} + \frac{1}{r}\frac{\partial\sigma_{r\phi}}{\partial\phi} + \frac{1}{r}\left(2\sigma_{rr} - \sigma_{\theta\theta} + \sigma_{r\phi}\cot\phi - \sigma_{\phi\phi}\right) + f_r = \rho\frac{\partial^2 u_r}{\partial t^2}$$

$$\tag{A.2.25.1}$$

Classical Mechanics and the Physics of Continua

$$\frac{\partial \sigma_{\theta r}}{\partial r} + \frac{1}{r \sin \phi} \frac{\partial \sigma_{\theta \theta}}{\partial \theta} + \frac{1}{r} \frac{\partial \sigma_{\theta \phi}}{\partial \phi} + \frac{2}{r}\left(\sigma_{\theta \phi}\right)\cot \phi + \frac{3}{r}\left(\sigma_{\theta r}\right) + f_\theta = \rho \frac{\partial^2 u_\theta}{\partial t^2} \qquad (A.2.25.2)$$

$$\frac{\partial \sigma_{\phi r}}{\partial r} + \frac{1}{r \sin \phi} \frac{\partial \sigma_{\phi \theta}}{\partial \theta} + \frac{1}{r} \frac{\partial \sigma_{\phi \phi}}{\partial \phi} + \frac{1}{r}\left(3\sigma_{\phi r} + \cot \theta \left(\sigma_{\phi \phi} - \sigma_{\theta \theta}\right)\right) + f_\phi = \rho \frac{\partial^2 u_\phi}{\partial t^2} \qquad (A.2.25.3)$$

2.17 SUMMARY

The fundamentals of continuum mechanics or mechanics of deformable bodies are reviewed in this chapter, tailored to the understanding of wave propagation in deformable bodies. Eulerian and Lagrangian coordinate systems are reviewed to provide better understanding of what coordinate system should be used in explaining the wave propagation. The fundamental elastodynamic equation is derived from the fundamentals. Thermodynamics of continua, internal energies, and strain energy densities are explained. Wave propagation in different material types requires understanding of material coefficients if a particular material type is anisotropic, depending on their degree of anisotropy. Hence, with different planes of symmetry, different materials are introduced with their respective necessary material coefficients. In the appendix, elastodynamic equations and stain-displacement relations are briefly introduced in cylindrical and spherical coordinate systems for broader applications.

REFERENCES

1. Morin, D., *Introduction to Classical Mechanics*. 2007, Cambridge: Cambridge University Press.
2. Jha, P., *Indian mathematics in the Dark Age*. Ganita-Bharati, 1995, **17**(1–4): p. 75–79.
3. Calinger, R., *A Contextual History of Mathematics*. 1999, Upper Saddle River, New Jersey: Prentice Hall PTR.
4. Hayashi, T., Kusuba, T., *Twenty-one algebraic normal forms of Citrabhanu*. Historia Mathematica, 1998, **25**(1): p. 1–21.
5. Sarma, U.K.V., Bhat, V., Pai, V., Ramasubramanian, K., *The discovery of Madhava series by Whish: An episode in historiography of science*. Ganita Bharati, 2010, **32**(1): p. 115–126.
6. Katz, V.J., ed. *The Mathematics of Egypt, Mesopotamia, China, India, and Islam: A Sourcebook*. 2007, Princeton, NJ: Princeton University Press.
7. Franco, A.B., *Avempace, projectile motion, and impetus theory*. Journal of the History of Ideas, 2003, **64**(4): p. 521–546.
8. Pines, S., *Abu'l-Barakāt al-Baghdādī, Hibat allah*. Vol. 1. 1970, New York: Charles Scribner's Sons.
9. Newton, I., *Philosophiæ Naturalis Principia Mathematica*. 1687, Edinburg, Scotland: Encyclopædia Britannica.
10. Lagrange, J.L., *Mécanique analytique*. 1788, Paris: Chez la Veuve Desaint.
11. Banerjee, S., Leckey, A.C.C., *Computational Nondestructive Evaluation Handbook*. 2020, Boca Raton, FL: CRC Press Taylor and Francis Group, p. 560.
12. Castillo, G.F.T., *An Introduction to Hamiltonian Mechanics*. 2018, New York City, NY: Springer.

13. Capaldi, F., *Continuum Mechanics: Constitutive Modeling of Structural and Biological Materials.* 2012, Cambridge: Cambridge University Press.
14. Eringen, A.C., *Mechanics of Continua.* 1980, Huntington, New York: Robert E. Krieger Publishing Company.
15. Mase, G.T., Smelser, R.E., Mase, G.E., *Continuum Mechanics for Engineers.* 3rd ed. 2009, Boca Raton, FL: CRC Press.
16. Lai, W.M., Rubin, D., Krempl, E., *Introduction to Continuum Mechanics.* 1974, New York: Pergamon Press Inc.
17. Shabana, A.A., *Computational Continuum Mechanics.* 3rd ed. 2018, Hoboken, NJ: John Wiley & Sons, Ltd.
18. Cho, C., Yu, S., Park, N., *Elastic Hamiltonian for quantum analog application.* Physical Review B, 2020, **101**(13): p. 134107.
19. Pavelka, M., Peshkov, I., Klika, V., *On Hamiltonian continuum mechanics.* Physica D: Nonlinear Phenomena, 2020, **408**: p. 132510.
20. Wiklund, K., *Wave interaction by Hamiltonian methods*, in *Department of Plasma Physics.* 2002, Umea, Sweden: Umea University, p. 55.

3 Acoustics and Elastic Wave Propagation in Fluids and Anisotropic Solids

3.1 WAVE PROPAGATION IN FLUID MEDIA

3.1.1 GOVERNING DIFFERENTIAL EQUATION WITH PRESSURE IN FLUID

Wave propagation in periodic metamaterial concerns fluid and solid media with fluid-solid interfaces. Hence, acoustic waves in fluid media are explained first in this chapter. Only incompressible fluids are considered. When elastic and acoustic wave energy is excited in incompressible fluid (air or liquid), the energy propagates in all possible directions with a constant wave velocity, resulting spherical wave fronts. A perfect fluid is a nondispersive medium, which means that the dispersion relation between the angular frequency (ω) and wave number (k) is linear with a constant slope. The slope is the wave velocity. Please note that an incompressible fluid cannot support any shear stresses. Pressure at a point in the fluid is always equal and compressive from all possible directions. This gives the normal stresses (discussed in Chapter 2) $\sigma_{11} = \sigma_{22} = \sigma_{33} = -p$, where $p(\mathbf{x}, t)$ is the pressure at a point at an instance in a fluid medium. The negative sign is to designate the compressive stress. By virtue of its material characteristics, fluid cannot take any tensile stresses either. Recalling the wave propagation equation or the governing partial differential equations of motion in any elastic media described in Chapter 2 and eliminating the shear stresses from Eq. A.2.1.1 – A2.1.3, we can write

$$\frac{\partial \sigma_{11}}{\partial x_1} + f_1 = \rho \frac{\partial^2 u_1}{\partial t^2} \ or \ -\frac{\partial p}{\partial x_1} + f_1 = \rho \frac{\partial^2 u_1}{\partial t^2} \tag{3.1.1}$$

$$\frac{\partial \sigma_{22}}{\partial x_2} + f_2 = \rho \frac{\partial^2 u_2}{\partial t^2} \ or \ -\frac{\partial p}{\partial x_2} + f_2 = \rho \frac{\partial^2 u_2}{\partial t^2} \tag{3.1.2}$$

$$\frac{\partial \sigma_{33}}{\partial x_3} + f_3 = \rho \frac{\partial^2 u_3}{\partial t^2} \ or \ -\frac{\partial p}{\partial x_3} + f_3 = \rho \frac{\partial^2 u_3}{\partial t^2} \tag{3.1.3}$$

Using the notation of the gradient of a scalar field (here the pressure field), we can write the above equations in the following form

$$-\nabla p + \mathbf{f} = \rho \ddot{\mathbf{u}} \ or \ \left\langle -p_{,i} + f_i = \rho \ddot{u}_i \right\rangle \widehat{\mathbf{e}_i} \ or \ simply \ -p_{,i} + f_i = \rho \ddot{u}_i \tag{3.2}$$

DOI: 10.1201/9781003225751-3

63

64   Metamaterials in Topological Acoustics

This shows that the above equations are analogous to a vector field that has three orthogonal independent components in three directions, x_1, x_2, and x_3, respectively. $\nabla \mathrm{p}$ is the gradient of the scalar pressure field. Gradient of a scalar field is always a vector field. As in Eq. 2.72, the ∇ is a nabla Hamiltonian operator or simply a gradient operator. Wave propagation is conceptually related to the divergence of a vector field. To find how the radius of the wave front changes over time, we need to find the divergence of the pressure field, $\nabla \mathrm{p}$. Hence, we apply the divergence operator on Eq. 3.2 and we get

$$-\nabla . \nabla \mathrm{p} + \nabla . \mathbf{f} = \rho \nabla . \ddot{\mathbf{u}} \ \ or \ \ -\mathrm{p}_{,ii} + f_{i,i} = \rho \ddot{u}_{i,i} \tag{3.3}$$

Due to the absence of shear modulus, the isotropic constitutive relation presented in Chapter 2, Eq. 2.93 and Eq. 2.103 can be modified for fluid and can be written as $\sigma_{ij} = \mathbb{L}\delta_{ij}e_{kk}$. The constitutive equation in fluid can be further modified considering the discussions on stress-strain relationship introduced in Chapter 2.

$$-\mathrm{p} = \mathbb{L}\left(e_{11} + e_{22} + e_{33}\right) = \mathbb{L}\left(u_{1,1} + u_{2,2} + u_{3,3}\right) = \mathbb{L}u_{i,i} = \mathbb{L}\nabla . \mathbf{u} \tag{3.4}$$

Substituting Eq. 3.4 into Eq. 3.3 we get

$$-\mathrm{p}_{,ii} - \frac{\rho}{\mathbb{L}}\ddot{\mathrm{p}} + f_{i,i} = 0 \ \ or \ \ -\mathrm{p}_{,ii} - \frac{1}{c_f^2}\ddot{\mathrm{p}} + f_{i,i} = 0 \tag{3.5}$$

where $c_f = \sqrt{\mathbb{L}/\rho}$ is the phase wave velocity in the fluid media, which is the square root of the ratio of the modulus ($\mathbb{L}$) and the density (ρ) of the fluid. Further neglecting body force, we can write the final form of the equation of wave propagation in fluid in the following form

$$\nabla^2 \mathrm{p} - \frac{1}{c_f^2}\frac{\partial^2 \mathrm{p}}{\partial t^2} = 0 \tag{3.6.1}$$

$$or \ \left(\frac{\partial^2 \mathrm{p}}{\partial x_1{}^2} + \frac{\partial^2 \mathrm{p}}{\partial x_2{}^2} + \frac{\partial^2 \mathrm{p}}{\partial x_3{}^2}\right) - \frac{1}{c_f^2}\frac{\partial^2 \mathrm{p}}{\partial t^2} = 0 \tag{3.6.2}$$

$$c_f^2\left(\frac{\partial^2}{\partial x_1{}^2} + \frac{\partial^2}{\partial x_2{}^2} + \frac{\partial^2}{\partial x_3{}^2}\right)\mathrm{p} = \mathcal{P}\mathrm{p} \tag{3.6.3}$$

where $\nabla^2 = \nabla . \nabla$ signifies the divergence of the gradient of a scalar field or squared nabla Hamiltonian operator. $\mathcal{P}$ is the eigenvalue of the differential operator on the left-hand side.

3.1.2 Pressure Velocity Relation

The above equation can be alternatively written in a pressure-velocity form. The velocity of a fluid particle along any arbitrary direction i can be assumed v_i. Then the acceleration in Eq. 3.2 will substitute $\dot{v}_i = \ddot{u}_i$. The governing equation will be

$$-\nabla \mathrm{p} + \mathbf{f} = \rho \dot{\mathbf{v}} \ \ or \ \ -\mathrm{p}_{,i} + f_i = \rho \dot{v}_i \tag{3.7}$$

Acoustics and Elastic Wave Propagation in Fluids and Anisotropic Solids

Please recollect the continuity equation Eq. 2.87 in Chapter 2. Axioms of continuum mechanics say that in addition to globally satisfying boundary conditions, it is necessary for the local mass to be conserved. Hence, here we reiterate the continuity equation as follows, starting from Eq. 2.29.

$$\int_v \frac{D}{Dt}(\rho dv) = 0 \ \ or \ \ \int_v \left[\frac{\partial \rho}{\partial t} + \nabla(\rho \mathbf{v})\right] dv = 0$$

$$i.e., \ \frac{\partial \rho}{\partial t} + \nabla(\rho \mathbf{v}) = 0 \ \ or \ \ \frac{\partial \rho}{\partial t} + \mathbf{v}.\nabla \rho + \rho \nabla.\mathbf{v} = 0 \tag{3.8}$$

In index notation, as in Eq. 2.87, the above equation reiterates to

$$\frac{\partial \rho}{\partial t} + (\rho v_i)_{,i} = 0 \ \ i.e., \ \frac{\partial \rho}{\partial t} + v_i \rho_{,i} + \rho v_{i,i} \tag{3.9}$$

Here, divergence of the velocity profile of the fluid particles appears in the equation. Referring to Eq. 3.4, we can write the bulk constitutive equation in terms of velocity as follows

$$-\mathrm{p} = \mathbb{L} \nabla.\mathbf{u} \ \ or \ \ -\frac{\partial \mathrm{p}}{\partial t} = \mathbb{L} \nabla.\mathbf{v} \tag{3.10}$$

where $\mathbb{L}$ is the first Lamé constant equivalent to the bulk modulus $\mathbb{K}$ for fluid. The bulk modulus of a material in terms of Lamé constants can be expressed as $\mathbb{K} = \mathbb{L} + 2/3(\mathrm{m})$. In fluid, the second Lamé constant is zero due to the incapability of the incompressible fluid to take shear stresses (i.e., $\mathbb{K} = \mathbb{L}$). Here thermal and mechanical effects are assumed to be uncoupled during wave propagation. Thus, the constitutive equation for perfect fluid can be written as

$$\nabla.\mathbf{v} = -\frac{1}{\mathbb{K}} \frac{\partial \mathrm{p}}{\partial t} \tag{3.11}$$

For an incompressible fluid, $\nabla \rho = 0$. Hence, evaluating Eq. 3.10 and Eq. 3.8, and referring to Eq. 3.6.1 we can write

$$\frac{1}{\mathbb{K}} \frac{\partial \mathrm{p}}{\partial t} = \frac{1}{\rho} \frac{\partial \rho}{\partial t} \ \ or \ \ \frac{\partial \mathrm{p}}{\partial t} = c_f^2 \frac{\partial \rho}{\partial t} \tag{3.12}$$

3.1.3 PRESSURE AND VELOCITY POTENTIAL IN FLUID MEDIA

Let's assume that a harmonic wave is propagating in a fluid with a frequency ω. The plane wave potential of a harmonic wave will consist of the spatial and temporal phases. Due to the oscillatory nature of the wave, the wave potential will take exponential form with imaginary arguments. Please note that for wave propagation in a three-dimensional homogeneous isotropic medium like fluid, the wave number

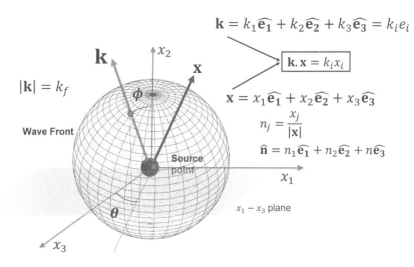

FIGURE 3.1 A spherical wave front in fluid and necessary mathematical terms to calculate phase $k_i x_i$.

along any direction of wave propagation (i.e., along the ray of the wave) orthogonal to the wave front will be the primary wave number that will follow the plane harmonic wave equations in the medium. Let's assume k_f is the wave number in the isotropic fluid. Hence, $k_f = c_f \omega = \dfrac{2\pi}{\lambda_f}$. Refer to Fig. 3.1, where a spherical wave front is shown. In the figure, $\mathbf{k} = k_i \hat{\mathbf{e}}_i$ is a vector pointing to the direction of wave propagating with magnitude of the wave number k_f. k_i are the three components of the wave number in Cartesian coordinate system. $\mathbf{x} = x_i \hat{\mathbf{e}}_i$ is a vector pointing to any arbitrary direction passing through a point x_i, where the pressure wave field is desired to be found. Please note that the phase of a propagating wave is the multiplication of the wave numbers and the distance traveled by the wave. In the same light, the spatial wave phase could be described as a dot product of $\mathbf{k}$ and $\mathbf{x}$ in Fig. 3.1. Hence, the pressure plane wave potential function in three dimensions will be

$$\mathrm{p}(x_j,t) = \overline{\mathrm{p}} e^{i(\mathbf{k}\cdot\mathbf{x}-\omega t)} = \overline{\mathrm{p}} e^{i(k_i x_i - \omega t)} \tag{3.13}$$

Alternatingly Eq. 3.13 can be written by separating the spatial and temporal parts of the harmonic wave as follows

$$\mathrm{p}(x_j,t) = \overline{\mathrm{p}} e^{i(k_i x_i)} e^{-i\omega t} = \mathrm{p}(x_j) e^{-i\omega t} \tag{3.14}$$

Due to the harmonic pressure field, the velocity field in the fluid will also take similar harmonic form, and the harmonic plane wave form for velocity can be written as

$$\mathbf{v}(x_j,t) = \overline{\mathbf{v}} e^{i(k_i x_i)} e^{-i\omega t} = \overline{\mathbf{v}}(x_j) e^{-i\omega t} = \overline{v}_i(x_j) e^{-i\omega t} \tag{3.15}$$

Acoustics and Elastic Wave Propagation in Fluids and Anisotropic Solids   67

Considering nontrivial condition $e^{-i\omega t} \neq 0$, while substituting Eqs. 3.14 and 3.15 in Eq. 3.7, we get

$$-\nabla p(x_j) + \mathbf{f} = -i\omega\rho\bar{\mathbf{v}}(x_j) \tag{3.16}$$

In the absence of body force, which is a general case for wave propagation, Eq. 3.16 will become

$$\nabla p(x_j) - i\omega\rho\bar{\mathbf{v}}(x_j) = 0 \quad \text{or} \quad p_{,i} - i\omega\rho\bar{v}_i = 0 \tag{3.17}$$

Similarly, the constitutive equation in Eq. 3.11 will be

$$\nabla.\bar{\mathbf{v}}(x_j) - \frac{i\omega p}{\mathbb{K}} = 0 \quad \text{or} \quad \bar{v}_{i,i} - \frac{i\omega p}{\mathbb{K}} = 0 \tag{3.18}$$

Alternatively, performing the second-order derivative of Eq. 3.14 with respect to time and eliminating the temporal harmonic part of the wave potential, Eq. 3.6.1 can be re written as

$$\nabla^2 p + k_f^2\, p = 0 \tag{3.19.1}$$

$$\text{or} \left(\frac{\partial^2}{\partial x_1^{\,2}} + \frac{\partial^2}{\partial x_2^{\,2}} + \frac{\partial^2}{\partial x_3^{\,2}} \right) p = k_f^2\, p \tag{3.19.2}$$

where $k_f = \omega/c_f$, and the equation is called a typical homogeneous Helmholtz equation [1]. For a specific frequency ω, k_f represents the eigenvalue of the differential operator on the left-hand side.

3.1.4   WAVE POTENTIAL IN FLUID

To solve Eq. 3.19 omitting the time harmonic, part let's assume an arbitrary spatial wave potential function $\phi(x_j)$ and gradient of which can be described as the displacement function in the fluid media. We can write

$$\phi = Ae^{i(k_i x_i)} \quad \text{and} \quad \mathbf{u} = \nabla\phi \tag{3.20}$$

Referring to Eq. 3.4, substituting Eq. 3.20 and the expression for c_f we can directly write the spatial expression for pressure as

$$p = -\mathbb{L}\nabla.(\nabla\phi) = -\mathbb{L}\nabla^2\phi \tag{3.21}$$

From Eq. 3.19 it can be realized that for a linear system, $\nabla^2\phi = -k_f^2\,\phi$. Thus, substituting $\nabla^2\phi$ value in Eq. 3.21, the pressure potential will revise to

$$p = \mathbb{L}k_f^2\,\phi = \mathbb{L}\frac{\omega^2}{c_f^2}\,\phi = \rho\omega^2\phi \tag{3.22}$$

This will automatically satisfy Eq. 3.19. Please note that Eq. 3.22 is a homogeneous solution of the homogeneous governing differential equation without any active source or body force in a fluid medium. The solution will be modified for Eq. 3.16 based on the type of body force considered in the medium. Considering a point source in a fluid medium, we could solve the pressure field analytically as an analytical Green's function in fluid. However, acoustic, and ultrasonic fields generated by a finite size source cannot be solved analytically. For such cases, one must rely on numerical computation. For an analytical solution to analyze and activate topological waves, here all wave incidences are assumed to be harmonic plane waves.

3.2 WAVE PROPAGATION IN SOLID MEDIA

3.2.1 NAVIER'S EQUATION OF MOTION

The governing differential equation of motion, also known as the equilibrium equation in a solid medium described in Chapter 2 Eq. 2.87, can be modified without body force as written in Eq. 3.23. Elimination of the body force from Eq. 2.87 may not be appropriate if there is an active wave source in the solid medium. However, propagation of the acoustic or ultrasonic wave due to plane wave incidence without any active wave source will follow the following equation of equilibrium, reiterating Eq. 2.72.

$$\sigma_{ij,j} - \rho(x_j)\ddot{u}_i = 0 \quad \text{or} \quad \nabla.\sigma = \rho(\mathbf{x})\ddot{\mathbf{u}} \tag{3.23}$$

where σ_{ij} is the stress and u_i is the displacement at a point $(\mathbf{x} = x_i\hat{e}_i)$ in a bulk isotropic solid. For inhomogeneous solid media, the density $(\rho(\mathbf{x}))$ may not remain constant over space; stress (σ_{ij}) and displacement (u_i) are, in general, functions of both space (x_i) and time (t).

Recollecting Eq. 2.92, the constitutive law for any linear material can be written as

$$\sigma_{ij} = \mathbb{C}_{ijkl}(x_p)e_{kl} \tag{3.24}$$

where e_{kl} are the strains in the isotropic solid, and p is another index that takes values 1, 2, and 3. For any generalized inhomogeneous (i.e., spatially changing materials) linear elastic isotropic or anisotropic material, $\mathbb{C}_{ijkl}(x_p)$ is the matrix of elastic constants written in Eqs. 2.95 and 2.96. Here the material properties represented by the elastic constants are changing over space. For isotropic materials, the stress-strain relation modifies to the following form

$$\sigma_{ij} = \mathbb{L}(x_p)\delta_{ij}e_{kk} + 2\mathrm{m}(x_p)e_{ij} \tag{3.25}$$

Acoustics and Elastic Wave Propagation in Fluids and Anisotropic Solids   69

where $\mathbb{L}$ and m are the two Lamé constants for isotropic material and δ_{ij} is the Kronecker delta ($\delta_{ij} = 1$ for $i = j$ else, $= 0$). After substituting the expression of strain, $e_{kl} = \dfrac{1}{2}\left(u_{k,l} + u_{l,k}\right)$ written in Eq. 2.8 into Eq. 3.25, one can get

$$\sigma_{ij} = \mathbb{L}\left(x_p\right)\delta_{ij}u_{k,k} + \mathrm{m}\left(x_p\right)\left(u_{i,j} + u_{j,i}\right) \tag{3.26}$$

Please note that the above equation is for inhomogeneous isotropic materials. In the following sections, the equations for homogeneous and inhomogeneous materials are explained and analyzed separately. The metamaterials and periodic phononic crystals are inherently inhomogeneous. Hence, in section 3.2.3 they are treated differently than homogeneous materials in section 3.2.2.

3.2.2 HOMOGENEOUS ISOTROPIC AND ANISOTROPIC MATERIALS

For homogeneous materials, we can assume constant material properties as follows

$$\mathbb{C}_{ijkl}\left(x_p\right) \equiv \mathbb{C}_{ijkl};\ \mathbb{L}\left(x_p\right) = \mathbb{L};\ \mathrm{m}\left(x_p\right) = \mathrm{m} \tag{3.27}$$

Performing the derivatives of σ_{ij} with respect to x_j and substituting in Eq. 3.23, we get

$$\mathbb{L}\delta_{ij}u_{k,kj} + \mathrm{m}\left(u_{i,jj} + u_{j,ij}\right) = \rho\ddot{u}_i \tag{3.28}$$

Applying the Kronecker delta identity $\delta_{ij}x_j = x_i$, and after successive index operation of the above equation, we get

$$\mathbb{L}u_{k,ki} + \mathrm{m}\left(u_{i,jj} + u_{j,ij}\right) = \rho\ddot{u}_t \tag{3.29}$$

or

$$\left(\mathbb{L} + \mathrm{m}\right)u_{j,ji} + \mathrm{m}\left(u_{i,jj}\right) = \rho\ddot{u}_t \tag{3.30}$$

Equation 3.30 can be written in vector form:

$$\left(\mathbb{L} + \mathrm{m}\right)\nabla\left(\nabla.\mathbf{u}\right) + \mathrm{m}\nabla^2\mathbf{u} = \rho\ddot{\mathbf{u}} \tag{3.31}$$

where ∇ is again the nabla Hamiltonian operator. Displacement at a point in a solid medium consists of three components (u_1, u_2, u_3) in three orthogonal directions (x_1, x_2, x_3), and the components of the displacement changes from point to point ($u_i\left(x_j,t\right)$) due to the wave propagation. Hence, by definition, displacement is a vector field. The following identity of the nabla Hamiltonian operation (see Appendix) on a vector field will be very useful to further simplify Eq. 3.31. Hence, applying

$$\nabla^2\mathbf{u} = \nabla\left(\nabla.\mathbf{u}\right) - \nabla\times\left(\nabla\times\mathbf{u}\right) \tag{3.32}$$

we get

$$(\mathbb{L}+m)\nabla(\nabla.\mathbf{u})+m\nabla(\nabla.\mathbf{u})-m\nabla\times(\nabla\times\mathbf{u})=\rho\ddot{\mathbf{u}} \quad (3.33.1)$$

$$\text{Or } (\mathbb{L}+2m)\nabla(\nabla.\mathbf{u})-m\nabla\times(\nabla\times\mathbf{u})=\rho\ddot{\mathbf{u}} \quad (3.33.2)$$

The above equation is called Navier's equation of motion [2, 3] without body force. Applying the permutation symbol (see Appendix), Navier's equation of motion without body force in index notation can be further written as follows

$$(\mathbb{L}+2m)u_{j,ji}-m\epsilon_{ijk}\epsilon_{kpq}u_{q,pj}=\rho\ddot{u}_i \quad (3.34)$$

If we have kept the body force in Eq. 3.34 intact, Navier's equation of motion in Eq. 3.33 would take the following form in vector and index notations, respectively

$$(\mathbb{L}+2m)\nabla(\nabla.\mathbf{u})-m\nabla\times(\nabla\times\mathbf{u})+\mathbf{f}=\rho\ddot{\mathbf{u}} \quad (3.35)$$

$$(\mathbb{L}+2m)u_{j,ji}-m\epsilon_{ijk}\epsilon_{kpq}u_{q,pj}+f_i=\rho\ddot{u}_i \quad (3.36)$$

3.2.2.1 Solving Navier's Equation of Motion in Homogeneous Isotropic Materials

3.2.2.1.1 Helmholtz Decomposition

Acoustic and ultrasonic wave propagation in a bulk isotropic solid medium can be solved assuming zero active source or the body force, and Eq. 3.33 would be sufficient to proceed. Pressure fields in Eqs. 3.14 and 3.21 are the scalar fields. The displacement field in fluid is expressed only as a gradient of a scalar potential Eq. 3.20. This is because the fluid cannot support any shear waves. On the contrary, waves in solids can support shear stresses, and the shear stresses give an additional part to the displacement contributed by the shear deformation. We know that the gradient of a scalar field (here a scalar wave potential) produces a vector field, which could represent the displacement field we need. Without any moment, normal stresses could produce such deformation. However, the shear stresses will produce moment about the center of a finite small volume and will cause shear deformation. To accommodate such shear displacement fields, the curl of a vector field may contribute additional displacement due to the shear stresses. During the wave propagation, there are possibilities of having three pairs of shear stresses in a finite volume of a solid. Thus, a vector potential with three different components is a reasonable assumption to accommodate the additional displacement components. The curl of a vector field (vector wave potential with three individual components) can produce a vector field. This additional field can be superposed with the gradient of a scalar potential, which is also a vector field. This new vector field could also potentially represent the additional displacement field required due to the shear stresses. Thus, it is assumed that the displacement wave field in an isotropic solid is a superposition of the gradient of a scalar wave potential (ϕ) and the curl of a vector wave potential (Ψ). Further, it could be said that any sufficiently smooth vector field ($\mathbf{u}$) could be resolved into

Acoustics and Elastic Wave Propagation in Fluids and Anisotropic Solids   71

a sum of an irrotational (i.e., curl-free) vector field (which could be obtained from the gradient of a scalar potential) and a solenoidal (i.e., divergence-free) vector filed (which could be obtained by taking the curl of a vector potential). Hence, the displacement wave field is decomposed into both scalar and vector potentials. Such decomposition was first proposed by Hermann von Helmholtz (1821–1894) and is called the Helmholtz decomposition [2, 3]. However, this decomposition may not result in a unique relationship between u_i, ϕ, and ψ_j. To achieve a unique relationship, an auxiliary condition was imposed and is called the gauge condition. It is said that the decomposition of a vector field into one scalar and three components of one vector potential will result in a unique relationship only if the divergence of the vector potential is zero (i.e., $\nabla.\Psi = 0$). This means that the vector potential must be non-divergent and should produce only a vortex field of displacements. This also means that the displacement field contributed by the vector potential is due only to the shear stresses but not due to the normal stresses. After the Helmholtz decomposition, we can write the displacement wave field as

$$\mathbf{u} = \nabla\phi + \nabla \times \Psi \tag{3.37.1}$$

$$u_i = \phi_{,i} + \epsilon_{ijk}\psi_{k,j} \tag{3.37.2}$$

3.2.2.1.2 Navier's Equation of Motion with Helmholtz Decomposition

Navier's equation of motion can be simplified further by implementing the Helmholtz decomposition (i.e., substituting Eq. 3.37 for Eq. 3.33) will modify to

$$(\mathbb{L}+2\mathrm{m})\nabla\left(\nabla^2\phi + \nabla.\left(\nabla\times\Psi\right)\right) - \mathrm{m}\nabla\times\left(\nabla\times\left(\nabla\phi + \nabla\times\Psi\right)\right) = \rho\left(\nabla\ddot{\phi} + \nabla\times\ddot{\Psi}\right) \tag{3.38}$$

According to the vector field identities (refer to Appendix) the divergence of the curl of a vector field in zero and, equivalently, the curl of the gradient of a scalar field is zero. Applying the following identities

$$\nabla.\left(\nabla\times\Psi\right) = 0; \ \nabla\times\left(\nabla\phi\right) = 0; \ \nabla\times\nabla\times\Psi = \nabla\left(\nabla.\Psi\right) - \nabla^2\Psi \tag{3.39}$$

Equation 3.38 could be modified to

$$(\mathbb{L}+2\mathrm{m})\nabla\left(\nabla^2\phi\right) - \mathrm{m}\nabla\times\left(\nabla\left(\nabla.\Psi\right) - \nabla^2\Psi\right) = \rho\left(\nabla\ddot{\phi} + \nabla\times\ddot{\Psi}\right) \tag{3.40}$$

$$(\mathbb{L}+2\mathrm{m})\nabla\left(\nabla^2\phi\right) - \mathrm{m}\nabla\times\left(\nabla\left(\nabla.\Psi\right) - \nabla^2\Psi\right) = \rho\left(\nabla\ddot{\phi} + \nabla\times\ddot{\Psi}\right) \tag{3.41}$$

And applying the gauge condition ($\nabla.\Psi = 0$), we get

$$(\mathbb{L}+2\mathrm{m})\nabla\left(\nabla^2\phi\right) - \mathrm{m}\nabla\times\left(-\nabla^2\Psi\right) = \rho\left(\nabla\ddot{\phi} + \nabla\times\ddot{\Psi}\right) \tag{3.42}$$

Applying the commutability of the del operator,

$$\nabla\left[(\mathbb{L}+2\mathrm{m})\left(\nabla^2\phi\right) - \rho\ddot{\phi}\right] + \nabla\times\left[\mathrm{m}\left(\nabla^2\Psi\right) - \rho\ddot{\Psi}\right] = 0 \tag{3.43}$$

72 Metamaterials in Topological Acoustics

Equation 3.43 appears like the Helmholtz decomposition in Eq. 3.37.1 and represents that the superposition of the gradient of a scalar field and the curl of a vector field is equal to zero. However, this is possible only when the scalar and vector fields are nonexistent, independently. Hence, the sufficient conditions for the Eq. 3.43 to be valid are

$$\left(\mathbb{L}+2\mathrm{m}\right)\left(\nabla^2\phi\right)-\rho\ddot{\phi}=0 \tag{3.44}$$

$$\mathrm{m}\left(\nabla^2\Psi\right)-\rho\ddot{\Psi}=0 \tag{3.45}$$

Equations 3.44 and 3.45 can be written in the form of wave equations as follows

$$\nabla^2\phi-\frac{1}{c_p^2}\ddot{\phi}=0 \tag{3.46}$$

$$\nabla^2\Psi-\frac{1}{c_s^2}\ddot{\Psi}=0 \tag{3.47}$$

where $c_p=\sqrt{\dfrac{(\mathbb{L}+2\mathrm{m})}{\rho}}$ and $c_s=\sqrt{\dfrac{\mathrm{m}}{\rho}}$ are the P wave, or longitudinal, and S wave, or shear wave, velocities in isotropic solid media.

As the solution of a typical wave equation in linear media can be described as a superposition of several plane harmonic waves, we could assume to solve Eqs. 3.46 and 3.47 like it was assumed in Eq. 3.13. A wave potential is composed of a spatial phase part and a time harmonic temporal part. Thus we could assume $\phi\left(x_j,t\right)=\varphi\left(x_j\right)e^{-i\omega t}$ and $\psi_i\left(x_j,t\right)=\Psi_i\left(x_j\right)e^{-i\omega t}$ or $\Psi\left(x_j,t\right)=\Psi(x_j)e^{-i\omega t}$. Substituting them in Eqs. 3.46 and 3.47, the wave equations transform into two Helmholtz equations as follows

$$\nabla^2\varphi+k_p^2\varphi=0 \tag{3.48}$$

$$\nabla^2\Psi+k_s^2\Psi=0 \tag{3.49}$$

where $k_p^2=\omega^2/c_p^2$ and $k_s^2=\omega^2/c_s^2$ are the square of the longitudinal wave number and the shear wave number, respectively.

3.2.2.1.3 Generalized Wave Potentials in Isotropic Solids

In Eq. 3.20 we found how to write a generalized wave potential for Helmholtz-type equations (Eqs. 3.19, 3.48, and 3.49). Ignoring the time harmonic part ($e^{-i\omega t}$), which is easily separable from both sides of the equation, we will further express the spatial part of the wave potential using a phase component as follows

$$\varphi=Ae^{i\left(\mathbf{k}^\mathbf{P}\cdot\mathbf{x}\right)}=Ae^{i\left(k_j^p x_j\right)} \ and \ \Psi=Be^{i\left(\mathbf{k}^\mathbf{s}\cdot\mathbf{x}\right)}=Be^{i\left(k_j^s x_j\right)} \tag{3.50}$$

where $\mathbf{k}^\mathbf{P}$ and $\mathbf{k}^\mathbf{s}$ describe the wave vectors pointed towards the propagation of a P wave and an S wave, respectively. As they are assumed to be different, it implies that

Acoustics and Elastic Wave Propagation in Fluids and Anisotropic Solids

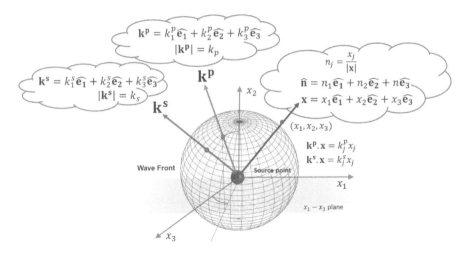

FIGURE 3.2 Spherical wave fronts in isotropic solid: Two wave directions – one for P wave (k^p) and another for S wave (k^s) – are shown to describe the mathematical equations for their respective phases $k_j^p x_j$ and $k_j^s x_j$.

the propagation direction of both the P wave and S wave will not be same in isotropic bulk media. Please note that $\mathbf{k}^p \cdot \mathbf{x} = k_j^p x_j$, $|\mathbf{k}^p| = k_p$, $\mathbf{k}^s \cdot \mathbf{x} = k_j^s x_j$, and $|\mathbf{k}^s| = k_s$. Please keep a note of the superscript p and s and the subscript j. Superscripts are for the type of wave, and subscripts designate the direction. Hence, k_j^s signifies the magnitude of the shear wave vector $\mathbf{k}^s$ along the j direction. Alternatively, it is the shear wave number along the j direction. Similarly, k_j^p is the longitudinal wave number along the j direction. $\mathcal{A}$ and $\mathcal{B}$ in Eq. 3.50 are the amplitude of the scalar and vector potentials, respectively. Additionally, please note that the amplitude vector **B** has three different amplitudes (B_1, B_2, and B_3 designated as B_k) for three corresponding orthogonal vector potentials. Please refer to Fig. 3.2, where two different wave directions, one for P wave ($\mathbf{k}^p$) and another for S wave ($\mathbf{k}^s$) are presented. If we are interested in investigating the wave field at an arbitrary spatial point (x_1, x_2, x_3) then the dot product between the wave direction vector and the position vector $\mathbf{x} = x_j \hat{e}_j$ will describe the phases of the P wave and the S wave, respectively, which is actually written in Eq. 3.50.

3.2.2.1.4 Phase Accumulation by a Traveling Wave

Phase is the critical concept for the study of topology. We call it the topological phases. Without which it is impossible to realize the wave propagation in topologically protected media as described in Chapter 1. Later in this chapter and in the following subsequent chapters the topological phases are described in detail. This section describes how to calculate the phase of a traveling wave. It is always challenging to visualize the phase. Hence, it is necessary to understand the phase from a fundamental understanding. In equation Eq. 3.50 and in Fig. 3.2, the phase is **k.x**, irrespective of P wave or S wave.

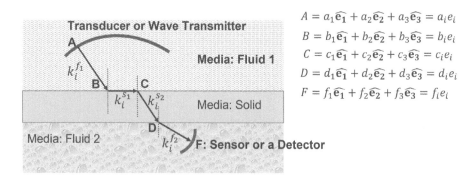

FIGURE 3.3 Schematic for an example showing the calculation of the phase incurred by a propagating wave.

Let's assume a scenario in which a wave starts from a transmitter and travels along path A and F (Fig. 3.3). First the wave (a general arbitrary wave, either P or an S wave) traveled along the path AB in fluid 1, then it traveled along the path BC on the surface of a solid or along the interface solid-fluid 1. Due to critical refraction, the wave travels along path CD in the solid. At last, the wave took the path along DF in fluid 2 due to refraction of the wave at the solid–fluid 2 interface. F is the point where the wave detector receives and senses the wave. We would like to find what would be the phase difference between A and F. To do so, we need the wave numbers along the individual wave paths and the position vector of points A, B, C, D, and F. Position vectors of points A, B, C, D, and F are written in Fig. 3.3. $\mathbf{k}^{f_1}$ is the wave vector in fluid 1 along the path AB. $\mathbf{k}^{s_1}$ is the wave vector along the path BC in solid. $\mathbf{k}^{s_2}$ is the wave vector along the path CD in solid. And at the last $\mathbf{k}^{f_2}$ is the wave vector along the path DF in fluid 2. Please note that fluid 1 and fluid 2 are two different fluids and thus the magnitude of wave vectors $\mathbf{k}^{f_1}$ and $\mathbf{k}^{f_2}$ namely $|\mathbf{k}^{f_1}|$ and $|\mathbf{k}^{f_2}|$ are different. Due to the change in directions, components of the wave vectors along the Cartesian coordinate directions are also different. Similarly, the components of $\mathbf{k}^{s_1}$ and $\mathbf{k}^{s_2}$ wave vectors are not the same. If the solid is an isotropic material, the magnitude of the wave vectors $\mathbf{k}^{s_1}$ and $\mathbf{k}^{s_2}$, namely $|\mathbf{k}^{s_1}|$ and $|\mathbf{k}^{s_2}|$, are equal. However, they are not the same if the material is anisotropic. In anisotropic material, the wave numbers are different in different directions due to their direction-dependent wave velocities. Next, let's find the vector along the directions AB, BC, CD, and DF.

$$\vec{AB} = \vec{B} - \vec{A} = (b_1 - a_1)\hat{e}_1 + (b_2 - a_2)\hat{e}_2 + (b_3 - a_3)\hat{e}_3 \quad (3.51.1)$$

$$\vec{BC} = \vec{C} - \vec{B} = (c_1 - b_1)\hat{e}_1 + (c_2 - b_2)\hat{e}_2 + (c_3 - b_3)\hat{e}_3 \quad (3.51.2)$$

$$\vec{CD} = \vec{D} - \vec{C} = (d_1 - c_1)\hat{e}_1 + (d_2 - c_2)\hat{e}_2 + (d_3 - c_3)\hat{e}_3 \quad (3.51.3)$$

$$\vec{DF} = \vec{F} - \vec{D} = (f_1 - d_1)\hat{e}_1 + (f_2 - d_2)\hat{e}_2 + (f_3 - d_3)\hat{e}_3 \quad (3.51.4)$$

Acoustics and Elastic Wave Propagation in Fluids and Anisotropic Solids 75

Now the change in phases along $\overrightarrow{AB}$, $\overrightarrow{BC}$, $\overrightarrow{CD}$, and $\overrightarrow{DF}$ can be written as the dot product of wave vector $\mathbf{k}$ and the directional vector $\mathbf{x}$. Let's describe the phase by α. Then the phases along each path can be written as

$$\alpha_{AB} = \mathbf{k}^{f1}.\overrightarrow{AB} = k_i^{f1}\left(b_i - a_i\right) \tag{3.52.1}$$

$$\alpha_{BC} = \mathbf{k}^{s1}.\overrightarrow{BC} = k_i^{s1}\left(c_i - b_i\right) \tag{3.52.2}$$

$$\alpha_{CD} = \mathbf{k}^{s2}.\overrightarrow{CD} = k_i^{s2}\left(d_i - c_i\right) \tag{3.52.3}$$

$$\alpha_{DF} = \mathbf{k}^{f2}.\overrightarrow{DF} = k_i^{f2}\left(f_i - d_i\right) \tag{3.52.4}$$

The total phase difference between A and F will be

$$\alpha = k_i^{f1}\left(b_i - a_i\right) + k_j^{s1}\left(c_j - b_j\right) + k_m^{s2}\left(d_m - c_m\right) + k_n^{f2}\left(f_n - d_n\right) \tag{3.53}$$

where a same index signifies the summation over the values it takes. The indices i, j, m and n take values 1, 2, and 3.

One very important aspect we have overlooked so far is the physical nature of the P waves and S waves described by the scalar potential φ and the vector potential Ψ. In calculation of phase, it should be mentioned whether the P wave or S wave are considered. This is because the wave numbers, or the magnitude of the wave vector for the P wave and the S wave, are different. To describe them in relation to the phase, first we need to understand the P waves and the S waves in isotropic solids.

3.2.2.1.5 Longitudinal Waves and Shear Waves in Homogeneous Isotropic Solids

Helmholtz decomposition of the displacement wave field gives scalar and vector potentials in isotropic solids that have very specific displacement patterns. To understand their pattern, let's consider one type of potential at a time. If a wave field is developed such a way that it has no rotational filed (i.e., the field is a curl-free wave field, then we can assume $\Psi = 0$). Expanding Eq. 3.37.1 with generalized wave potentials expressed in Eq. 3.50, we can write

$$\mathbf{u} = \nabla\varphi e^{-i\omega t} \quad or \quad \mathbf{u} = \mathbf{k}^{\mathbf{p}}.Ae^{i\left(\mathbf{k}^{\mathbf{P}}.\mathbf{x}\right)}e^{-i\omega t} \tag{3.54.1}$$

$$u_i = \varphi_{,i}e^{-i\omega t} \tag{3.54.2}$$

or

$$u_i = i\delta_{ij}k_j^p Ae^{i\left(k_j^p x_j\right)}e^{-i\omega t} = ik_i^p Ae^{i\left(k_j^p x_j\right)}e^{-i\omega t} = Ak_i^p e^{i\left(k_j^p x_j\right)}e^{-i\omega t} \tag{3.54.3}$$

where the Kronecker delta rule is used and the complex amplitude is implied $A = i\mathcal{A}$. The direction of the particle displacement (u_i) in the wave field is along the direction of the wave propagation (k_i^p). That means the particle of the media will move along the direction of the wave propagation as shown in Fig. 3.4. This is

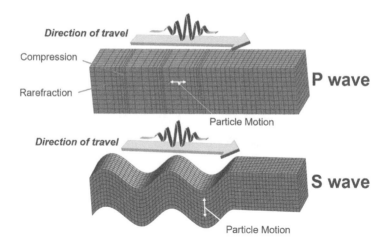

FIGURE 3.4 Schematic description of P wave and S wave.

called the P wave or the longitudinal wave, which is mathematically represented by a scalar potential.

Next, if a wave field is expressed in such a way that it has no divergence (i.e., the field is a divergence-free wave field), then we can assume $\phi = 0$. Expanding Eq. 3.37.1 with generalized wave potentials expressed in Eq. 3.50, we can write

$$\mathbf{u} = \nabla \times \boldsymbol{\Psi} e^{-i\omega t} \tag{3.55}$$

Using the permutation symbol used in index notation, and assuming the amplitudes (B_k) in Eq. 3.50 are constants, we can express Eq. 3.55 in index notation as follows

$$u_k = \hat{\mathbf{e}}_k \epsilon_{kij} \frac{\partial \left(B_j e^{i(k_l^s x_l)} \right)}{\partial x_i} e^{-i\omega t} = i\, \hat{\mathbf{e}}_k \epsilon_{kij} B_j k_l^S \delta_{li} e^{i(k_l^s x_l)} e^{-i\omega t} \tag{3.56}$$

Applying the Kronecker delta rule and consuming the imaginary part inside the amplitude, we get

$$u_k = \hat{\mathbf{e}}_k \epsilon_{kij} C_j k_i^s e^{i(k_l^s x_l)} e^{-i\omega t} \tag{3.57}$$

where $C_j = iB_j$. Expanding the displacements in three orthogonal directions, we can write

$$\begin{aligned} u_1 &= \left[C_3 k_2^s e^{i(k_l^s x_l)} - C_2 k_3^s e^{i(k_l^s x_l)} \right] e^{-i\omega t} \\ u_2 &= \left[C_1 k_3^s e^{i(k_l^s x_l)} - C_3 k_1^s e^{i(k_l^s x_l)} \right] e^{-i\omega t} \\ u_3 &= \left[C_2 k_1^s e^{i(k_l^s x_l)} - C_1 k_2^s e^{i(k_l^s x_l)} \right] e^{-i\omega t} \end{aligned} \tag{3.58}$$

Acoustics and Elastic Wave Propagation in Fluids and Anisotropic Solids   77

Next, let's take a dot product of the above displacement field **u** with the wave propagation direction $\mathbf{k}^s$.

$$\mathbf{u}.\mathbf{k}^s = u_i k_i^s = u_1 k_1^s + u_2 k_2^s + u_3 k_3^s \tag{3.59}$$

Substituting Eq. 3.58 into Eq. 3.59, one can easily show that the sum in Eq. 3.59 will result zero. Hence, we can conclude that the displacement field of a divergence-free wave field is orthogonal to the direction of wave propagation. None of the particle displacements will be along the direction of the wave vector $\mathbf{k}^s$ as shown in Fig. 3.4. This is called the S wave, or the transverse wave, which is mathematically represented by a vector potential with three orthogonal components.

The vector potential will have three components. In the case of two dimensional in-plane problems, the survivability of the components of the vector potential will depend on the choice of the axes in the problem. For example, if the $x_1 - x_2$ plane is considered, then out-of-plane displacement u_3 is zero. According to Eq. 3.58, C_1 and C_2 should be zero, and hence, only the Ψ_3 component will survive with amplitude C_3. In-plane displacements only due to the shear wave will take the following form.

$$
\begin{aligned}
u_1 &= \left[C_3 k_2^s e^{i\left(k_l^s x_l\right)} \right] e^{-i\omega t} \\
u_2 &= \left[-C_3 k_1^s e^{i\left(k_l^s x_l\right)} \right] e^{-i\omega t}
\end{aligned}
\tag{3.60}
$$

On the same $x_1 - x_2$ plane, if a P wave is considered along the direction $x_1,$ then the total displacements along x_1 and x_2 will be

$$u_1 = A k_1^p e^{i\left(k_j^p x_j\right)} e^{-i\omega t} + C_3 k_2^s e^{i\left(k_l^s x_l\right)} e^{-i\omega t} \tag{3.61.1}$$

$$u_2 = -C_3 k_1^s e^{i\left(k_l^s x_l\right)} e^{-i\omega t} \tag{3.61.2}$$

If a P wave is considered along the direction $x_2,$ then the total displacements along x_1 and x_2 will be

$$u_1 = C_3 k_2^s e^{i\left(k_l^s x_l\right)} e^{-i\omega t} \tag{3.62.1}$$

$$u_2 = A k_2^p e^{i\left(k_j^p x_j\right)} e^{-i\omega t} - C_3 k_1^s e^{i\left(k_l^s x_l\right)} e^{-i\omega t} \tag{3.62.2}$$

Please note that in both the above cases, one or the other displacement is coupled due to the nature of the P wave and the S wave. In the above cases, we have two S waves propagating orthogonally, one along the direction x_1 and another along the direction x_2. Hypothetically, if we have a situation where a P wave and a single S wave is propagating only along the direction x_1 or x_2, then we will get a complete decoupled displacements on the $x_1 - x_2$ plane as follows

- P and S waves traveling along the x_1 direction with $u_3 = 0$

$$u_1 = Ak_1^p e^{i\left(k_j^p x_j\right)} e^{-i\omega t} \tag{3.63.1}$$

$$u_2 = -C_3 k_1^s e^{i\left(k_l^s x_l\right)} e^{-i\omega t} \tag{3.63.2}$$

- P and S waves traveling along the x_2 direction with $u_3 = 0$

$$u_1 = C_3 k_2^s e^{i\left(k_l^s x_l\right)} e^{-i\omega t} \tag{3.64.1}$$

$$u_2 = Ak_2^p e^{i\left(k_j^p x_j\right)} e^{-i\omega t} \tag{3.64.2}$$

where the negative sign could be consumed inside the C_3 coefficient; however, not to make things complicated, it is kept as it is. Substituting these special displacement patterns into the governing differential equation, Eqs. 3.38 through 3.43 will be decoupled for P wave and S wave automatically, satisfying the curl-free part by the u_1 displacement and the divergence-free part by the u_2 displacement.

On the contrary if the $x_1 - x_3$ plane is considered, then out-of-plane displacement u_2 should be enforced to zero. According to Eq. 3.58, C_1 and C_3 should be zero, and hence, only the Ψ_2 component will survive. Like Eqs. 3.63.1 and 3.63.2, decoupled u_1 and u_3 displacements could be achieved when

- a P wave and an S wave both are traveling along the x_1 direction with $u_2 = 0$

$$u_1 = Ak_1^p e^{i\left(k_j^p x_j\right)} e^{-i\omega t} \tag{3.65.1}$$

$$u_3 = C_2 k_1^s e^{i\left(k_l^s x_l\right)} e^{-i\omega t} \tag{3.65.2}$$

- P and S waves both are traveling along the x_3 direction with $u_2 = 0$

$$u_1 = C_2 k_3^s e^{i\left(k_l^s x_l\right)} e^{-i\omega t} \tag{3.66.1}$$

$$u_3 = Ak_3^p e^{i\left(k_j^p x_j\right)} e^{-i\omega t} \tag{3.66.2}$$

Hence, note that the components of the vector potential in the Helmholtz decomposition of the displacement wave field written in Eq. 3.37.1 should be judicially selected based on a specific scenario of wave propagation and an appropriate derivate should be used based on the selection of the coordinate system.

This situation is particularly helpful to present the governing equation in Hamiltonian form. Elastic waves for their counter quantum mechanical applications will require the understanding of the system by deriving their Hamiltonian equations. Such scenarios will be further discussed under inhomogeneous isotropic material and in later chapters.

3.2.2.1.6 In-Plane and Out-of-Plane Shear Waves in Isotropic Solids

A P wave with a wave velocity c_p has particle motions along the principal direction of the wave vector $\mathbf{k}^p$, which is the direction of the wave propagation. However, S waves with wave velocity c_s have the particle motions orthogonal to the direction of wave vector $\mathbf{k}^s$, or the wave propagation direction. Referring to Fig. 3.5, where $\mathbf{k}^p$ and $\mathbf{k}^s$ are in the same arbitrary direction, it can be visualized that the particles have several options in choosing the direction of motion to carry the S wave along the dotted line, as long as they are orthogonal to the wave vectors $\mathbf{k}^p$ and $\mathbf{k}^s$. In Fig. 3.5, a Cartesian coordinate system is used. Instead of considering infinite possibilities, which are redundant, we will consider only two possible orthogonal directions. One is called in-plane motion and the other is called out-of-plane motion. One of them is coupled with the P wave, and another is independent and decoupled from the P wave. In Fig. 3.5, $\mathbf{k}^p$ and $\mathbf{k}^s$ wave vectors are pointed in an arbitrary direction. A circular plane, orthogonal to the wave vectors, can be imagined. The particle displacements of the P wave are along the wave vector $\mathbf{k}^p$, designated as P. However, the particle displacements of the S waves have multiple possibilities (dotted arrows in Fig. 3.5), out of which only two directions are marked, along SV and SH, while the wave vectors are pointing in the direction of $\mathbf{k}^p$ and $\mathbf{k}^s$, which signifies the wave propagation direction. Particle displacement along SV is orthogonal to the P wave vector and considered in-plane with the P wave, which is coupled. Considering the same P wave vector, particle motion along SH is also orthogonal to the P wave but out of plane to the P–SV plane as indicated in Fig. 3.5. It is immediately evident that the choice of SV and SH purely depends on a problem and the respective coordinate system. If the wave propagation of interest is confined on a $x_1 - x_2$ plane, then the

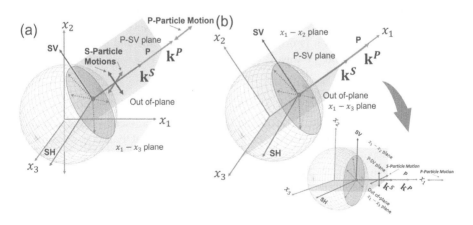

FIGURE 3.5 a) Showing possible particle motions when the wave vectors for P wave and S wave are along the same arbitrary direction in Cartesian coordinate system, b) showing possible particle motions when the wave vectors for P wave and S wave are along the x_1 axis.

particle motions that result in u_1 and u_2 displacements will create the P–SV duo, and the SH wave will contribute to u_3 displacement. Similarly, if the wave propagation of interest is confined on a $x_1 - x_3$ plane, then the particle motions that result in u_1 and u_3 displacements will create the P–SV duo, and the SH wave will contribute to u_2 displacements. To state it in a general form, there is no specific SV or SH wave mode as such; there are two shear wave modes and one longitudinal wave mode. If the shear wave propagation with particle motion is orthogonal to the longitudinal motion and lies on a plane of interest, then that shear wave is marked as an SV wave and the other shear wave causing out-of-plane motion is marked as an SH wave. Generally speaking, the particle motion of an SV wave is in-plane with the P wave but results in a shear wave with vertical polarization. Please note again that the vertical direction does not always refer to the x_3 axis or the direction of the gravity. The shear vertical wave, or SV wave, which has the wave vector $\mathbf{k}^s$, has the same wave velocity c_s as the shear horizontal, or SH wave. Similarly, The particle motion in the SH direction is out of plane with respect to the P–SV plane but results in a shear wave with horizontal polarization. Please note again that the horizontal direction does not always refer to the x_2 or x_1 axis or orthogonal to the gravity. A shear horizontal wave, or SH wave, which has the wave vector $\mathbf{k}^s$, has the wave velocity c_s as does the shear vertical or SV wave.

In most common cases, the wave propagation direction is aligned with the x_1 axis. Thus, the x_2 axis, which is vertical to the x_1 axis, designates the polarization of the SV waves in a right-hand coordinate system. Alternatively, the x_3 axis, which is horizontal and orthogonal to the x_1 axis, designates the polarization of the SH waves. Figure 3.5b shows the same situation presented in Fig. 3.5a, with the x_1 axis intentinally aligned to the wave vector for our convenience. Without the loss of the generality, in most of our analysis this convenience is used almost all the time. With the x_1 axis pointed along the $\mathbf{k}^p$ and $\mathbf{k}^s$ vectors, the x_2 axis is the direction of particle motion for the SV waves and the x_3 axis is the direction of the particle motion for the SH waves. From this section it is clear that in isotropic infinite media, three wave modes – P wave, SV wave, and SH wave – propagate with three orthogonal polarizations. It is also necessary to mention here that the SH wave is independent from P and SV wave because it is orthogonal to the P–SV plane and doesn't carry any components of the P and SV waves.

3.2.2.1.7 *Wave Potentials for P, SV, and SH Waves and their Relation*

It is evident that the P wave, or the dilatational wave, is expressed by a scalar potential φ. Similarly, an S wave is expressed by the vector potential Ψ. But as we consider SV and SH separately based on the particle motions in two mutually orthogonal directions, each could be expressed by its own potentials similar to a scalar potential. As the particle motion for the SH wave is orthogonal to the SV and P wave, it does not have any component on the P–SV plane and demands separate treatment. Considering the coordinate alignment used in Fig. 3.5b, let's consider a case where there is no in-plane motion of the particles on the P–SV plane, (i.e., u_1 and u_2 are zero). Only u_3 mothion survives. Let's consider the wave equation in Eq. 3.35, which could be expanded with all displacement components before applying the Helmholtz

Acoustics and Elastic Wave Propagation in Fluids and Anisotropic Solids 81

decomposition. Further plugging $u_1 = u_2 = 0$ into the equation, Eq. 3.36 could be rewritten only for the u_3 displacement as follows

$$\mathrm{m}\left(u_{3,11} + u_{3,22}\right) - \rho \ddot{u}_3 = 0 \tag{3.67.1}$$

$$\nabla^2 u_3 - \frac{1}{c_s^2} \ddot{u}_3 = 0 \tag{3.67.2}$$

which is similar to Eq. 3.46 or 3.47. Next, substituting plane wave potential for the SH wave,

$$u_3\left(x_j,t\right) = \Psi_{SH}\left(x_j\right) e^{-i\omega t} \tag{3.68}$$

Eq. 3.67.2 will be revised to

$$\nabla^2 \Psi_{SH} + k_s^2 \Psi_{SH} = 0 \tag{3.69.1}$$

$$\text{or } \left(\frac{\partial^2}{\partial x_1^2} + \frac{\partial^2}{\partial x_2^2} + \frac{\partial^2}{\partial x_3^2}\right) \Psi_{SH} = -k_s^2 \Psi_{SH} \tag{3.69.2}$$

where $k_s^2 = \omega^2 / c_s^2$ is the square of the shear wave number discussed previously. For a specific frequency ω, k_s represents the eigenvalue of the differential operator on the left-hand side. Based on generalized plane wave potential discussed in section 3.2.2.1.3, the $\Psi_{SH}\left(x_j\right)$ potential can be further modified to

$$\Psi_{SH}\left(x_j\right) = B_{SH} e^{i\left(k_j^s x_j\right)} \tag{3.70}$$

where B_{SH} is the amplitude of the SH wave along the x_3 axis, with wave propagation along the x_1 axis with velocity c_s and wave number k_s at radial frequency ω.

Keeping our discussion along the same line, we can express the wave potentials for the P wave and SV wave as follows

$$\varphi_P\left(x_j\right) = A_P e^{i\left(k_j^P x_j\right)} \tag{3.71}$$

$$\Psi_{SV}\left(x_j\right) = B_{SV} e^{i\left(k_j^s x_j\right)} \tag{3.72}$$

SV wave and P wave propagate on the same plane. Hence, they are coupled. Thus, when P wave is incident on an isotropic interface, it will break down into both P wave and SV wave when reflected from an interface and will similarly break down when transmitted to another isotropic media beyond the interface. However, when SH wave is incident on an isotropic interface, it will not break down into P wave or SV wave but will remain SH wave in the reflection and remain SH wave in transmission to another isotropic media beyond the interface. To express these multiple waves generated due to the interaction of the waves at the interfaces, different wave potentials with different wave amplitudes should be assumed and appropriate interface conditions should be applied. The objective of this book is to make a transition of the fundamental understanding of waves to the understanding of the topological effects. Hence, classical analysis of waves at the interfaces is considered trivial. In this book all such scenarios are omitted and could be found in books on wave propagation in solids [2–4].

3.2.2.2 Solving Navier's Equation of Motion in Homogeneous Anisotropic Materials

Wave fronts in fluid and isotropic materials are spherical because the wave velocities in all directions are equal. Although P waves and S waves have different wave velocities, their wave fronts are still spherical, as shown in Fig. 3.5a. Waves in anisotropic solid media [5], however, do not propagate with a spherical wave front as in isotropic materials. This situation is shown in Fig. 3.6 for an anisotropic material GaAs. Due to constant amplitude in all directions in an isotropic solid, a wave velocity plot in three dimensions generates a spherical plot of the wave front. Similarly, considering a source at the origin in GaAs, an anisotropic material, quasi shear wave velocities are different in different directions. Due to different amplitudes of velocities along different directions, wave velocity plot in three dimensions (3D) generates specific architectures (as in Fig. 3.6) of the wave fronts, and these architectures are uniquely dependent on the material property matrix or the $\mathbb{C}_{mn}$ matrix in Eq. 2.96. Composite materials are good examples of anisotropic solid material with different degrees. For example, triclinic, monoclinic, orthotropic, and transversely isotropic materials are example of

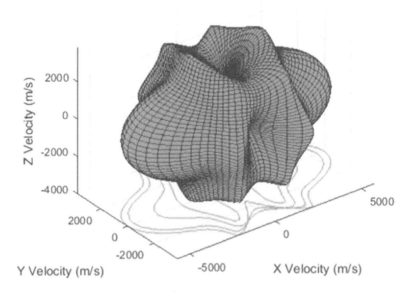

FIGURE 3.6 3D wave velocity profile of quasi-shear wave in anisotropic media (GaAs), showing non spherical wave front.

Acoustics and Elastic Wave Propagation in Fluids and Anisotropic Solids 83

generalized anisotropic material with different degrees of anisotropy or with different plane and/or axis of symmetry as discussed in section 2.14 in Chapter 2.

3.2.2.2.1 Wave Polarization and Slowness Surface

Anisotropy, together with homogeneous or inhomogeneous material properties, is the key ingredient for topological wave behavior. Different characteristics of wave dispersion in different directions may trigger topological waves. In anisotropic material, polarization of the waves, or the particle motion while the wave propagates, plays a key role. Understanding of topological waves in anisotropic material is an open field of research that has not been fully explored yet. We have barely scratched the surface in the past two decades. While detailed discussions are reserved for later chapters, here, propagation in homogeneous anisotropic media is discussed.

First, to understand polarization, it is recommended to refer to the previous sections on isotropic materials. We briefly introduced the particle motions of a wave system in an isotropic solid in section 3.2.2.1.6. It shows that the polarization (same as the particle motion) in isotropic solids consists of three directions. P wave or the longitudinal wave has the particle motion along the direction of the wave propagation. An S wave causes lateral particle motion orthogonal to the direction of the wave propagation. In both cases the direction of the wave energy is aligned with the direction of the wave vector (Fig. 3.5) (i.e., perpendicular to the surface of the wave front, which is spherical due to isotropy). This scenario completely changes when anisotropy is considered. Anisotropy sometimes causes anomalous wave polarization [6, 7]. Finding the wave polarization in an anisotropic solid is not straightforward. This requires a bit more explanation and understanding of wave energy direction, particle motion, wave propagation direction, and wave fronts, as discussed below.

In this section, wave propagation in bulk anisotropic solids with different degrees of anisotropy will be discussed. If an acoustic emitter or a transducer is placed in an anisotropic media, it emits the mechanical energy in the forward direction, and a wave front is created. Say we are interested in investigating the wave behavior along a wave vector $\mathbf{k} = k_i \hat{\mathbf{e}}_i$, as shown in Fig. 3.7a. Unlike in isotropic material, the energy propagation or the wave polarity will not necessarily be along the direction of the wave vector $\mathbf{k}$. The unit vector $\hat{\mathbf{n}} = n_i \hat{\mathbf{e}}_i$ in Fig. 3.7a is along the $\mathbf{k}$ vector, where n_i are the direction cosines of the unit vector $\hat{\mathbf{n}}$ (i.e., cosines of the respective angles made by the $\mathbf{k}$ vector with the coordinate axes x_1, x_2, and x_3 respectively). The profile of the wave front was achieved by joining the points with equal phases around the wave source. Wave fronts are equiphasic surfaces. Figure 3.7a shows the wave front generated in an arbitrary media with an arbitrary shape for now. The velocity of the wave energy c_E along the unit vector $\hat{\mathbf{N}} = N_i \hat{\mathbf{e}}_i$ is at an angle θ_2 with respect to the wave vector $\mathbf{k}$. Here, N_i are the direction cosines of the unit vector $\hat{\mathbf{N}}$ (i.e., cosines of the respective angles made by the unit vector $\hat{\mathbf{N}}$ with the coordinate axes x_1, x_2, and x_3 respectively). The energy velocity or the wave group velocity c_E is not necessarily along the wave vector $\mathbf{k}$. Anisotropy results in wave energy propagation along a different direction compared to phase velocity c, marked with a unit vector $\hat{\mathbf{n}}$ along the direction of $\mathbf{k}$. Although the $\hat{\mathbf{N}}$ and $\mathbf{k}$ vectors are known on a same plane in Fig. 3.7a,

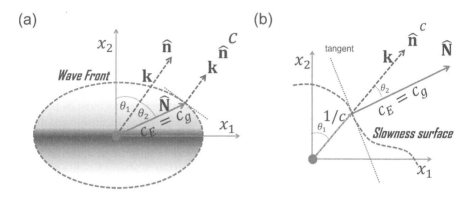

FIGURE 3.7 A schematic showing the wave field inside a bulk anisotropic medium: a) wave front surface and energy propagation b) slowness surface and energy propagation.

note that the plane of energy propagation may not be on the same plane as the incident wave. This will depend on the degree of anisotropy and material coefficients. It means that the unit vector $\hat{\mathbf{N}} = N_i \hat{\mathbf{e}}_i$ in Fig. 3.7a may not be on the $x_1 - x_2$ plane as shown. $\hat{\mathbf{N}}$ could potentially be along any direction in the 3D coordinate system based on the material properties and the direction of the $\mathbf{k}$ vector. However, in most materials they are on the same plane. For example, if the material is transversely isotropic, $\hat{\mathbf{N}}$ will definitely lie on the same plane as $\hat{\mathbf{n}}$.

It is to be noted that the wave vector $\mathbf{k}$ is perpendicular to the wave front, where the direction of the wave energy propagation and the wave front intersects, shown in Fig. 3.7a. Figure 3.7b shows a velocity slowness surface in an anisotropic material. Slowness surface is the most vital component to discuss wave propagation in anisotropic material, which was not necessary for an isotropic material. Slowness surface is a generalized concept for all types of materials. Slowness surface is a surface in 3D created by plotting the inverse of the phase velocities along different directions of the wave propagation. In isotropic material slowness surfaces are spherical due to its isotropy. But in anisotropic material, the architecture of the slowness surfaces could be very different.

Along the $\mathbf{k}$ vector or along the direction of the wave propagation, the inverse of the phase wave velocities results in the slowness. If the phase velocities are plotted in the same way, it will result in a wave velocity surface. They are reciprocal in terms of numerical values. However, in wave theory, velocity surfaces and slowness surfaces are not called reciprocal. Wave front surfaces and slowness surfaces are called reciprocal surfaces. The reason behind this reciprocal nature is discussed below.

Interestingly, a normal ($\hat{\mathbf{N}}$) to the slowness surface at the intersection of the surface and the intended direction wave propagation along the wave vector $\mathbf{k}$ actually shows the direction of the propagation of the wave energy. $\hat{\mathbf{N}}$ can be defined as the wave energy vector. Wave energy velocity and wave group velocity are synonyms. Hence, the normal vector $\hat{\mathbf{N}}$ points towards the direction of the group velocity of the wave.

Next, it is necessary to summarize three different surfaces in anisotropic media. They exist in isotropic media but do not play much importance in analyzing the wave propagation. This is because they are all concentric and never intersect each other in isotropic material. In anisotropic media these three surfaces (wave front surface, wave velocity surface, and wave slowness surface) are not necessarily concentric.

Wave front surfaces are created by joining the particles that are in same phase in the material when the wave propagates. It shows the direction of propagation of wave energy. Wave velocity surfaces are created by joining the tip of the wave vectors that represent the respective amplitude of the phase wave velocities in the direction of the wave vector. Slowness surfaces are created by joining the tip of the wave vectors with magnitude that represents the respective amplitude of the inverse of the phase velocities in the direction of the wave vector **k**.

To present the relationship between these surfaces, let's refer to Fig. 3.8. Here, three different but very important possible intersecting surfaces are shown. At any arbitrary point on the slowness surface if a normal is drawn, the normal $\hat{\mathbf{N}}$ will represent the direction of the energy propagation of the wave at that point. If at the same point a tangent is drawn, say $\hat{\mathbf{T}}$, the tangent vector will essentially intersect orthogonally with the vector $\widehat{\mathbf{N}_g}$ if drawn from the origin (wave source) parallel to the vector $\hat{\mathbf{N}}$. Further, if the group velocity magnitude ($c_E = c_g$) is plotted along the direction of the $\widehat{\mathbf{N}_g}$ vector, the point will lie on the wave front surface, or the energy surface as shown in Fig. 3.8. At the newly found point, the normal ($\hat{\mathbf{n}}$) to the wave front surface will show the direction of the wave vector **k** along the unit normal $\hat{\mathbf{n}}$. The tangent to the wave front surface will share the tangent of the velocity surface shown in Fig. 3.8. The point on the phase velocity surface can be found by mapping the magnitude of the phase velocity c along the direction of the wave vector **k**. The angle made by the unit normal $\widehat{\mathbf{N}_g}$ and $\hat{\mathbf{n}}$ will contribute to the magnitude of the phase velocity and group velocity of the wave mode and vice versa. From Fig. 3.8 We can write

$$c = c_g \left(\widehat{\mathbf{N}_g} \cdot \hat{\mathbf{n}} \right) \tag{3.73}$$

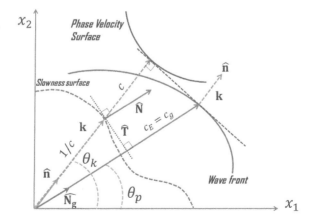

FIGURE 3.8 Schematics of the wave surfaces and their relationship.

To facilitate the discussion in the following chapters, please note that the slowness is equivalent to the wave number for a monochromatic wave with a known frequency ω. Slowness is $1/c$; however, the magnitude of the wave vector $\mathbf{k}$ is $|\mathbf{k}| = k = \omega/c$. Thus, if the slowness surfaces are drawn for specific frequencies ω, then the slowness would generate equifrequency surfaces (EFSs) of the respective wave modes on the wave number planes $(k_x - k_y - k_z)$. If at a specific frequency, a 2D slowness surface is drawn for a specific wave mode on the $x - y$ plane, the slowness surface will represent the equifrequency contours (EFCs) of that wave mode on the $k_x - k_y$ plane. Similarly, this understanding is equivalent for $k_y - k_z$ or $k_x - k_z$ planes. Thus, the slowness and the wave number will be used synonymously in this book for a specific monochromatic wave mode.

The three surfaces discussed above do not necessarily tell how the wave modes are polarized in anisotropic media. Particle motions in three orthogonal directions for the three unique wave modes needs to be further realized. Unlike P, SV, and SH waves in isotropic media, anisotropic media have quasi-longitudinal (qP), quasi-fast shear (qFS), and quasi-slow shear (qSS) wave modes. The surfaces in Figs. 3.7 and 3.8 are presented for only one such wave mode. Each individual wave mode has its own polarization (direction of the particle motion). These directions are orthogonal to each other but cannot be found intuitively. To find the wave polarization for each quasi-wave mode in an anisotropic solid, the following solution steps are necessary.

3.2.2.2.2 Governing Elastodynamic Equation in Anisotropic Media

It is assumed that readers are familiar with index notation and with the basics of continuum mechanics discussed in Chapter 2. Let's recall the fundamental elastodynamic equation, or the equation of motion in a solid body derived in Eq. 2.87 in Chapter 2.

$$\sigma_{ij,j} + F_i = \rho \ddot{u}_i \tag{3.74}$$

where $F_i = \rho f_i$ is the body force, σ_{ij} is the stress, and u_i is the displacement at a point $(\mathbf{x} = x_i \hat{e}_i)$ in a bulk anisotropic solid. For homogeneous anisotropic solid media, the density (ρ) remains constant; however, stress (σ_{ij}) and displacement (u_i) are the general functions of both space (x_i) and time (t). Recollecting Eq. 3.24 for homogeneous anisotropic media, the constitutive law for any linear material and the strain displacement relation in linear elastic material written in Eq. 2.88 can be written as

$$\sigma_{ij} = \mathbb{C}_{ijkl} e_{kl}; \; e_{kl} = \frac{1}{2}\left(u_{k,l} + u_{l,k}\right) \tag{3.75}$$

where e_{kl} is the strain in the anisotropic solid. For any generalized linear elastic anisotropic material, $\mathbb{C}_{ijkl}$ is the matrix of elastic constants written in Eqs. 2.95 and 2.96. Assuming the material is linear, and assuming a very short exposure time of the wave compared to the loading history of the material, the above equations are valid in anisotropic material. Substituting the geometrically linear strain-displacement

Acoustics and Elastic Wave Propagation in Fluids and Anisotropic Solids   87

relation and the constitutive law in Eq. 3.74 into the elastodynamic equation in Eq. 3.73, the equation takes the form in terms of displacement [8].

$$\mathbb{C}_{ijml}\frac{\partial^2 u_m}{\partial x_j\,\partial x_l} + F_i = \rho\ddot{u}_i \tag{3.76}$$

The above equation is solved using a monochromatic harmonic displacement function. Such solutions can be commutable to the other neighboring frequencies generated by wave sources. Scalar and vector potentials in isotropic media were assumed to decompose the displacement field. They were $\phi(x_j,t)=\varphi(x_j)e^{-i\omega t}$ and $\psi_i(x_j,t)=\Psi_i(x_j)e^{-i\omega t}$. Please note that the wave propagation problem is solved at one unique frequency at a time. The time harmonic part of the potential functions $e^{-i\omega t}$ is separable and is omitted due to its appearance on either side of the equation (i.e., $e^{-i\omega t}\neq 0$). The wave number ($k=\omega/c$) is calculated using a specific angular frequency ω, which is the same frequency that appeared in this time harmonic part $e^{-i\omega t}$. The potential is monochromatic. The magnitude of the wave vector is the inverse of the phase wave velocity c at a constant frequency. Wave vector is synonymous to the slowness vector discussed in the previous section.

Helmholtz decomposition will not be suitable for anisotropic media, and thus there is no point in assuming those potentials separately. This is because in anisotropic media the three wave surfaces are not concentric. To solve the wave propagation problem monochromatically, let's assume monochromatic potential function directly imposed on each component of the displacement wave field in anisotropic solid.

$$u_m = Ap_m e^{i(\mathbf{k}.\mathbf{x}-\omega t)} \tag{3.77}$$

In the above equation, A is the scalar wave amplitude. Monochromatic wave frequency is denoted as ω. Wave vector is $\mathbf{k}$, and $\mathbf{x}$ is the position vector. $\mathbf{k}.\mathbf{x}$ is the dot product between $\mathbf{k}$ and $\mathbf{x}$ represents the phase component of the wave as discussed in section 3.2.2.1.4. The most important part in Eq. 3.77 is the parameter p_m. Here, $\hat{\mathbf{p}} = p_m\hat{\mathbf{e}}_\mathbf{m}$ is called the polarization vector. It is a unit vector pointing to the direction of the particle motion of the respective wave modes. After substituting Eq. 3.77 into Eq. 3.76 and following a few mathematical simplifications [8] we get the following equation.

$$\left[\mathbb{C}_{ijml}k_jk_l - \rho\delta_{im}\omega^2\right]Ap_m = -F_i \tag{3.78}$$

The above system of equations is subjected to external force and depends on the quality and magnitude of the force field. However, to extract the nature of the eigen-wave field of an uninterrupted system, it is necessary to solve the homogeneous system of equations. Hence, without the body force, the Navier's equation in anisotropic media and the nontrivial solution of the equation can be written as

$$\left[\mathbb{C}_{ijml}n_jn_l - \rho\delta_{im}c^2\right]p_m = 0 \text{ or } \left[\Gamma_{im} - \rho\delta_{im}c^2\right]p_m \tag{3.79}$$

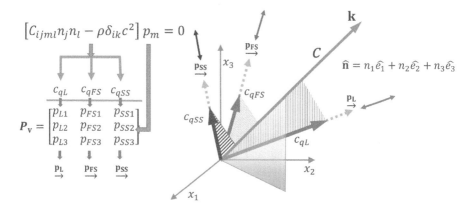

FIGURE 3.9 Graphical understanding of the Christoffel solution: phase velocities of quasi wave modes and their polarities. Phase velocity (c) along the direction of wave vector k is the superposition of the contribution from the particle motions of each wave modes along the k vector.

$\Gamma_{im} = \mathbb{C}_{ijml} n_j n_l$. Equation 3.79 is an eigenvalue problem, where $c^2 = \omega^2/k^2$ is the square of the phase velocity of a wave mode along the direction of **k** vector. The n_j parameters are the direction cosines of the unit vector along the wave propagation direction (along the **k** vector). The solution of this equation gives the fundamental wave modes that are propagating in the material with material constants $\mathbb{C}_{ijml}$. Equation 3.79 is the Christoffel equation [5]. The solution of this equation will provide three eigenmodes with wave velocities c_{qL}, c_{qFS}, and c_{qSS}, respectively. The directions found from the eigenvector matrix **P**$_v$ in Fig. 3.9 are composed of polarization vectors of each mode p_m. The figure graphically explains the meaning of the solution of Eq. 3.79. The solution is performed for a specific direction of wave propagation along the **k** vector with direction normal $\hat{n}$. It can be realized that the superposition of the projections of the phase velocities of the wave modes (c_{qL}, c_{qFS}, and c_{qSS}) on the **k** vector gives the phase velocity of the wave along the **k** or the wave propagation direction. Direction of the wave energy will be different, like discussed in Fig. 3.8. There are three wave modes: quasi-longitudinal (qL), quasi-fast shear (qFS) or quasi shear 2, and quasi-slow shear (qSS) or quasi shear 1 wave modes with velocities c_{qL}, c_{qFS}, and c_{qSS}, respectively. These three wave modes have polarization along the directions $\mathbf{p}_L$, $\mathbf{p}_{FS}$, and $\mathbf{p}_{SS}$, respectively. $\mathbf{p}_L$, $\mathbf{p}_{FS}$, and $\mathbf{p}_{SS}$, are the eigenvectors with their respective components along x_1, x_2, and x_3.

$$\mathbf{p}_L = p_{L1}\hat{e}_1 + p_{L2}\hat{e}_2 + p_{L3}\hat{e}_3 \qquad (3.80.1)$$

$$\mathbf{p}_{FS} = p_{FS1}\hat{e}_1 + p_{FS2}\hat{e}_2 + p_{FS3}\hat{e}_3 \qquad (3.80.2)$$

$$\mathbf{p}_{SS} = p_{SS1}\hat{e}_1 + p_{SS2}\hat{e}_2 + p_{SS3}\hat{e}_3 \qquad (3.80.3)$$

The above equations are graphically explained in Fig. 3.9. For each wave propagation direction **k** with varying angles, the wave velocity values of c_{qL}, c_{qFS}, and c_{qSS} and

Acoustics and Elastic Wave Propagation in Fluids and Anisotropic Solids   89

the elements in $\mathbf{P_v}$ matrix will be different. If the magnitude of the wave velocities c_{qL}, c_{qFS}, and c_{qSS} are plotted for all possible wave propagation directions (**k** along $\hat{\mathbf{n}}$) in three dimensions by discretizing a sphere (unit normal $\hat{\mathbf{n}}$ at each point on the surface of the sphere are calculated from the basic equations used in spherical coordinate system), the velocity surfaces are generated for all three modes separately. Such wave velocity surfaces can be generated with different material properties $\mathbb{C}_{ijml}$ (Eq. 3.79) and will produce different wave velocity surfaces. By taking the inverse of the velocity surfaces, slowness surfaces can be created. If a specific frequency is considered, the slowness will result EFC or EFS in the wave number domain.

3.2.2.3 Understanding Wave Modes with Normal and Anomalous Polarity

To explore different situations and explanation for the topological waves in metamaterials requires understanding of the particle motion that includes polarization and spin. Thus, wave polarization will be discussed in detail in this section. Spin and relevant discussions are reserved for the later chapters. Fundamental wave modes in isotropic homogeneous media are discussed in section 3.2.2.1. P, SV, and SH modes are the three primary modes present in isotropic media. P waves are the normally polarized waves with particle motion along the direction of the wave vector. SH and SV wave modes are orthogonally polarized with the direction of the wave vector. Wave modes in anisotropic media are discussed in section 3.2.2.2. Longitudinal and shear wave modes in anisotropic media are not always longitudinal or orthogonal to the direction of the wave propagation as they are for isotropic. Hence, the wave modes in anisotropic media are called quasi modes, as they were introduced in the previous section, namely qL, qFS, and qSS modes.

Next, to characterize the wave polarity with respect to the direction of the wave vector **k**, a new parameter called the polarity differential (PoDi) is created. The PoDi (θ_{pd}), or the polarity differential, for each wave mode (qL, qFS, and qSS) is calculated by subtracting the wave propagation angle (θ_k) (i.e., the angle made by the wave vector **k** with respect to a reference plane) from the angle of the direction of the particle motion or simply the polarization angle (θ_p) (obtained from the polarization vectors $\mathbf{p_L}$, $\mathbf{p_{FS}}$, $\mathbf{p_{SS}}$ for the respective wave modes) with respect to the same reference plane. Please refer to Fig. 3.8 to understand the θ_k and θ_p, where they are explicitly marked to find the PoDi. The PoDi can be expressed as

$$\left[\theta_{pd}\right]^{qm} = \left|\theta_p - \theta_k\right|^{qm} \tag{3.81}$$

where qm stands for quasi wave modes in anisotropic media. m can take identification for the longitudinal wave mode as L, fast shear as FS and slow shear as SS. In isotropic media, qm can take the identification for the longitudinal mode as P, shear vertical as SV and shear horizontal as SH. From fundamentals, it is known that the wave polarization (i.e., particle motion) is along the direction of the wave vector for P waves in isotropic media. Thus, the PoDi for P waves will be a constant value for different direction of the wave vector **k**. With increasing θ_k from 0° to 90°, the $\left[\theta_{pd}\right]^{P}$ will be a constant value of 0 signifying a purely longitudinal mode. Similarly due to the nature of orthogonal polarity of SV and SH waves, with increasing θ_k from 0°

to 90° the $\left[\theta_{pd}\right]^{SV} = \left[\theta_{pd}\right]^{SH}$ will be a constant value of $\pi/2$. $\theta_{pd} = \pi/2$, signifying a pure shear mode. Thus, the difference in angle between P wave polarization and SV or SH wave polarization is always $\pi/2$ over a range of wave propagation angle (θ_k), described above. On the contrary in anisotropic media, the angles of polarities (θ_p) for different wave modes are not constant over a range of wave propagation angle (θ_k). Thus the difference in angle of polarity between the quasi-longitudinal and quasi-shear wave modes are also neither constant nor $\pi/2$ over a range of θ_k. Although the difference is not always $\pi/2$, it does not mean that the wave modes are not orthogonal. Mutual dot products of the polarities $\mathbf{p}_L$, $\mathbf{p}_{FS}$, and $\mathbf{p}_{SS}$ are zero. Specifically, $\mathbf{p}_L \cdot \mathbf{p}_{FS} = 0$, $\mathbf{p}_L \cdot \mathbf{p}_{SS} = 0$, and $\mathbf{p}_{SS} \cdot \mathbf{p}_{FS} = 0$. In the following sections, a few case studies are presented to visualize and understand the polarity of the quasi-wave modes. It is also explained how and when the polarities are normal and when the polarity are anomalous.

Anomalous polarity is not a naturally occurring phenomenon but a physically admissible phenomenon. Based on fundamental physics, anomalous polarity is possible. Hence, anomalous polarity may result in artificial metamaterial if designed with proper understanding. How such design of metamaterials could cause topological behavior is an open field of research.

3.2.2.3.1 Normal Polarization in Isotropic and Anisotropic Media

<u>Case I: Isotropic Material:</u> Figure 3.10a shows the 3D phase wave velocity profile of all the wave modes (P, SV, and SH) in an isotropic material. Aluminum, an isotropic material, is used for this study. The constitutive property matrix of aluminum

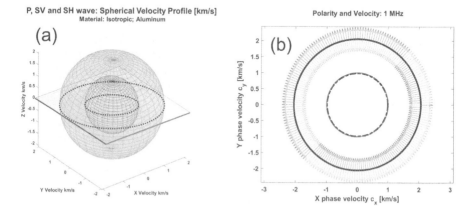

FIGURE 3.10 Case I: Phase wave velocity profile and the wave polarities of different wave modes in isotropic solid aluminum. a) Equifrequency surfaces (EFS) for P, SV, and SH wave modes at 1 MHz, schematic showing a x-y cut plane resulted circular equifrequency contours (EFC) b) Showing x-y plane of the EFC at 1 MHz for P, SV, and SH wave modes with their respective polarities. On x-y plane the SV wave polarities in the out of plane direction, whereas the SV wave polarities are tangent to the SH EFC (barely visible).

Acoustics and Elastic Wave Propagation in Fluids and Anisotropic Solids

with Lamé constants ($\mathbb{L} = 6.13\ Gpa$; $\mathrm{m} = 2.64\ GPa$) and density $2700\ kg/m^3$ can be written as

$$\mathbb{C}_{mn} = \begin{bmatrix} 11.41 & 6.13 & 6.13 & 0 & 0 & 0 \\ & 11.41 & 6.13 & 0 & 0 & 0 \\ & & 11.41 & 0 & 0 & 0 \\ & & & 2.64 & 0 & 0 \\ & Sym & & & 2.64 & 0 \\ & & & & & 2.64 \end{bmatrix} GPa \qquad (3.82)$$

Here the constitutive matrix is written according to the convention presented in Eq. 2.100.3 in Chapter 2. Figure 3.10b shows the cross section of the wave velocity profiles of each wave mode on the $x_1 - x_2$ plane at $x_3 = 0$, with their respective polarity identified with small vectors. A firm circle is used to indicate a P wave, and dashed lines are used to indicate an S wave. Next, the wave slowness, or the EFCs for the P, SV, and SH wave modes in wave number domain are shown in Fig. 3.11a. In this figure, the wave polarities are also identified with small lines overlaid with the slowness profile. Figure 3.11b shows the polarity differential $\left[\theta_{pd}\right]$ of different wave modes (longitudinal and shear wave modes). In isotropic media, θ_{pd} is constant and equal to 0 for the P wave, and equal to $\pi/2$ for the shear wave. The polarity differential below $\pi/4$ is identified as the P zone, and the differential between $\pi/4 - \pi/2$ is identified as the S zone. It is generally assumed that the P wave differential polarities belong to the P zone and the S wave differential polarities belong to the S zone. It can be found later that it is not always true.

Case II: Orthotropic Material: Figure 3.12 shows the 3D phase wave velocity profile of all the wave modes (qL, qFS, and qSS, in Fig. 3.12a, b, and c, respectively) in orthotropic material with nine independent coefficients. The constitutive matrix used for a sample orthotropic material with density $1500\ kg/m^3$ is

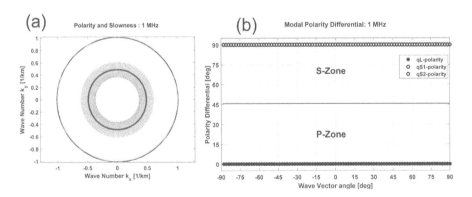

FIGURE 3.11 Case I: Wave slowness profile or the ESCs on a wave number plane $(k_x - k_y)$ are shown for each wave mode with their respective modal polarities. a) EFC on $k_x - k_y$ plane for P, SV and SH wave modes at 1 MHz, b) showing the change in polarity differential (PoDi) of each mode over the range of wave vector angles between −90° and 90°. Isotropic material shows constant PoDi of P, SV, and SH modes.

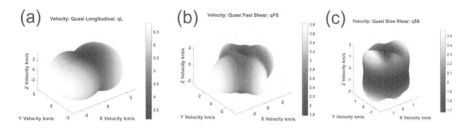

FIGURE 3.12 3D velocity profile in orthotropic material: a) quasi longitudinal mode, b) quasi fast shear (qFS) mode, c) quasi slow shear (qSS) mode.

$$\mathbb{C}_{mn} = \begin{bmatrix} 70 & 23.9 & 6.2 & 0 & 0 & 0 \\ 23.9 & 33 & 6.8 & 0 & 0 & 0 \\ 6.2 & 6.8 & 14.7 & 0 & 0 & 0 \\ 0 & 0 & 0 & 4.2 & 0 & 0 \\ 0 & 0 & 0 & 0 & 4.7 & 0 \\ 0 & 0 & 0 & 0 & 0 & 21.9 \end{bmatrix} GPa \quad (3.83)$$

Figure 3.13a shows the cross sections of the wave velocity profiles of each wave mode on the $x_1 - x_2$ plane with its respective polarity. A firm line is used to indicate qL mode, and dashed lines are used to indicate qFS and qSS modes. Next the wave slowness, alternatively the EFCs for the qL, qFS, and qSS wave modes in wave number domain, are shown in Fig. 3.13b. Like earlier, the wave polarities are also identified with small lines overlaid with the slowness profile. Figure 3.13c shows the polarity differential $\left[\theta_{pd}\right]$ of different wave modes (quasi-longitudinal and quasi-shear). The polarity differential below $\pi/4$ is identified as the P zone, and the differential between $\pi/4 - \pi/2$ is identified as the S-zone. It can be seen that the θ_{pd} for the qL mode is mostly confined to the P zone, and for the qFS mode it is confined to the S zone. qFS mode has polarity components on the $x - y$ plane but the polarity components for qSS mode are orthogonal to the $x - y$ plane (i.e., out of plane). This is an example of the normal polarization in a non-isotropic media.

Case III: Transversely Isotropic Material: Barium sodium niobate ($Ba_2NaNb_5O_{15}$) is chosen to demonstrate the normal polarity condition is transversely isotropic media. Here 3D profiles of wave velocities are not explicitly shown for all the wave modes (qL, qFS, and qSS). Transversely isotropic media may have four to five independent coefficients. Here, for barium sodium niobate four unique coefficients are used. The constitutive matrix of $Ba_2NaNb_5O_{15}$ with density 5300 kg/m^3 is written below.

$$\mathbb{C}_{mn} = \begin{bmatrix} 23.9 & 5 & 5 & 0 & 0 & 0 \\ 5 & 13.5 & 5 & 0 & 0 & 0 \\ 5 & 5 & 13.5 & 0 & 0 & 0 \\ 0 & 0 & 0 & 9.45 & 0 & 0 \\ 0 & 0 & 0 & 0 & 6.6 & 0 \\ 0 & 0 & 0 & 0 & 0 & 6.6 \end{bmatrix} GPa \quad (3.84)$$

Acoustics and Elastic Wave Propagation in Fluids and Anisotropic Solids

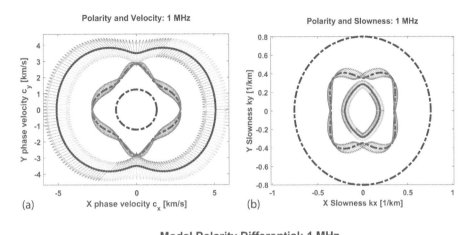

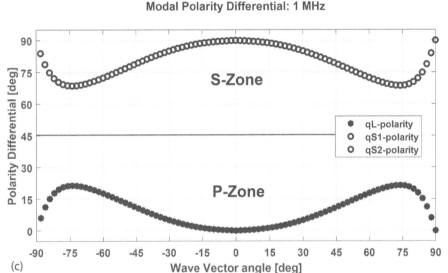

FIGURE 3.13 Orthotropic material: a) velocity profile EFC with polarity at 1 MHz, b) slowness profile EFC in wave number domain at 1 MHz, c) polarity differential over a range of direction of wave propagation, polarity differential says that the qL and qFS no essential overlap or crossing between their polarities.

Figure 3.14a shows the cross section of the wave velocity profiles of each wave mode on the $x_1 - x_2$ plane at $x_3 = 0$, with their respective polarity identified with small individual vectors. A firm circle is used to indicate the P wave, and dashed lines are used to indicate the S wave. Next the wave slowness, or the EFCs for the qL, qFS, and qSS wave modes in wave number domain, is shown in Fig. 3.14b. Figure 3.14c shows the polarity differential $[\theta_{pd}]$ of different wave modes (quasi-longitudinal and quasi-shear). It can be seen that in barium sodium niobate the θ_{pd} for the qL mode is mostly

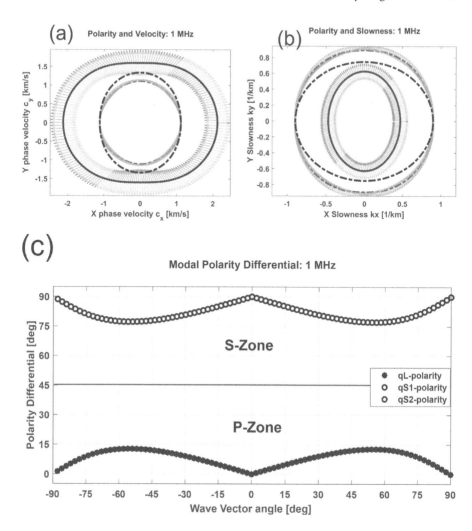

FIGURE 3.14 Transversely isotropic material: Barium sodium niobate ($Ba_2NaNb_5O_{15}$). a) Velocity profile EFC with polarity at 1 MHz, b) slowness profile EFC in wave number domain at 1 MHz, c) polarity differential over a range of direction of wave propagation, polarity differential says that the qL and qFS no essential overlap or crossing between their polarities.

confined to the P zone and for the qFS mode it is confined to the S zone. The qFS mode has polarity components on the $x - y$ plane, but the polarity components for qSS mode are orthogonal to the $x - y$ plane (i.e., out of plane). This is also an example of the normal polarization in transversely isotropic media.

3.2.2.3.2 Abnormal Polarization in Anisotropic Media

Case IV: Transversely Isotropic Material: Calcium formate is chosen to demonstrate the abnormal polarity condition in transversely isotropic media. Calcium

Acoustics and Elastic Wave Propagation in Fluids and Anisotropic Solids **95**

formate is stable at room temperature. It is inflammable and forms orthorhombic crystals. The constitutive matrix of calcium formate with density 2020 kg/m^3 is written below.

$$\mathbb{C}_{mn} = \begin{bmatrix} 2.44 & 2.48 & 2.48 & 0 & 0 & 0 \\ 2.48 & 4.92 & 2.48 & 0 & 0 & 0 \\ 2.48 & 2.48 & 4.92 & 0 & 0 & 0 \\ 0 & 0 & 0 & -0.02 & 0 & 0 \\ 0 & 0 & 0 & 0 & 2.82 & 0 \\ 0 & 0 & 0 & 0 & 0 & 2.82 \end{bmatrix} GPa \qquad (3.85)$$

Figure 3.15a shows a cross-section of the wave velocity profiles of each wave mode on the $x_1 - x_2$ plane at $x_3 = 0$. All the wave modes are presented, with their respective polarity identified with small vectors. Please note that there are two overlapping circular profiles for the qSS mode in the figure. This mode has the orthogonal polarity to the $x_1 - x_2$ plane. Next, the wave slowness, or the EFCs for the qL and qFS, wave modes in wave number domain, is shown in Fig. 3.15b. Due to high wave number values for qSS along the k_y direction, compared to qFS, the slowness profile for the qSS wave mode is not presented in Fig. 3.15b. The wave polarities are also identified with small lines overlaid with the slowness profile. An inset in Fig. 3.15b shows a zoomed view of the polarities of the qL and qFS wave modes near θ_k close to zero. Figure 3.15c shows the polarity differential $\left[\theta_{pd}\right]$ of qL and qFS modes. The polarity differential below $\pi/4$ is identified as the P zone, and the differential between $\pi/4 - \pi/2$ is identified as the S zone. In calcium formate, the θ_{pd} for the qL mode near $\theta_k = 0$ is equal to $\pi/2$. According to the definition of PoDi, $\theta_{pd} = \pi/2$ signifies a shear wave mode. Hence, it is noted that the qL mode has the polarity much like the shear wave polarity, and the polarity differential is $\pi/2$. On the contrary, the θ_{pd} for the qFS mode near $\theta_k = 0$ is equal to zero. According to the definition of PoDi, $\theta_{pd} = 0$ signifies a longitudinal wave mode. Hence, the qFS mode has polarity like a longitudinal wave. Mutually, qL and qFS modes tend to share opposite polarities and have very close wave velocities. Thus, they share anomalous behavior (lack of wave identity) till $\theta_k = \sim 10°$ (Fig. 3.15c). Beyond $\theta_k = \sim 10°$, the qL wave polarity tend to comes back to the P zone, and qFS wave polarity tends to go back to the S zone. Along the wave propagation direction, if θ_k lies between $\sim -10° < \theta_k < \sim 10°$, the wave velocities of qL and qFS modes are very comparable and close to each other. This is an example of the abnormal polarization in transversely isotropic media. It seems that for a certain wave propagation direction $\sim -10° < \theta_k < \sim 10°$ wave modes tend to switch their identity and may have opportunity to convert to one another. This behavior is useful for specific wave propagation phenomena and for devising new metamaterial for topological wave behavior. Recently, researchers [9] conceptualized a new metamaterial that can have similar abnormal polarity and showed unique wave propagation with orthogonal bending of wave using a wedge structure.

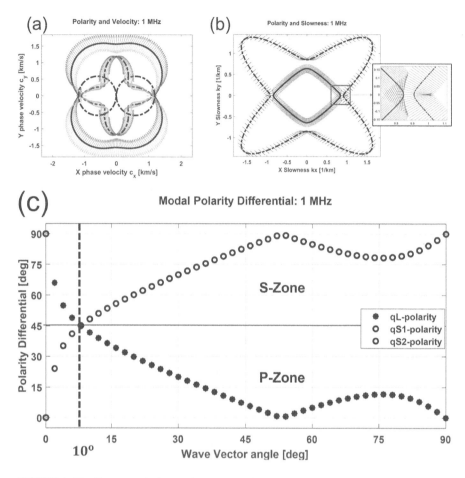

FIGURE 3.15 Transversely isotropic material: Calcium formate. a) Velocity profile EFC with polarity at 1 MHz, b) slowness profile EFC in wave number domain at 1 MHz (except the qSS mode, with out of plane polarity), c) polarity differential over a range of direction of wave propagation, polarity differential says that qL and qFS has significant overlap or crossing between their polarities.

<u>Case V: Monoclinic Material:</u> The constitutive matrix of a monoclinic material with density 1560 kg/m^3 is written below.

$$\mathbb{C}_{mn} = \begin{bmatrix} 102.6 & 24.1 & 6.3 & 0 & 0 & 40 \\ 24.1 & 18.7 & 6.4 & 0 & 0 & 10 \\ 6.3 & 6.4 & 13.3 & 0 & 0 & -0.1 \\ 0 & 0 & 0 & 3.8 & 0.9 & 0 \\ 0 & 0 & 0 & 0.9 & 5.3 & 0 \\ 40 & 10 & -0.1 & 0 & 0 & 23.6 \end{bmatrix} GPa \quad (3.86)$$

Figure 3.16a shows a cross-section view of the wave velocity profiles of each wave mode in monoclinic material on the $x_1 - x_2$ plane. The wave modes are presented with their respective polarity identified with small vectors. Next, the EFCs for the qL and qFS wave modes in wave number domain are shown in Fig. 3.16b. The wave polarities are identified with small lines overlaid with the slowness profile. Figure 3.16c shows the polarity differential $[\theta_{pd}]$ of qL and qFS modes. In monoclinic material, the θ_{pd} for qL mode is mostly confined to the P zone, and the polarity differential for the qFS mode is confined to the S zone except for $\theta_k > \sim 76°$. This could

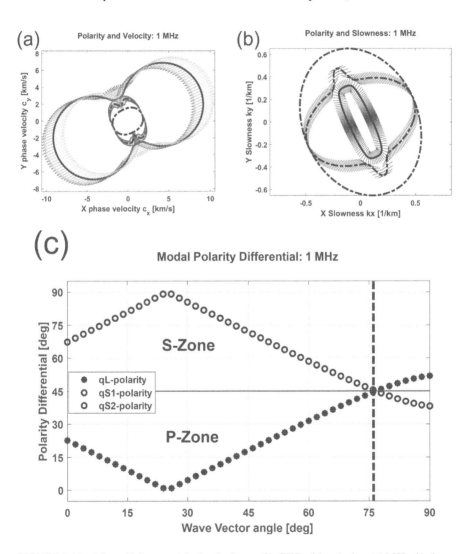

FIGURE 3.16 Monoclinic material: a) velocity profile EFC with polarity at 1 MHz, b) slowness profile EFC in wave number domain at 1 MHz, c) polarity differential over a range of direction of wave propagation, polarity differential says that qL and qFS has some overlaps or crossing between their polarities.

be an example of abnormal polarity. However, please note that the wave velocities of qL and qFS modes in this abnormal zone are close but not close enough. Thus, ambiguity in modal behavior does not exist at ~76°. With close observation, we can say that near the θ_k = ~90° the qL and qFS mode velocities are comparable where the differential polarities of these modes are one opposite zone (qL polarity in S zone and qFS polarity in P zone). Possibly, wave propagation along the x_2 axis may resolve to an abnormal polarity and wave behavior might be ambiguous.

Case V: Orthotropic Material: Gallium Arsenide (GaAs) is chosen for this study. GaAs with the following material constants is used in getting the EFCs.

$$
\mathbb{C}_{mn} = \begin{bmatrix} 72.3 & 10.2 & 10.3 & 0.00 & 0.00 & 0.00 \\ 10.2 & 78.3 & 24.4 & 0.00 & 0.00 & 0.00 \\ 10.3 & 24.4 & 78.3 & 0.00 & 0.00 & 0.00 \\ 0.00 & 0.00 & 0.00 & 5.00 & 0.00 & 0.00 \\ 0.00 & 0.00 & 0.00 & 0.00 & 175.2 & 0.00 \\ 0.00 & 0.00 & 0.00 & 0.00 & 0.00 & 175.2 \end{bmatrix} GPa \quad (3.87)
$$

Figure 3.17a shows a cross-section view of the wave velocity profiles of each wave mode in GaAs on the $x_1 - x_2$ plane. The wave modes are presented with their respective polarity, identified with small vectors. EFCs for the qL and qFS wave modes in wave number domain are shown in Fig. 3.17b. Figure 3.17c shows the polarity differential $\left[\theta_{pd} \right]$ of qL and qFS modes. In GaAs the θ_{pd} for qL and qFS modes has significant overlap in terms of polarity. Unlike Case IV for calcium formate, the wave velocities in GaAs for the qL and qFS modes are not comparable. Hence, although they are not inseparable in terms of polarity, they are quite separable due to their difference in wave velocities. This could be an example of abnormal polarity, but GaAs does not exhibit any ambiguous wave behavior.

It is therefore understood that to exploit abnormal polarity in a material, the quasi-wave modes must have opposite polarities (qL mode: θ_{pd} lies in S zone, and qS mode θ_{pd} resides in P zone) with very close wave velocities. This is a unique and accidental critical state where abnormal polarities could be exploited for topological wave behavior.

3.2.2.4 Exploring Abnormal Polarities

Wave velocities (c), slowness ($1/c$ or the wave numbers k at a specific frequency), and the polarities ($\mathbf{p_L}$, $\mathbf{p_{FS}}$, and $\mathbf{p_{SS}}$) are obtained from the eigensolution of the Christoffel equation in Eq. 3.79. The Christoffel equation has the Kelvin-Christoffel matrix Γ, or Γ_{im} [4]. The Kelvin-Christoffel matrix is a 3x3 matrix composed of elements obtained from the material property matrix and the unit vector along the intended wave propagation direction. Please note that $\Gamma_{im} = \mathbb{C}_{ijml} n_j n_l$. Elements of the symmetric Γ matrix can be expanded as follows.

$$
\Gamma_{11} = \mathbb{C}_{11} n_1^2 + \mathbb{C}_{66} n_2^2 + \mathbb{C}_{55} n_3^2 + 2\mathbb{C}_{56} n_2 n_3 + 2\mathbb{C}_{15} n_3 n_1 + 2\mathbb{C}_{16} n_1 n_2 \quad (3.88.1)
$$

$$
\Gamma_{22} = \mathbb{C}_{66} n_1^2 + \mathbb{C}_{22} n_2^2 + \mathbb{C}_{44} n_3^2 + 2\mathbb{C}_{24} n_2 n_3 + 2\mathbb{C}_{46} n_3 n_1 + 2\mathbb{C}_{26} n_1 n_2 \quad (3.88.2)
$$

Acoustics and Elastic Wave Propagation in Fluids and Anisotropic Solids 99

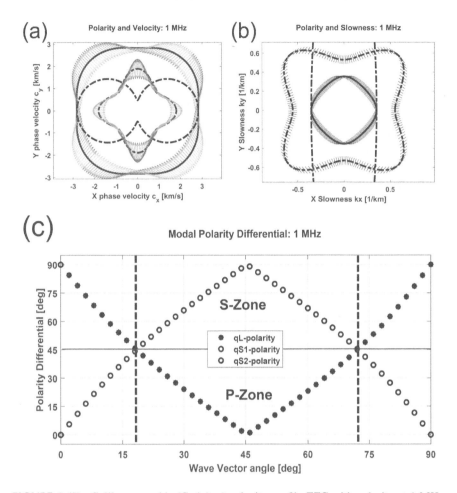

FIGURE 3.17 Gallium arsenide (GaAs): a) velocity profile EFC with polarity at 1 MHz, b) slowness profile EFC in wave number domain at 1 MHz, c) polarity differential over a range of direction of wave propagation, polarity differential says that qL and qFS has some overlaps or crossing between their polarities.

$$\Gamma_{33} = \mathbb{C}_{55}n_1^2 + \mathbb{C}_{44}n_2^2 + \mathbb{C}_{33}n_3^2 + 2\mathbb{C}_{34}n_2n_3 + 2\mathbb{C}_{35}n_3n_1 + 2\mathbb{C}_{45}n_1n_2 \quad (3.88.3)$$

$$\Gamma_{12} = \mathbb{C}_{16}n_1^2 + \mathbb{C}_{26}n_2^2 + \mathbb{C}_{45}n_3^2 + (\mathbb{C}_{46} + \mathbb{C}_{25})n_2n_3 \\ + (\mathbb{C}_{56} + \mathbb{C}_{14})n_3n_1 + (\mathbb{C}_{66} + \mathbb{C}_{12})n_1n_2 \quad (3.88.4)$$

$$\Gamma_{13} = \mathbb{C}_{15}n_1^2 + \mathbb{C}_{46}n_2^2 + \mathbb{C}_{35}n_3^2 + (\mathbb{C}_{45} + \mathbb{C}_{36})n_2n_3 \\ + (\mathbb{C}_{55} + \mathbb{C}_{13})n_3n_1 + (\mathbb{C}_{56} + \mathbb{C}_{14})n_1n_2 \quad (3.88.5)$$

$$\Gamma_{23} = \mathbb{C}_{56}n_1^2 + \mathbb{C}_{24}n_2^2 + \mathbb{C}_{34}n_3^2 + (\mathbb{C}_{44} + \mathbb{C}_{23})n_2n_3 \\ + (\mathbb{C}_{36} + \mathbb{C}_{45})n_3n_1 + (\mathbb{C}_{46} + \mathbb{C}_{25})n_1n_2 \quad (3.88.6)$$

The convention for the constitutive matrix presented in Eq. 2.100.3 in Chapter 2 is used here. Referring to the equations in Eqs. 3.88.1 through 3.88.6, it can be seen that the matrix elements in the Kelvin-Christoffel matrix are primarily governed by the values of the material constants. Essentially the matrix is symmetric. Hence, the unique elements in the matrix are the elements for which the equations are written above. Eigenvalue s and eigenvectors of the matrix give the wave slowness and the wave polarities. However, they are not unique. It is possible to have situations where dissimilar stiffness matrices could give similar Kelvin-Christoffel matrices. Eigensolution may give a similar slowness profile but may have different polarities. Such media are kinematically identical but dynamically different [6]. It is noted that identification or classification of different anisotropic solids only by their slowness or velocity profile is neither sufficient nor unique [6, 7]. This is counterintuitive. Hence, a solid medium must be ranked with both its velocity and polarization.

3.2.2.4.1 Anomalous Companions

Two anisotropic media (media I and media II) with different stiffness matrices are called "anomalous companions [6, 7]" if the elements of the Kelvin-Christoffel matrix are identical ($\Gamma^I = \Gamma^{II}$). This means the matrix invariants are also identical. This is possible only when a) the diagonal terms Γ_{ii} are identical, b) the square of the off diagonal (Γ_{ij}^2) terms are identical, c) product of three mutually exclusive off diagonal terms ($\Gamma_{ij}\Gamma_{ik}\Gamma_{jk}$) is identical. In most general anisotropic media, the material coefficients follow a typical trend.

$$\{\mathbb{C}_{11}, \mathbb{C}_{22}, \mathbb{C}_{33}\} > \{\mathbb{C}_{44}, \mathbb{C}_{55}, \mathbb{C}_{66}\} \tag{3.89}$$

which means that the primary elastic moduli along any specific axes are always greater than the shear moduli of the media. (Please refer to Eq. 2.99 for visualizing this statement.) Generally speaking, the materials following the rule in Eq. 3.89 have a distinct qL wave mode, well separated from qFS and qSS wave modes. These materials belong to the category of "normally polarized" media. This condition was satisfied in Eqs. 3.82 through 3.84. Please note that the condition, however, was not satisfied for the constitutive matrix in Eqs. 3.85 through 3.87. The later media demonstrated abnormal polarity. These situations are well demonstrated in section 3.2.2.3.2.

Further, in articles by Helbig et al [6, 7], three conditions are articulated to achieve "anomalous companions" that demonstrate abnormal polarities.

<u>Condition 1:</u> First let's assume that the Kelvin-Christoffel matrix for two anomalous companions with similar eigenvalues must have identical elements ($\Gamma^I = \Gamma^{II}$). With careful observation of Eqs. 3.88.1 through 3.88.6, it can be found that the diagonal terms of the Kelvin-Christoffel matrix for all possible wave vector direction must be equal in order to achieve $\Gamma^I = \Gamma^{II}$, if only if, they share the constitutive coefficients, $\mathbb{C}_{11}, \mathbb{C}_{22}, \mathbb{C}_{33}, \mathbb{C}_{44}, \mathbb{C}_{55}, \mathbb{C}_{66}, \mathbb{C}_{15}, \mathbb{C}_{16}, \mathbb{C}_{56}, \mathbb{C}_{24}, \mathbb{C}_{26}, \mathbb{C}_{46}, \mathbb{C}_{34}, \mathbb{C}_{35}, \mathbb{C}_{45}.$ (Refer to Eq. 2.96.) Beyond these coefficients, two "anomalous companions" can have different $\mathbb{C}_{23}, \mathbb{C}_{13}, \mathbb{C}_{12}, \mathbb{C}_{14}, \mathbb{C}_{25}, \mathbb{C}_{36}$ coefficients. Please refer to Fig. 3.18 to visualize the matrix.

Acoustics and Elastic Wave Propagation in Fluids and Anisotropic Solids · 101

(a)

$\mathbb{C}_{11}$	$\mathbb{C}_{12}$	$\mathbb{C}_{13}$	$\mathbb{C}_{14}$	$\mathbb{C}_{15}$	$\mathbb{C}_{16}$
$\mathbb{C}_{21}$	$\mathbb{C}_{22}$	$\mathbb{C}_{23}$	$\mathbb{C}_{24}$	$\mathbb{C}_{25}$	$\mathbb{C}_{26}$
$\mathbb{C}_{31}$	$\mathbb{C}_{32}$	$\mathbb{C}_{33}$	$\mathbb{C}_{34}$	$\mathbb{C}_{35}$	$\mathbb{C}_{36}$
$\mathbb{C}_{41}$	$\mathbb{C}_{42}$	$\mathbb{C}_{43}$	$\mathbb{C}_{44}$	$\mathbb{C}_{45}$	$\mathbb{C}_{46}$
$\mathbb{C}_{51}$	$\mathbb{C}_{52}$	$\mathbb{C}_{53}$	$\mathbb{C}_{54}$	$\mathbb{C}_{55}$	$\mathbb{C}_{56}$
$\mathbb{C}_{61}$	$\mathbb{C}_{62}$	$\mathbb{C}_{63}$	$\mathbb{C}_{64}$	$\mathbb{C}_{65}$	$\mathbb{C}_{66}$

(b)

$\mathbb{C}_{11}$	$\mathbb{C}_{12}$	$\mathbb{C}_{13}$	$\mathbb{C}_{14}$	0	0
$\mathbb{C}_{21}$	$\mathbb{C}_{22}$	$\mathbb{C}_{23}$	0	$\mathbb{C}_{25}$	0
$\mathbb{C}_{31}$	$\mathbb{C}_{32}$	$\mathbb{C}_{33}$	0	0	$\mathbb{C}_{36}$
$\mathbb{C}_{41}$	0	0	$\mathbb{C}_{44}$	0	0
0	$\mathbb{C}_{52}$	0	0	$\mathbb{C}_{55}$	0
0	0	$\mathbb{C}_{63}$	0	0	$\mathbb{C}_{66}$

FIGURE 3.18 Constitutive matrix and visualization of the material coefficients that creates 'anomalous companion': a) condition 1: coefficients in the light boxes must be equal but the coefficients in the dark boxes may differ, b) condition 2: coefficients in gray boxes must be equal to zero.

<u>Condition 2</u>: With careful observation of the eigensolutions from the Kelvin-Christoffel equation and Eqs. 3.88.1 through 3.88.6, it can be seen that the eigenvalues of the system will be indifferent if two of the three off-diagonal coefficients (Γ_{12}, Γ_{23}, Γ_{31}) in the Kelvin-Christoffel matrix are of equal magnitude but carry opposite signs for all possible wave vector directions. Hence, the necessary condition for similar eigenvalue s of two anomalous companions will be

$$\Gamma_{12}{}^I = \pm\Gamma_{12}{}^{II}; \; \Gamma_{13}{}^I = \pm\Gamma_{13}{}^{II}; \; \Gamma_{23}{}^I = \pm\Gamma_{23}{}^{II} \tag{3.90}$$

To achieve Eq. 3.90, the following statement must hold true. Referring to Fig. 3.18, the coefficients in light boxes, except the diagonal terms (i.e., nine coefficients: $\mathbb{C}_{15}$, $\mathbb{C}_{16}$, $\mathbb{C}_{56}$, $\mathbb{C}_{24}$, $\mathbb{C}_{26}$, $\mathbb{C}_{46}$, $\mathbb{C}_{34}$, $\mathbb{C}_{35}$, $\mathbb{C}_{45}$) contribute to the off-diagonal terms in the Kelvin-Christoffel matrix (Eqs. 3.88.4, 3.88.5, and 3.88.6) in such a way that they must be equal to zero. With this condition Eqs. 3.88.4 through 3.88.6 for Γ^I will be

$$\Gamma_{12}{}^I = \left(\mathbb{C}_{25}{}^I\right)n_2 n_3 + \left(\mathbb{C}_{14}{}^I\right)n_3 n_1 + \left(\mathbb{C}_{66} + \mathbb{C}_{12}{}^I\right)n_1 n_2 \tag{3.91.1}$$

$$\Gamma_{13}{}^I = \left(\mathbb{C}_{36}{}^I\right)n_2 n_3 + \left(\mathbb{C}_{55} + \mathbb{C}_{13}{}^I\right)n_3 n_1 + \left(\mathbb{C}_{14}{}^I\right)n_1 n_2 \tag{3.91.2}$$

$$\Gamma_{23}{}^I = \left(\mathbb{C}_{44} + \mathbb{C}_{23}{}^I\right)n_2 n_3 + \left(\mathbb{C}_{36}{}^I\right)n_3 n_1 + \left(\mathbb{C}_{25}{}^I\right)n_1 n_2 \tag{3.91.3}$$

From Condition 1, it was concluded that the anomalous companions may have different $\mathbb{C}_{23}$, $\mathbb{C}_{13}$, $\mathbb{C}_{12}$, $\mathbb{C}_{14}$, $\mathbb{C}_{25}$, $\mathbb{C}_{36}$ coefficients. Thus, Eqs. 3.88.4 through 3.88.6 for Γ^{II} will be

$$\Gamma_{12}{}^{II} = \left(\mathbb{C}_{25}{}^{II}\right)n_2 n_3 + \left(\mathbb{C}_{14}{}^{II}\right)n_3 n_1 + \left(\mathbb{C}_{66} + \mathbb{C}_{12}{}^{II}\right)n_1 n_2 \tag{3.92.1}$$

$$\Gamma_{13}{}^{II} = \left(\mathbb{C}_{36}{}^{II}\right)n_2 n_3 + \left(\mathbb{C}_{55} + \mathbb{C}_{13}{}^{II}\right)n_3 n_1 + \left(\mathbb{C}_{14}{}^{II}\right)n_1 n_2 \tag{3.92.2}$$

$$\Gamma_{23}{}^{II} = \left(\mathbb{C}_{44} + \mathbb{C}_{23}{}^{II}\right)n_2 n_3 + \left(\mathbb{C}_{36}{}^{II}\right)n_3 n_1 + \left(\mathbb{C}_{25}{}^{II}\right)n_1 n_2 \tag{3.92.3}$$

102 Metamaterials in Topological Acoustics

Please note that the specific identified coefficients under Condition 1 are different in media I and media II.

<u>Condition 3:</u> Eq. 3.79 reads $\left[\Gamma_{im} - \rho\delta_{im}c^2\right] = 0$, which can also read $\left[\Gamma - \lambda\mathbf{I}\right] = 0$, having eigenvalues λ_i $i = 1, 2,$ and 3. The characteristic equation of the eigenvalue problem can be written as

$$\lambda^3 - \Upsilon_1\lambda^2 + \Upsilon_2\lambda - \Upsilon_3 = 0 \tag{3.93}$$

where

$$\Upsilon_1 = \Gamma_{11} + \Gamma_{22} + \Gamma_{33} \tag{3.94.1}$$

$$\Upsilon_2 = \Gamma_{22}\Gamma_{33} - \Gamma_{23}{}^2 + \Gamma_{11}\Gamma_{33} - \Gamma_{13}{}^2 + \Gamma_{11}\Gamma_{22} - \Gamma_{12}{}^2 \tag{3.94.2}$$

$$\Upsilon_3 = \Gamma_{11}\Gamma_{22}\Gamma_{33} + 2\,\Gamma_{12}\Gamma_{13}\Gamma_{23} - \Gamma_{11}\Gamma_{23}{}^2 - \Gamma_{22}\Gamma_{13}{}^2 - \Gamma_{33}\Gamma_{12}{}^2 \tag{3.94.3}$$

Υ_1, Υ_2, and Υ_3 are the three invariants of the Kelvin-Christoffel matrix. With close observation, there are four effective coefficients (Γ_{12}, Γ_{13}, Γ_{23} and $\Gamma_{12}\Gamma_{13}\Gamma_{23}$) with eight sign combinations possible. Together they will result in the same invariants and the same three eigenvalues (slowness or velocities). This indicates that eight conditions for slowness surface can arise. However, implying Condition 2, mentioned above, $(\Gamma_{12}\Gamma_{13}\Gamma_{23})' = (\Gamma_{12}\Gamma_{13}\Gamma_{23})''$ will further assemble the eight slowness profiles in to two groups ($\Gamma_{12}\Gamma_{13}\Gamma_{23}$ is positive or negative). These groups will be from different individual signs of the coefficients Γ_{12}, Γ_{13}, and Γ_{23} but result in the same product value $\Gamma_{12}\Gamma_{13}\Gamma_{23}$ as elaborated in Table 3.1.

If two anomalous companion media are taken, they must differ in their algebraic signs of only two off-diagonal terms in the Kelvin–Christoffel matrix, while they share the same magnitude. Now, implying Condition 2, it can be further stated that to achieve the above state in Condition 3, one must have $\mathbb{C}_{14} = \mathbb{C}_{25} = \mathbb{C}_{36} = 0$ or any two of these stiffness coefficients that are equal to zero (i.e., $\mathbb{C}_{14} = \mathbb{C}_{25} = 0$ or $\mathbb{C}_{25} = \mathbb{C}_{36} = 0$ or $\mathbb{C}_{14} = \mathbb{C}_{36} = 0$). Figure 3.19 shows these conditions to visualize the material coefficients with their necessary symmetries. These materials will give normal polarizations in their symmetry plane but will demonstrate abnormal polarization in other orthogonal planes. $\mathbb{C}_{14} = \mathbb{C}_{25} = \mathbb{C}_{36} = 0$ satisfies the case for orthotropic and transversely isotropic media as presented in the previous section.

TABLE 3.1

Sign Combination of Off-Diagonal Terms in the Kelvin-Christoffel Matrix

Γ_{12}	+	+	−	−	−	−	+	+
Γ_{13}	+	−	−	+	−	+	+	−
Γ_{23}	+	−	+	−	−	+	−	+
$\Gamma_{12}\Gamma_{13}\Gamma_{23}$	+	+	+	+	−	−	−	−

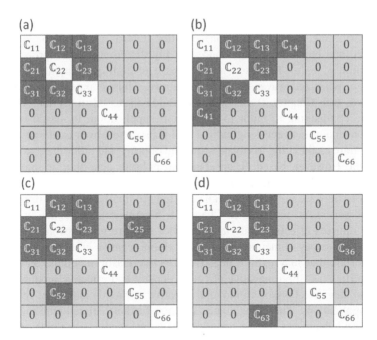

FIGURE 3.19 Constitutive matrix and visualization of the material coefficients that creates 'anomalous companion' based on condition 3: a) $\mathbb{C}_{14} = \mathbb{C}_{25} = \mathbb{C}_{36} = 0$, b) $\mathbb{C}_{25} = \mathbb{C}_{36} = 0$; $\mathbb{C}_{14} \neq 0$, y-z symmetry, c) $\mathbb{C}_{14} = \mathbb{C}_{36} = 0$; $\mathbb{C}_{25} \neq 0$ x-z symmetry, d) $\mathbb{C}_{14} = \mathbb{C}_{25} = 0$; $\mathbb{C}_{36} \neq 0$ x-y symmetry.

3.2.2.4.2 Metamaterials with Abnormal Polarity

In the above sections the conditions on the material properties are concluded purely based on the digital values of the material coefficients and elements of the Kelvin-Christoffel matrix. However, it is not necessary that all conditions will result in stable constitutive behavior. Based on the discussion in Chapter 2, we see that the constitutive matrix must be positive definite and must have positive strain energy under any deformation field. It is also necessary that all principal minors of the constitutive matrix must be positive. Considering such a scenario, most natural materials except a few may not exhibit abnormal polarity. However, if artificially created, a metamaterial (refer to Chapter 1) can exhibit such a unique material state. Designing metamaterials with unique material coefficients to explore various abnormal polarity conditions is an open field of research.

Applications of polarized waves were realized with electromagnetic waves in electromagnetism and optics. Liquid crystal displays (LCDs) that are dominating our society of the 21st century, telecommunication devices, application of photoelasticity, and devices like the polariscope are all the result of tailored applications of polarized electromagnetic waves. On the other hand, controlled and polarized elastic waves are rare and even challenging to achieve. They are crucial for the study of seismology and its application to mitigate earthquake-related damage to civil structures. Additionally, they may have tremendous applications in wave guiding, wave tunneling, acoustic computing, selective wave transport, acoustic clocking, energy harvesting, impact mitigation, vibration control, ultrasonic nondestructive evaluation, and biomedical ultrasound. For any direction of wave propagation or the direction of vector $\mathbf{k}$, elastic waves consist of both qL (longitudinal) and qS (transverse) polarized waves as discussed above. Thus, compared to electromagnetic waves (see Chapter 4) and waves in fluid (discussed in section 3.1), elastic waves are challenging to tailor to any specific application. In elastic waves, the modal conversion between longitudinal and transverse waves is universal, which makes it more challenging to isolate a specific polarity and tailor them to a specific application. However, based on an abnormal polarity condition, it can be imagined that a specific metamaterial could be devised to create a tuned wave polarizer. An elastic wave polarizer can tune qL (longitudinally) and/or q, (transversely) polarized waves. As discussed in Chapter 1, metamaterial gives the opportunity to create unusual effective material properties that are not possible naturally. At resonance frequency, a metamaterial may exhibit negative effective mass density. Metamaterials that may exhibit negative effective elastic modulus, double negative properties (i.e., both negative effective density and effective modulus, simultaneously), negative Poisson's ratio, and negative refraction index, are realized for various applications. These metamaterials were further realized with their unique polarization behavior to achieve complete shear-free media (like fluid) and tunable directional band gaps for unique applications [10–18].

It is outlined in the previous section that most general materials follow Eq. 3.89. These materials result in confined polarity of the qL modes into P zone $(0 - \pi/4)$ and qS modes into S zone $(\pi/4 - \pi/2)$ for all wave propagation directions, as shown in Figs. 3.11, 3.13, and 3.14. Abnormal polarization (discussed in section 3.2.2.3.2) shows that if, by any means, the shear modulus is higher than the modulus along the primary normal direction, material may exhibit abnormal polarity. If $\mathbb{C}_{11} < \mathbb{C}_{66} < \mathbb{C}_{22}$ condition is satisfied and agrees with the conditions discussed in section 3.2.1.4.1, the condition will always result an abnormal polarization condition. This is a unique anisotropy requirement that is hard to find in natural materials. However, if a metamaterial is devised with this requirement, the metamaterial will always guarantee abnormal polarity.

Only recently, Lee et. al [9]. created a metamaterial by creating slits in a metallic plate. No locally resonant nondispersive metamaterial was used, which can only give zero to almost zero group velocity of the wave modes. It was mentioned that the key requirements were as follows: 1) Effective normal stiffness must be lower than the shear stiffness for a specific material axis (e.g., $\mathbb{C}_{11}$). 2) Effective normal stiffness along the orthogonal directions must be higher than the shear stiffness. This satisfies the $\mathbb{C}_{11} < \mathbb{C}_{66} < \mathbb{C}_{22}$ requirement. It is indeed a challenging task. Figure 3.20 shows the design of the material that they conceptualized to achieve the above requirements. The effective stiffness was calculated from the stress-strain relation obtained from a simulation. For example, to obtain $\mathbb{C}_{11}$, stress was calculated using equation $\sigma_{11} = \dfrac{F_1}{3L_y t}$, and the strain was calculated using $e_{11} = \delta_{11}/3L_x$, where F_1 was the applied force and δ_{11} was the measured elongation from the simulation. $\mathbb{C}_{11} = \sigma_{11}/e_{11}$. L_x, L_y, and t are the length and width of the sample material along

Acoustics and Elastic Wave Propagation in Fluids and Anisotropic Solids 105

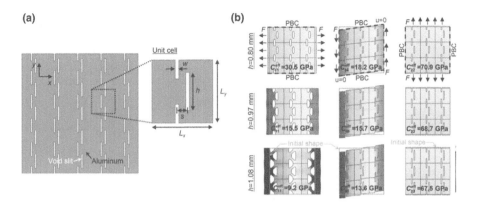

FIGURE 3.20 Off-centered double-slit metamaterial: a) schematic configuration, b) static deformation behavior under x-directional dilatational load (first column), shear load (second column), y-directional dilatational load (third column). The figure is reprinted from Ref [9] with permission.

x and y directions and the thickness, respectively. Similarly, $\mathbb{C}_{22}$ stiffness was also calculated. To obtain $\mathbb{C}_{66}$, shear stress was calculated using equation $\sigma_{12} = F/3L_y t$, and the strain was calculated using $e_{12} = 0.5\gamma_{12}$, where γ_{12} is the shear deformation or the angular deformation obtained from the simulation.

Figure 3.21 shows the EFCs (first column), polarization differential or PoDi (Eq. 3.81) of the qL and qFS wave modes (second column), frequency-dependent effective density (ρ_{eff}) (third column), and frequency-dependent effective modulus $\mathbb{C}_{11}^{\text{eff}}$, $\mathbb{C}_{66}^{\text{eff}}$, $\mathbb{C}_{22}^{\text{eff}}$ (fourth column) for three different sets of design parameters presented in Fig. 3.20 along the three rows, (a), (b) and (c). It can be seen that when $\mathbb{C}_{11} < \mathbb{C}_{66} < \mathbb{C}_{22}$ is satisfied, abnormal polarity prevails (Fig. 3.21b and c). Figure 3.21 is directly taken from the article by Lee et. al [9]. However, to make a connection between our forgoing discussion, definition of P zone and S zone for the PoDi remains intact. Hence, Figs. 3.21b and c show switching of the qL and qFS wave modes with their polarities for specific directions of wave propagation. Further it was demonstrated that, using abnormal polarity, the designed metamaterial exhibits wave-mode conversion capability. It was able to convert a longitudinal wave mode to a transverse wave mode using the following physical logic.

Researchers created a wedge design of the metamaterial embedded in the host structure made from an aluminum plate. The design presented in Fig. 3.21c was chosen to create the wedge. A critical angle was found using the ratio of the quasi-shear wave velocity and quasi-longitudinal wave velocities as $\theta_c = tan^{-1}\left(\sqrt{c_S/c_L}\right) = tan^{-1}\left(\sqrt{\mathbb{C}_{66}/\mathbb{C}_{11}}\right)$. This was to ensure modal conversion in the metamaterial plate. A specific angle was created to find the wedge angle $\theta_s = 90° - \theta_c$ as shown in Fig. 3.22a. The rest of the part of the plate remained isotropic. A wave vector along the x direction was sent with a longitudinally polarized wave number k_0^L. At the interface between the isotropic aluminum plate and the metamaterial wedge along the y-axis, the wave number should modify to k_x^L to keep the longitudinal polarity of the incident wave indicated in Fig. 3.22a and b. Please note that the longitudinal polarity along the x direction is practiced by the qFS wave slowness

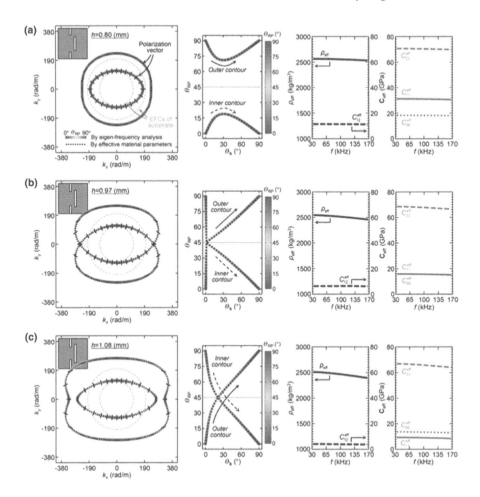

FIGURE 3.21 Schematic shows the equifrequency contours for in-plane wave modes along the first column, polarization differential or PoDi along the middle column, and frequency dependent effective material properties along the third and fourth columns a) with slit height 0.8 mm, b) with slit height 0.97 mm, and c) with slit height 1.08 mm. The figure is reprinted from Ref [9] with permission.

(refer to Fig. 3.21c) but not by the qL wave mode. Hence, a P wave in an aluminum plate will convert to qFS wave mode to keep the polarization within P zone owing to the abnormal polarization. Next, as the wave propagates along the metamaterial shaped like a wedge, the wave incident on the inclined surface of the wedge can be represented with a wave number k_x^L. Following the wave number conversion rule or based on the conservation of the tangential momentum [4, 5] the wave number along the wedge boundary must be unique. A dotted line in Fig. 3.22b and c shows the projection of the wave number k_x^L on the wedge interface (θ_s line) and then further projection on the orthogonal reflection plane. The wave number k_w propagating along the interface is unique (Fig. 3.22c). Projection of the k_w wave number along the y-axis will give the wave number k_y^S along the y-direction of the same

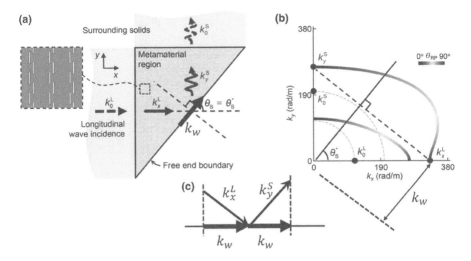

FIGURE 3.22 Schematics showing the wave number conversion rule based on the conservation of the tangential momentum: a) mode conversion and polarization flip, b) explanation through slowness curves. The figure is reprinted from Ref [9] with permission.

wave mode but with different slowness or wave number value. Owing to the abnormal polarity, although the wave mode (qFS mode) remains the same, the polarity of the mode switches to transverse polarization. This shows that it is not necessary to find the θ_s or the wedge angle using the equation $\theta_s = 90° - tan^{-1}\left(\sqrt{\mathbb{C}_{66}/\mathbb{C}_{11}}\right)$; it can be obtained from pure geometry. Geometrical considerations are as follows. It is necessary for the wave number along the x-axis to have a value $k_x{}^L$ due to the longitudinally polarized wave input from the left at a specific frequency obtained from the EFC. It was intended to obtain the same wave mode along the orthogonal direction but with different polarity owing to the abnormal polarization of the similar mode. So, it is inevitable that the wave number along the y-axis at the same frequency must be the best available wave number from the EFC, which is $k_y{}^S$. Once these two wave numbers on the two orthogonal directions are found, the values ($k_x{}^L$, $k_y{}^S$) on the graph can be joined using a dotted line as shown in Fig. 3.22b and c. A line orthogonal to the dotted line passing through the reference origin will give the angle θ_s. Please refer to the firm line in Fig. 3.22b. The θ_s is the design angle for the wedge. Hence, if a metamaterial is created with tailored material properties, EFC at multiple frequencies can be obtained for that metamaterial. By observing EFCs and polarization of the wave modes simultaneously, abnormal polarization directions/orientations can be found, and polarization differential plots can be visualized. Next, based on EFCs and polarization differential plots wave guiding and wave tuning effect can be exploited predictively.

Figure 3.23 shows the evidence and application of the abnormal polarization. Figure 3.23a shows the velocity field along the x and y directions on the entire geometry, including the wedge meta structure. Similar velocity fields are shown in Fig. 3.23b, replacing the metamaterial wedge by an equivalent anisotropic material

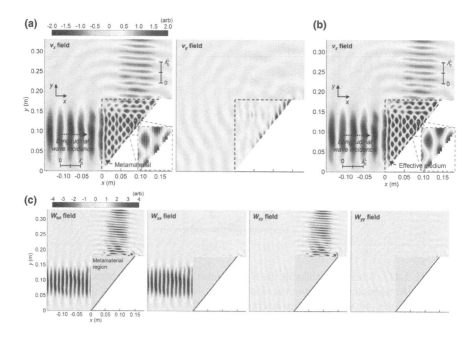

FIGURE 3.23 Orthogonal bending of wave energy: a) demonstration of velocity field through metamaterial, b) demonstration of velocity field through effective medium, c) demonstration of strain energy density of the respective polarized wave through effective medium. The figure is reprinted from Ref [9] with permission.

(without discontinuity) with material coefficients presented in the fourth column of Fig. 3.21c. This proves that a continuous homogeneous material with equivalent unique material properties where the $\mathbb{C}_{11} < \mathbb{C}_{66} < \mathbb{C}_{22}$ condition is satisfied would demonstrate similar abnormal polarization behavior as was designed for the metamaterial. Figure 3.23c shows the strain energy density fields on the entire geometry. The first wave field shows the total energy density $\varepsilon_t = \varepsilon_{xx} + \varepsilon_{yy} + \varepsilon_{xy}$. Second, third, and fourth wave fields show the energy densities ε_{xx}, ε_{xy}, and ε_{yy}, respectively. Comparing the second and third wave fields, it is evident that the longitudinally polarized wave propagated along the x direction and the transversely polarized wave propagated along the y direction.

Please note that the above problem is just a single demonstration of exploiting abnormal polarity in metamaterials. There is much other literature that explored metamaterials for multiple unique behaviors, but it is not explicitly discussed herein. This research field is vast, and there are multiple possibilities of the design of novel metamaterials. In this chapter, an attempt has been made to describe fundamental physics and provide the physical guidelines/requirements for the design of new metamaterials from the perspective of effective material properties. One could exploit a complete design envelope for different metamaterial structures with necessary design parameters using numerical study. One could also create surrogate model(s) to obtain the design parameter(s) for metamaterials with specific design objective and tailored applications.

Acoustics and Elastic Wave Propagation in Fluids and Anisotropic Solids **109**

Please note that these wave behaviors are not topological in general sense; however, topological wave behavior significantly depends on the wave polarization and their phases. Understanding from this chapter on wave phases and wave polarization will be helpful for understanding the topological wave behaviors in the later chapters.

3.2.3 Nonhomogeneous Isotropic and Anisotropic Materials

Wave propagation in nonhomogeneous media will be the key for understanding unique topological wave behaviors. Please note, here all materials that are not homogeneous in spatial and temporal sense (e.g., material properties are the function of space and time) are considered nonhomogeneous, or inhomogeneous. In Chapter 1, while discussing metamaterials, the heterogeneity of material was introduced. A heterogeneous material is composed of multiple dissimilar material constituents, and they may result in special variability of the material properties. From here onward in this book, in the context of metamaterials, nonhomogeneous and heterogeneous materials carry similar meaning.

In section 3.2.2.4, metamaterial design is introduced to obtain a unique wave behavior exploiting the abnormal polarity. However, the requirements for material properties are a bit unnatural. The new material needs an artificial design of the material properties, introducing local geometric perturbations. Geometric perturbation could be the removal or addition of another material of the same or different properties to the host material. Metamaterials are generally periodic in nature for easy manufacturability. Once the metamaterials or the articulated materials are designed, the new materials do not remain homogeneous due to their local architectures and geometric perturbations. They might be nonhomogeneous in the classical physical sense or could be a heterogeneous composition of different material constituents. For example, the slits created in an aluminum plate to achieve certain wave behavior, as discussed in the previous section, makes the plate nonhomogeneous. In that example, specific constitutive properties were the objective to achieve abnormal polarity in the material. It was shown that an effective material (homogeneous) with similar material properties can result in similar intended behavior. But the material is inhomogeneous. Thus, the following sections are specifically dedicated to understanding the wave equations in nonhomogeneous media.

For nonhomogeneous materials, we use material properties as follows:

$$\mathbb{C}_{ijkl} \equiv \mathbb{C}_{ijkl}\left(x_p\right); \ \mathbb{L} = \mathbb{L}\left(x_p\right); \ \mathrm{m} = \mathrm{m}\left(x_p\right); \ \rho = \rho\left(x_p\right) \tag{3.95}$$

3.2.3.1 Wave Equations for Nonhomogeneous Isotropic Media

Here we start with the governing differential equation written in Eq. 3.23 $\nabla.\sigma = \rho(\mathbf{x})\ddot{\mathbf{u}}$. The equations with all the terms are written in Eqs. A.2.1.1 through A.2.1.3. In an isotropic nonhomogeneous media, the constitutive relation (stress-strain relation) is written in Eqs. 3.25 and 3.26. In matrix form those equations can be expanded to

$$\sigma = \begin{bmatrix} \sigma_{11} & \sigma_{21} & \sigma_{31} \\ \sigma_{12} & \sigma_{22} & \sigma_{32} \\ \sigma_{13} & \sigma_{23} & \sigma_{33} \end{bmatrix} =$$

$$\begin{bmatrix} \mathbb{L}(x_p)(\nabla.\mathbf{u})+2\mathrm{m}(x_p)\mathrm{e}_{11} & 2\mathrm{m}(x_p)(\mathrm{e}_{12}) & 2\mathrm{m}(x_p)(\mathrm{e}_{13}) \\ 2\mathrm{m}(x_p)(\mathrm{e}_{21}) & \mathbb{L}(x_p)(\nabla.\mathbf{u})+2\mathrm{m}(x_p)(\mathrm{e}_{22}) & 2\mathrm{m}(x_p)(\mathrm{e}_{32}) \\ 2\mathrm{m}(x_p)(\mathrm{e}_{31}) & 2\mathrm{m}(x_p)(\mathrm{e}_{32}) & \mathbb{L}(x_p)(\nabla.\mathbf{u})+2\mathrm{m}(x_p)\mathrm{e}_{33} \end{bmatrix} \cdots$$

$$(3.96)$$

where the (e_{ij}) elements and the expressions for $\nabla.\mathbf{u}$ are written in Eqs. A.3.7 and A.3.2, respectively.

The stress matrix can be further written as

$$\sigma = \mathbb{L}(x_p)(\nabla.\mathbf{u}) \begin{bmatrix} 1 & 0 & 0 \\ 0 & 1 & 0 \\ 0 & 0 & 1 \end{bmatrix} + 2\mathrm{m}(x_p) \begin{bmatrix} \mathrm{e}_{11} & \mathrm{e}_{12} & \mathrm{e}_{13} \\ \mathrm{e}_{21} & \mathrm{e}_{22} & \mathrm{e}_{23} \\ \mathrm{e}_{31} & \mathrm{e}_{32} & \mathrm{e}_{33} \end{bmatrix} \qquad (3.97)$$

or

$$\sigma = \mathbb{L}(x_p)(\nabla.\mathbf{u}).\mathbf{I} + 2\mathrm{m}(x_p)\epsilon \qquad (3.98.1)$$

Referring to Eq. A.3.7 in Appendix,

$$\sigma = \mathbb{L}(x_p)(\nabla.\mathbf{u}).\mathbf{I} + \mathrm{m}(x_p)(\nabla\mathbf{u}+\nabla\mathbf{u}^T) \qquad (3.98.2)$$

Next, operating the nabla Hamiltonian on the stress matrix will result the expressions written in Eq. A.3.12, which can be further elaborated using Eq. 3.96 as follows

$$\nabla.\sigma = \begin{bmatrix} \frac{\partial}{\partial x_1}\left[\mathbb{L}(x_p)(\nabla.\mathbf{u})+2\mathrm{m}(x_p)\mathrm{e}_{11}\right]+\frac{\partial}{\partial x_2}\left(2\mathrm{m}(x_p)(\mathrm{e}_{21})\right)+\frac{\partial}{\partial x_3}\left(2\mathrm{m}(x_p)(\mathrm{e}_{31})\right) \\ \frac{\partial}{\partial x_1}\left(2\mathrm{m}(x_p)(\mathrm{e}_{12})\right)+\frac{\partial}{\partial x_2}\left[\mathbb{L}(x_p)(\nabla.\mathbf{u})+2\mathrm{m}(x_p)(\mathrm{e}_{22})\right]+\frac{\partial}{\partial x_3}\left(2\mathrm{m}(x_p)(\mathrm{e}_{32})\right) \\ \frac{\partial}{\partial x_1}\left(2\mathrm{m}(x_p)(\mathrm{e}_{13})\right)+\frac{\partial}{\partial x_2}\left(2\mathrm{m}(x_p)(\mathrm{e}_{23})\right)+\frac{\partial}{\partial x_3}\left[\mathbb{L}(x_p)(\nabla.\mathbf{u})+2\mathrm{m}(x_p)\mathrm{e}_{33}\right] \end{bmatrix} \cdots$$

$$(3.99)$$

If the material is homogeneous, the derivative of the Lamé constants will result in zero. However, metamaterials are nonhomogeneous, and thus the derivatives of the material constants cannot be ignored. They must contribute to the governing differential equation. Further, expanding the expressions term by term in Eq. 3.100 will

Acoustics and Elastic Wave Propagation in Fluids and Anisotropic Solids 111

result in long expressions as follows. For simplicity $\mathbb{L}(x_p)$ is written as λ and $\mathrm{m}(x_p)$ is written as μ where their spatial inhomogeneity is implied.

$$\nabla.\sigma = \begin{bmatrix} \lambda\left(\dfrac{\partial}{\partial x_1}(\nabla.\mathbf{u})\right) + (\nabla.\mathbf{u})\dfrac{\partial\lambda}{\partial x_1} + 2\mu\dfrac{\partial}{\partial x_1}\left(\dfrac{\partial u_1}{\partial x_1}\right) + 2\dfrac{\partial u_1}{\partial x_1}\left(\dfrac{\partial\mu}{\partial x_1}\right) \\[2mm] \lambda\left(\dfrac{\partial}{\partial x_2}(\nabla.\mathbf{u})\right) + (\nabla.\mathbf{u})\dfrac{\partial\lambda}{\partial x_2} + 2\mu\dfrac{\partial}{\partial x_2}\left(\dfrac{\partial u_2}{\partial x_2}\right) + 2\dfrac{\partial u_2}{\partial x_2}\left(\dfrac{\partial\mu}{\partial x_2}\right) \\[2mm] \lambda\left(\dfrac{\partial}{\partial x_3}(\nabla.\mathbf{u})\right) + (\nabla.\mathbf{u})\dfrac{\partial\lambda}{\partial x_3} + 2\mu\dfrac{\partial}{\partial x_3}\left(\dfrac{\partial u_3}{\partial x_3}\right) + 2\dfrac{\partial u_3}{\partial x_3}\left(\dfrac{\partial\mu}{\partial x_3}\right) \end{bmatrix}$$

$$\tag{3.100}$$

$$\begin{bmatrix} + \mu\dfrac{\partial}{\partial x_2}\left(\dfrac{\partial u_2}{\partial x_1} + \dfrac{\partial u_1}{\partial x_2}\right) + \mu\dfrac{\partial}{\partial x_3}\left(\dfrac{\partial u_3}{\partial x_1} + \dfrac{\partial u_1}{\partial x_3}\right) + 2e_{21}\dfrac{\partial\mu}{\partial x_2} + 2e_{31}\dfrac{\partial\mu}{\partial x_3} \\[2mm] + \mu\dfrac{\partial}{\partial x_1}\left(\dfrac{\partial u_1}{\partial x_2} + \dfrac{\partial u_2}{\partial x_1}\right) + \mu\dfrac{\partial}{\partial x_3}\left(\dfrac{\partial u_3}{\partial x_2} + \dfrac{\partial u_2}{\partial x_3}\right) + 2e_{12}\dfrac{\partial\mu}{\partial x_1} + 2e_{32}\dfrac{\partial\mu}{\partial x_3} \\[2mm] + \mu\dfrac{\partial}{\partial x_1}\left(\dfrac{\partial u_1}{\partial x_3} + \dfrac{\partial u_3}{\partial x_1}\right) + \mu\dfrac{\partial}{\partial x_2}\left(\dfrac{\partial u_3}{\partial x_2} + \dfrac{\partial u_2}{\partial x_3}\right) + 2e_{23}\dfrac{\partial\mu}{\partial x_2} + 2e_{13}\dfrac{\partial\mu}{\partial x_3} \end{bmatrix}$$

Writing $2\mu\dfrac{\partial}{\partial x_1}\left(\dfrac{\partial u_1}{\partial x_1}\right)$ as a summation of two similar terms as $\mu\dfrac{\partial}{\partial x_1}\left(\dfrac{\partial u_1}{\partial x_1}\right) + \mu\dfrac{\partial}{\partial x_1}\left(\dfrac{\partial u_1}{\partial x_1}\right)$ and likewise for other terms, collecting similar terms by category the above expression simplifies to the following

$$\nabla.\sigma = \begin{bmatrix} \dfrac{\partial\lambda}{\partial x_1}(\nabla.\mathbf{u}) + 2\left[\dfrac{\partial\mu}{\partial x_1}e_{11} + \dfrac{\partial\mu}{\partial x_2}e_{12} + \dfrac{\partial\mu}{\partial x_3}e_{13}\right] + \lambda\dfrac{\partial}{\partial x_1}(\nabla.\mathbf{u}) \\[2mm] \dfrac{\partial\lambda}{\partial x_2}(\nabla.\mathbf{u}) + 2\left[\dfrac{\partial\mu}{\partial x_1}e_{21} + \dfrac{\partial\mu}{\partial x_2}e_{22} + \dfrac{\partial\mu}{\partial x_3}e_{23}\right] + \lambda\dfrac{\partial}{\partial x_2}(\nabla.\mathbf{u}) \\[2mm] \dfrac{\partial\lambda}{\partial x_3}(\nabla.\mathbf{u}) + 2\left[\dfrac{\partial\mu}{\partial x_1}e_{31} + \dfrac{\partial\mu}{\partial x_2}e_{32} + \dfrac{\partial\mu}{\partial x_3}e_{33}\right] + \lambda\dfrac{\partial}{\partial x_3}(\nabla.\mathbf{u}) \end{bmatrix}$$

$$\tag{3.101}$$

$$\begin{bmatrix} + \left(\mu\dfrac{\partial^2 u_1}{\partial x_1^2} + \mu\dfrac{\partial^2 u_1}{\partial x_2^2} + \mu\dfrac{\partial^2 u_1}{\partial x_3^2}\right) + \mu\dfrac{\partial}{\partial x_1}(\nabla.\mathbf{u}) \\[2mm] + \left(\mu\dfrac{\partial^2 u_2}{\partial x_1^2} + \mu\dfrac{\partial^2 u_2}{\partial x_2^2} + \mu\dfrac{\partial^2 u_2}{\partial x_3^2}\right) + \mu\dfrac{\partial}{\partial x_2}(\nabla.\mathbf{u}) \\[2mm] + \left(\mu\dfrac{\partial^2 u_3}{\partial x_1^2} + \mu\dfrac{\partial^2 u_3}{\partial x_2^2} + \mu\dfrac{\partial^2 u_3}{\partial x_3^2}\right) + \mu\dfrac{\partial}{\partial x_3}(\nabla.\mathbf{u}) \end{bmatrix}$$

Again, collecting similar terms, the $\nabla.\sigma$ in matrix form will be

$$\nabla.\sigma = \begin{bmatrix} (\lambda+\mu)\dfrac{\partial}{\partial x_1}(\nabla.\mathbf{u})+\mu\left(\dfrac{\partial^2}{\partial x_1^2}+\dfrac{\partial^2}{\partial x_2^2}+\dfrac{\partial^2}{\partial x_3^2}\right)u_1+(\nabla.\mathbf{u})\dfrac{\partial\lambda}{\partial x_1} \\[2mm] (\lambda+\mu)\dfrac{\partial}{\partial x_2}(\nabla.\mathbf{u})+\mu\left(\dfrac{\partial^2}{\partial x_1^2}+\dfrac{\partial^2}{\partial x_2^2}+\dfrac{\partial^2}{\partial x_3^2}\right)u_2+(\nabla.\mathbf{u})\dfrac{\partial\lambda}{\partial x_2} \\[2mm] (\lambda+\mu)\dfrac{\partial}{\partial x_3}(\nabla.\mathbf{u})+\mu\left(\dfrac{\partial^2}{\partial x_1^2}+\dfrac{\partial^2}{\partial x_2^2}+\dfrac{\partial^2}{\partial x_3^2}\right)u_3+(\nabla.\mathbf{u})\dfrac{\partial\lambda}{\partial x_3} \\[4mm] +2\left[\dfrac{\partial\mu}{\partial x_1}\left(\dfrac{\partial u_1}{\partial x_1}\right)+\dfrac{1}{2}\dfrac{\partial\mu}{\partial x_2}\left(\dfrac{\partial u_1}{\partial x_2}+\dfrac{\partial u_2}{\partial x_1}\right)+\dfrac{1}{2}\dfrac{\partial\mu}{\partial x_3}\left(\dfrac{\partial u_1}{\partial x_3}+\dfrac{\partial u_3}{\partial x_1}\right)\right] \\[4mm] +2\left[\dfrac{1}{2}\dfrac{\partial\mu}{\partial x_1}\left(\dfrac{\partial u_2}{\partial x_1}+\dfrac{\partial u_1}{\partial x_2}\right)+\dfrac{\partial\mu}{\partial x_2}\dfrac{\partial u_2}{\partial x_2}+\dfrac{1}{2}\dfrac{\partial\mu}{\partial x_3}\left(\dfrac{\partial u_2}{\partial x_3}+\dfrac{\partial u_3}{\partial x_2}\right)\right] \\[4mm] +2\left[\dfrac{1}{2}\dfrac{\partial\mu}{\partial x_1}\left(\dfrac{\partial u_3}{\partial x_1}+\dfrac{\partial u_1}{\partial x_3}\right)+\dfrac{1}{2}\dfrac{\partial\mu}{\partial x_2}\left(\dfrac{\partial u_3}{\partial x_2}+\dfrac{\partial u_2}{\partial x_3}\right)+\dfrac{\partial\mu}{\partial x_3}\dfrac{\partial u_3}{\partial x_3}\right] \end{bmatrix} \tag{3.102}$$

Applying Eq. A.3.11 and further breaking $2\dfrac{\partial\mu}{\partial x_1}\left(\dfrac{\partial u_1}{\partial x_1}\right)=\dfrac{\partial\mu}{\partial x_1}\left(\dfrac{\partial u_1}{\partial x_1}\right)+\dfrac{\partial\mu}{\partial x_1}\left(\dfrac{\partial u_1}{\partial x_1}\right)$ for the similar terms, Eq. 3.102 modifies to

$$\nabla.\sigma = \begin{bmatrix} (\lambda+\mu)\dfrac{\partial}{\partial x_1}(\nabla.\mathbf{u})+\mu\nabla^2 u_1+(\nabla.\mathbf{u})\dfrac{\partial\lambda}{\partial x_1} \\[2mm] (\lambda+\mu)\dfrac{\partial}{\partial x_2}(\nabla.\mathbf{u})+\mu\nabla^2 u_2+(\nabla.\mathbf{u})\dfrac{\partial\lambda}{\partial x_2} \\[2mm] (\lambda+\mu)\dfrac{\partial}{\partial x_3}(\nabla.\mathbf{u})+\mu\nabla^2 u_3+(\nabla.\mathbf{u})\dfrac{\partial\lambda}{\partial x_3} \\[4mm] +\left(\dfrac{\partial\mu}{\partial x_1}\dfrac{\partial u_1}{\partial x_1}+\dfrac{\partial\mu}{\partial x_2}\dfrac{\partial u_1}{\partial x_2}+\dfrac{\partial\mu}{\partial x_3}\dfrac{\partial u_1}{\partial x_3}\right)+\left(\dfrac{\partial\mu}{\partial x_1}\dfrac{\partial u_1}{\partial x_1}+\dfrac{\partial\mu}{\partial x_2}\dfrac{\partial u_2}{\partial x_1}+\dfrac{\partial\mu}{\partial x_3}\dfrac{\partial u_3}{\partial x_1}\right) \\[4mm] +\left(\dfrac{\partial\mu}{\partial x_1}\dfrac{\partial u_2}{\partial x_1}+\dfrac{\partial\mu}{\partial x_2}\dfrac{\partial u_2}{\partial x_2}+\dfrac{\partial\mu}{\partial x_3}\dfrac{\partial u_2}{\partial x_3}\right)+\left(\dfrac{\partial\mu}{\partial x_1}\dfrac{\partial u_1}{\partial x_2}+\dfrac{\partial\mu}{\partial x_2}\dfrac{\partial u_2}{\partial x_2}+\dfrac{\partial\mu}{\partial x_3}\dfrac{\partial u_3}{\partial x_2}\right) \\[4mm] +\left(\dfrac{\partial\mu}{\partial x_1}\dfrac{\partial u_3}{\partial x_1}+\dfrac{\partial\mu}{\partial x_2}\dfrac{\partial u_3}{\partial x_2}+\dfrac{\partial\mu}{\partial x_3}\dfrac{\partial u_3}{\partial x_3}\right)+\left(\dfrac{\partial\mu}{\partial x_1}\dfrac{\partial u_1}{\partial x_3}+\dfrac{\partial\mu}{\partial x_2}\dfrac{\partial u_2}{\partial x_3}+\dfrac{\partial\mu}{\partial x_3}\dfrac{\partial u_3}{\partial x_3}\right) \end{bmatrix} \tag{3.103}$$

Acoustics and Elastic Wave Propagation in Fluids and Anisotropic Solids 113

Please refer to Appendix for a few standard forms of matrices with μ. Using Eqs. A.3.16.1 and A.3.16.2, the above equation will read

$$
\nabla.\sigma = \begin{bmatrix} (\lambda+\mu)\dfrac{\partial}{\partial x_1}(\nabla.\mathbf{u})+\mu\nabla^2 u_1' +(\nabla.\mathbf{u})\dfrac{\partial\lambda}{\partial x_1} \\[2mm] (\lambda+\mu)\dfrac{\partial}{\partial x_2}(\nabla.\mathbf{u})+\mu\nabla^2 u_2 +(\nabla.\mathbf{u})\dfrac{\partial\lambda}{\partial x_2} \\[2mm] (\lambda+\mu)\dfrac{\partial}{\partial x_3}(\nabla.\mathbf{u})+\mu\nabla^2 u_3 +(\nabla.\mathbf{u})\dfrac{\partial\lambda}{\partial x_3} \end{bmatrix} + \underline{\nabla}\mu^1 + \underline{\nabla}\mu^2 \quad (3.104)
$$

With a minor mathematical treatment, Eq. 3.104 will take the following form where Eq. A.3.16.4 will further be simplified to

$$
\nabla.\sigma = \begin{bmatrix} (\lambda+\mu)\dfrac{\partial}{\partial x_1}(\nabla.\mathbf{u})+\mu\nabla^2 u_1 +(\nabla.\mathbf{u})\dfrac{\partial\lambda}{\partial x_1} \\[2mm] (\lambda+\mu)\dfrac{\partial}{\partial x_2}(\nabla.\mathbf{u})+\mu\nabla^2 u_2 +(\nabla.\mathbf{u})\dfrac{\partial\lambda}{\partial x_2} \\[2mm] (\lambda+\mu)\dfrac{\partial}{\partial x_3}(\nabla.\mathbf{u})+\mu\nabla^2 u_3 +(\nabla.\mathbf{u})\dfrac{\partial\lambda}{\partial x_3} \end{bmatrix} + 2\underline{\nabla}\mu^1 + \left(\underline{\nabla}\mu^2 - \underline{\nabla}\mu^1\right) \quad (3.105)
$$

Please note that the expression for $\underline{\nabla}\mu^1$ can be further written as $\left(\nabla\mu.\underline{\nabla}\right)\mathbf{u}$ using Eq. A.3.17.3. Hence, $\nabla.\sigma$ could be further written as

$$
\nabla.\sigma = (\lambda+\mu)\underline{\nabla}(\nabla.\mathbf{u})+\mu\nabla^2 u+\underline{\nabla}\lambda(\nabla.\mathbf{u})+2\left(\nabla\mu.\underline{\nabla}\right)\mathbf{u}+\underline{\nabla}\mu\times\underline{\nabla}\times\mathbf{u} \quad (3.106)
$$

The final governing differential equation in a nonhomogeneous isotropic material will read

$$
(\lambda+\mu)\underline{\nabla}(\nabla.\mathbf{u})+\mu\nabla^2 u+\underline{\nabla}\lambda(\nabla.\mathbf{u})+2\left(\nabla\mu.\underline{\nabla}\right)\mathbf{u}+\underline{\nabla}\mu\times\underline{\nabla}\times\mathbf{u}+\mathbf{f} = \rho(\mathbf{x})\ddot{u} \quad (3.107)
$$

Please note the difference between the two nabla operators ∇ and $\underline{\nabla}$, the former being a matrix and the latter being a column vector as discussed in the appendix of this chapter: $\underline{\nabla} = \nabla.\mathbf{I}$. Equation 3.107 is the elastodynamic wave equation for any nonhomogeneous isotropic bulk medium. If a specific structure has its own boundary, appropriate boundary conditions should be applied to solve the above equation. As discussed in section 3.2.2.1, the following steps are generally adopted to solve the wave equations:

- First wave potentials are assumed to satisfy the governing equation.
- Further displacements and stresses are formulated using the wave potentials.
- Appropriate boundary conditions are applied to get eigen equation(s)
- The eigen problem is formulated and eigensolutions are found that represent wave modes.

114 Metamaterials in Topological Acoustics

However, in case of nonhomogeneous media, it is required to know the functions that explicitly describe how the material properties are varying over a spatial and temporal domain. By knowing explicit functions, their derivatives can be found. If the material is periodic their Fourier transforms and Fourier coefficients are used in the equation. These steps are discussed in Chapter 6 on periodic media.

3.2.3.2 Wave Equations for Nonhomogeneous Anisotropic Media

Here we reiterate the governing wave equation from Chapter 2 Appendix as follows, which is the same as the equation $\nabla.\sigma = \rho(\mathbf{x})\ddot{\mathbf{u}}$ with the body force.

$$\frac{\partial \sigma_{11}}{\partial x_1} + \frac{\partial \sigma_{12}}{\partial x_2} + \frac{\partial \sigma_{13}}{\partial x_3} + f_1 = \rho \frac{\partial^2 u_1}{\partial t^2} \tag{3.108.1}$$

$$\frac{\partial \sigma_{21}}{\partial x_1} + \frac{\partial \sigma_{22}}{\partial x_2} + \frac{\partial \sigma_{23}}{\partial x_3} + f_2 = \rho \frac{\partial^2 u_2}{\partial t^2} \tag{3.108.2}$$

$$\frac{\partial \sigma_{31}}{\partial x_1} + \frac{\partial \sigma_{32}}{\partial x_2} + \frac{\partial \sigma_{33}}{\partial x_3} + f_3 = \rho \frac{\partial^2 u_3}{\partial t^2} \tag{3.108.3}$$

A complete generic solution in anisotropic media with spatially varying material coefficients is extremely challenging. Hence, a few simpler cases will be used to demonstrate the governing equations needed. A plane wave on a two-dimensional plane (e.g., $x - y$ plane or according to our nomenclature $x_1 - x_2$ plane) will not have any z or x_3 variation. Thus, it is safe to assume $\dfrac{\partial}{\partial x_3} = 0$. Hence, the above equations on a 2D plane can be written as

$$\frac{\partial \sigma_{11}}{\partial x_1} + \frac{\partial \sigma_{12}}{\partial x_2} + f_1 = \rho \frac{\partial^2 u_1}{\partial t^2} \tag{3.109.1}$$

$$\frac{\partial \sigma_{21}}{\partial x_1} + \frac{\partial \sigma_{22}}{\partial x_2} + f_2 = \rho \frac{\partial^2 u_2}{\partial t^2} \tag{3.109.2}$$

$$\frac{\partial \sigma_{31}}{\partial x_1} + \frac{\partial \sigma_{32}}{\partial x_2} + f_3 = \rho \frac{\partial^2 u_3}{\partial t^2} \tag{3.109.3}$$

From sections 3.2.2.3 and 3.2.2.4, the discussion on normal and abnormal polarity, it is evident that the polarity of the two quasi wave modes (one longitudinal and one transverse) always remains on an $x_1 - x_2$ plane, whereas the remaining quasi-transverse wave mode is vertically polarized and always has out-of-plane displacement. Using this understanding, we can break the above equations in to two categories: a) horizontally polarized wave equations on the $x_1 - x_2$ plane and b) vertically polarized wave equations on the $x_1 - x_2$ plane. Additionally, another category we could consider is a plane-stress state.

Acoustics and Elastic Wave Propagation in Fluids and Anisotropic Solids

a. Horizontally polarized waves with coupled u_1 and u_2 displacements

$$\frac{\partial \sigma_{11}}{\partial x_1} + \frac{\partial \sigma_{12}}{\partial x_2} + f_1 = \rho \frac{\partial^2 u_1}{\partial t^2} \tag{3.110.1}$$

$$\frac{\partial \sigma_{21}}{\partial x_1} + \frac{\partial \sigma_{22}}{\partial x_2} + f_2 = \rho \frac{\partial^2 u_2}{\partial t^2} \tag{3.110.2}$$

b. Vertically polarized wave with u_3 displacement only on the $x_1 - x_2$ plane

$$\frac{\partial \sigma_{31}}{\partial x_1} + \frac{\partial \sigma_{32}}{\partial x_2} + f_3 = \rho \frac{\partial^2 u_3}{\partial t^2} \tag{3.111}$$

Alternatively, referring to Eqs. 3.108.1 through 3.108.3, if a medium can sustain only the vertically polarized wave with $u_1 = 0$ and $u_2 = 0$ displacements with no dependency on the $x_1 - x_2$ plane, the wave equation will read

$$\frac{\partial \sigma_{33}}{\partial x_3} + f_3 = \rho \frac{\partial^2 u_3}{\partial t^2} \tag{3.112}$$

c. Considering Eqs. 3.108.1 through 3.108.3, the plane-stress state can be substituted using the expression $\sigma_{33} = 0$. This state, however, does not eliminate the displacement u_3. Hence, the plane-stress and $\frac{\partial}{\partial x_3} = 0$ states are not the same.

Although this discussion is presented under anisotropic media, the equations above are equally valid for isotropic systems.

Next the stresses can be substituted with their respective space-dependent constitutive equations as follows. Here we primarily use the orthotropic material as a standard anisotropic medium. The constitutive material properties written in Eq. 2.100.3 are reiterated here with spatial dependency.

$$\begin{Bmatrix} \sigma_{11} \\ \sigma_{22} \\ \sigma_{33} \\ \sigma_{12} \\ \sigma_{23} \\ \sigma_{31} \end{Bmatrix} = \begin{bmatrix} \mathbb{C}_{11}(x_p) & \mathbb{C}_{12}(x_p) & \mathbb{C}_{13}(x_p) & 0 & 0 & 0 \\ \mathbb{C}_{21}(x_p) & \mathbb{C}_{22}(x_p) & \mathbb{C}_{23}(x_p) & 0 & 0 & 0 \\ \mathbb{C}_{31}(x_p) & \mathbb{C}_{32}(x_p) & \mathbb{C}_{33}(x_p) & 0 & 0 & 0 \\ 0 & 0 & 0 & \mathbb{C}_{44}(x_p) & 0 & 0 \\ 0 & 0 & 0 & 0 & \mathbb{C}_{55}(x_p) & 0 \\ 0 & 0 & 0 & 0 & 0 & \mathbb{C}_{66}(x_p) \end{bmatrix} \begin{Bmatrix} \dfrac{\partial u_1}{\partial x_1} \\[2mm] \dfrac{\partial u_2}{\partial x_2} \\[2mm] \dfrac{\partial u_3}{\partial x_3} \\[2mm] \dfrac{1}{2}\left(\dfrac{\partial u_1}{\partial x_2} + \dfrac{\partial u_2}{\partial x_1} \right) \\[2mm] \dfrac{1}{2}\left(\dfrac{\partial u_2}{\partial x_3} + \dfrac{\partial u_3}{\partial x_2} \right) \\[2mm] \dfrac{1}{2}\left(\dfrac{\partial u_3}{\partial x_1} + \dfrac{\partial u_1}{\partial x_3} \right) \end{Bmatrix} \cdots \tag{3.113}$$

116   Metamaterials in Topological Acoustics

Substituting the material properties $\mathbb{C}_{ij}(x_p) \equiv C_{ij}$, $\rho(x_p) \equiv \rho$ and the strain expressions in terms of displacements, the wave equation will take the following forms:

a. Horizontally polarized waves with coupled u_1 and u_2 displacements

$$\frac{\partial}{\partial x_1}\left[C_{11}\frac{\partial u_1}{\partial x_1} + C_{12}\frac{\partial u_2}{\partial x_2}\right] + \frac{\partial}{\partial x_2}\left[\frac{1}{2}C_{44}\left(\frac{\partial u_1}{\partial x_2} + \frac{\partial u_2}{\partial x_1}\right)\right] + f_1 = \rho\frac{\partial^2 u_1}{\partial t^2} \quad (3.114.1)$$

$$\frac{\partial}{\partial x_1}\left[\frac{1}{2}C_{44}\left(\frac{\partial u_1}{\partial x_2} + \frac{\partial u_2}{\partial x_1}\right)\right] + \frac{\partial}{\partial x_2}\left[C_{21}\frac{\partial u_1}{\partial x_1} + C_{22}\frac{\partial u_2}{\partial x_2}\right] + f_2 = \rho\frac{\partial^2 u_2}{\partial t^2} \quad (3.114.2)$$

b. Vertically polarized wave with u_3 displacement only on the $x_1 - x_2$ plane

$$\frac{\partial}{\partial x_1}\left[\frac{1}{2}C_{66}\left(\frac{\partial u_3}{\partial x_1} + \frac{\partial u_1}{\partial x_3}\right)\right] + \frac{\partial}{\partial x_2}\left[\frac{1}{2}\mathbb{C}_{55}\left(\frac{\partial u_2}{\partial x_3} + \frac{\partial u_3}{\partial x_2}\right)\right] + f_3 = \rho\frac{\partial^2 u_3}{\partial t^2} \quad (3.115)$$

or

$$\frac{\partial}{\partial x_1}\left[\frac{1}{2}C_{66}\left(\frac{\partial u_3}{\partial x_1}\right)\right] + \frac{\partial}{\partial x_2}\left[\frac{1}{2}C_{55}\left(\frac{\partial u_3}{\partial x_2}\right)\right] + f_3 = \rho\frac{\partial^2 u_3}{\partial t^2} \quad (3.116)$$

When the wave is vertically polarized, please note that displacements $u_1 = u_2 = 0$.

c. In case of plane-stress condition $\sigma_{33} = 0$, and thus from Eq. 3.113 we can write

$$\frac{\partial u_3}{\partial x_3} = \frac{1}{C_{33}}\left[C_{31}\frac{\partial u_1}{\partial x_1} + C_{32}\frac{\partial u_2}{\partial x_2}\right] \quad (3.117)$$

Hence, σ_{11} and σ_{22} revise to

$$\sigma_{11} = C_{11}\frac{\partial u_1}{\partial x_1} + C_{12}\frac{\partial u_2}{\partial x_2} + \left[C_{31}\frac{\partial u_1}{\partial x_1} + C_{32}\frac{\partial u_2}{\partial x_2}\right] \quad (3.118.1)$$

$$\sigma_{22} = C_{21}\frac{\partial u_1}{\partial x_1} + C_{22}\frac{\partial u_2}{\partial x_2} + \left[C_{31}\frac{\partial u_1}{\partial x_1} + C_{32}\frac{\partial u_2}{\partial x_2}\right] \quad (3.118.2)$$

Wave equations under Plane-stress condition will be

$$\frac{\partial}{\partial x_1}\left[C_{11}\frac{\partial u_1}{\partial x_1} + C_{12}\frac{\partial u_2}{\partial x_2} + \frac{C_{13}}{C_{33}}\left[C_{31}\frac{\partial u_1}{\partial x_1} + C_{32}\frac{\partial u_2}{\partial x_2}\right]\right]$$
$$+ \frac{\partial}{\partial x_2}\left[\frac{1}{2}C_{44}\left(\frac{\partial u_1}{\partial x_2} + \frac{\partial u_2}{\partial x_1}\right)\right] + f_1 = \rho\frac{\partial^2 u_1}{\partial t^2}\cdots \quad (3.119.1)$$

Acoustics and Elastic Wave Propagation in Fluids and Anisotropic Solids   117

$$\frac{\partial}{\partial x_1}\left[\frac{1}{2}C_{44}\left(\frac{\partial u_1}{\partial x_2}+\frac{\partial u_2}{\partial x_1}\right)\right]+\frac{\partial}{\partial x_2}\left[C_{21}\frac{\partial u_1}{\partial x_1}+C_{22}\frac{\partial u_2}{\partial x_2}+\frac{C_{23}}{C_{33}}\left[C_{31}\frac{\partial u_1}{\partial x_1}+C_{32}\frac{\partial u_2}{\partial x_2}\right]\right]$$

$$+f_2=\rho\frac{\partial^2 u_2}{\partial t^2}\cdots \tag{3.119.2}$$

$$\frac{\partial}{\partial x_1}\left[\frac{1}{2}C_{66}\left(\frac{\partial u_3}{\partial x_1}+\frac{\partial u_1}{\partial x_3}\right)\right]+\frac{\partial}{\partial x_2}\left[\frac{1}{2}\mathbb{C}_{55}\left(\frac{\partial u_2}{\partial x_3}+\frac{\partial u_3}{\partial x_2}\right)\right]+f_3=\rho\frac{\partial^2 u_3}{\partial t^2}\cdots \tag{3.119.3}$$

3.2.3.3 Solution of Wave Equations for Nonhomogeneous Media

It is impossible to solve the above equations in nonhomogeneous media unless the spatial and/or temporal variation of the material properties are explicitly described. This requires the description of the constituents of the materials. In this book only periodic metamedia made of discrete elements or periodic arrangements of continuum superlattices are considered. Such media are made of either phononic crystals (PnCs) or metamaterials. The solution of wave equations in periodic media is discussed in later chapters.

For topological wave applications, it is necessary to revisit the above wave equations from the quantum mechanics perspective. Many topological behaviors are explained by quantum phenomena. Similar phenomena were also observed with acoustics and elastic waves. Thus, it is automatically realized that these waves must have quantum explanation for their topological behaviors. Under engineering curriculum, quantum mechanics is not formally taught to the students, and hence, sometimes it is hard for the students to understand this specific direction of research and contribute simultaneously. To overcome this challenge and to realize the quantum analogous applications of acoustic and elastic waves from the fundamentals of quantum mechanics, a chapter on quantum mechanics is introduced in this book. How quantum mechanics helps understand topological phenomena and helps tailor the above wave equations for quantum analog applications is discussed in the following chapters. Terminology related to the quantum effects, (e.g., quantum Hall effect (QHE), quantum spin Hall effect (QSHE), quantum valley Hall effect (QVHE), Dirac cone and Dirac-like cone behaviors, pseudo-spin, topological phases and geometric phase) are introduced to make a connection between quantum physics and engineering for seamless development of the concepts.

3.3   APPENDIX: UNDERSTANDING NABLA HAMILTONIAN OPERATIONS

Many operations used in this chapter are explained here in matrix form.

3.3.1   Nabla Hamiltonian

- ∇ is a nabla Hamiltonian operator or simply a gradient operator.

In matrix form for three-dimensional system

$$\nabla = \begin{bmatrix} \dfrac{\partial}{\partial x_1} & \dfrac{\partial}{\partial x_2} & \dfrac{\partial}{\partial x_3} \\[2ex] \dfrac{\partial}{\partial x_1} & \dfrac{\partial}{\partial x_2} & \dfrac{\partial}{\partial x_3} \\[2ex] \dfrac{\partial}{\partial x_1} & \dfrac{\partial}{\partial x_2} & \dfrac{\partial}{\partial x_3} \end{bmatrix} = \dfrac{\partial}{\partial x_i}\hat{\mathbf{e}}_{\mathbf{i}} \tag{A.3.1}$$

Here, three rows are added for three equations that are present in the governing differential equation. Refer to Eqs. A.2.1.1 through A.2.1.3 in Chapter 2.

Using a nabla operator when operated on the displacement field $\mathbf{u} = [u_1 \quad u_2 \quad u_3]^T \equiv u_i\hat{\mathbf{e}}_{\mathbf{i}}$, we get

$$\nabla.\mathbf{u} = \begin{bmatrix} \dfrac{\partial u_1}{\partial x_1} + \dfrac{\partial u_2}{\partial x_2} + \dfrac{\partial u_3}{\partial x_3} \\[2ex] \dfrac{\partial u_1}{\partial x_1} + \dfrac{\partial u_2}{\partial x_2} + \dfrac{\partial u_3}{\partial x_3} \\[2ex] \dfrac{\partial u_1}{\partial x_1} + \dfrac{\partial u_2}{\partial x_2} + \dfrac{\partial u_3}{\partial x_3} \end{bmatrix} = \begin{bmatrix} u_{i,i} \\[1ex] u_{i,i} \\[1ex] u_{i,i} \end{bmatrix} \text{ or simply } u_{i,i} \tag{A.3.2}$$

- $\nabla.(\nabla.\mathbf{u})$ is not the same as $\nabla(\nabla.\mathbf{u})$

$\nabla.(\nabla.\mathbf{u})$ means the divergence of the divergence of the displacement field. Substituting Eq. A.3.2 we get

$$\nabla.(\nabla.\mathbf{u}) = \nabla.(u_{i,i}) = \begin{bmatrix} \dfrac{\partial}{\partial x_1} & \dfrac{\partial}{\partial x_2} & \dfrac{\partial}{\partial x_3} \\[2ex] \dfrac{\partial}{\partial x_1} & \dfrac{\partial}{\partial x_2} & \dfrac{\partial}{\partial x_3} \\[2ex] \dfrac{\partial}{\partial x_1} & \dfrac{\partial}{\partial x_2} & \dfrac{\partial}{\partial x_3} \end{bmatrix} \cdot \begin{bmatrix} \dfrac{\partial u_1}{\partial x_1} + \dfrac{\partial u_2}{\partial x_2} + \dfrac{\partial u_3}{\partial x_3} \\[2ex] \dfrac{\partial u_1}{\partial x_1} + \dfrac{\partial u_2}{\partial x_2} + \dfrac{\partial u_3}{\partial x_3} \\[2ex] \dfrac{\partial u_1}{\partial x_1} + \dfrac{\partial u_2}{\partial x_2} + \dfrac{\partial u_3}{\partial x_3} \end{bmatrix} = \begin{bmatrix} \dfrac{\partial}{\partial x_1} & \dfrac{\partial}{\partial x_2} & \dfrac{\partial}{\partial x_3} \\[2ex] \dfrac{\partial}{\partial x_1} & \dfrac{\partial}{\partial x_2} & \dfrac{\partial}{\partial x_3} \\[2ex] \dfrac{\partial}{\partial x_1} & \dfrac{\partial}{\partial x_2} & \dfrac{\partial}{\partial x_3} \end{bmatrix} \cdot \begin{bmatrix} u_{i,i} \\[1ex] u_{i,i} \\[1ex] u_{i,i} \end{bmatrix}$$

$$= \begin{bmatrix} \dfrac{\partial}{\partial x_1}(u_{i,i}) + \dfrac{\partial}{\partial x_2}(u_{i,i}) + \dfrac{\partial}{\partial x_3}(u_{i,i}) \\[2ex] \dfrac{\partial}{\partial x_1}(u_{i,i}) + \dfrac{\partial}{\partial x_2}(u_{i,i}) + \dfrac{\partial}{\partial x_3}(u_{i,i}) \\[2ex] \dfrac{\partial}{\partial x_1}(u_{i,i}) + \dfrac{\partial}{\partial x_2}(u_{i,i}) + \dfrac{\partial}{\partial x_3}(u_{i,i}) \end{bmatrix} \tag{A.3.3}$$

Acoustics and Elastic Wave Propagation in Fluids and Anisotropic Solids   119

On the contrary, when there is no dot product symbol outside the bracket, e.g., $\nabla(\nabla.\mathbf{u})$, this means the gradient of the divergence of the displacement field. A gradient operator is like the nabla Hamiltonian operator, but only the trace elements are used to calculate the gradient. Hence, we write

$$
\nabla.\mathbf{I} =
\begin{bmatrix}
\dfrac{\partial}{\partial x_1} & \dfrac{\partial}{\partial x_2} & \dfrac{\partial}{\partial x_3} \\[2mm]
\dfrac{\partial}{\partial x_1} & \dfrac{\partial}{\partial x_2} & \dfrac{\partial}{\partial x_3} \\[2mm]
\dfrac{\partial}{\partial x_1} & \dfrac{\partial}{\partial x_2} & \dfrac{\partial}{\partial x_3}
\end{bmatrix}
.
\begin{bmatrix}
1 & 0 & 0 \\
0 & 1 & 0 \\
0 & 0 & 1
\end{bmatrix}
=
\begin{bmatrix}
\dfrac{\partial}{\partial x_1} & 0 & 0 \\[2mm]
0 & \dfrac{\partial}{\partial x_2} & 0 \\[2mm]
0 & 0 & \dfrac{\partial}{\partial x_3}
\end{bmatrix}
\tag{A.3.4}
$$

Thus, to avoid confusion, whenever we have a dot product symbol, the nabla Hamiltonian will take the matrix form in Eq. A.3.1. If the same operator indicates the gradient operation, then only the trace elements are used, as written in Eq. A.3.4. In some literature, $\nabla.\mathbf{I}$ is written as $\underline{\nabla}$. In some places, it might read $\underline{\nabla}(\nabla.\mathbf{u})$.

Hence, we can say the $\nabla.(\nabla.\mathbf{u})$ and $\nabla(\nabla.\mathbf{u})$ are not the same, but we can write,

$$
\nabla.\mathbf{I}(\nabla.\mathbf{u}) = \nabla(\nabla.\mathbf{u})
\tag{A.3.5.1}
$$

$$
\begin{bmatrix}
\dfrac{\partial}{\partial x_1} & \dfrac{\partial}{\partial x_2} & \dfrac{\partial}{\partial x_3} \\[2mm]
\dfrac{\partial}{\partial x_1} & \dfrac{\partial}{\partial x_2} & \dfrac{\partial}{\partial x_3} \\[2mm]
\dfrac{\partial}{\partial x_1} & \dfrac{\partial}{\partial x_2} & \dfrac{\partial}{\partial x_3}
\end{bmatrix}
\begin{bmatrix}
1 & 0 & 0 \\
0 & 1 & 0 \\
0 & 0 & 1
\end{bmatrix}
\begin{bmatrix}
\dfrac{\partial u_1}{\partial x_1} + \dfrac{\partial u_2}{\partial x_2} + \dfrac{\partial u_3}{\partial x_3} \\[2mm]
\dfrac{\partial u_1}{\partial x_1} + \dfrac{\partial u_2}{\partial x_2} + \dfrac{\partial u_3}{\partial x_3} \\[2mm]
\dfrac{\partial u_1}{\partial x_1} + \dfrac{\partial u_2}{\partial x_2} + \dfrac{\partial u_3}{\partial x_3}
\end{bmatrix}
\tag{A.3.5.2}
$$

$$
\begin{bmatrix}
\dfrac{\partial}{\partial x_1} & 0 & 0 \\[2mm]
0 & \dfrac{\partial}{\partial x_2} & 0 \\[2mm]
0 & 0 & \dfrac{\partial}{\partial x_3}
\end{bmatrix}
\begin{bmatrix}
\dfrac{\partial u_1}{\partial x_1} + \dfrac{\partial u_2}{\partial x_2} + \dfrac{\partial u_3}{\partial x_3} \\[2mm]
\dfrac{\partial u_1}{\partial x_1} + \dfrac{\partial u_2}{\partial x_2} + \dfrac{\partial u_3}{\partial x_3} \\[2mm]
\dfrac{\partial u_1}{\partial x_1} + \dfrac{\partial u_2}{\partial x_2} + \dfrac{\partial u_3}{\partial x_3}
\end{bmatrix}
=
\begin{bmatrix}
\dfrac{\partial}{\partial x_1}\left(u_{i,i}\right) \\[2mm]
\dfrac{\partial}{\partial x_2}\left(u_{i,i}\right) \\[2mm]
\dfrac{\partial}{\partial x_3}\left(u_{i,i}\right)
\end{bmatrix}
= \nabla(\nabla.\mathbf{u})
\tag{A.3.5.3}
$$

3.3.2 STRAIN MATRIX USING NABLA HAMILTONIAN FORM

- How $\nabla\mathbf{u}$ or $\nabla\mathbf{u}^T$ are interpreted in three-dimension governing differential equations

$$\nabla \mathbf{u} = \nabla . \mathbf{I}\left(\mathbf{u}^T\right) = \begin{bmatrix} \dfrac{\partial}{\partial x_1} \\[8pt] \dfrac{\partial}{\partial x_2} \\[8pt] \dfrac{\partial}{\partial x_3} \end{bmatrix} \begin{bmatrix} u_1 & u_2 & u_3 \end{bmatrix} = \begin{bmatrix} \dfrac{\partial u_1}{\partial x_1} & \dfrac{\partial u_2}{\partial x_1} & \dfrac{\partial u_3}{\partial x_1} \\[8pt] \dfrac{\partial u_1}{\partial x_2} & \dfrac{\partial u_2}{\partial x_2} & \dfrac{\partial u_3}{\partial x_2} \\[8pt] \dfrac{\partial u_1}{\partial x_3} & \dfrac{\partial u_2}{\partial x_3} & \dfrac{\partial u_3}{\partial x_3} \end{bmatrix} \tag{A.3.6.1}$$

In some literature, researchers use a notation for the outer product. An outer product is defined using the symbol $\otimes$. The above equation is the example of an outer product. Hence, we can also write

$$\nabla \mathbf{u} = \begin{bmatrix} \dfrac{\partial}{\partial x_1} \\[8pt] \dfrac{\partial}{\partial x_2} \\[8pt] \dfrac{\partial}{\partial x_3} \end{bmatrix} \begin{bmatrix} u_1 & u_2 & u_3 \end{bmatrix} = \underline{\nabla} \otimes \mathbf{u}^T \tag{A.3.6.2}$$

where $\mathbf{u}$ is a column vector as defined earlier. Transposing the matrix in Eq. A.3.6.1, we get

$$\nabla \mathbf{u}^T = \begin{bmatrix} \dfrac{\partial u_1}{\partial x_1} & \dfrac{\partial u_2}{\partial x_1} & \dfrac{\partial u_3}{\partial x_1} \\[8pt] \dfrac{\partial u_1}{\partial x_2} & \dfrac{\partial u_2}{\partial x_2} & \dfrac{\partial u_3}{\partial x_2} \\[8pt] \dfrac{\partial u_1}{\partial x_3} & \dfrac{\partial u_2}{\partial x_3} & \dfrac{\partial u_3}{\partial x_3} \end{bmatrix}^T = \begin{bmatrix} \dfrac{\partial u_1}{\partial x_1} & \dfrac{\partial u_1}{\partial x_2} & \dfrac{\partial u_1}{\partial x_3} \\[8pt] \dfrac{\partial u_2}{\partial x_1} & \dfrac{\partial u_2}{\partial x_2} & \dfrac{\partial u_2}{\partial x_3} \\[8pt] \dfrac{\partial u_3}{\partial x_1} & \dfrac{\partial u_3}{\partial x_2} & \dfrac{\partial u_3}{\partial x_3} \end{bmatrix} \tag{A.3.6.3}$$

Using an outer product symbol $\otimes$, we can rewrite Eq. A.3.6.3 as

$$\nabla \mathbf{u}^T = \begin{bmatrix} u_1 \\ u_2 \\ u_3 \end{bmatrix} \begin{bmatrix} \dfrac{\partial}{\partial x_1} & \dfrac{\partial}{\partial x_2} & \dfrac{\partial}{\partial x_3} \end{bmatrix} = \mathbf{u} \otimes \left(\nabla . \mathbf{I}\right)^T = \mathbf{u} \otimes \underline{\nabla}^T \tag{A.3.6.4}$$

Referring to the strain equations Eq. A.2.2.1 through A.2.2.6 in Chapter 2, we can write

$$\epsilon = \begin{bmatrix} e_{11} & e_{12} & e_{13} \\ e_{21} & e_{22} & e_{23} \\ e_{31} & e_{32} & e_{33} \end{bmatrix} = \frac{1}{2}\left(\nabla \mathbf{u} + \nabla \mathbf{u}^T\right) = \frac{1}{2} \begin{bmatrix} 2\dfrac{\partial u_1}{\partial x_1} & \left(\dfrac{\partial u_2}{\partial x_1} + \dfrac{\partial u_1}{\partial x_2}\right) & \left(\dfrac{\partial u_3}{\partial x_1} + \dfrac{\partial u_1}{\partial x_3}\right) \\[10pt] \left(\dfrac{\partial u_1}{\partial x_2} + \dfrac{\partial u_2}{\partial x_1}\right) & 2\dfrac{\partial u_2}{\partial x_2} & \left(\dfrac{\partial u_3}{\partial x_2} + \dfrac{\partial u_2}{\partial x_3}\right) \\[10pt] \left(\dfrac{\partial u_1}{\partial x_3} + \dfrac{\partial u_3}{\partial x_1}\right) & \left(\dfrac{\partial u_2}{\partial x_3} + \dfrac{\partial u_3}{\partial x_2}\right) & 2\dfrac{\partial u_3}{\partial x_3} \end{bmatrix}$$

$$\tag{A.3.7}$$

Acoustics and Elastic Wave Propagation in Fluids and Anisotropic Solids **121**

Alternatively, using the outer product symbol the strain matrix will read

$$\epsilon = \frac{1}{2}\left(\underline{\nabla}\otimes\mathbf{u}^T + \mathbf{u}\otimes\underline{\nabla}^T\right) \tag{A.3.8}$$

3.3.3 LAPLACIAN USING NABLA HAMILTONIAN

- **How $\nabla.(\nabla\mathbf{u})$ will read in matrix form**

Using Eq. 3.1 and 3.6.1, the above expression will be

$$\nabla.(\nabla\mathbf{u}) = \begin{bmatrix} \dfrac{\partial}{\partial x_1} & \dfrac{\partial}{\partial x_2} & \dfrac{\partial}{\partial x_3} \\[2mm] \dfrac{\partial}{\partial x_1} & \dfrac{\partial}{\partial x_2} & \dfrac{\partial}{\partial x_3} \\[2mm] \dfrac{\partial}{\partial x_1} & \dfrac{\partial}{\partial x_2} & \dfrac{\partial}{\partial x_3} \end{bmatrix} \cdot \begin{bmatrix} \dfrac{\partial u_1}{\partial x_1} & \dfrac{\partial u_2}{\partial x_1} & \dfrac{\partial u_3}{\partial x_1} \\[2mm] \dfrac{\partial u_1}{\partial x_2} & \dfrac{\partial u_2}{\partial x_2} & \dfrac{\partial u_3}{\partial x_2} \\[2mm] \dfrac{\partial u_1}{\partial x_3} & \dfrac{\partial u_2}{\partial x_3} & \dfrac{\partial u_3}{\partial x_3} \end{bmatrix}$$

$$= \begin{bmatrix} \left(\dfrac{\partial^2 u_1}{\partial x_1^2} + \dfrac{\partial^2 u_1}{\partial x_2^2} + \dfrac{\partial^2 u_1}{\partial x_3^2}\right) & \left(\dfrac{\partial^2 u_2}{\partial x_1^2} + \dfrac{\partial^2 u_2}{\partial x_2^2} + \dfrac{\partial^2 u_2}{\partial x_3^2}\right) & \left(\dfrac{\partial^2 u_3}{\partial x_1^2} + \dfrac{\partial^2 u_3}{\partial x_2^2} + \dfrac{\partial^2 u_3}{\partial x_3^2}\right) \\[3mm] \left(\dfrac{\partial^2 u_1}{\partial x_1^2} + \dfrac{\partial^2 u_1}{\partial x_2^2} + \dfrac{\partial^2 u_1}{\partial x_3^2}\right) & \left(\dfrac{\partial^2 u_2}{\partial x_1^2} + \dfrac{\partial^2 u_2}{\partial x_2^2} + \dfrac{\partial^2 u_2}{\partial x_3^2}\right) & \left(\dfrac{\partial^2 u_3}{\partial x_1^2} + \dfrac{\partial^2 u_3}{\partial x_2^2} + \dfrac{\partial^2 u_3}{\partial x_3^2}\right) \\[3mm] \left(\dfrac{\partial^2 u_1}{\partial x_1^2} + \dfrac{\partial^2 u_1}{\partial x_2^2} + \dfrac{\partial^2 u_1}{\partial x_3^2}\right) & \left(\dfrac{\partial^2 u_2}{\partial x_1^2} + \dfrac{\partial^2 u_2}{\partial x_2^2} + \dfrac{\partial^2 u_2}{\partial x_3^2}\right) & \left(\dfrac{\partial^2 u_3}{\partial x_1^2} + \dfrac{\partial^2 u_3}{\partial x_2^2} + \dfrac{\partial^2 u_3}{\partial x_3^2}\right) \end{bmatrix}$$

$$\tag{A.3.9}$$

If the matrix in Eq. A.3.9 is multiplied with an identity matrix, we get

$$\nabla.(\nabla\mathbf{u}).\mathbf{I} = \begin{bmatrix} \left(\dfrac{\partial^2 u_1}{\partial x_1^2} + \dfrac{\partial^2 u_1}{\partial x_2^2} + \dfrac{\partial^2 u_1}{\partial x_3^2}\right) & 0 & 0 \\[3mm] 0 & \left(\dfrac{\partial^2 u_2}{\partial x_1^2} + \dfrac{\partial^2 u_2}{\partial x_2^2} + \dfrac{\partial^2 u_2}{\partial x_3^2}\right) & 0 \\[3mm] 0 & 0 & \left(\dfrac{\partial^2 u_3}{\partial x_1^2} + \dfrac{\partial^2 u_3}{\partial x_2^2} + \dfrac{\partial^2 u_3}{\partial x_3^2}\right) \end{bmatrix}$$

$$\tag{A.3.10}$$

Thus, it can be concluded that the above expression is a Laplace operator.

$$\nabla.(\nabla\mathbf{u}).\mathbf{I} = \nabla^2\mathbf{u} \tag{A.3.11}$$

122 Metamaterials in Topological Acoustics

where $\nabla^2 = \dfrac{\partial^2}{\partial x_1^2} + \dfrac{\partial^2}{\partial x_2^2} + \dfrac{\partial^2}{\partial x_3^2}$ is called the Laplace operator. It will be used in Chapter 4 whenever a second-order spatial derivative appears.

3.3.4 NABLA OPERATION ON STRESS MATRIX AND SPATIALLY VARYING MATERIAL CONSTANTS

- How the nabla operator in the governing equation $\nabla.\sigma = \rho(\mathbf{x})\ddot{\mathbf{u}}$ will read in matrix form

$$
\nabla.\sigma =
\begin{bmatrix}
\dfrac{\partial}{\partial x_1} & \dfrac{\partial}{\partial x_2} & \dfrac{\partial}{\partial x_3} \\[2ex]
\dfrac{\partial}{\partial x_1} & \dfrac{\partial}{\partial x_2} & \dfrac{\partial}{\partial x_3} \\[2ex]
\dfrac{\partial}{\partial x_1} & \dfrac{\partial}{\partial x_2} & \dfrac{\partial}{\partial x_3}
\end{bmatrix}
\cdot
\begin{bmatrix}
\sigma_{11} & \sigma_{21} & \sigma_{31} \\
\sigma_{12} & \sigma_{22} & \sigma_{32} \\
\sigma_{13} & \sigma_{23} & \sigma_{33}
\end{bmatrix}
=
\begin{bmatrix}
\dfrac{\partial \sigma_{11}}{\partial x_1} + \dfrac{\partial \sigma_{12}}{\partial x_2} + \dfrac{\partial \sigma_{13}}{\partial x_3} \\[2ex]
\dfrac{\partial \sigma_{21}}{\partial x_1} + \dfrac{\partial \sigma_{22}}{\partial x_2} + \dfrac{\partial \sigma_{23}}{\partial x_3} \\[2ex]
\dfrac{\partial \sigma_{31}}{\partial x_1} + \dfrac{\partial \sigma_{32}}{\partial x_2} + \dfrac{\partial \sigma_{33}}{\partial x_3}
\end{bmatrix}
\qquad \text{(A.3.12)}
$$

The expressions are similar to those before derived before and written in Eqs. A.2.1.1 through A.2.1.3.

- The gradient of material coefficients found by operating the nabla Hamiltonian on any material constant as needed

Let's take the nabla operator $(\underline{\nabla})$ of a spatially varying material constant. For example, the spatially varying second Lamé constant $\mathrm{m}(\mathbf{x}) = \mu$ is used herein.

$$
\underline{\nabla}\mu = (\nabla.\mathbf{I})\mu =
\begin{bmatrix}
\dfrac{\partial \mu}{\partial x_1} & \dfrac{\partial \mu}{\partial x_2} & \dfrac{\partial \mu}{\partial x_3} \\[2ex]
\dfrac{\partial \mu}{\partial x_1} & \dfrac{\partial \mu}{\partial x_2} & \dfrac{\partial \mu}{\partial x_3} \\[2ex]
\dfrac{\partial \mu}{\partial x_1} & \dfrac{\partial \mu}{\partial x_2} & \dfrac{\partial \mu}{\partial x_3}
\end{bmatrix}
\cdot
\begin{bmatrix}
1 & 0 & 0 \\
0 & 1 & 0 \\
0 & 0 & 1
\end{bmatrix}
=
\begin{Bmatrix}
\dfrac{\partial \mu}{\partial x_1} \\[2ex]
\dfrac{\partial \mu}{\partial x_2} \\[2ex]
\dfrac{\partial \mu}{\partial x_3}
\end{Bmatrix}
\qquad \text{(A.3.13)}
$$

- Curl of the displacement field $\underline{\nabla} \times \mathbf{u}$

$$
\underline{\nabla} \times \mathbf{u} =
\begin{bmatrix}
\widehat{\mathbf{e}}_1 & \widehat{\mathbf{e}}_2 & \widehat{\mathbf{e}}_3 \\[1ex]
\dfrac{\partial}{\partial x_1} & \dfrac{\partial}{\partial x_2} & \dfrac{\partial}{\partial x_3} \\[2ex]
u_1 & u_2 & u_3
\end{bmatrix}
=
\begin{bmatrix}
\left(\dfrac{\partial u_3}{\partial x_2} - \dfrac{\partial u_2}{\partial x_3} \right) \\[2ex]
-\left(\dfrac{\partial u_3}{\partial x_1} - \dfrac{\partial u_1}{\partial x_3} \right) \\[2ex]
\left(\dfrac{\partial u_2}{\partial x_1} - \dfrac{\partial u_1}{\partial x_2} \right)
\end{bmatrix}
\qquad \text{(A.3.14)}
$$

Acoustics and Elastic Wave Propagation in Fluids and Anisotropic Solids 123

- Cross product between the gradient of the material constant $\underline{\nabla}\mu$ and the curl of the displacement field $\underline{\nabla}\times\mathbf{u}$ to explore the transverse components of the wave modes

$$\underline{\nabla}\mu\times\underline{\nabla}\times\mathbf{u}=\begin{bmatrix} \widehat{\mathbf{e}}_1 & \widehat{\mathbf{e}}_2 & \widehat{\mathbf{e}}_3 \\ \dfrac{\partial\mu}{\partial x_1} & \dfrac{\partial\mu}{\partial x_2} & \dfrac{\partial\mu}{\partial x_3} \\ \left(\dfrac{\partial u_3}{\partial x_2}-\dfrac{\partial u_2}{\partial x_3}\right) & \left(\dfrac{\partial u_1}{\partial x_3}-\dfrac{\partial u_3}{\partial x_1}\right) & \left(\dfrac{\partial u_2}{\partial x_1}-\dfrac{\partial u_1}{\partial x_2}\right) \end{bmatrix}$$

$$=\begin{bmatrix} \left[\dfrac{\partial\mu}{\partial x_2}\dfrac{\partial u_2}{\partial x_1}-\dfrac{\partial\mu}{\partial x_2}\dfrac{\partial u_1}{\partial x_2}-\dfrac{\partial\mu}{\partial x_3}\dfrac{\partial u_1}{\partial x_3}+\dfrac{\partial\mu}{\partial x_3}\dfrac{\partial u_3}{\partial x_1}\right] \\ -\left[\dfrac{\partial\mu}{\partial x_1}\dfrac{\partial u_2}{\partial x_1}-\dfrac{\partial\mu}{\partial x_1}\dfrac{\partial u_1}{\partial x_2}-\dfrac{\partial\mu}{\partial x_3}\dfrac{\partial u_3}{\partial x_2}+\dfrac{\partial\mu}{\partial x_3}\dfrac{\partial u_2}{\partial x_3}\right] \\ \left[\dfrac{\partial\mu}{\partial x_1}\dfrac{\partial u_1}{\partial x_3}-\dfrac{\partial\mu}{\partial x_1}\dfrac{\partial u_3}{\partial x_1}-\dfrac{\partial\mu}{\partial x_2}\dfrac{\partial u_3}{\partial x_2}+\dfrac{\partial\mu}{\partial x_2}\dfrac{\partial u_2}{\partial x_3}\right] \end{bmatrix} \quad\text{(A.3.15)}$$

Referring to Eq. 3.103 and two components related to the gradient of the material, constant $\underline{\nabla}\mu$ are collected and written as follows with two new variable names.

$$\underline{\nabla}\mu^1=\begin{bmatrix} \left(\dfrac{\partial\mu}{\partial x_1}\dfrac{\partial u_1}{\partial x_1}+\dfrac{\partial\mu}{\partial x_2}\dfrac{\partial u_1}{\partial x_2}+\dfrac{\partial\mu}{\partial x_3}\dfrac{\partial u_1}{\partial x_3}\right) \\ \left(\dfrac{\partial\mu}{\partial x_1}\dfrac{\partial u_2}{\partial x_1}+\dfrac{\partial\mu}{\partial x_2}\dfrac{\partial u_2}{\partial x_2}+\dfrac{\partial\mu}{\partial x_3}\dfrac{\partial u_2}{\partial x_3}\right) \\ \left(\dfrac{\partial\mu}{\partial x_1}\dfrac{\partial u_3}{\partial x_1}+\dfrac{\partial\mu}{\partial x_2}\dfrac{\partial u_3}{\partial x_2}+\dfrac{\partial\mu}{\partial x_3}\dfrac{\partial u_3}{\partial x_3}\right) \end{bmatrix} \quad\text{(A.3.16.1)}$$

$$\underline{\nabla}\mu^2=\begin{bmatrix} \left(\dfrac{\partial\mu}{\partial x_1}\dfrac{\partial u_1}{\partial x_1}+\dfrac{\partial\mu}{\partial x_2}\dfrac{\partial u_2}{\partial x_1}+\dfrac{\partial\mu}{\partial x_3}\dfrac{\partial u_3}{\partial x_1}\right) \\ \left(\dfrac{\partial\mu}{\partial x_1}\dfrac{\partial u_1}{\partial x_2}+\dfrac{\partial\mu}{\partial x_2}\dfrac{\partial u_2}{\partial x_2}+\dfrac{\partial\mu}{\partial x_3}\dfrac{\partial u_3}{\partial x_2}\right) \\ \left(\dfrac{\partial\mu}{\partial x_1}\dfrac{\partial u_1}{\partial x_3}+\dfrac{\partial\mu}{\partial x_2}\dfrac{\partial u_2}{\partial x_3}+\dfrac{\partial\mu}{\partial x_3}\dfrac{\partial u_3}{\partial x_3}\right) \end{bmatrix} \quad\text{(A.3.16.2)}$$

With careful observation, it can be seen that

$$
\underline{\nabla}\mu^2 - \underline{\nabla}\mu^1 = \begin{bmatrix} \left(\dfrac{\partial \mu}{\partial x_1}\dfrac{\partial u_1}{\partial x_1} + \dfrac{\partial \mu}{\partial x_2}\dfrac{\partial u_2}{\partial x_1} + \dfrac{\partial \mu}{\partial x_3}\dfrac{\partial u_3}{\partial x_1} \right) - \left(\dfrac{\partial \mu}{\partial x_1}\dfrac{\partial u_1}{\partial x_1} + \dfrac{\partial \mu}{\partial x_2}\dfrac{\partial u_1}{\partial x_2} + \dfrac{\partial \mu}{\partial x_3}\dfrac{\partial u_1}{\partial x_3} \right) \\[2mm] \left(\dfrac{\partial \mu}{\partial x_1}\dfrac{\partial u_1}{\partial x_2} + \dfrac{\partial \mu}{\partial x_2}\dfrac{\partial u_2}{\partial x_2} + \dfrac{\partial \mu}{\partial x_3}\dfrac{\partial u_3}{\partial x_2} \right) - \left(\dfrac{\partial \mu}{\partial x_1}\dfrac{\partial u_2}{\partial x_1} + \dfrac{\partial \mu}{\partial x_2}\dfrac{\partial u_2}{\partial x_2} + \dfrac{\partial \mu}{\partial x_3}\dfrac{\partial u_2}{\partial x_3} \right) \\[2mm] \left(\dfrac{\partial \mu}{\partial x_1}\dfrac{\partial u_1}{\partial x_3} + \dfrac{\partial \mu}{\partial x_2}\dfrac{\partial u_2}{\partial x_3} + \dfrac{\partial \mu}{\partial x_3}\dfrac{\partial u_3}{\partial x_3} \right) - \left(\dfrac{\partial \mu}{\partial x_1}\dfrac{\partial u_3}{\partial x_1} + \dfrac{\partial \mu}{\partial x_2}\dfrac{\partial u_3}{\partial x_2} + \dfrac{\partial \mu}{\partial x_3}\dfrac{\partial u_3}{\partial x_3} \right) \end{bmatrix}
$$

$$(A.3.16.3)$$

will result to the exact expression obtained from Eq. A.3.15. Hence, it can be said that

$$
\underline{\nabla}\mu^2 - \underline{\nabla}\mu^1 = \underline{\nabla}\mu \times \underline{\nabla} \times \mathbf{u} \tag{A.3.16.4}
$$

- Dot product between the gradient of the material constant $\underline{\nabla}\mu$ and the nabla Hamiltonian

$$
\left(\nabla\mu.\underline{\nabla} \right) = \begin{bmatrix} \dfrac{\partial \mu}{\partial x_1} & \dfrac{\partial \mu}{\partial x_2} & \dfrac{\partial \mu}{\partial x_3} \\[2mm] \dfrac{\partial \mu}{\partial x_1} & \dfrac{\partial \mu}{\partial x_2} & \dfrac{\partial \mu}{\partial x_3} \\[2mm] \dfrac{\partial \mu}{\partial x_1} & \dfrac{\partial \mu}{\partial x_2} & \dfrac{\partial \mu}{\partial x_3} \end{bmatrix} . \begin{Bmatrix} \dfrac{\partial}{\partial x_1} \\[2mm] \dfrac{\partial}{\partial x_2} \\[2mm] \dfrac{\partial}{\partial x_3} \end{Bmatrix} = \begin{bmatrix} \left(\dfrac{\partial \mu}{\partial x_1}\dfrac{\partial}{\partial x_1} + \dfrac{\partial \mu}{\partial x_2}\dfrac{\partial}{\partial x_2} + \dfrac{\partial \mu}{\partial x_3}\dfrac{\partial}{\partial x_3} \right) \\[2mm] \left(\dfrac{\partial \mu}{\partial x_1}\dfrac{\partial}{\partial x_1} + \dfrac{\partial \mu}{\partial x_2}\dfrac{\partial}{\partial x_2} + \dfrac{\partial \mu}{\partial x_3}\dfrac{\partial}{\partial x_3} \right) \\[2mm] \left(\dfrac{\partial \mu}{\partial x_1}\dfrac{\partial}{\partial x_1} + \dfrac{\partial \mu}{\partial x_2}\dfrac{\partial}{\partial x_2} + \dfrac{\partial \mu}{\partial x_3}\dfrac{\partial}{\partial x_3} \right) \end{bmatrix}
$$

$$(A.3.17.1)$$

$$
\left(\nabla\mu.\underline{\nabla} \right)\mathbf{u} = \begin{bmatrix} \left(\dfrac{\partial \mu}{\partial x_1}\dfrac{\partial}{\partial x_1} + \dfrac{\partial \mu}{\partial x_2}\dfrac{\partial}{\partial x_2} + \dfrac{\partial \mu}{\partial x_3}\dfrac{\partial}{\partial x_3} \right) \\[2mm] \left(\dfrac{\partial \mu}{\partial x_1}\dfrac{\partial}{\partial x_1} + \dfrac{\partial \mu}{\partial x_2}\dfrac{\partial}{\partial x_2} + \dfrac{\partial \mu}{\partial x_3}\dfrac{\partial}{\partial x_3} \right) \\[2mm] \left(\dfrac{\partial \mu}{\partial x_1}\dfrac{\partial}{\partial x_1} + \dfrac{\partial \mu}{\partial x_2}\dfrac{\partial}{\partial x_2} + \dfrac{\partial \mu}{\partial x_3}\dfrac{\partial}{\partial x_3} \right) \end{bmatrix} \begin{bmatrix} u_1 \\[2mm] u_2 \\[2mm] u_3 \end{bmatrix} = \begin{bmatrix} \left(\dfrac{\partial \mu}{\partial x_1}\dfrac{\partial u_1}{\partial x_1} + \dfrac{\partial \mu}{\partial x_2}\dfrac{\partial u_1}{\partial x_2} + \dfrac{\partial \mu}{\partial x_3}\dfrac{\partial u_1}{\partial x_3} \right) \\[2mm] \left(\dfrac{\partial \mu}{\partial x_1}\dfrac{\partial u_2}{\partial x_1} + \dfrac{\partial \mu}{\partial x_2}\dfrac{\partial u_2}{\partial x_2} + \dfrac{\partial \mu}{\partial x_3}\dfrac{\partial u_2}{\partial x_3} \right) \\[2mm] \left(\dfrac{\partial \mu}{\partial x_1}\dfrac{\partial u_3}{\partial x_1} + \dfrac{\partial \mu}{\partial x_2}\dfrac{\partial u_3}{\partial x_2} + \dfrac{\partial \mu}{\partial x_3}\dfrac{\partial u_3}{\partial x_3} \right) \end{bmatrix}
$$

$$(A.3.17.2)$$

It is easily comparable to say

$$
\left(\nabla\mu.\underline{\nabla} \right)\mathbf{u} = \underline{\nabla}\mu^1 \tag{A.3.17.3}
$$

3.4 SUMMARY

Wave propagation in fluid and elastic homogeneous and nonhomogeneous media are discussed with easy-to-follow mathematical steps and physical examples. Wave polarity and its effect on wave propagation is the key for many topological behaviors, and thus it is necessary to understand the polarity of waves in anisotropic media. Additional conditions that cause abnormal polarity in material are discussed. To explore anomalous polarity by the design of new metamaterials for specific application, this chapter provides both physical and mathematical insight to the reader for further research and exploration.

REFERENCES

1. Haberman, R., *Elementary Applied Partial Differential Equations with Fourier Series and Boundary Value Problems.* 3rd ed. 1997, New Jersey: Prentice Hall.
2. Achenbach, J.D., *Wave Propagation in Elastic Solids.* 1999, New York: Elsevier.
3. Graff, K.F., *Wave Motion in Elastic Solids.* 1975, New York: Dover Publication.
4. Banerjee, S., Leckey, A.C.C., *Computational Nondestructive Evaluation Handbook.* 2020, Boca Raton, FL: CRC Press Taylor and Francis Group. 560.
5. Auld, B.A., *Acoustic Fields and Waves in Solids.* 1990, Hoboken, NJ: Wiley.
6. Helbig, K., Carcione, J.M., Anomalous polarization in anisotropic media. European Journal of Mechanics A/Solids, 2009, **28**(4): p. 704–711.
7. Helbig, K., Schoenberg, M., Anomalous polarization of elastic waves in transversely isotropic media. Journal of the Acoustical Society of America, 1987, **81**(5): p. 1235–1245.
8. Banerjee, S., Kundu, T., *Advanced application of distributed point source method: ultrasonic field modeling in solid media*, in *DPSM for Modeling Engineering Problems*, T. Kundu, D. Placko, Editors. 2007, Hoboken, New Jersey, USA: John & Willey Publication.
9. Lee, H.J., Lee, J.-R., Moon, S.H., Je, T.-J., Jeon, E.-chae, Kim, K., Kim, Y.Y., *Off-centered double-slit metamaterial for elastic wave polarization anomaly.* Scientific Reports, 2017, **7**(15378).
10. Zheng, L.-Y., et al., Acoustic cloaking by a near-zero-index phononic crystal. *Applied Physics Letters*, 2014, **104**(16): p. 161904.
11. Hajian, H., Ozbay, E., Caglayan, H., Enhanced transmission and beaming via a zero-index photonic crystal. Applied Physics Letters, 2016, **109**(3): p. 031105.
12. Geng, Z.-G., et al., Acoustic delay-line filters based on largely distorted topological insulators. Applied Physics Letters, 2018, **113**(3): p. 033503.
13. Caballero, D., et al., *Large two-dimensional sonic band gaps.* Physical Review E, 1999, **60**(6): p. R6316.
14. Lai, Y., Zhang, X., Zhang, Z.-Q., *Large sonic band gaps in 12-fold* quasicrystals. Journal of Applied Physics, 2002, **91**(9): p. 6191–6193.
15. Li, X., et al., *Large acoustic band gaps created by rotating square rods in two-dimensional periodic composites.* Journal of Applied Physics D, 2002, **36**(1): p. L15.
16. Wu, F., Liu, Z., Liu, Y.J.P.R.E., *Splitting and tuning characteristics of the point defect modes in two-dimensional* phononic crystals. Physical Review E: Statistical, Nonlinear, and Soft Matter Physics, 2004, **69**(6): p. 066609.
17. Song, A., et al., *Band* structures *in a* two-dimensional phononic crystal with rotational multiple scatterers. International Journal of Modern Physics B, 2017, **31**(6): p. 1750038.
18. Hyun, J., et al., *Systematic realization of double-zero-index* phononic crystals with hard inclusions. Scientific Reports, 2018, **8**(1): p. 7288.

4 Electromagnetic Wave Propagation

4.1 FIELD AND FIELD THEORIES

The concept of field was first introduced by Michael Faraday in 1849 [1]. Despite having only basic school education and being a self-trained scientist, Faraday was one of the most influential scientists of the 19th century, at least in my opinion. Concept of field was in the making since the period of Sir Isaac Newton. Classical mechanics has the concept of a discrete system and its behavior using the position and momentum of an object. However, when the degrees of freedom of a system are very large, then the concept of field is inevitable. *Field is the representation of any scalar, vector, or tensor quantity over a permeable space that is invisible but can be felt, measured, or altered by or through external causes.* From the classical point of view, field is a map of a quantity (tensor of zero to any degree). In modern physics, this concept is no different.

The concept of the field fundamentally started in the 19th century with electromagnetism. Electric fields and magnetic fields are coupled, and it was identified that they influence each other through an induction described by Faraday [2], called Faraday's induction law. When a charge is placed in an environment with existing charges, the new charge must feel the force according to Coulomb's law [1]. If the charge is moved from one point to another, the force acting on the free charge changes and gradually decays as it goes away from the source, following a distance square law [3]. A similar rule applies to the gravitational force field. The force is real, but it is invisible. An invisible push over the entire space is created not only by the charges but also by the magnetic materials. This fascinated Faraday, and he imagined that the entire space is filled with small force vectors [1]. This imagination triggered the concept of the field theory. The concept of electric fields and magnetic fields were first conceived in physics. Like an electrical charge, if there is magnetic monopole moving in a magnetic field, the monopole also experiences the force, called magnetic force. It is inevitable that an electrical charge in an electric field and a magnetic pole in a magnetic field will always experience a force; however, Faraday also found that there is an induction effect between these two fields [1]. That means if a magnetic pole is placed in an electric field, or an electric charge is placed in a magnetic field with a differential motion, they will also feel some degree of additional force that could be derived from the induction law [3].

In general, when particles are moving in a space, they create a trajectory of motion due the force they feel. Now, if there are several trajectories present in a space, representing the motion of all the particles in a system, the small tangential vectors, or the arrows signify the force along the trajectories, creating a map of a playground where all the particles are playing in the field. Such a field is called a vector field, and all mathematical treatments related to vector calculus must be valid. Motion of a particle, deformations, velocities, or the forces acting on a particle are

DOI: 10.1201/9781003225751-4

127

the vector quantities. They all create the vector field due to their nonstationary or non-static operations. Similarly, temperature, density, pressure, potential, etc. – the scalar quantities distributed over a space – are also the examples of fields. Continuum mechanics, discussed in Chapter 2, results in a 'continuum field' when stress, strains, and material properties (tensors) are presented over an entire space of interest. A continuum field gives the stress field or the strain field distribution on the entire structure or the material of interest. Stress or strain and their respective field distribution may change from point to point and even change over time at a single point when the wave propagates (Chapter 3).

Figure 4.1 shows a few examples of fields that we frequently encounter in science and engineering. Wind velocity around an airfoil, fluid flow inside a venturi meter, a magnetic field due to a magnet, an electric field due to charges, etc. are the examples of fields. A field in most cases is manifested by the local motion of a particle or the response of a local particle due to a local force. A field is a spatial or spatio-temporal variation of the direction of the force itself. Irrespective of being scalar, vector, or tensor quantities, the field is a map of local parameters with direction and/or magnitudes as shown in Fig. 4.1.

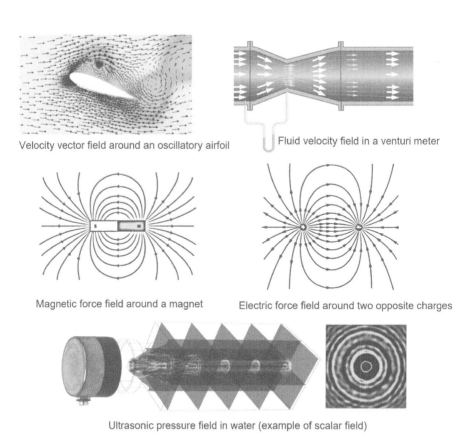

FIGURE 4.1 A few example of fields as written below the respective figures.

Electromagnetic Wave Propagation   129

Like classical mechanics, quantum mechanics (discussed in the next chapter) also uses the concept of fields. Quantization of classical electrodynamics or quantum electrodynamics [4] uses the concept of quantum fields. The same is true for quantum chromodynamics [5]. A quantum field is comparatively a modern concept in quantum mechanics due to the recognition of the presence of angular momentum and energy of the electromagnetic field.

Field theory is a theory to describe the field within a physical framework. How a field can dynamically change (spatially or spatio-temporally) due to the internal or external physical effects is generally described under the framework of a field theory. A field quantity, or the dependent variable, changes in space and time with respect to the internal independent physical quantities. Irrespective of a classical or quantum mechanical system, the physical phenomena related to fields are predictable by writing the Lagrangian function or the Hamiltonian principle of the system (refer to Chapter 2) and treating them with an infinite number of degrees of freedom. This will result a field theory. The resulting field theories are referred to as classical or quantum field theories. A theory is a set of rules, laws, and processes that predictively describe the dynamics of a field. Hence, field and field theories are mutually inclusive in physics.

4.2 ELECTRIC AND MAGNETIC FIELDS

Current through a wire is essentially the rate of flow of charges through the wire. In 1820 Danish physicist Hans Christian Øersted [1, 2] discovered that an electric current flowing through a wire caused deflection of the needle of a compass next to a wire carrying a current. The deflection was highest when the needle was perpendicular to the wire. He concluded that when current flows through a wire, the rate of change of charges causes a magnetic field ($\mathbf{H}$) to develop around the wire. Later, Andre-Marie Ampere in 1826 [1, 3] quantified the magnetic field generated by a wire carrying an electric current, experimentally observed by Øersted. Alternatively, if an electrical wire is moved with a certain velocity near a magnet, the wire starts to carry current due to the induction. This signifies that if the magnetic field is changed over time, the rate of change of the magnetic field can cause generation of electric field in a conductive wire. This phenomenon was first reported by Faraday in 1831 conducting an experiment with a metal ring with two independent wires wrapped around on either side. This was an ingenious idea, which changed the course of physics in a completely new direction, giving birth to electromagnetism. The induction phenomenon was also independently discovered by Joseph Henry in 1832 [2]. Please note that the concept of mutual interaction between electric and magnetic fields came much later, in 1849, at which time it was rejected. The concept of field by Faraday was resurrected by James Clerk Maxwell in 1861 [2] giving proper mathematical framework using Oliver Heaviside's [2] mathematical formulation of Faradays law. In the following subsections, the mathematical treatment is presented not in chronological order but in a conceptual order of use, the way Maxwell's equations are presented in today's literature on electromagnetism.

4.2.1 Conservative and Nonconservative Fields

Although electric ($\mathbf{E}$) and magnetic fields ($\mathbf{H}$) are complementary in nature, it was found that an electric field is conservative, but a magnetic field is not. As a conservative field, an electric field ($\mathbf{E}(\mathbf{x})$), which is a vector field, can be presented as a gradient of a scalar field of electric potential ($V(\mathbf{x})$). However, as a nonconservative field, the magnetic field ($\mathbf{H}(\mathbf{x})$) can be presented only as the curl of another vector field and will be

$$\mathbf{E}(x_j) = -\nabla V(x_j); \ \mathbf{H}(x_j) = \nabla \times \mathbf{A}(x_j) \tag{4.1}$$

4.2.2 Coulomb's Law and Gauss's Law

In 1785 Charles Augustine de Coulomb [3] described the force felt by a charge when a similar charge was placed closed to it. Using a torsional wire, measuring its deformation, Coulomb calculated the repulsive electrostatic force and discovered that the square law of gravitation force is also valid when two charges are present in space. Although the concept of field theory was not present during Coulomb's time, here the field theory is used for easier description of the Coulomb's law. The electric field generated by a charge q at rest at any point in space at a radial distance of r from the charge q can be written as

$$\mathbf{E}(r) = \frac{q}{4\pi\epsilon_0 r^2} \hat{\mathbf{e}}_r \tag{4.2}$$

where ϵ_0 is the permittivity of the free space or vacuum. Along the radial direction $\hat{\mathbf{e}}_r$ is the unit normal vector (Fig. A2.2). Referring to the spherical coordinate system in Fig. A2.2, the electric field in Cartesian coordinate system will read

$$\mathbf{E}(x_j) = \sqrt{\frac{q}{12\pi\epsilon_0 x_i^2}} \ \hat{\mathbf{e}}_1 + \sqrt{\frac{q}{12\pi\epsilon_0 x_i^2}} \hat{\mathbf{e}}_2 + \sqrt{\frac{q}{12\pi\epsilon_0 x_i^2}} \hat{\mathbf{e}}_3 \tag{4.3}$$

where index notation is implied. If the charge is infinitesimal with a spatial charge density of ρ, the total charge q can be written as the volume integral of the charge density.

$$q = \iiint_\Omega \rho \, d\Omega \tag{4.4}$$

On the other hand, according to the law by Carl Fredrick Gauss [3], proposed in 1853, the surface integral of the electric field is the result of the total charge present in the space enclosed by the surface. Mathematically this statement could be written as

$$\iint_\Gamma \mathbf{E}.d\vec{\Gamma} = \frac{1}{\epsilon_0} \iiint_\Omega \rho \, d\Omega \tag{4.5}$$

Electromagnetic Wave Propagation

where ρ is the charge density and $d\vec{\Gamma}$ is the elemental surface vector with area and direction (see Appendix Fig. A4.1). Applying the Gauss divergence theorem (transferring the surface integral to a volume integral) discussed in Appendix, Gauss's law be

$$\iiint_\Omega \nabla.\mathbf{E}\, d\Omega = \frac{1}{\epsilon_0} \iiint_\Omega \rho\, d\Omega;\ \nabla.\mathbf{E} = \frac{\rho}{\epsilon_0} \tag{4.6}$$

Similarly, Gauss's law for a magnetic field will not have any charge in an enclosed space, and the surface integral of the magnetic field will read

$$\iint_\Gamma \mathbf{H}.d\vec{\Gamma} = 0 \tag{4.7}$$

Applying the Gauss divergence theorem,

$$\iiint_\Omega \nabla.\mathbf{H}\, d\Omega = 0 \text{ or } \nabla.\mathbf{H} = 0 \tag{4.8}$$

If in a space there is no charge, then the divergence of both the electric and magnetic field will be equal to zero.

$$\nabla.\mathbf{E} = 0;\ \nabla.\mathbf{H} = 0 \tag{4.9}$$

These two equations derived from the Gauss's law are considered an integral part of the Maxwell's equations.

4.2.3 Ampere's Law and Ampere-Maxwell Equation

The magnetic field ($\mathbf{H}$) presented in the previous sections is the result of an electrical wire carrying current, proposed by Øersted. However, if a material of interest is placed in this magnetic field $\mathbf{H}$, some degrees of magnetic field lines will be carried by the material. That means that an external magnetic field will arise in the material due to the external electric field, not due to its intrinsic property. This magnetic intensity is measured by the magnetic flux density ($\mathbf{B}$) in the material. The intensity of the magnetic field inside the material is either enhanced or worsened by the magnetic constitutive property. For example, the induced magnetic field ($\mathbf{B}$) inside an iron rod will be significantly higher than the magnetic field induced inside a wooden block. Iron is a magnetic material, and thus the magnetic permeability is high. In a free space, or in vacuum, it is obvious that the generated magnetic field ($\mathbf{H}$) by a wire carrying current and induced magnetic field ($\mathbf{B}$) in the vacuum are equal, and their ratio should be equal to 1. The induced magnetic field is also called the magnetic flux density as mentioned above. In SI units, due to specific respective dimensionality of $\mathbf{B}$ and $\mathbf{H}$ fields, the constitutive relation for vacuum can be written as $\mathbf{B} = \mu_0\mathbf{H}$, where μ_0 is the magnetic permeability of vacuum equal to $4\pi \times 10^{-7}$ Wb/Amp-m. The relative permeability (μ_r) is a dimensionless quantity and must be multiplied with μ_0 to get the magnetic flux density in any general material. Thus, the magnetic flux density in any material with relative magnetic permeability μ_r will read

$$\mathbf{B} = \mu_0\mu_r\mathbf{H} = \mu\mathbf{H} \tag{4.10}$$

Gauss's law on the divergence of a magnetic field (Eq. 4.9) will result in $\nabla.\mathbf{B} = 0$ for any general material. From here onwards in this chapter, $\mathbf{B}$ and $\mathbf{H}$ will be used synonymously, assuming the magnetic field is induced in a free space or in air.

Following the phenomena described by Øersted, Ampere [1] proposed a mathematical description of the magnetic field generated by a wire carrying an electric current. It says that the contour integral of the magnetic field generated by a wire, around the circumference of the electrical wire carrying electric current, is equal to the current carried by the wire. Current is the rate of change of charge per unit time. Again, it was found that the net electric current I through a conductive wire is equal to the area integral of the electric current density $\mathbf{J}$ flowing through the wire. In mathematical term it can be written as

$$I = \iint_\Gamma \mathbf{J}.d\vec{\Gamma} \tag{4.11}$$

Thus, mathematically the Ampere's law reads

$$\oint \mathbf{B}.d\vec{\mathbf{c}} = \mu I = \mu \iint_\Gamma \mathbf{J}.d\vec{\Gamma} \tag{4.12}$$

Applying Stokes' theorem (see Appendix Eq. A4.5), Ampere's law will be

$$\iint_\Gamma (\nabla \times \mathbf{B}).d\vec{\Gamma} = \mu \iint_\Gamma \mathbf{J}.d\vec{\Gamma} \tag{4.13}$$

According to the identities of vector calculus, the divergence of the curl of a magnetic field, i.e., $\nabla.(\nabla \times \mathbf{B})$, must be equal to zero. That signifies that divergence of current $\nabla.\mathbf{J} = 0$. However, $\nabla.\mathbf{J} = -\dfrac{\partial \rho}{\partial t}$, ensures the time-varying charge density is non-zero. Alternatively, $\mathbf{J} = 0$ will result $\nabla \times \mathbf{B} = 0$. This contradicts Faraday's induction law described in the next section. Hence, later Maxwell in 1861 [2] generalized Ampere's equation, applying time-varying currents by adding displacement current term, $\mathbf{J}_D$. Overall, the basic current density generated in a material is directly proportional to the time-varying change of the electric displacement $\mathbf{D}$ induced by an electric field $\mathbf{E}$. Like the magnetic field ($\mathbf{B}$), the electric displacement also follows

Electromagnetic Wave Propagation

a constitutive law of the medium and is directly proportional to the applied electric field $\mathbf{E}$. Dielectric permittivity ϵ is the proportionality constant and the electric displacement can be written as

$$\mathbf{D} = \epsilon \mathbf{E} \tag{4.14.1}$$

If the material has prior polarization (polarization density $\mathbf{P}$), then the electric displacement $\mathbf{D}$ modifies to

$$\mathbf{D} = \epsilon \mathbf{E} + \mathbf{P} \text{ or } \mathbf{D} = \epsilon_0 \mathbf{E} + \mathbf{P} \tag{4.14.2}$$

where ϵ_0 is the permittivity of the free space or vacuum as it is used in the Gauss's law in Eq. 4.6. Next, to satisfy the Gauss's law, the electric displacement field also satisfy

$$\nabla.\mathbf{D} = \epsilon_0 \ \nabla.\mathbf{E} + \nabla.\mathbf{P} = \nabla.(\epsilon_0 \mathbf{E} + \mathbf{P}) \tag{4.14.3}$$

Substituting Eq. 4.6,

$$\nabla.\mathbf{D} = \rho + \nabla.\mathbf{P} = \rho - \rho_p = \rho_f \tag{4.14.4}$$

where ρ is the total charge used in the Gauss's law, ρ_p is the charge due to the polarized dipoles present or induced in the material, and ρ_f is the free charges. Hence,

$$\nabla.\mathbf{P} = -\rho_p; \ \nabla.(\epsilon_0 \mathbf{E} + \mathbf{P}) = \rho_f \tag{4.14.5}$$

Generally speaking, a vacuum may not have any prior polarization, and $\mathbf{P}$ could be zero. However, for a transversely polarized wave in photonic metamaterials that is not true, where $\nabla.\mathbf{D} = 0$, but $\nabla.\mathbf{E} \neq 0$. Adding the displacement current term proposed by Maxwell with the rate of change of electric displacement $\mathbf{D}$, the total current density $\mathbf{J}$ in the wire in a magnetic field can be written as $\mathbf{J} = \dfrac{\partial \mathbf{D}}{\partial t}$. Substituting Eq. 4.14.2 for any material with permittivity ϵ

$$\mathbf{J} = \epsilon \frac{\partial \mathbf{E}}{\partial t} + \frac{\partial \mathbf{P}}{\partial t} = \mathbf{J}_D + \epsilon \frac{\partial \mathbf{E}}{\partial t} \tag{4.14.6}$$

Substituting Eq. 4.14 into Eq. 4.12, the Ampere-Maxwell equation will be derived as follows

$$\oint \mathbf{B}.d\vec{c} = \iint_\Gamma \mu \mathbf{J}_D.d\vec{\Gamma} + \epsilon\mu \iint_\Gamma \frac{\partial \mathbf{E}}{\partial t}.d\vec{\Gamma} \tag{4.15.1}$$

Applying Stokes' theorem (see Appendix, Eq. A.4.5), Eq. 4.15.1 modifies to a surface integral

$$\iint_\Gamma (\nabla \times \mathbf{B}).d\vec{\Gamma} = \iint_\Gamma \left(\mu \mathbf{J}_D + \epsilon\mu \frac{\partial \mathbf{E}}{\partial t} \right).d\vec{\Gamma} \tag{4.15.2}$$

Thus, at every point on the surface of a material of interest, the curl of the magnetic field will be

$$\nabla \times \mathbf{B} = \mu \mathbf{J}_D + \epsilon\mu \frac{\partial \mathbf{E}}{\partial t} \tag{4.15.3}$$

Assuming no dipole moment ($\mathbf{P}$) in material and thus no displacement current in a free space, the curl of the induced magnetic field or the effective magnetic flux density will read

$$\nabla \times \mathbf{B} = \epsilon\mu \frac{\partial \mathbf{E}}{\partial t} \tag{4.15.4}$$

where $\epsilon\mu$ is the product of the magnetic permeability (μ) and the dielectric permittivity (ϵ) of the space or the material of interest. In free space this product will result in to $\epsilon_0\mu_0$ where ϵ_0 is the dielectric permittivity of the free space equal to 8.854188×10^{-12} F/m and μ_0 is the magnetic permeability of free space ($4\pi \times 10^{-7}$Wb/Amp-m mentioned earlier).

4.2.4 FARADAY'S INDUCTION LAW

Faraday found that when the magnetic field or the magnetic flux density in a material is changed, then the electric field is generated. Later Heaviside mathematically explained this phenomenon and described that the rate of change of the magnetic flux density over the entire surface of the material will be equal to the contour integral of the electric field induced. This was used and further mathematically developed by Maxwell. Later it was called the Maxwell-Faraday equation [2].

$$\oint \mathbf{E}.d\vec{c} = -\frac{\partial}{\partial t} \iint_\Gamma \mathbf{B}.d\vec{\Gamma} \tag{4.16.1}$$

Applying Stokes' theorem again, the equation modifies to

$$\iint_\Gamma (\nabla \times \mathbf{E}).d\vec{\Gamma} = -\iint_\Gamma \frac{\partial \mathbf{B}}{\partial t}.d\vec{\Gamma} \tag{4.16.2}$$

Hence, at every point on the surface the following equation hold true for any material

$$\nabla \times \mathbf{E} = -\frac{\partial \mathbf{B}}{\partial t} \tag{4.16.3}$$

Electromagnetic Wave Propagation

135

4.3 MAXWELL'S ELECTROMAGNETIC WAVE EQUATION

In summary, there are two interconnected fields following coupled governing equations written in Eqs. 4.15.4 and 4.16.3 reiterated here again in Table 4.1.

Following step-by-step operations will lead to the two wave equations representative of the electromagnetic wave equations, described in a tabular form in Table 4.2.

4.3.1 SOLUTION OF ELECTROMAGNETIC WAVE EQUATIONS

In Table 4.3 the solutions for both the electric and magnetic fields are presented.

TABLE 4.1

Gauss's Law, Ampere-Maxwell and Maxwell-Faraday Equations

Electric Field Equation	**Magnetic Field Equation**
Gauss's law	Gauss's law
$$\mathbf{\nabla.D} = \rho_f$$	$$\mathbf{\nabla.H} = 0$$
$$\rho_f = \rho - \rho_p$$	$$\mathbf{\nabla.B} = 0$$
$$\mathbf{\nabla.E} = \frac{\rho}{\epsilon}$$	
Constitutive equation	Constitutive equation
$$\mathbf{D} = \epsilon_0\epsilon_r\mathbf{E} + \mathbf{P}$$	$$\mathbf{B} = \mu_0\mu_r\mathbf{H}$$
With $\mathbf{P} = 0$ in free space	In free space
$$\mathbf{D} = \epsilon_0\mathbf{E}$$	$$\mathbf{B} = \mu_0\mathbf{H}$$
$$\mathbf{\nabla} \times \mathbf{E} = -\frac{\partial \mathbf{B}}{\partial t}$$	$$\mathbf{\nabla} \times \mathbf{B} = \epsilon\mu\frac{\partial \mathbf{E}}{\partial t}$$

TABLE 4.2

Electromagnetic Wave Equations

Electric Field Equation	**Magnetic Field Equation**
Faraday's Law	Ampere's Law
$$\mathbf{\nabla} \times \mathbf{E} = -\frac{\partial \mathbf{B}}{\partial t}$$	$$\mathbf{\nabla} \times \mathbf{B} = \epsilon\mu\frac{\partial \mathbf{E}}{\partial t}$$
Taking curl on both sides	Taking curl on both sides
$$\mathbf{\nabla} \times (\mathbf{\nabla} \times \mathbf{E}) = -\mathbf{\nabla} \times \frac{\partial \mathbf{B}}{\partial t} = -\frac{\partial}{\partial t}(\mathbf{\nabla} \times \mathbf{B})$$	$$\mathbf{\nabla} \times \mathbf{\nabla} \times \mathbf{B} = \epsilon\mu \ \mathbf{\nabla} \times \frac{\partial \mathbf{E}}{\partial t}$$
Refer to Eq. 3.32 for any vector field identity	Refer to Eq. 3.32 for any vector field identity
$$\nabla^2\mathbf{E} = \mathbf{\nabla}(\mathbf{\nabla.E}) - \mathbf{\nabla} \times (\mathbf{\nabla} \times \mathbf{E})$$	$$\nabla^2\mathbf{B} = \mathbf{\nabla}(\mathbf{\nabla.B}) - \mathbf{\nabla} \times (\mathbf{\nabla} \times \mathbf{B})$$
$$\therefore \mathbf{\nabla} \times (\mathbf{\nabla} \times \mathbf{E}) = \mathbf{\nabla}(\mathbf{\nabla.E}) - \nabla^2\mathbf{E}$$	$$\therefore \mathbf{\nabla} \times (\mathbf{\nabla} \times \mathbf{B}) = \mathbf{\nabla}(\mathbf{\nabla.B}) - \nabla^2\mathbf{B}$$

(Continued)

136 Metamaterials in Topological Acoustics

TABLE 4.2 *(Continued)*
Electromagnetic Wave Equations

Electric Field Equation	**Magnetic Field Equation**
Substituting the above identity	Substituting the above identity

$$\nabla(\nabla.\mathbf{E}) - \nabla^2 \mathbf{E} = -\frac{\partial}{\partial t}(\nabla \times \mathbf{B})$$

$$\nabla(\nabla.\mathbf{B}) - \nabla^2 \mathbf{B} = \epsilon\mu \, \nabla \times \frac{\partial \mathbf{E}}{\partial t}$$

Applying Eq. 4.9 : divergence of an electric field is zero, $\nabla.\mathbf{E} = 0$

Applying Eq. 4.9: divergence of magnetic field is zero, $\nabla.\mathbf{H} = 0$; $\mathbf{B} = \mu\mathbf{H} \therefore \nabla.\mathbf{B} = 0$

$$-\nabla^2 \mathbf{E} = -\frac{\partial}{\partial t}(\nabla \times \mathbf{B})$$

$$-\nabla^2 \mathbf{B} = \epsilon\mu \, \frac{\partial}{\partial t}(\nabla \times \mathbf{E})$$

Substituting Eq. 4.15.4

Substituting Eq. 4.16.3

$$\nabla^2 \mathbf{E} = \epsilon\mu \frac{\partial^2 \mathbf{E}}{\partial t^2}$$

$$\nabla^2 \mathbf{B} = \epsilon\mu \frac{\partial^2 \mathbf{B}}{\partial t^2}$$

If the divergence of the electric field is nonzero but follows the Gauss's law, using Eq. 4.6 and substituting Eq. 4.15.4

With no such situation, the equation remains

$$\nabla^2 \mathbf{B} = \epsilon\mu \frac{\partial^2 \mathbf{B}}{\partial t^2}$$

$$\nabla(\nabla.\mathbf{E}) - \nabla^2 \mathbf{E} = -\epsilon\mu \frac{\partial^2 \mathbf{E}}{\partial t^2}$$

$$\nabla^2 \mathbf{E} = \epsilon\mu \frac{\partial^2 \mathbf{E}}{\partial t^2} + \nabla\left(\frac{\rho}{\epsilon}\right)$$

$$\nabla^2 \mathbf{E} = \frac{1}{c^2} \frac{\partial^2 \mathbf{E}}{\partial t^2}$$

$$\nabla^2 \mathbf{B} = \frac{1}{c^2} \frac{\partial^2 \mathbf{B}}{\partial t^2}$$

Electric field propagates with velocity

Magnetic field propagates with velocity

$$c = \frac{1}{\sqrt{\epsilon\mu}}$$

$$c = \frac{1}{\sqrt{\epsilon\mu}}$$

In free space $c = \frac{1}{\sqrt{\epsilon_0\mu_0}} = 3 \times 10^8$ m/s

In free space $c = \frac{1}{\sqrt{\epsilon_0\mu_0}} = 3 \times 10^8$ m/s

Speed of light

Speed of light

But $\nabla \times \mathbf{E} = -\frac{\partial \mathbf{B}}{\partial t}$; thus, $\mathbf{E} \perp \mathbf{B}$; electric field is orthogonal to the magnetic field.
Electric and magnetic are mutually orthogonal fields.

But $\nabla \times \mathbf{B} = \epsilon\mu \frac{\partial \mathbf{E}}{\partial t}$; thus $\mathbf{B} \perp \mathbf{E}$; magnetic field is orthogonal to the electric field
Magnetic and electric are mutually orthogonal fields.

TABLE 4.3
Electromagnetic Wave Equations

Solution of the Electric Field Equation	**Solution of Magnetic Field Equation**
As in Eq. 3.77, which is used to solve acoustic wave fields in solids or fluids, a similar wave potential is assumed for the electric field	As in Eq. 3.77, which is used to solve acoustic wave fields in solids or fluids, a similar wave potential is assumed for the magnetic field

(Continued)

Electromagnetic Wave Propagation

TABLE 4.3 *(Continued)*
Electromagnetic Wave Equations

Solution of the Electric Field Equation	Solution of Magnetic Field Equation

$$\mathbf{E} = \mathcal{E}_m \hat{\mathbf{e}}_m e^{i(\mathbf{k} \cdot \mathbf{x} - \omega t)}$$

$$\mathbf{B} = \mathcal{B}_n \hat{\mathbf{e}}_n e^{i(\mathbf{k} \cdot \mathbf{x} - \omega t)}$$

where $\hat{\mathbf{e}}_m$ is the unit vectors in three different directions, $\mathcal{E}_m$ is the magnitudes of the electric field along the m-th direction, and all components together indicate the polarity of the electric field. m takes values 1, 2, and 3. $\mathcal{E} = \sqrt{\mathcal{E}_m^2}$ is the amplitude of the electric wave field.

where $\hat{\mathbf{e}}_n$ is the unit vectors in three different directions, $\mathcal{B}_n$ is the magnitudes of the magnetic field along the n-th direction, and all components together indicate the polarity of the magnetic field. n takes values 1, 2, and 3. $\mathcal{B} = \sqrt{\mathcal{B}_n^2}$ is the amplitude of the magnetic wave field.

$|\mathbf{k}|$ is the wave number; ω is the frequency; wave velocity of the electric field $c = \dfrac{\omega}{|\mathbf{k}|}$

$|\mathbf{k}|$ is the wave number; ω is the frequency; wave velocity of the magnetic field $c = \dfrac{\omega}{|\mathbf{k}|}$

See Fig. 4.2a to visualize the electric field direction and wave propagation direction.

See Fig. 4.2a to visualize the electric field direction and wave propagation direction.

Dot product between the electric and magnetic fields would read:

Dot product between the magnetic and electric fields would read:

$$\mathbf{E} \cdot \mathbf{B} = \mathcal{E}_m \hat{\mathbf{e}}_m e^{i(\mathbf{k} \cdot \mathbf{x} - \omega t)} \cdot \mathcal{B}_n \hat{\mathbf{e}}_n e^{i(\mathbf{k} \cdot \mathbf{x} - \omega t)}$$

$$\mathbf{E} \cdot \mathbf{B} = \mathcal{E}_m \mathcal{B}_n e^{2i(\mathbf{k} \cdot \mathbf{x} - \omega t)} \left(\hat{\mathbf{e}}_m \cdot \hat{\mathbf{e}}_n \right)$$

$$\because \mathbf{E} \perp \mathbf{B} \; ; \; \therefore \; \hat{\mathbf{e}}_m \cdot \hat{\mathbf{e}}_n = 0$$

$$\mathbf{E} \cdot \mathbf{B} = 0$$

$$\mathbf{B} \cdot \mathbf{E} = \mathcal{B}_n \hat{\mathbf{e}}_n e^{i(\mathbf{k} \cdot \mathbf{x} - \omega t)} \cdot \mathcal{E}_m \hat{\mathbf{e}}_m e^{i(\mathbf{k} \cdot \mathbf{x} - \omega t)}$$

$$\mathbf{B} \cdot \mathbf{E} = \mathcal{B}_n \mathcal{E}_m e^{2i(\mathbf{k} \cdot \mathbf{x} - \omega t)} \left(\hat{\mathbf{e}}_n \cdot \hat{\mathbf{e}}_m \right)$$

$$\because \mathbf{B} \perp \mathbf{E} \; ; \; \therefore \; \hat{\mathbf{e}}_n \cdot \hat{\mathbf{e}}_m = 0$$

$$\mathbf{B} \cdot \mathbf{E} = 0$$

Wave propagation direction is orthogonal to the electric field.

Wave propagation direction is orthogonal to the electric field.

The cross product between these two fields represents the direction of wave propagation.
Operate: $\mathbf{E} \times \mathbf{B}$

The cross product between these two fields represents the direction of wave propagation.
Operate: $\mathbf{B} \times \mathbf{E}$

$$\mathbf{E} \times \mathbf{B} = \begin{vmatrix} \hat{\mathbf{e}}_1 & \hat{\mathbf{e}}_2 & \hat{\mathbf{e}}_3 \\ \mathcal{E}_1 & \mathcal{E}_2 & \mathcal{E}_3 \\ \mathcal{B}_1 & \mathcal{B}_2 & \mathcal{B}_3 \end{vmatrix}$$

$$\mathbf{B} \times \mathbf{E} = \begin{vmatrix} \hat{\mathbf{e}}_1 & \hat{\mathbf{e}}_2 & \hat{\mathbf{e}}_3 \\ \mathcal{B}_1 & \mathcal{B}_2 & \mathcal{B}_3 \\ \mathcal{E}_1 & \mathcal{E}_2 & \mathcal{E}_3 \end{vmatrix}$$

$$= \hat{\mathbf{e}}_1 \left(\mathcal{E}_2 \mathcal{B}_3 - \mathcal{B}_2 \mathcal{E}_3 \right) - \hat{\mathbf{e}}_2 \left(\mathcal{E}_1 \mathcal{B}_3 - \mathcal{B}_1 \mathcal{E}_3 \right) + \hat{\mathbf{e}}_3 \left(\mathcal{E}_1 \mathcal{B}_2 - \mathcal{B}_1 \mathcal{E}_2 \right)$$

$$= \hat{\mathbf{e}}_1 \left(\mathcal{B}_2 \mathcal{E}_3 - \mathcal{E}_2 \mathcal{B}_3 \right) - \hat{\mathbf{e}}_2 \left(\mathcal{B}_1 \mathcal{E}_3 - \mathcal{E}_1 \mathcal{B}_3 \right) + \hat{\mathbf{e}}_3 \left(\mathcal{B}_1 \mathcal{E}_2 - \mathcal{E}_1 \mathcal{B}_2 \right)$$

Example: Fig. 4.2b
An electric field polarized along x_2 axis will have only $\hat{\mathbf{e}}_2$ component of the field, and $\mathcal{E}_2 = \mathcal{E} \neq 0; \mathcal{E}_1 = \mathcal{E}_3 = 0,$

Example: Fig. 4.2b
A magnetic field polarized along x_3 axis will have only $\hat{\mathbf{e}}_3$ component of the field, and $\mathcal{B}_3 = \mathcal{B} \neq 0$; $\mathcal{B}_1 = \mathcal{B}_2 = 0,$

$$\mathbf{E} = \mathcal{E}_2 \hat{\mathbf{e}}_2 = \mathcal{E} \hat{\mathbf{e}}_2$$

$$\mathbf{B} = \mathcal{B}_3 \hat{\mathbf{e}}_3 = \mathcal{B} \hat{\mathbf{e}}_3$$

This simultaneously satisfies the $\mathbf{E} \perp \mathbf{B}$ condition, and $\mathbf{E} \cdot \mathbf{B} = 0$

This simultaneously satisfies the $\mathbf{B} \perp \mathbf{E}$ condition, and $\mathbf{B} \cdot \mathbf{E} = 0$

$$\mathbf{E} \times \mathbf{B} = \hat{\mathbf{e}}_1 \left(\mathcal{E}_2 \mathcal{B}_3 \right)$$

$$\mathbf{B} \times \mathbf{E} = \hat{\mathbf{e}}_1 \left(-\mathcal{E}_2 \mathcal{B}_3 \right)$$

(Continued)

TABLE 4.3 *(Continued)*
Electromagnetic Wave Equations

Solution of the Electric Field Equation
Faraday's law

$\nabla \times \mathbf{E} = -\dfrac{\partial \mathbf{B}}{\partial t}$ will results

$$\dfrac{\partial \mathcal{E}_2}{\partial x_1} = -\dfrac{\partial \mathcal{B}_3}{\partial t}$$

Applying wave potential along the $\hat{\mathbf{e}}_2$ direction,
$\mathbf{E} = \mathcal{E} e^{i(\mathbf{k}\cdot\mathbf{x} - \omega t)}$
Substituting above potentials,

$$\mathcal{E} i k_1 e^{i(\mathbf{k}\cdot\mathbf{x}-\omega t)} = -\mathcal{B}(-i\omega) e^{i(\mathbf{k}\cdot\mathbf{x}-\omega t)}$$

$$\dfrac{\mathcal{E}}{\mathcal{B}} = \dfrac{\omega}{k_1} = c$$

Here the wave number along x_1 is the wave vector along the x_1 axis, thus $k_1 = |\mathbf{k}| = k$

The ratio of the magnitudes of the electric and magnetic fields is equal to the wave velocity of the electromagnetic wave (c), along the orthogonal direction to the plane of the electric and magnetic field.

Solution of Magnetic Field Equation
Ampere's Law

$\nabla \times \mathbf{B} = \epsilon\mu \dfrac{\partial \mathbf{E}}{\partial t}$ will results

$$\dfrac{\partial \mathcal{B}_3}{\partial x_1} = \epsilon\mu \dfrac{\partial \mathcal{E}_2}{\partial t}$$

Applying wave potential along the $\hat{\mathbf{e}}_3$ direction,
$\mathbf{B} = \mathcal{B} e^{i(\mathbf{k}\cdot\mathbf{x} - \omega t)}$
Substituting above potentials,

$$\mathcal{B} i k_1 e^{i(\mathbf{k}\cdot\mathbf{x}-\omega t)} = \epsilon\mu \mathcal{E}(-i\omega) e^{i(\mathbf{k}\cdot\mathbf{x}-\omega t)}$$

$$\dfrac{\mathcal{B}}{\mathcal{E}} = \epsilon\mu \dfrac{\omega}{k_1} = \epsilon\mu c$$

Referring to $c = \dfrac{1}{\sqrt{\epsilon\mu}}$

$$\dfrac{\mathcal{B}}{\mathcal{E}} = \dfrac{1}{c}$$

The ratio of the magnitudes of the electric and magnetic fields is equal to the wave velocity of the electromagnetic wave (c), along the orthogonal direction to the plane of the electric and magnetic field.

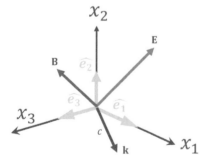
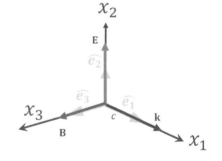

FIGURE 4.2 Generic electromagnetic field and wave propagation.

4.4 COMPARISON OF ELECTROMAGNETIC AND ELASTIC ACOUSTIC WAVE EQUATIONS

Electromagnetic wave equation, elastic wave equation, and acoustic wave equations are compared in a tabular form depicted in Table 4.4.

Electromagnetic Wave Propagation

TABLE 4.4
Electromagnetic Waves and Elastic Waves in Isotropic Media

Electromagnetic Wave Chapter 4	Acoustic Waves Chapter 3	Elastic Waves Chapter 3
$$\nabla^2 \mathbf{E} = \frac{1}{c^2}\frac{\partial^2 \mathbf{E}}{\partial t^2}$$ $$\nabla^2 \mathbf{B} = \frac{1}{c^2}\frac{\partial^2 \mathbf{B}}{\partial t^2}$$	$$\nabla^2 \mathrm{p} - \frac{1}{c_f^2}\frac{\partial^2 \mathrm{p}}{\partial t^2} = 0$$	$$(\lambda + \mu)\underline{\nabla}(\nabla.\mathbf{u}) + \mu\nabla^2 u$$ $$+\underline{\nabla}\lambda(\nabla.\mathbf{u}) + (\nabla\mu.\underline{\nabla})\mathbf{u}$$ $$+\underline{\nabla}\mu \times \underline{\nabla} \times \mathbf{u} + \mathbf{f} = \rho(\mathbf{x})\ddot{\mathbf{u}}$$ $\mathbf{u}$ = displacement field Considering only the out of plane shear wave $$\nabla^2 u_3 - \frac{1}{c_s^2}\frac{\partial^2 u_3}{\partial t^2} = 0$$
Polarization	**Polarization**	**Polarization**
Transverse and longitudinal	Only longitudinal	Both transverse and longitudinal, coupled
Simultaneously electric field and magnetic field waves propagate with wave velocity c, keeping orthogonality condition to each other. Hence, the magnetic field $\mathbf{B}$ could be used to manipulate, electric field $\mathbf{E}$. Wave vortex is one such example.	A pressure wave propagates in fluid media with longitudinal wave velocity c_f. Fluid can not hold shear waves. No external field is present inherent to the wave field like magnetic field for electromagnetic waves, thus manipulation or control of pressure is challenging.	An elastic wave propagates in solid media (for simplicity here on, isotopic materials are discussed; refer to Chapter 3 for anisotropic material) with longitudinal and shear wave velocities c_p and c_s, respectively. No inherent field to pose control of the displacement field. Hence, challenging.
Constitutive equation	**Constitutive equation**	**Constitutive equation**
Homogeneous	Homogeneous	Homogeneous
$$\mathbf{D} = \epsilon_0\epsilon_r\mathbf{E} + \mathbf{P}$$ $$\mathbf{B} = \mu_0\mu_r\mathbf{H}$$	$$\nabla.\mathbf{v} = -\frac{1}{\mathbb{K}}\frac{\partial \mathrm{p}}{\partial t}$$	$$\sigma = \mathbb{L}(\nabla.\mathbf{u}).\mathbf{I} + 2\mathrm{m}\epsilon$$
Inhomogeneous	Inhomogeneous	Inhomogeneous (Eq. 3.98)
$$\mathbf{D} = \epsilon_0\epsilon_r(\mathbf{x})\mathbf{E} + \mathbf{P}$$ $$\mathbf{B} = \mu_0\mu_r(\mathbf{x})\mathbf{H}$$	$$\nabla.\mathbf{v} = -\frac{1}{\mathbb{K}(\mathbf{x})}\frac{\partial \mathrm{p}}{\partial t}$$ $\mathbf{v}$ = velocity matrix	$$\sigma = \mathbb{L}(x)(\nabla.\mathbf{u}).\mathbf{I}$$ $$+2\mathrm{m}(\mathbf{x})\epsilon$$ ϵ = strain matrix σ = stress matrix
Wave potentials	**Wave potential**	**Wave POTENTIALS**
Direct potentials for the fields	Direct potential for pressure	Helmholtz decomposition
$$\mathbf{E} = \mathcal{E}_m\hat{\mathbf{e}}_m e^{i(\mathbf{k}.\mathbf{x}-\omega t)}$$ $$\mathbf{B} = \mathcal{B}_n\hat{\mathbf{e}}_n e^{i(\mathbf{k}.\mathbf{x}-\omega t)}$$	$$\mathrm{p} = \bar{\mathrm{p}}e^{i(\mathbf{k}.\mathbf{x}-\omega t)}$$ $$\mathbf{v} = \bar{v}_m\hat{\mathbf{e}}_m e^{i(\mathbf{k}.\mathbf{x}-\omega t)}$$	$$\mathbf{u} = \nabla\phi + \nabla \times \Psi$$ Indirect longitudinal and shear potentials $$\phi = \mathcal{A}e^{i(\mathbf{k}^P.\mathbf{x}-i\omega t)}$$ $$\Psi = B_n\hat{\mathbf{e}}_n e^{i(\mathbf{k}^s.\mathbf{x}-i\omega t)}$$

(Continued)

TABLE 4.4 *(Continued)*
Electromagnetic Waves and Elastic Waves in Isotropic Media

Electromagnetic Wave	Acoustic Waves	Elastic Waves
Poynting vector and energy flux density	Poynting vector and energy flux density	Poynting vector and energy flux density
$\mathbf{P}_{El} = Re\left[\dfrac{\mathbf{E} \times \mathbf{H}^*}{2}\right]$	$\mathbf{P}_A = -\dfrac{1}{2} Re\left[p^* \mathbf{v}\right]$	$\mathbf{P}_E = -\dfrac{1}{2} Re\left[\sigma^* . \mathbf{v}\right]$
	$\mathbf{v}$ = velocity field in pressure acoustics	$\mathbf{v}$ = velocity field in elastic solid

4.5 APPENDIX

4.5.1 DIVERGENCE THEOREM

In any arbitrary vector field, an imaginary surface (e.g., spherical, cuboidal, ellipsoidal, cylindrical etc.) could be assumed. Sometimes it is imperative in science and engineering to find how much flux has passed through that surface. The surface integral of a vector field is called the flux integral. The surface is shown in Fig. A.4.1 with an example of a spherical surface and an arbitrary field $\mathbf{E}(x_j) = x_1 x_2 \hat{\mathbf{e}}_1 + x_2 x_3 \hat{\mathbf{e}}_2 + x_3 x_1 \hat{\mathbf{e}}_3$.

The vector field also passes through an incremental surface element $d\vec{\Gamma}$ taken from the spherical surface. To find the flux flowing through that incremental surface, one should take a dot product between the field and the direction normal to that incremental surface. The dot product will signify the component of the field along the normal to the incremental local surface area. The integral of the product over the entire surface will give the total flux flowing away or going in through the surface Γ.

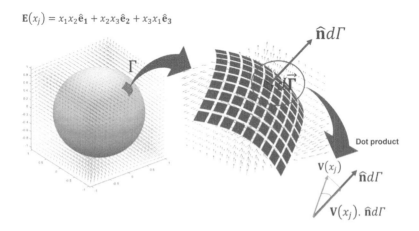

FIGURE A.4.1 Schematic representation of a vector field passing through a surface, projection of the vector field on the surface normal at each point on the surface.

Electromagnetic Wave Propagation   141

Mathematically, if the $\mathbf{E}$ field is known, then the total flux (Φ) passing through the surface Γ can be written as

$$\Phi = \oiint \mathbf{E}.d\vec{\Gamma} = \oiint \mathbf{E}.\hat{n}\,d\Gamma \tag{A.4.1}$$

Factually, the net charge enclosed by a spherical surface in an electric field $\mathbf{E}(x_j,t)$ can be written as

$$q = \epsilon_0 \mathbf{E};\ Q = \epsilon_0 \oiint \mathbf{E}.d\vec{\Gamma} = \epsilon_0 \oiint \mathbf{E}.\hat{n}\,d\Gamma \tag{A.4.2}$$

where ϵ_0 is the permittivity of the free surface.

Often the surface integrals in Eqs. A.4.1 and A.4.2 are difficult to perform. The surface through which the flux passes needs to be parameterized, and $\hat{n}d\Gamma$ must be properly calculated. The unit normal to a surface at any arbitrary point ($\hat{n}$) can be calculated by taking the cross product of the respective tangential vectors towards the change of the parameters that defines the curvature of the surface. For example, a spherical surface (Fig. 3.1) is defined by its radius (r), horizontal angle (θ), and the azimuthal angle (ϕ). The Gauss divergence theorem omits the requirement of the calculation of the unit normal for a specific surface. The Gauss divergence theorem is stated as follows. The total flux of a vector field passes through a surface supposedly calculated through a surface integral and can be simply calculated by the volume integral of the divergence of the vector field over the entire volume enclosed by the surface. Mathematically Eq. A.4.2 can then be written using divergence theorem as follows

$$\epsilon_0 \oiint \mathbf{E}.\hat{n}\,d\Gamma = \int_\Omega \nabla.\mathbf{E}\ d\Omega \tag{A.4.3}$$

where $\nabla.\mathbf{E}$ is the divergence of the electric field ($\mathbf{E}$) and is true for any vector fields.

4.5.2 Stokes' Theorem

In a vector field, an imaginary surface (e.g., spherical, cuboidal, ellipsoidal, cylindrical, etc.) can be assumed to be cut by a plane shown in Fig. A.4.2. Sometimes it is imperative in electromagnetism to find how much flux is flowing along the boundary of an arbitrary surface if the surface is cut by a plane. The contour integral of a field is called the contour integral along a boundary of an object, for example, an electrical wire. The surface shown in Fig. A.4.2 is an arbitrary surface and an arbitrary velocity field $\mathbf{E}(x_j) = x_1 x_2 \hat{e}_1 + x_2 x_3 \hat{e}_2 + x_3 x_1 \hat{e}_3$. The vector field passes through an incremental contour element $d\vec{c}$ taken along the boundary of the surface Γ cut by a plane. To find the total field along the incremental contour, one should take a dot product between the vector field and the direction normal to that incremental contour vector as shown in Fig. A.4.2. The dot product signifies the component of the field along the normal to the contour vector. Integral of the product over the entire contour

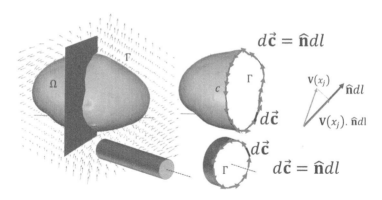

FIGURE A.4.2 Schematic representation of a vector field passing through a surface, projection of the vector field on the tangent of a curve which is a contour enclosing the surface.

line will then give the total flux flowing along the boundary. Mathematically, if the field (**E**) is known then the total flow can be written as

$$F = \oint \mathbf{E}.d\vec{c} = \oint \mathbf{E}.\hat{\mathbf{n}}\,dl \qquad (A.4.4)$$

where $\hat{\mathbf{n}}$ is the unit normal vector pointing along the tangent to the contour of small length segment dl. Further Stokes' theorem says that the field along a contour line enclosing a boundary of a surface can be found by taking curl of the vector field over the entire surface. Hence, mathematically the Stokes' theorem can be written as

$$\oint \mathbf{E}.\hat{\mathbf{n}}\,dl = \int_\Gamma (\nabla \times \mathbf{E}).d\vec{\Gamma} \qquad (A.4.5)$$

4.6 SUMMARY

Understanding of electromagnetic waves is the key to the study of photonics. In this chapter electromagnetic wave equation is derived from the fundamentals. The electromagnetic wave equation is solved from the fundamentals, comparing the wave equations for acoustics and electromagnetic waves simultaneously. Phonons in acoustics and elastodynamics are equivalent to the photons in electrodynamics. The comparison is presented for equivalent application of phononics in acoustic metamaterials in a later chapter. Understanding of electromagnetic waves and its equivalency helps understand the fundamentals of quantum mechanics discussed in this book.

REFERENCES

1. Forbes, N., Mahon, B., *Faraday, Maxwell, and the Electromagnetic Field: How Two Men Revolutionized Physics*. 2014, Amherst, New York: Prometheus Book.
2. Fitzpatrick, R., *Maxwell's Equations and the Principles of Electromagnetism*. 2008, Jones & Bartlett Learning. 2008-01-22: https://www.jblearning.com/.
3. Gonzalez, G., *Advanced Electromagnetic Wave Propagation Methods*. 2021, Boca Raton, FL: CRC Press.
4. Feynman, R.P., *Quantum Electrodynamics*. 1971, Boca Raton, FL: CRC Press.
5. Greiner, W., Schramm, S., Stein, E., Bromley, D.A., *Quantum Chromodynamics*. 2006, Berlin, Germany: Springer.

5 Quantum Mechanics for Engineers

5.1 PARTICLE WAVES AND THE SCHRÖDINGER EQUATION

5.1.1 QUANTIZED ENERGY

The objective of this chapter is to equip the reader with basic concepts of waves that pertain to the concept of quantum mechanics. The evolution of quantum mechanics started from the concept of photons proposed by Max Planck in 1900 [1] while theoretically explaining the 'Blackbody Radiation'. It was realized that the energy (E) of an electromagnetic wave like light is inherently linked to its wave frequency (v). More precisely, energy is directly proportional to the frequency of the electromagnetic waves when the energy can be released like a packet of energy. This gave the fundamental quantized energy equation [2]. As the energy is directly proportional to the frequency, different wavelengths of the waves will have different energy keeping the wave velocity constant. If a wave can have only a discrete number of wave numbers, the energy of the wave carrying the wavelength should also be discrete. This is called quantized energy. This breeds the idea that was postulated that electromagnetic energy can be emitted only in quantized form. In a certain system only a few discrete energy levels are possible. Those are equivalent to their packets of quanta or the photons. Thus, energy of a quantum or a single photon carrying wavelength λ can be written as [3]

$$E = hv = h\frac{c}{\lambda} = 2\pi\hbar\frac{c}{\lambda} \tag{5.1}$$

where h is Planck's constant carrying value $6.62607015 \times 10^{-35}$ Joule-sec (J-s), $\hbar$ is reduced Planck's constants, and c is the wave velocity for electromagnetic wave $c = 3 \times 10^8$ m/s.

Quantum means 'amount' in Latin. In the modern era quantum means the smallest discrete unit possible for any physical quantity such as energy or matter. Quanta of electromagnetic waves are called 'photons'; quanta for acoustic or elastic waves are called 'phonons'.

5.1.2 RELATIVISTIC PARTICLES

To build the quantum mechanics concepts from the ground up, it is necessary to bring the concept of a relativistic particle and its energy, introduced by Sir Albert Einstein [3, 4] in 1905. It was realized that a particle moving with a velocity of light (c) could not be seen by the observer observing the particle from its reference plane. All dimensions of the particle will shrink to null for the observer. However, hypothetically if the same observer is riding the particle, that observer will not see any change to the geometry of the particle. Alternatively, please note that the particle and

DOI: 10.1201/9781003225751-5

144   Metamaterials in Topological Acoustics

riding observer both will shrink to null for an observer fixed at the reference plane. This concept is similar to the Lagrangian and Eulerian coordinate systems described in Chapter 2 but not exactly the same. The observer stationary at a reference plane is referred to in a Lagrangian system, whereas the deformed coordinate system travels with the object in a Eulerian system. This concept of 'shrink to null' is nothing but conversion of the mass (m) of the object to its equivalent energy. This concept gives Einstein's legendary equation [2]

$$E = mc^2 \tag{5.2}$$

5.1.3 WAVE PARTICLE DUALITY

From section 5.1.2, it is apparent that a particle of mass m moving with a velocity of light c will be completely converted to its equivalent energy E as written in Eq. 5.2. From section 5.1.1, it is also evident that a quantum of a photon with velocity c and wavelength λ will have the energy E equivalent to the wave energy written in Eq. 5.1. The next stroke of a genius comes from a French aristocrat, Louis de Broglie in 1925 [5]. Are Eqs. 5.1 and 5.2 the same for a particle like an electron moving with electromagnetic wave velocity? With this question in mind, de Broglie derived the following equations equating the energy in Eqs. 5.1 and 5.2 to a relativistic particle. This ground-breaking concept fundamentally established the dual nature of wave and matter, known as 'wave particle duality'. Equating (5.1) and (5.2) will be

$$mc^2 = h\frac{c}{\lambda}; \ mc = \frac{h}{\lambda} = p; \ \lambda = \frac{2\pi\hbar}{p}; \ p = \frac{h}{\lambda} = \frac{2\pi\hbar}{\lambda} \tag{5.3}$$

where p is the momentum of the particle.

5.1.3.1 Momentum of a Relativistic Particle

The story does not end here. A particle with rest mass m moving with a speed of light has momentum $p = mc$, when the speed of the particle $c \neq 0$. These particles are relativistic particles. It appears that the rest of the mass is completely converted to energy. However, a relativistic particle with any arbitrary velocity ($v < c$) may not have all its mass disappeared as energy. The particle will have an effective mass m_e depending on its velocity v. The effective mass was derived by Einstein in 1905 [6] as follows

$$m_e = \frac{mv}{\sqrt{1 - \dfrac{v^2}{c^2}}} \tag{5.4.1}$$

In general, momentum p of a relativistic particle then can be written as [6]

$$p = \frac{mv}{\sqrt{1 - \dfrac{v^2}{c^2}}} = m_e v \tag{5.4.2}$$

Quantum Mechanics for Engineers

145

5.1.3.2 Relativistic Energy in Terms of Momentum

The energy from the effective mass will then read

$$E = m_e c^2 = \frac{mc^2 v}{\sqrt{1 - \frac{v^2}{c^2}}} \tag{5.4.3}$$

When the rest mass of a particle is zero, then it seems from the above equation that the energy is also zero. But we know that is not true for massless particles like electrons and protons. Hence, new equations are required for massless relativistic particles. Let's use the square of the momentum $\times$ speed of light as follows

$$(pc)^2 = \frac{(mv)^2}{1 - \frac{v^2}{c^2}} \cdot c^2 = \frac{m^2 \frac{v^2}{c^2}}{1 - \frac{v^2}{c^2}} \cdot c^4 \tag{5.4.4}$$

By adding and subtracting the term and substituting Eq. 5.4.3,

$$(pc)^2 = \frac{m^2 c^4 \left[\frac{v^2}{c^2} - 1 \right]}{1 - \frac{v^2}{c^2}} + \frac{m^2 c^4}{1 - \frac{v^2}{c^2}} = -m^2 c^4 + \left(m_e c^2 \right)^2 \tag{5.4.5}$$

$$(pc)^2 = -\left(mc^2 \right)^2 + E^2; \; E^2 = (pc)^2 + \left(mc^2 \right)^2; \; E = \sqrt{(pc)^2 + \left(mc^2 \right)^2} \tag{5.4.6}$$

Here, according to Eq. 5.4.6, for a massless relativistic particle $m = 0$, the energy will read

$$E = pc; \; p = \frac{h}{\lambda} \tag{5.4.7}$$

5.1.4 WAVE FUNCTION

If a particle can be described by an equivalent nature of a wave, using wave particle duality, it can be argued that the particle must propagate with four canonical characteristics of the wave, namely, frequency ($\omega = 2\pi v$), wave number ($k = 2\pi/\lambda$), amplitude ($\mathcal{A}$), and phase (Φ). A spatial and time varying potential function for the wave nature can thus be written as

$$\Psi = \mathcal{A} e^{i(k.x - \omega t)} \tag{5.5.1}$$

This wave potential has a similar pattern to that expressed in section 3.2.2.1.3 in Chapter 3 for acoustic and elastic waves. The wave function above is a one-dimensional wave function. The term $k.x$ in Eq. 5.5.1 describes the phase of the propagating wave as discussed in section 3.2.2.1.5 in Chapter 3. If the wave function is expressed in three dimensions, the wave number k will become a vector $\mathbf{k} = k_i \hat{\mathbf{e}}_i$ and the position of a particle x will become a position vector $\mathbf{x} = x_i \hat{\mathbf{e}}_i$. Thus, the phase will read $\mathbf{k} \cdot \mathbf{x} = k_i x_i$; a scalar quantity and wave function in three dimensions will be

$$\Psi = \mathcal{A} e^{i(\mathbf{k} \cdot \mathbf{x} - \omega t)} = \mathcal{A} e^{i(k_i x_i - \omega t)} \tag{5.5.2}$$

The one-dimensional wave potential can further be expressed with energy E and momentum p as follows. Please note the wave velocity $c = \omega / k$.

$$\Psi = \mathcal{A} \exp\left[-i\omega\left(t - \frac{k}{\omega}x\right)\right] = \mathcal{A} \exp\left[-i\left(2\pi v t - \frac{2\pi v x}{c}\right)\right] = \mathcal{A} \exp\left[-i2\pi\left(v t - \frac{x}{\lambda}\right)\right] \tag{5.5.3}$$

As, $E = 2\pi \hbar v$ from Eq. 5.1, and $p = 2\pi \hbar / \lambda$ from Eq. 5.3, then Eq. 5.5.1 will be

$$\Psi = \mathcal{A} \exp\left[-i\left(\frac{E}{\hbar}t - \frac{p}{\hbar}x\right)\right] = \mathcal{A} \exp\left[-\frac{i}{\hbar}(Et - px)\right] \tag{5.5.4}$$

Taking spatial derivatives of the wave function in Eq. 5.5.2 with respect to x would read

$$\frac{\partial}{\partial x}\Psi = \left(\frac{i}{\hbar}p\right)\mathcal{A} \exp\left[-\frac{i}{\hbar}(Et - px)\right] = \left(\frac{i}{\hbar}p\right)\Psi \tag{5.5.5}$$

Similarly, the second-order derivative will read

$$\frac{\partial^2}{\partial x^2}\Psi = \left(\frac{-1}{\hbar^2}p^2\right)\mathcal{A} \exp -\frac{i}{\hbar}(Et - px) = -\left(\frac{p}{\hbar}\right)^2 \Psi \tag{5.5.6}$$

The time derivative of the wave function will be

$$\frac{\partial}{\partial t}\Psi = -\left(\frac{i}{\hbar}E\right)\mathcal{A} \exp -\frac{i}{\hbar}(Et - px) = -\left(\frac{i}{\hbar}E\right)\Psi \tag{5.5.7}$$

Alternatively, as the wave function cannot disappear itself, or $\Psi \neq 0$, the energy (from Eq. 5.5.7) and the momentum (from Eq. 5.5.5) can further be written as and time and spatial derivative operators as follows, respectively [2].

$$\text{Energy operator: } E = -\frac{\hbar}{i}\frac{\partial}{\partial t} = i\hbar\frac{\partial}{\partial t} \tag{5.5.8}$$

$$\text{Momentum operator: } p^2 = -\hbar^2 \frac{\partial^2}{\partial x^2} \text{ and } p = -i\hbar\frac{\partial}{\partial x} \tag{5.5.9}$$

In classical mechanics these terms were never expressed as operators; however, from classical wave function they are obvious and are frequently used in quantum mechanics. Also, it is obvious that these operators are valid only when wave particle duality is enforced.

Quantum Mechanics for Engineers

147

5.1.5 THE SCHRÖDINGER EQUATION

Please refer to section 2.13 in Chapter 2. Starting from classical mechanics, where the Hamiltonian of a system is described by the summation of kinetic energy and potential energy, the total energy E of a particle can be written as

$$E = \mathcal{K} + \mathcal{U} \text{ or } E = \frac{1}{2}mc^2 + \mathcal{U} \tag{5.6.1}$$

Substituting the momentum expression obtained from Eq. 5.3 and multiplying the wave function Ψ in Eq. 5.5.1 on both sides of Eq. 5.6.1 will be

$$E\Psi = \frac{p^2}{2m}\Psi + \mathcal{U}\Psi \tag{5.6.2}$$

Substituting the operators in Eqs. 5.5.8 and 5.5.9, into Eq. 5.6.2 the wave equation is derived:

$$ih\frac{\partial}{\partial t}\Psi = -\frac{\hbar^2}{2m}\frac{\partial^2}{\partial x^2}\Psi + \mathcal{U}\Psi \tag{5.7}$$

Alternatively,

$$-\frac{\hbar^2}{2m}\frac{\partial^2}{\partial x^2}\Psi = \left(ih\frac{\partial}{\partial t}\Psi - \mathcal{U}\Psi\right) \tag{5.8.1}$$

$$\text{or } -\frac{\hbar^2}{2m}\frac{\partial^2}{\partial x^2}\Psi = (E - \mathcal{U})\Psi \tag{5.8.2}$$

$$\text{or } \left(-\frac{\hbar^2}{2m}\frac{\partial^2}{\partial x^2} + \mathcal{U}\right)\Psi = ih\frac{\partial}{\partial t}\Psi \tag{5.8.3}$$

For two- and three-dimensional systems the equation will be

$$\left[-\frac{\hbar^2}{2m}\left(\frac{\partial^2}{\partial x_1^2} + \frac{\partial^2}{\partial x_2^2}\right) + \mathcal{U}\right]\Psi = ih\frac{\partial}{\partial t}\Psi \tag{5.8.4}$$

$$\left[-\frac{\hbar^2}{2m}\left(\frac{\partial^2}{\partial x_1^2} + \frac{\partial^2}{\partial x_2^2} + \frac{\partial^2}{\partial x_3^2}\right) + \mathcal{U}\right]\Psi = ih\frac{\partial}{\partial t}\Psi \tag{5.8.5}$$

$$\text{or } H\Psi = E\Psi \tag{5.8.6}$$

where the differential form of the summation of kinetic energy and potential energy is substituted with a new notation H, called Hamiltonian. Hamiltonian is essentially the total energy of the system.

Please note that the above equations are one-dimensional or multi-dimensional equations. This equation is called the time dependent Schrödinger equation. This was not the original derivation step proposed by Schrödinger. However, for engineers it is easier to understand from the perspective of classical wave, and hence the above equations are derived from the perspective of engineering mechanics [3].

Next, if the energy operator in Eq. 5.5.6 is not substituted in the equation Eq. 5.6.2, then the energy of the system will become the eigenvalues of the system. Any differential form of an equation must have their eigenvalues. To an engineer, it is almost inevitable to know that a system may have multiple eigenvalues, which are that characteristics of the system and corresponding mode shapes are the eigenfunctions. Here the quantum system is also defined by the Hamiltonian. Hence, the system must have multiple eigenvalues and their corresponding mode shapes. Eq. 5.8.6 can be rewritten as

$$H\Psi_N = E_N\Psi_N \tag{5.8.7}$$

where N takes values, $1, 2, 3, \ldots$ and the eigenvalues are E_N for energy with values E_1, E_2, E_3, $\ldots$ Ψ_N is the eigenfunction for the N-th eigenvalue. Let's assume that Ψ_M is another eigenfunction of the Hamiltonian of the system in Eq. 5.8.6, where M also takes values, $1, 2, 3, \ldots$ Then Ψ_M must be orthogonal to the eigenfunction Ψ_N. If Ψ_M is operated on the Eq. 5.8.7, with Dirac notation it would read.

$$\langle \Psi_M \mid H \mid \Psi_N \rangle = \langle \Psi_M \mid E_N \mid \Psi_N \rangle \tag{5.8.8}$$

As the eigenvalues are real, E_N can be taken out from the right side and the above equation will be

$$\langle \Psi_M \mid H \mid \Psi_N \rangle = E_N \langle \Psi_M \mid \Psi_N \rangle \tag{5.8.9}$$

As Ψ_N and Ψ_M are orthogonal to each other, the above equation can be written in a matrix from where the inner product of Ψ_N and Ψ_M (i.e., $\langle \Psi_M \mid \Psi_N \rangle$) will result in Kronecker delta δ_{MN}. For a finite number of eigenfunctions, the equation will take a matrix form as follows

$$H_{MN} = E_N \delta_{MN} \tag{5.8.10}$$

$E_N \delta_{MN}$ is a diagonal matrix carrying the eigenvalues along the diagonal. This leads to a conclusion that the size of a Hamiltonian matrix will depend on the number of eigenfunctions considered in the solution. If only three eigenstates are considered for a Hamiltonian, then there will be only three eigenvalues (i.e., three energy states).

In mathematics, a Hamiltonian matrix ($\mathbf{H}$) is specifically defined as a $2n \times 2n$ matrix. If the $\mathbf{H}$ is pre-multiplied with a skew-symmetric matrix $\mathbf{M}$, the new matrix will always be symmetric. That means that the transpose or the conjugate transpose (if the matrix is complex) of the new matrix $\mathbf{JH}$, which is $(\mathbf{MH})^{\mathrm{T}}$ or $(\mathbf{MH})^{*}$, will be equal to $\mathbf{MH}$ as indicated in Eq. 5.8.11. If the Hamiltonian matrix has eigenvalues E, then $-E$, E^{*}, and $-E^{*}$ are also its eigenvalues.

$$\mathbf{M} = \begin{bmatrix} \mathbf{0}_{n \times n} & \mathbf{I}_{n \times n} \\ -\mathbf{I}_{n \times n} & \mathbf{0}_{n \times n} \end{bmatrix}; \ (\mathbf{MH})^{T} = \mathbf{MH}; \ (\mathbf{MH})^{*} = \mathbf{MH} \tag{5.8.11}$$

Quantum Mechanics for Engineers

Generally the Hamiltonian of a system represents a global behavior as well as the local behavior of the system. Sometimes it is difficult to comprehend the total Hamiltonian of a system to understand a specific phenomenon at a specific energy state. Then it becomes necessary to focus on the subsystem and write the Hamiltonian of the subsystem only. The Hamiltonian of a subsystem is alternatively called a 'reduced Hamiltonian' [7]. Later in this chapter, in section 5.3, the reduced Hamiltonian in light of **k.p** perturbation is discussed.

5.1.6 Time-Independent Schrödinger Equation

Without complicating the solution, let's assume the system has no potential energy (i.e., $\mathcal{U} = 0$). Then the eigenvalue problem in Eq. 5.8.5 for a one-dimensional system could be rewritten as

$$-\frac{\hbar^2}{2m}\frac{\partial^2 \Psi}{\partial x^2} = E\Psi \tag{5.9.1}$$

For two- and three-dimensional systems the equation will be

$$-\frac{\hbar^2}{2m}\left(\frac{\partial^2}{\partial x_1^2} + \frac{\partial^2}{\partial x_2^2}\right)\Psi = E\Psi \tag{5.9.2}$$

$$-\frac{\hbar^2}{2m}\left(\frac{\partial^2}{\partial x_1^2} + \frac{\partial^2}{\partial x_2^2} + \frac{\partial^2}{\partial x_3^2}\right)\Psi = E\Psi \tag{5.9.3}$$

Please note that Eq. 5.9.3 is no different from Eqs. 3.6.3, 3.19.2, or, 3.69.2 in Chapter 3, where wave velocities are the eigenvalues. To illuminate the comparative view, only the one-dimensional differential operators are considered. Please refer to Fig. 5.1 for this discussion.

Comparative eigenvalue solutions are presented in a tabular form, a) based on quantum mechanics perspective, to find position of an electron along a fixed, straight, narrow tube of length ℓ_1 (Fig. 5.1), and b) from engineering mechanics perspective, to find the wave solution for a vertically polarized wave along a two-end fixed rope of length ℓ_1 (Fig. 5.1). Their respective similarities and differences are identified in Table 5.1.

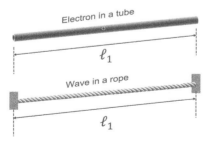

FIGURE 5.1 Examples from quantum mechanics and engineering mechanics.

TABLE 5.1
Eigenvalue Solution

Quantum Mechanics
Governing Differential Equation

$$\left(-\frac{\hbar^2}{2m}\frac{\partial^2}{\partial x^2}\right)\Psi = i\hbar\frac{\partial}{\partial t}\Psi$$

Eigenvalue problem for homogeneous differential equation

$$\left(\frac{\hbar^2}{2m}\frac{\partial^2}{\partial x^2}\right)\Psi = -E\Psi$$

Partial to ordinary differential equation
Hamiltonian

$$H = \frac{\hbar^2}{2m}\frac{\partial^2}{\partial x^2}$$

Solution in the form

$$\Psi = e^{mx}$$

Modified equation.

$$-\frac{\hbar^2}{2m}m^2 e^{mx} = E e^{mx}; \ e^{mx} \neq 0$$

Eigenvalues

$$m = \pm\frac{\sqrt{-2mE}}{\hbar} = \pm\frac{i\sqrt{2mE}}{\hbar}$$

Solution

$$\Psi = C_1 e^{\frac{i\sqrt{2mE}}{\hbar}x} + C_1 e^{-\frac{i\sqrt{2mE}}{\hbar}x}$$

C_1 and C_2 are arbitrary constants.
Boundary conditions

As we are interested in the position of the electron inside a linear tube, there is no possibility of the electron being present at the two ends of the tube. That says, at $x = 0$ and $x = \ell_1$; $\Psi = 0$

$$@ \ x = 0; \ \Psi = 0 \text{ gives}$$
$$C_1 + C_2 = 0; \ C_2 = -C_1$$
$$@ \ x = \ell_1; \ \Psi = 0 \text{ gives}$$
$$C_1 e^{\frac{i\sqrt{2mE}}{\hbar}\ell_1} + C_2 e^{-\frac{i\sqrt{2mE}}{\hbar}\ell_1} = 0 \text{ substituting } C_2 = -C_1$$
$$C_1\left(e^{\frac{i\sqrt{2mE}}{\hbar}\ell_1} - e^{-\frac{i\sqrt{2mE}}{\hbar}\ell_1}\right) = 0$$

Engineering Mechanics
Governing Differential Equation

$$c_s^2\frac{\partial^2}{\partial x^2}u_3 = \frac{\partial^2 u_3}{\partial t^2}$$

Wave to accommodate out-of-plane displacement only for simplicity
Eigenvalue problem for homogeneous differential equation

$$c_s^2\frac{\partial^2}{\partial x^2}u_3 = -\mathcal{W}u_3$$

Partial to ordinary differential equation
Hamiltonian

$$H = c_s^2\frac{\partial^2}{\partial x^2}$$

Solution in the form

$$u_3 = e^{mx}$$

Modified equation.

$$c_s^2 m^2 e^{mx} = -\mathcal{W}e^{mx}; \ e^{mx} \neq 0$$

Eigenvalues

$$m = \pm\frac{\sqrt{\mathcal{W}}}{c_s} = \pm\frac{i\omega}{c_s} = \pm ik_s$$

Solution

$$u_3 = C_1 e^{ik_s x} + C_1 e^{-ik_s x}$$

C_1 and C_2 are arbitrary constants.
Boundary conditions

As we are interested in propagation of the wave in a rope fixed at the two ends, there are no displacements at those end points. That says, at $x = 0$ and $x = \ell_1$; $u_3 = 0$

$$@ \ x = 0; \ u_3 = 0 \text{ gives}$$
$$C_1 + C_2 = 0; \ C_2 = -C_1$$
$$@ \ x = \ell_1; \ u_3 = 0 \text{ gives}$$
$$C_1 e^{ik_s\ell_1} + C_2 e^{-ik_s\ell_1} = 0$$

substituting $C_2 = -C_1$

$$C_1\left(e^{ik_s\ell_1} - e^{-ik_s\ell_1}\right) = 0$$

(Continued)

Quantum Mechanics for Engineers

151

TABLE 5.1 *(Continued)*
Eigenvalue Solution

Quantum Mechanics Governing Differential Equation	Engineering Mechanics Governing Differential Equation

Nontrivial solution

$$C_1 \neq 0$$

$$e^{\frac{i\sqrt{2mE}}{\hbar}\ell_1} - e^{-\frac{i\sqrt{2mE}}{\hbar}\ell_1} = 0$$

Applying Euler's formula

$$e^{i\theta} = \cos\theta + i\sin\theta$$

$$2i\sin\left(\frac{\sqrt{2mE}}{\hbar}\ell_1\right) = 0$$

$\because$ Energy cannot be negative

$$\sin\left(\frac{\sqrt{2mE}}{\hbar}\ell_1\right) = 0$$

$$\sin\left(\frac{\sqrt{2mE}}{\hbar}\ell_1\right) = \sin n\pi, \; n = 0,1,2,3\ldots$$

$$\frac{\sqrt{2mE}}{\hbar}\ell_1 = n\pi$$

$$k = \frac{n\pi}{\ell_1}$$

or $E = \dfrac{n^2\pi^2\hbar^2}{2m\ell_1^2}; \; n = 0,1,2,3\ldots$

Conclusion

Energies are discrete.

$k = \dfrac{\sqrt{2mE}}{\hbar}$ is the wave number

Energy (E) and wave number (k) relation are linear here, but not always. There could be dispersion ($E(k)$) for other complex quantum systems.

Referring to Eqs. 5.1 and 5.3, expressions for energy and momentum, it is apparent that

$$E = 2\pi\hbar\nu$$

$k = \dfrac{p}{\hbar}$; momentum divided by reduced Planck's constant is the quantum wave number.

Nontrivial solution

$$C_1 \neq 0$$

$$e^{ik_s\ell_1} - e^{-ik_s\ell_1} = 0$$

Applying Euler's formula

$$e^{i\theta} = \cos\theta + i\sin\theta$$

$$2i\sin(k_s\ell_1) = 0$$

$\because W$ cannot be negative

$$\sin(k_s\ell_1) = 0$$

$$\sin(k_s\ell_1) = \sin n\pi; \; n = 0,1,2,3\ldots$$

$$k_s = \frac{n\pi}{\ell_1}$$

$$W = k_s^2 c_s^2 = \omega^2 = \frac{n^2\pi^2 c_s^2}{\ell_1^2}; \; n = 0,1,2,3\ldots$$

Conclusion

Frequencies are discrete.

k_s is the wave number, and ω is the frequency

Frequency (ω) and wave number (k) relation are linear here, but not always. There could be dispersion ($\omega(k)$) for the other complex wave problems in complex geometry.

(Continued)

152 Metamaterials in Topological Acoustics

TABLE 5.1 *(Continued)*
Eigenvalue Solution

Quantum Mechanics
Governing Differential Equation
Possible solutions of the wave function, or the eigenfunctions are

$$\Psi_n = C\sin\left(\frac{n\pi}{\ell_1}x\right); n = 0,1,2,3\ldots \text{ so}$$

$$\Psi_1 = C\sin\left(\frac{\pi}{\ell_1}x\right); \Psi_2 = C\sin\left(\frac{2\pi}{\ell_1}x\right)$$

$$\Psi_3 = C\sin\left(\frac{3\pi}{\ell_1}x\right); \Psi_5 = C\sin\left(\frac{5\pi}{\ell_1}x\right)$$

Eigenfunctions are all discrete; each eigenfunction is a possibility of finding the electron in the tube.

Total Solution

$$\Psi = \sum_{n=0}^{\infty} C_n \sin\left(\frac{n\pi}{\ell_1}x\right)$$

Differences

Under any circumstances, without any further boundary conditions, it is necessary that an electron to be present anywhere along that tube between the two end boundaries. There must be a certain possibility of finding the electron along the tube. Hence, the probability of finding the particle should be equal to 1 with any order of eigenfunction. Each eigensolution above must be normalized such that the probability of finding the electron is integrated over all the possible positions equate to 1. This governs the following criteria. The function inner product (see Appendix for understanding) of two wave functions must be equal to 1 to ensure the orthogonality condition.

$$\langle\Psi_n \mid \Psi_n\rangle = \int_0^{\ell_1} |C|^2 \sin^2\left(\frac{n\pi}{\ell_1}x\right)dx = 1$$

$$\therefore C = \sqrt{\frac{2}{\ell_1}}$$

Infinite eigenvalues and eigenfunctions

$$E_n = \frac{n^2\pi^2\hbar^2}{2m\ell_1^2}; \Psi_n = \sqrt{\frac{2}{\ell_1}}\sin\left(\frac{n\pi}{\ell_1}x\right)$$

Engineering Mechanics
Governing Differential Equation
Possible solutions of the displacement wave function u_3, or the eigenfunctions are

$$u_3^n = C\sin\left(\frac{n\pi}{\ell_1}x\right); n = 0,1,2,3\ldots \text{ so}$$

$$u_3^1 = C\sin\left(\frac{\pi}{\ell_1}x\right); u_3^2 = C\sin\left(\frac{2\pi}{\ell_1}x\right)$$

$$u_3^3 = C\sin\left(\frac{3\pi}{\ell_1}x\right); u_3^5 = C\sin\left(\frac{5\pi}{\ell_1}x\right)$$

Eigenfunctions are all discrete. Superposition of these possible mode shapes (eigenfunctions) will give the total solution.

Total Solution

$$u_3 = \sum_{n=0}^{\infty} C_n \sin\left(\frac{n\pi}{\ell_1}x\right)$$

Differences

All eigenvalues are orthogonal to each other. The function inner product of two eigenfunctions will be

$$I = \int_0^{\ell_1} \sin\left(\frac{m\pi}{\ell_1}x\right)\sin\left(\frac{n\pi}{\ell_1}x\right)dx$$

$I = 1$ when $m = n$

$I = 0$ when $m \neq n$

The actual solution of C_n will depend on the given force or displacement to the system. An additional initial condition in time will express the coefficient in a Fourier series, and applying the orthogonality condition will solve u_3.

(Continued)

Quantum Mechanics for Engineers

TABLE 5.1 *(Continued)*
Eigenvalue Solution

Quantum Mechanics	Engineering Mechanics
Governing Differential Equation	**Governing Differential Equation**
Visualization of eigenfunctions and the square of their magnitude is shown in	Visualization of eigenfunctions and corresponding wavelength that fits the rope is shown in
Fig. 5.2 left column	Fig. 5.2 Right column
Note the difference between	No such complex understating is required.
Ψ_n and $\lvert\Psi_n\rvert^2$	
Eigenvalues and eigenfunction in two dimensions (to find an electron in a box of length ℓ_1, width ℓ_2)	Eigenvalues and eigenfunction in two dimensions (to find the out of plane displacement of a plate of length ℓ_1, width ℓ_2)
Energy:	Frequency:

$$E_{nr} = \frac{n^2\pi^2\hbar^2}{2m\ell_1^2} + \frac{r^2\pi^2\hbar^2}{2m\ell_2^2}$$

$$\omega_{nr}^2 = \frac{n^2\pi^2 c_s^2}{\ell_1^2} + \frac{r^2\pi^2 c_s^2}{\ell_2^2}$$

Modal wave function

Modal eigenfunction or mode shape

$$\Psi_{nr} = \sqrt{\frac{4}{\ell_1\ell_2}}\sin\!\left(\frac{n\pi}{\ell_1}x_1\right)\sin\!\left(\frac{r\pi}{\ell_1}x_2\right)$$

$$u_3^{\,nr} = C_{nr}\sin\!\left(\frac{n\pi}{\ell_1}x_1\right)\sin\!\left(\frac{r\pi}{\ell_2}x_2\right)$$

$$n,r = 0,1,2,3\ldots$$

$$n,r = 0,1,2,3\ldots$$

Eigenvalues and eigenfunction in three dimensions (to find an electron in a box of length ℓ_1, width ℓ_2 and height ℓ_3)

Eigenvalues and eigenfunction in three dimensions (u_3 displacement in 3D solid body of length ℓ_1, width ℓ_2, and height ℓ_3)

Energy:

Frequency:

$$E_{nrq} = \frac{n^2\pi^2\hbar^2}{2m\ell_1^2} + \frac{r^2\pi^2\hbar^2}{2m\ell_2^2} + \frac{q^2\pi^2\hbar^2}{2m\ell_3^2}$$

$$\omega_{nrq}^2 = \frac{n^2\pi^2 c_s^2}{\ell_1^2} + \frac{r^2\pi^2 c_s^2}{\ell_2^2} + \frac{q^2\pi^2 c_s^2}{\ell_3^2}$$

Modal wave function

Modal eigenfunction or mode shape

$$\Psi_{nrq} = \sqrt{\frac{8}{\ell_1\ell_2\ell_3}}\sin\!\left(\frac{n\pi}{\ell_1}x_1\right)\sin\!\left(\frac{r\pi}{\ell_1}x_2\right)\sin\!\left(\frac{q\pi}{\ell_1}x_3\right)$$

$$u_3^{\,nrq} = C_{nrq}\sin\!\left(\frac{n\pi}{\ell_1}x_1\right)\sin\!\left(\frac{r\pi}{\ell_2}x_2\right)\sin\!\left(\frac{q\pi}{\ell_3}x_3\right)$$

$$n,r,q = 0,1,2,3\ldots$$

$$n,r,q = 0,1,2,3\ldots$$

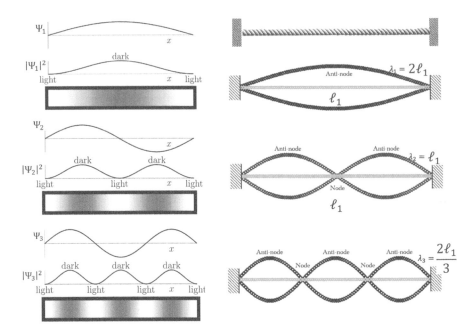

FIGURE 5.2 Comparison and meaning of the wave functions in quantum mechanics and engineering mechanics, respectively.

5.1.7 Time-Dependent Schrödinger Equation

The time-dependent Schrödinger equation is reiterated as

$$\left[-\frac{\hbar^2}{2m}\left(\frac{\partial^2}{\partial x_1^2}+\frac{\partial^2}{\partial x_2^2}+\frac{\partial^2}{\partial x_3^2}\right)+\mathcal{U}\right]\Psi=i\hbar\frac{\partial}{\partial t}\Psi \qquad (5.9.4)$$

where the solution of the wave function should be a function of both time and space, $\Psi(x_j,t)$. Applying the rule of separation of variable $\Psi(x_j,t)=\psi(x_j)f(t)$, one can get the time-dependent part of equation as follows:

$$i\hbar\frac{df}{dt}=E \qquad (5.9.5)$$

The solution could be written in the form $f(t)=\exp(Et/i\hbar)=\exp(-iEt/\hbar)$. Thus, the complete solution of the eigenfunctions for the time-dependent Schrödinger equation (referring to the wave potential written in Eq. 5.5.5) in one, two, and three dimensions, will be visualized as follows. Although it seems complicated, referring to Eq. 5.5.1 and the solutions presented in Table 5.1 are obvious.

One dimension:

$$\Psi(x_j,t)=\psi(x_1)e^{-iEt/\hbar}=\sum_{n=0}^{\infty}\mathcal{A}_n\,e^{ip_{1n}x_1/\hbar}e^{-iE_nt/\hbar} \qquad (5.9.6)$$

Quantum Mechanics for Engineers

155

Two dimensions:

$$\Psi(x_j,t) = \psi(x_1,x_2)e^{-iEt/\hbar} = \sum_{r=0}^{\infty}\sum_{n=0}^{\infty} A_{nr}\, e^{\frac{i(p_{nr1}x_1+p_{nr2}x_2)}{\hbar}} e^{-iE_{nr}t/\hbar} \qquad (5.9.7)$$

Three dimensions:

$$\Psi(x_j,t) = \psi(x_1,x_2,x_3)e^{-iEt/\hbar}$$

$$= \sum_{q=0}^{\infty}\sum_{r=0}^{\infty}\sum_{n=0}^{\infty} A_{nrq}\, e^{\frac{i(p_{nrq1}x_1+p_{nrq2}x_2+p_{nrq3}x_3)}{\hbar}} e^{-\frac{iE_{nrq}t}{\hbar}} \qquad (5.9.8)$$

$$= \sum_{q=0}^{\infty}\sum_{r=0}^{\infty}\sum_{n=0}^{\infty} A_{nrq}e^{-i\mathbf{p}_{nrq}\cdot\mathbf{x}/\hbar} e^{-iE_{nrq}t/\hbar}$$

where p_1, p_2, and p_3 are the momentum of the particle in three directions along x_1, x_2, and x_3, respectively, and are described in the following sections as operators. n, r, and q take values 0 to ∞ and are the modal number. They are introduced in Table 5.1. Here in this solution the potential energy function $\mathcal{U}$ did not receive any special attention. However, it deserves special attention when waves in periodic potentials are discussed. Waves in periodic media will be central discussions in this book in the following chapters.

Here, another very important discussion should take place. Please refer to Eq. 5.9.5. Also refer to Table 4.4, where electromagnetic, acoustic, and elastic waves are compared. It is immediately evident that Eq. 5.9.5 has a first-order time derivative, whereas other wave equations have second-order time derivatives. When the first-order partial time derivative is present in a physical differential equation, it is considered a diffusion equation, not a wave equation. Decaying exponential functions with real eigenvalues are the solutions for such a system with a first-order time derivative. When exponential eigenfunctions have complex eigenvalues representing oscillatory response, it would be the appropriate eigenfunctions for the wave equations having second-order partial derivative with respect to time. On the contrary, the Schrödinger equation, however, does not have a second-order partial time derivative, but still the exponential eigenfunctions have complex eigenvalues. Why is it so?

Unlike other wave equations, which were derived from the first principle of balance equations and solved using wave potentials, Schrödinger arrived at the quantum wave equation assuming a wave potential for a particle – conceptualized based on wave particle duality. A complex momentum operator is responsible for this situation, which resulted in the complex first-order partial time derivative of the wave function in Eq. 5.9.5. Irrespective of this disparity, many quantum phenomena were found to be equivalent for electromagnetic, acoustic, and elastic waves, pointing towards underlying physical parity of wave behavior.

156 Metamaterials in Topological Acoustics

5.1.8 Schrödinger Equation for a Particle in an Electromagnetic Field

Let's assume a particle with charge q has a feature of wave particle duality and the motion of the particle can be expressed by the Schrödinger equation. Repeating the time-dependent Schrödinger equation in Eq. 5.9.5, the momentum operator is further explicitly written as

$$\left[-\frac{\hbar^2}{2m}\nabla^2 + \mathcal{U}\right]\Psi = i\hbar\frac{\partial}{\partial t}\Psi; \left[-\frac{p^2}{2m} + \mathcal{U}\right]\Psi = i\hbar\frac{\partial}{\partial t}\Psi \tag{5.9.9}$$

Usually the function $\mathcal{U}$ for a charge q can be expressed by the electrical potential $q\phi$ when the charge is placed in an electric field with a potential ϕ. However, the charged particle q in an electromagnetic field experiences an electric force due to the electric field $\mathbf{E}$ that cause an additional momentum of the particle. The force experienced by the particle is

$$\mathbf{F} = q\mathbf{E} \tag{5.9.10}$$

Referring to Chapter 4 on electromagnetic fields, with the curl of an electric field, Faraday's law of induction can be written as depicted in Eq. 4.16.3: $\nabla \times \mathbf{E} = -\frac{\partial \mathbf{B}}{\partial t}$. It is obvious that any vector field could be written as a curl of another vector field. Thus Faraday's law could be written as

$$\nabla \times \mathbf{E} = -\frac{\partial}{\partial t}(\nabla \times \mathbf{A}); \ \mathbf{E} = -\frac{\partial \mathbf{A}}{\partial t} \tag{5.9.11}$$

where $\mathbf{B} = \nabla \times \mathbf{A}$.

Thus, the force and additional momentum on a charged particle can be deduced to

$$\mathbf{F} = q\mathbf{E} = m\frac{\partial \mathbf{v}}{\partial t}; \ \mathbf{F} = -q\frac{\partial \mathbf{A}}{\partial t} = m\frac{\partial \mathbf{v}}{\partial t}; \ m\mathbf{v} = -q\mathbf{A} \tag{5.9.12}$$

Modifying the total moment to p to $p - q\mathbf{A}$, the Schrödinger equation will read

$$\left[\frac{1}{2m}\left(i\hbar\nabla + q\mathbf{A}\right)^2 + q\phi\right]\Psi = i\hbar\frac{\partial}{\partial t}\Psi \tag{5.9.13}$$

Or in alternate form

$$\left[\frac{1}{2m}\left(\frac{\hbar}{i}\nabla - q\mathbf{A}\right)^2 + q\phi\right]\Psi = -i\hbar\frac{\partial}{\partial t}\Psi \tag{5.9.14}$$

5.2 QUANTUM OPERATORS

In quantum mechanics the operators are used to define eigenvalue problems. Each operator described below has its own eigenvalues. At each eigenvalue, their unique wave functions are recorded as their respective eigenfunctions. Eigenvalues of a

Quantum Mechanics for Engineers

quantum operators give the values of the quantity when the corresponding quantity is measured during an experiment. The eigenvectors are the quantum states of the quantity at that specific well-defined value of the quantity. In terms of structural dynamics, it means that at a specific frequency (eigenvalue) of vibration, the structure has a specific mode shape (eigenvector), and they together define the eigenstate of the structure. An operator (A) is called a Hermitian if it has the same effect on the inner product with its all its eigenvector state. If $\Psi_1(x_j,t)$ and $\Psi_2(x_j,t)$ are two eigenvectors of the operator A, then the following equation is true:

$$\langle A \ \Psi_1 | \Psi_2 \rangle = \langle \Psi_1 | A \ \Psi_2 \rangle \qquad (5.9.15)$$

Hermitian operators have two unique characteristics.

1. All the eigenvalues of a Hermitian operators are real.
2. All the eigenvectors are distinct and orthogonal to each other.

It is known that these two conditions are satisfied for acoustic and elastic waves, and thus it means that their corresponding Hamiltonian operator is Hermitian.

Based on section 5.1 it is clear that the states of a system actually correspond to a wave function of the system [4]. This is also obvious for acoustic and elastic waves. In this section it is emphasized that the operators are the observable quantities, or the 'observable quantities correspond to the operators' [4]. Please note that any algebraic addition and subtraction of multiple operators could also act as an operator. When the addition and subtraction are conducted in complex plane, then some new unique behavior of the operators emerge. This is due to the uncertainty of a few quantities in quantum mechanics.

Unlike in classical mechanics, in quantum mechanics, all operators are not uniquely solvable, and this results in uncertainty. The Heisenberg principle [2–4] describes the uncertainty between the linear momentum and position of the particle. This means that for a particle, the position and the linear momentum cannot be uniquely determined. Similarly, all components of the angular momentum could not be uniquely known for certain. As depicted in Table 5.1, the solution patterns from the quantum equation and the elastic wave equation are quite similar, and the conclusions from quantum solutions may be equation applicable for acoustic and elastic waves. Quantum mechanics offer the understanding of the operators, which is valuable for studying topological behavior of acoustic and elastic waves.

5.2.1 Hamiltonian Operator

In three-dimensional space, Eq. 5.9.1 will require a wave function to be expressed in the form written in Eq. 5.5.2. Following similar derivation steps, the second-order spatial derivatives of the wave function will have three parts, as appeared in the Laplace operator. The Laplace operator is discussed in section 3.3.3 in Chapter 3. The Laplace operators appeared in acoustics and elastic wave equations in Eqs. 3.6.1, 3.6.2, 3.56, 3.57, and 3.67.2 when similar wave potentials (here wave function) were used. Based on similarity, skipping the detailed derivation, the time-dependent

158 Metamaterials in Topological Acoustics

Schrödinger equation for a particle behaving like a wave in three dimensions will read

$$-\frac{\hbar^2}{2m}\left(\frac{\partial^2}{\partial x_1^2}+\frac{\partial^2}{\partial x_2^2}+\frac{\partial^2}{\partial x_3^2}\right)\Psi+\mathcal{U}\Psi=i\hbar\frac{\partial}{\partial t}\Psi \tag{5.10.1}$$

or

$$\left(-\frac{\hbar^2}{2m}\nabla^2+\mathcal{U}\right)\Psi=i\hbar\frac{\partial}{\partial t}\Psi \tag{5.10.2}$$

Eq. 5.10.2 has the operator on the left-hand side that operates on the wave function and is equal to the equivalent first-order time derivative form of the energy operator (Eq. 5.5.6). Referring to Eq. 5.5, the operator on the left-hand side signifies the energy that is the Hamiltonian of the system. Considering position x_j or $\mathbf{r}$ as an operator, except angular momentum operators (discussed in the next section), major operators in quantum mechanics are summarized as follows:

$$\text{Energy operator: } E=-\frac{\hbar}{i}\frac{\partial}{\partial t}=i\hbar\frac{\partial}{\partial t} \tag{5.11.1}$$

$$\text{Momentum operator: } p_1^2=-\hbar^2\frac{\partial^2}{\partial x_1^2}\text{ and } p_1=-i\hbar\frac{\partial}{\partial x_1} \tag{5.11.2}$$

$$p_2^2=-\hbar^2\frac{\partial^2}{\partial x_2^2}\text{ and } p_2=-i\hbar\frac{\partial}{\partial x_2} \tag{5.11.3}$$

$$p_3^2=-\hbar^2\frac{\partial^2}{\partial x_3^2}\text{ and } p_3=-i\hbar\frac{\partial}{\partial x_3} \tag{5.11.4}$$

$$\text{Kinetic Energy operator: } \mathcal{K}=\frac{p_1^2+p_2^2+p_3^2}{2m}=\frac{p^2}{2m}=-\frac{\hbar^2}{2m}\nabla^2 \tag{5.11.5}$$

$$\text{Hamiltonian operator: } H=-\frac{\hbar^2}{2m}\nabla^2+\mathcal{U} \tag{5.11.6}$$

$$H=\frac{p^2}{2m}+\mathcal{U} \tag{5.11.7}$$

5.2.2 Ladder Operators and Properties

This operator is particularly useful and has more significance when the angular momentum is introduced as an operator. Although it is early in this chapter, an attempt has been made to make sense of the ladder operator using a simple example with two basic operators: position $\mathbf{x}$ and momentum $\mathbf{p}$. For a simple harmonic

Quantum Mechanics for Engineers

oscillator of one degree of freedom, with position x, momentum p, and frequency of oscillation ω, the Hamiltonian operator can be written as

$$H = \frac{p^2}{2m} + \frac{1}{2}m\omega^2 x^2 \qquad (5.11.8)$$

Next, adapting the complex operators from quantum mechanics, let's assume two new operators.

$$A^- = \frac{1}{\sqrt{2}}(x+ip) \text{ and } A^+ = \frac{1}{\sqrt{2}}(x-ip) \qquad (5.11.9)$$

It is obvious that these two new operators are not Hermitian and thus do not represent any observable quantity. *Please note the alternate signs in Eq. 5.11.9.* We are not interested in the eigenvalues or eigenfunctions produced by these operators but are interested in finding what it does to the eigenfunction Ψ if operated (e.g., $A^+\Psi$ or $A^-\Psi$). First let's find the product of these operators for their later use.

$$A^+ A^- = \frac{1}{2}(x-ip)(x+ip) = \frac{1}{2}\left(x^2 + p^2 + i(xp - px)\right) \qquad (5.11.10)$$

$$A^- A^+ = \frac{1}{2}(x+ip)(x-ip) = \frac{1}{2}\left(x^2 + p^2 - i(xp - px)\right) \qquad (5.11.11)$$

With a unitary system, where m and ω are normalized to 1,

$$A^+ A^- = H + \frac{i}{2}[x,p] = H - \frac{1}{2} \qquad (5.11.12)$$

$$A^- A^+ = H - \frac{i}{2}[x,p] = H + \frac{1}{2} \qquad (5.11.13)$$

where $[x,p] = (xp - px) = i\hbar$ is used from the commutator relation discussed in section 5.2.4.

Now let's see how any of the complex operators A^+ or A^- changes the eigenstate (eigenvalues and eigenfunction) E and Ψ. From Eq. 5.8.6 $H\Psi = E\Psi$ is valid for this simple system. Let's test the following operations using Eqs. 5.11.12 and 5.11.13.

$$H\left(A^+\Psi\right) = \left(A^+ A^- + \frac{1}{2}\right)\left(A^+\Psi\right) \qquad (5.11.14)$$

$$H\left(A^+\Psi\right) = A^+\left(A^- A^+ + \frac{1}{2}\right)(\Psi) \qquad (5.11.15)$$

$$H\left(A^+\Psi\right) = A^+ (H+1)(\Psi) \qquad (5.11.16)$$

$$H\left(A^+\Psi\right) = (H+1)\left(A^+\Psi\right) \qquad (5.11.17)$$

$$H\left(A^+\Psi\right) = (E+1)\left(A^+\Psi\right) \qquad (5.11.18)$$

It says that the new eigenfunction $A^+\Psi$ has an eigenvalue one level above the original eigenvalue for the eigenfunction Ψ when the Hamiltonian is well defined. Similarly

$$H\left(A^-\Psi\right) = \left(A^- A^+ - \frac{1}{2}\right)\left(A^-\Psi\right) \qquad (5.11.19)$$

$$H\left(A^-\Psi\right) = A^- \left(A^+ A^- - \frac{1}{2}\right)(\Psi) \qquad (5.11.20)$$

$$H\left(A^-\Psi\right) = A^- (H-1)(\Psi) \qquad (5.11.21)$$

$$H\left(A^-\Psi\right) = (H-1)\left(A^-\Psi\right) \qquad (5.11.22)$$

$$H\left(A^-\Psi\right) = (E-1)\left(A^-\Psi\right) \qquad (5.11.23)$$

It says that the new eigenfunction $A^-\Psi$ has an eigenvalue one level below the original eigenvalue for the eigenfunction Ψ when the Hamiltonian is well defined. Hence, the complex plus and minus operators can increase or decrease the eigenstates going up and down the ladder. This ladder is of course not infinite in either direction. Although shown for the Hamiltonian, this basic ladder concept is true for angular momentum operators discussed in section 5.2.4.4.

5.2.3 ANGULAR MOMENTUM OPERATORS

In classical mechanics, the angular momentum about a specific axis is described by the moment of the momentum about that axis. The linear momentum of an object is described by the product of its mass and velocity. The moment is taken to describe the rotational motion of an object. Moment is described by a vector cross product and indicates the rotational plane that is orthogonal to the axis, about which the moment is taken. That means that it is necessary to know the position vector $\mathbf{r} \equiv (x_1, x_2, x_3) \equiv x_i\hat{\mathbf{e}}_i$ of the object, the mass m of the object and the velocity $\mathbf{v} \equiv (v_{x_1}, v_{x_2}, v_{x_3}) \equiv v_i\hat{\mathbf{e}}_i$ of the object in a space to find the angular momentum. Angular momentum is required to describe the motion of the celestial bodies like planets moving around their respective suns. Similarly, in quantum mechanics, motion of an electron around a nucleus requires angular momentum. This momentum is called the orbital angular momentum (OAM). In quantum mechanics the motion of an electron can also manifest its own axis. Such motion is called the 'spin'. As a planet rotates about its sun, it also spins about its own axis. Thus, the particles may have additional angular momentum called spin angular momentum (SAM). Together OAM and SAM give the definition of the total angular momentum (TAM). The eigenvalues of angular momentum and their quantum numbers are critically important in quantum mechanics. Thus, is it inevitable that the quantum analogue applications of acoustic elastic waves will have equal importance.

Quantum Mechanics for Engineers

For the photon particles (e.g., light or electromagnetic waves) the orbital angular momentum could be divided into two types, internal OAM and external OAM. Orbital angular momentum for electromagnetic waves is unrelated to polarization. External OAM purely depends on the reference coordinate system or the origin. External OAM is calculated using spatial distribution of the parameters of the electromagnetic field (e.g., the electric field $\mathbf{E}$ or the magnetic field $\mathbf{B}$). If the field distribution of an electromagnetic field is symmetric about the central axis of a propagating wave, and the reference axis (about which the OAM is calculated) is selected along the wave vector direction, then the orbital angular momentum will vanish. This will not be the case if the wave vector direction is not aligned with the axis of interest about which the OAM is calculated. This OAM is called the external OAM.

In contrast, internal OAM is more valuable when the electromagnetic field is not symmetric about the central axis but propagates like a helical wave front. Here, polarity of the wave field is an important parameter, where the polarity of the wave field changes continuously. When the wave amplitude remains constant, but the polarity rotates at a constant rate on a plane perpendicular to the direction of the wave propagation, a wave vortex is created. This is called circularly polarized waves. Circular polarization is manifested when electric ($\mathbf{E}$) and magnetic field ($\mathbf{B}$) continuously rotates around the wave propagation axis. The change in polarity is caused by the difference in the phase between the two planar components. This is possible only when the two orthogonal components of the fields are at a phase difference of $\pi/2$ for an anticlockwise rotation (left-handed) or at a phase difference of $-\pi/2$ for a clockwise rotation (right-handed). Hence, please note that the phasal differences and the change in polarity are synonymous.

A wave front in a helical shape is possible only when the wave field has singularity at the center but the phase of the wave field circularly changes from 0 to 2π over one wavelength (λ) or the integer multiple of wavelength ($|m|\lambda$), where m is an integer discussed in the next paragraph. If the phasal change of 0 to 2π is accommodated over one wavelength, then there will be only one helical wavefront. However, if the phasal change of 0 to 2π is spread over an integer multiple of the wavelength, then there are possibilities of multiple helices intertwined at a longitudinal spacing of single wavelength (λ). Please refer to Fig. 5.3a–e. Single-helix or multiple-helix wave fields tend to generate OAM of different charges (or topological charges) based on the number of helical wave fronts present in the wave field. Here, 'charge' should not be confused with the electrical charge. It is nothing but the quantum number associated with the angular momentum; at least, this is how it is called in quantum mechanics. Next, the rotation of the helix could be anticlockwise (left-handed) or clockwise (right-handed), which will make this charge positive (+) or negative (−), respectively. Irrespective of the number of helices or the topological charge, left-handedness or right-handedness, each helix generates a phase difference of 0 to 2π. Overall, this phenomenon is called the photonic quantum vortex, electromagnetic vortex, or simply the optical vortex for the light waves of specific topological charge, and internal OAM does not vanish.

Figure 5.3a–e show a well-known schematic of the situations with different OAMs. Figure 5.4 shows a typical wave form and corresponding phases in a grayscale that is used in Fig. 5.3. Generally, the external OAM is called the OAM with

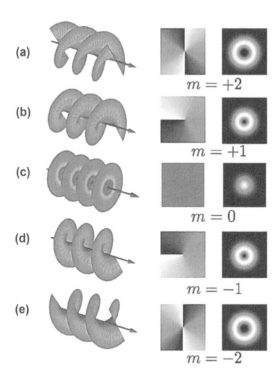

FIGURE 5.3 Schematic showing situations for external and internal OAM for electromagnetic waves: a) optical vortex with topological charge +2, b) with topological charge +1, c) external OAM is zero, with topological charge zero, d) optical vortex with topological charge −1, e) with topological charge −2. m is the quantum number for the OAM.

topological charge zero. The *m* parameter in Fig. 5.3 is called the 'topological charge' and is a quantum number for angular momentum. Please recollect that the external OAM vanishes with the choice of the momentum axis (i.e., when it is aligned with the direction of the wave propagation). In Fig. 5.3c the field is symmetric about the central axis. This is not the case when $m \neq 0$. When the topological charge is $m = +1$

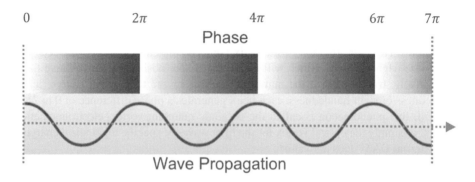

FIGURE 5.4 Wave form and corresponding phase.

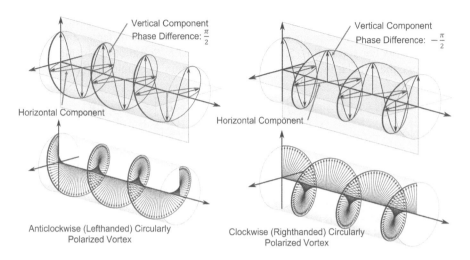

FIGURE 5.5 Wave vortex.

(Fig. 5.3b), there is only one helical wave front, and the phase of the wave field circularly changes from 0 to 2π over one wavelength (λ). It is caused by two orthogonal components of the wave fields that are at a phase difference of $\pi/2$ with anticlockwise rotation as shown in Fig. 5.3a. This is opposite when the topological charge is $m = -1$ (Fig. 5.3 d). There is only one helical wave front, and the phase of the wave field circularly changes from 0 to 2π over one wavelength (λ). Further, it is caused by two orthogonal components of the wave field that are at a phase difference of $\pi/2$ with clockwise rotation as shown in Fig. 5.3 b. When the topological charge is $m = +2$ (Fig. 5.3a), then two helical wave fronts are generated at a phase difference of π. But the phase of the wave field of each helix circularly changes from 0 to 2π in counterclockwise direction over two wavelengths (2λ). In the case of topological charge $m = -2$ (Fig. 5.3e), two helical wave fronts are started at a phase difference of π, with each circularly changing its phase in a clockwise direction between 0 and 2π over the two wavelengths (2λ). Figure 5.5 shows the wave vortex in graphical form.

Above, OAM is described in terms of electromagnetic waves, which is easier to explain because the waves of photons are with spin 1. However, in quantum mechanics, the eigenstates for the angular momentum of the particles may not always have integer eigenvalues or integer topological charges. In quantum mechanics, particles called fermions will have non-integer eigenstates. Particles that rotate about their own axes, resulting in angular momentum due to their spin (spin angular momentum) but having non-integer eigenstates, are unrealistic. Thus, these discussions would take the next leap in this chapter to introduce quantum concept of angular momentum. In the following sections these concepts are mathematically treated for final rigor.

5.2.3.1 Orbital Angular Momentum (OAM)

Orbital angular momentum $\mathbf{L} \equiv (L_1, L_2, L_3) \equiv L_i \hat{\mathbf{e}}_i$ is described by

$$\mathbf{L} = \mathbf{r} \times \mathbf{p} = \mathbf{r} \times (m\mathbf{v}) \tag{5.12}$$

where $\mathbf{p} \equiv (p_1, p_2, p_3) \equiv p_i \hat{\mathbf{e}}_i$ is the liner momentum of the object. In quantum mechanics however, the momentum (p_i) of a particle in three different directions is the operators derived in Eqs. 5.11.2, 5.11.3, and 5.11.5. Substituting the momentum operators, the angular momentum operator in quantum mechanics will read

$$\mathbf{L} = \mathbf{r} \times \mathbf{p} = x_i \hat{\mathbf{e}}_i \times p_i \hat{\mathbf{e}}_i = -i\hbar \left(x_i \hat{\mathbf{e}}_i \times \frac{\partial}{\partial x_j} \hat{\mathbf{e}}_j \right) = -i\hbar \left(\mathbf{r} \times \nabla \right) \tag{5.13}$$

$$L_i = -i\hbar \left(\epsilon_{ijk} \, x_j \frac{\partial}{\partial x_k} \hat{\mathbf{e}}_i \right) \tag{5.14}$$

where ∇ is the gradient operator or the nabla Hamiltonian operator discussed in Chapters 2 and 3. ϵ_{ijk} is called the permutation symbol, or Livi-Civita symbol [8]. Most of the time, the angular momentum is described in spherical coordinates $(\mathbf{L} \equiv (L_r, L_\theta, L_\phi))$ due to the rotational nature of the particles, like electrons (fermions) or photons (bosons). Utilizing the definition of the gradient operator in spherical coordinate system in Chapter 2, Eq. A.2.20, and referring to Fig. A.2.2, the position vector $\mathbf{r}$ will simply be the radial distance from the origin and will be equal to $r = \sqrt{x_i^2}$, and thus the angular momentum will be

$$\nabla = \left(\frac{\partial}{\partial r} \hat{\mathbf{e}}_r + \frac{1}{r\sin\phi} \frac{\partial}{\partial \theta} \hat{\mathbf{e}}_\theta + \frac{1}{r} \frac{\partial}{\partial \phi} \hat{\mathbf{e}}_\phi \right) \tag{5.15.1}$$

$$\mathbf{L} = -i\hbar \left(\mathbf{r} \times \nabla \right) = -i\hbar \left(|\mathbf{r}| \nabla \right) = -i\hbar \, r \left(\frac{\partial}{\partial r} \hat{\mathbf{e}}_r + \frac{1}{r\sin\phi} \frac{\partial}{\partial \theta} \hat{\mathbf{e}}_\theta + \frac{1}{r} \frac{\partial}{\partial \phi} \hat{\mathbf{e}}_\phi \right) \tag{5.15.2}$$

$$L_r = -i\hbar \, r \frac{\partial}{\partial r}; \; L_\theta = -i\hbar \, \frac{1}{\sin\phi} \frac{\partial}{\partial \theta}; \; L_\phi = -i\hbar \, \frac{\partial}{\partial \phi} \tag{5.15.3}$$

In classical mechanics, the x_3 component of the motion of a body describes the motion along the x_3 axis. However, L_3, the x_3 component of the angular momentum $\mathbf{L}$, describes the motion about or around the x_3 axis. In quantum mechanics, the x_3 component of the linear momentum p_3 is written as $-i\hbar \dfrac{\partial}{\partial x_3}$ (Eq. 5.11.5), where the derivative with respect the x_3 coordinate is taken along the x_3-axis. Hence, logically the equivalent for x_3 angular momentum is $-i\hbar$ times the derivative with respect to the angle ϕ around the x_3-axis. So, sometimes L_ϕ is equivalently written as L_3. Similar orbital angular momentum is described in Chapter 6 for elastic and acoustic waves.

$$\text{OAM Operator: } L_3 = L_\phi = -i\hbar \, \frac{\partial}{\partial \phi} \tag{5.15.4}$$

In quantum mechanics, there will never be a case where an exact zero OAM vector exists. It is obvious that there are three orbital angular momentum operators namely L_1, L_2, and L_3.

Quantum Mechanics for Engineers

As in Eq. 5.11.9, let's introduce two new complex OAM operators using the L_1, L_2.

$$L^+ = L_1 + iL_2 \tag{5.15.5}$$

$$L^- = L_1 - iL_2 \tag{5.15.6}$$

As discussed in section 5.2.2, these operators will also result in a raised or lowered eigenstate when operated with the current OAM eigenstate, respectively.

5.2.3.2 Spin Angular Momentum (SAM)

Spin angular momentum is noted as $\mathbf{S} \equiv (S_1, S_2, S_3) \equiv S_i \hat{\mathbf{e}}_i$. Spin is an intrinsic property of a particle. An electron in quantum mechanics spinning about its axis has either $+1/2$ or $-1/2$ spin. Two electrons in same quantum levels are entangled by their opposite spins. The reason for $+1/2$ or $-1/2$ comes from Pauli's exclusion principle [2], which says that all quantum numbers for two particles cannot be the same. As they are the eigenvalues, two spin states of a particle are orthogonal to each other. It was mentioned in Table 5.1 that the wave functions written in Eqs. 5.9.6 to 5.9.8 do not give the exact location of a particle, but the inner product must be equal 1 to signify that the probability of finding the particle at a point is 1. In acoustic and elastic waves, the velocity field and the displacement field have their inherent polarity explicitly present in the wave function. Polarity of a wave function can be defined as a direction of the amplitude vector. The quantum wave function $\Psi(x_j, t)$ is not the polarity-defining function but a probability density function. It cannot give any polarity or direction. Spin, on the other hand, has direction because it is necessary to mention which axis the particle is rotating about. Quantum particles, mostly fermions, exist in pairs, with spin up $+1/2$ ($\uparrow$) or spin down $-1/2$ ($\downarrow$) spins. This information is not included in the wave function $\Psi(x_j, t)$. But it is now required to be included. Hence the wave function modifies to

$$\Psi = \Psi(x_j, S_3, t) \tag{5.16.1}$$

$|\Psi|^2$ is the square magnitude of the probability density function, if integrated over the entire space or volume (i.e., $\int |\Psi(x_j, t)|^2 \, d\Omega$), the integral will result in the probability of finding the particle at location x_j. If $|\Psi|^2$ derived from Eq. 5.16.1 is integrated over the entire space or volume (i.e., $\int |\Psi(x_j, S_3, t)|^2 \, d\Omega ds$), the integral will result the probability of finding the particle with spin angular momentum S_3 about the x_3 axis at location x_j. Similarly, other components of spin angular momentum can also be included in the wave function, as needed. As spin quantum numbers are not continuous but just two discrete numbers $+1/2$ ($\uparrow$) or spin down $-1/2$ ($\downarrow$), the integral space for spin is not continuous but just two points. Integral space is not possible unless a vector of wave functions with two separate spin angular momentum are considered to ensure the summation. This modifies the wave function as follows.

$$\Psi = \Psi(x_j, S_3, t) = \begin{bmatrix} \Psi\left(x_j, \frac{1}{2}\hbar, t\right) \\ \Psi\left(x_j, -\frac{1}{2}\hbar, t\right) \end{bmatrix} = \begin{bmatrix} \Psi^+(x_j, t) \\ \Psi^-(x_j, t) \end{bmatrix} \quad or \quad \begin{bmatrix} \Psi^\uparrow(x_j, t) \\ \Psi^\downarrow(x_j, t) \end{bmatrix} \tag{5.16.2}$$

where the square magnitude integral of Ψ^+ or $\Psi^\uparrow$ wave function gives the probability of finding a particle at location x_j with spin up, and the square magnitude integral of Ψ^- or $\Psi^\downarrow$ wave function gives the probability of finding a particle at location x_j with spin down. This two-dimensional vector in three-dimensional space is called 'spinor' [9]. Using the spinor, the wave function will read.

$$\Psi\left(x_j, S_3, t\right) = \Psi^+\left(x_j, t\right) + \Psi^-\left(x_j, t\right) \qquad (5.16.3)$$

These two spin eigenstates are orthogonal and are related to the spin unit vectors as are used in the Cartesian coordinate system $\hat{\mathbf{e}}_i$. Spin unit vectors $\hat{\mathbf{s}}_1$ and $\hat{\mathbf{s}}_2$ are noted respectively $\left(\frac{1}{2}\hbar\right)$ and $\left(-\frac{1}{2}\hbar\right)$. For $S_3(\uparrow)$ and $S_3(\downarrow)$ they will respectively be

$$\hat{\mathbf{s}}_1 = 1, \hat{\mathbf{s}}_2 = 0; \hat{\mathbf{s}}_1 = 0, \hat{\mathbf{s}}_2 = 1 \qquad (5.16.4)$$

Spin distinguishes two types of elementary particles: a) fermions and b) bosons [10]. Fermions have half-integer spins, but bosons have integer spins. Photons are called gauge bosons with spin 1. Photon spins are connected to the circular polarization discussed in section 5.2.3 (Fig. 5.3). Additionally, as with Eq. 5.15.6, let's introduce two additional complex operators for SAM.

$$S^+ = S_1 + iS_2 \text{ and } S^- = S_1 - iS_2 \qquad (5.16.5)$$

As discussed in section 5.2.2, these operators will also result in a raised or lowered eigenstate when operated with the current SAM eigenstate, respectively.

5.2.3.3 Total Angular Momentum (TAM)

Total angular momentum is noted as $\mathbf{J} \equiv (J_1, J_2, J_3) \equiv J_i \hat{\mathbf{e}}_i$.

$$\mathbf{J} = \mathbf{L} + \mathbf{S} \qquad (5.17.1)$$

For a system, the total angular momentum must always be conserved, but it is not necessary that the orbital angular momentum or the spin angular momentum are conserved individually. In the case of nonzero angular momentum, the orbital angular momentum about any arbitrary axis is always less than the total angular momentum about that axis, even when the orbital angular momentum vector is aligned with the arbitrary axis selected. This is due to the uncertainty of angular momentum in quantum mechanics. If one component of the angular momentum is known for certain, the other two components will be uncertain.

Due to the uncertainty principle and non-commutating behavior (see the next section), square total angular momenta are discrete in quantum mechanics, and take only the multiple of half integer values as per the rule below.

$$J^2 = j(j+1)\hbar^2; \; j = 0, \frac{1}{2}, 1, \frac{3}{2}, 2, \frac{5}{2}.... \qquad (5.17.2)$$

Quantum Mechanics for Engineers

where j is azimuthal quantum number. The J_3 operator, considering any arbitrary direction as x_3 with magnetic quantum number, m_j will be

$$J_3 = m_j \hbar;\ m_j = -j\ to + j\ \text{with integer increment} \qquad (5.17.3)$$

This relation is not true for J_1 and J_2, due to incommutability of the other two orthogonal components of total angular momentum as discussed in section 5.2.3. Here, it must be recognized that for a particle without any preferential direction, any arbitrary direction can be used as x_3 axis and other two orthogonal axes as x_1 and x_2. Additionally, as with Eq. 5.15.6, let's introduce two additional complex operators for TAM

$$J^+ = J_1 + iJ_2 \text{ and } J^- = J_1 - iJ_2 \qquad (5.17.4)$$

As is discussed in section 5.2.2, these operators will also result in a raised or lowered eigenstate when operated with the current TAM eigenstate, respectively.

5.2.3.4 Square Orbital Angular Momentum

When orbital angular momentum ($\mathbf{L}$) describes the momentum of the particle about any arbitrary axis, sometimes it is necessary to know only the magnitude of the OAM. OAM magnitude can be calculated as follows.

$$L^2 = \mathbf{L}.\mathbf{L} = -i\hbar\ (\mathbf{r} \times \nabla)\ . - i\hbar\ (\mathbf{r} \times \nabla) = -\hbar^2 \mathbf{r}.\left(\nabla \times (\mathbf{r} \times \nabla)\right) \qquad (5.17.5.1)$$

$$L^2 = -\hbar^2 x_i\ \epsilon_{ijk}\ \left(\frac{\partial}{\partial x_j} \left(\epsilon_{klm}\ x_l\ \frac{\partial}{\partial x_m} \right) \right) \qquad (5.17.5.2)$$

where ϵ_{ijk} is again the Livi-Civita symbol. Transforming L^2 to the spherical coordinate system, the square angular momentum operator will read

$$\text{Square AM Operator: } L^2 = -\hbar^2\ \frac{1}{\sin\theta}\ \frac{\partial}{\partial\theta}\left(\sin\theta\ \frac{\partial}{\partial\theta} \right) - -\hbar^2\ \frac{1}{\sin^2\theta}\ \frac{\partial^2}{\partial\phi^2} \qquad (5.17.6)$$

Due to the uncertainty principle and non-commutating behavior (see the next section), square orbital angular momentum is discrete in quantum mechanics and takes only the integer values as per the rule below

$$L^2 = l(l+1)\hbar^2;\ l = 0,1,2,3,\ldots \qquad (5.17.7)$$

where l is the azimuthal quantum number. The L_3 operator, considering any arbitrary direction as x_3 with magnetic quantum number, m_l will be

$$L_3 = m_l \hbar;\ m_l = -l\ to + l\ \text{with integer increment} \qquad (5.17.8)$$

This relation is not true for L_1 and L_2, due to incommutability of the other two orthogonal components of orbital angular momentum as discussed in section 5.2.3.

168 Metamaterials in Topological Acoustics

Here, it must be recognized that for a particle without any preferential direction, any arbitrary direction can be used as x_3 axis and other two orthogonal axes x_1 and x_2. Earlier it was discussed that an operator must have eigenvalues and a set of eigenfunctions to describe the eigenstates. As the eigenvalues are well defined for L_3 for specific L^2, the respective eigenfunctions $(Y_{m_l}^l)$ can be found from Eq. 5.15.4 when Eq. 5.17.8 is substituted.

$$L_3 Y_{m_l}^l = -i\hbar \frac{\partial}{\partial\phi} Y_{m_l}^l = m_l \hbar Y_{m_l}^l \qquad (5.17.9)$$

$$\text{or } -i\hbar \frac{\partial}{\partial\phi} Y_{m_l}^l = m_l \hbar Y_{m_l}^l \qquad (5.17.10)$$

$Y_{m_l}^l$ provides all the 'spherical harmonics' [2, 10] for specific l and a set of defined m_l from $-l$ to $+l$ as the eigenfunctions for all the possible eigenstates of OAM.

5.2.3.5 Square Spin Angular Momentum

Similarly, the magnitude of the spin angular momentum can be calculated as follows:

$$S^2 = \mathbf{S.S} = S_1^2 + S_2^2 + S_3^2 \qquad (5.17.11)$$

Due to the uncertainty principle and non-commutating behavior (see the next section), square spin angular momenta are discrete in quantum mechanics and take only the multiple of half integer values as per the rule below.

$$S^2 = s(s+1)\hbar^2; \ s = 0, \frac{1}{2}, 1, \frac{3}{2}, 2, \frac{5}{2}.... \qquad (5.17.12)$$

Multiples of half integers are for fermions, and multiples of integers are for bosons. s is an azimuthal quantum number for spin. The S_3 operator, considering any arbitrary direction as x_3 with spin magnetic quantum number, m_s will be

$$S_3 = m_s \hbar; \ m_s = -s \ to + s \ \text{with integer increment.} \qquad (5.17.13)$$

This relation is not true for S_1 and S_2, due to incommutability of the other two orthogonal components of spin angular momentum as discussed in section 5.2.3. Here, it must be recognized that for a particle without any preferential direction, any arbitrary direction can be used as x_3 axis and other two orthogonal axes as x_1 and x_2.

5.2.4 Observable Operators

Now it is necessary to be familiarized with the quantum mechanics terminologies. The next most important concept of the quantum mechanics is the uncertainty principle. Observable quantities are the physical or quantum quantities that can be measured through experiments in a laboratory. Operators mentioned in these sections are the

- total energy operator
- position operator

Quantum Mechanics for Engineers

- momentum operator
- kinetic energy operator
- Hamiltonian operator
- orbital angular momentum operator
- spin angular momentum operator
- total angular momentum operator
- square angular momentum operator

These operators are supposed to be the observables. It is classically recognized that if one quantity (operator) is observed, another quantity is in relation to the first quantity, and could be uniquely calculated without ambiguity. However, in quantum mechanics, if two observable operators do not commute, then they are called complementary observable operators. Here it is necessary to know what is meant by commuting operators. Let's assume $\mathbb{A}$ and $\mathbb{B}$ are the two operators of a system. For $\mathbb{A}$ and $\mathbb{B}$ to be commutable

$$[\mathbb{A}, \mathbb{B}] = \mathbb{AB} - \mathbb{BA} = 0 \tag{5.18}$$

where $[,]$ is a commutator and $\mathbb{A}$ and $\mathbb{B}$ are the two elements of ring theory in mathematics [11]. Details on ring theory are omitted here; however, the most valuable deduction from ring theory is used for further discussion. If $\mathbb{A}$ and $\mathbb{B}$ does not commute, then Eq. 5.18 will result in a nonzero value. In this situation, $\mathbb{A}$ and $\mathbb{B}$, two complementary observables, cannot be measured simultaneously. If one is measured, the other will result in some form of uncertainty in the measurement. The uncertainty, however, is not arbitrary. The commuting operator on these complementary observables carries a certain value and satisfies an uncertainty principle.

5.2.4.1 Heisenberg Uncertainty Principle

Among these principles, the Heisenberg uncertainty principle [2] is the most popular. Heisenberg principle was the first principle that actually revealed the manifestation of uncertainty in quantum mechanics. In the Heisenberg uncertainty principle, $\mathbb{A}$ is the position ($\mathbf{x}$) of a particle like an electron and $\mathbb{B}$ is the momentum ($\mathbf{p}$) of the particle. The Heisenberg uncertainty principle says that the position and the momentum of any particle cannot be observed simultaneously with absolute certainty. The inequality relation was later modified by Kennard and Weyl [12] where they used the standard deviation of position σ_x and the standard deviation of momentum σ_p to write the Heisenberg uncertainty principle as

$$\sigma_x \sigma_p \geq \frac{\hbar}{2} \tag{5.19}$$

Considering wave particle duality introduced in section 5.1.3, the position of a particle is described by Eq. 5.5.1. Omitting the time harmonic part, the spatial wave function in one dimension will read

$$\psi(x_1) = \mathcal{A} e^{i(kx_1)} \tag{5.20.1}$$

Recollecting Eq. 5.5.5, the spatial wave function can further be expressed in terms of momentum as

$$\psi(x_1) = A \exp\left(\frac{i}{\hbar}(px_1)\right) \tag{5.20.2}$$

From the solution of the Schrödinger equation, it is well-known that the wave function to describe the position of a particle is not absolute, but the summation over all possible energy states (Eqs. 5.9.6 through 5.9.8) must be considered. Hence, Eq. 5.20.1 in any dimension will read

$$\psi(\mathbf{x}) = \sum_{nrq} A_{nrq} \exp\left(\frac{i}{\hbar}(\mathbf{p}_{nrq}.\mathbf{x})\right) \tag{5.20.3}$$

By now it is understood that the wave function $\psi(\mathbf{x})$ describes the probability of finding a particle at the point $\mathbf{x}$. Hence, if the $\psi(\mathbf{x})$ function pinpoints a location of the particle with probability equal to 1, then that position is resulted from the summation over all possible momenta $\mathbf{p}_{nrq}$ of the particle. Considering a one-dimensional scenario, Fig. 5.6 shows the situation in a graphical form. In Fig. 5.6a shows a position-momentum plane on which the objective is to find the particle at the crosshair intersection. However, the position and momentum of the particle create a shaded blob with ambiguity. If the position is precisely detected as indicated in Fig. 5.6b and c, the momentum spreads across its axis with all possible values, as in Eq. 5.20.2. Similarly, if the objective is to find the momentum precisely, (Fig. 5.6d and e), then the blob spreads across the position axis. No matter how precisely the experiment is conducted, the size of the blob is always constant and is a function of Planck's

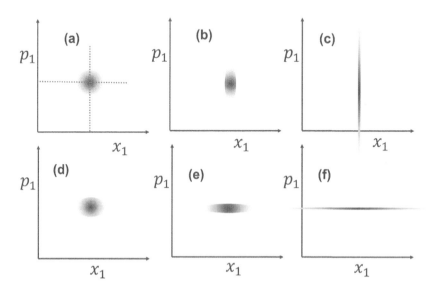

FIGURE 5.6 Schematics for Heisenberg uncertainty principle with ambiguity in precisely determining both the position and momentum simultaneously, (a-f) shows such different scenarios.

Quantum Mechanics for Engineers 171

constant as indicated in Eq. 5.19, which will never disappear, causing uncertainty. The commutator operator operated on the complementary observables position, and the momentum will result.

$$[x, p] = xp - px = i\hbar \qquad (5.20.4)$$

5.2.4.2 Uncertainty Principle for Angular Momentum

Like position and momentum, any two components of the orbital angular momentum of a particle do not commute. Considering three components of orbital angular momentum $\mathbf{L} \equiv (L_1, L_2, L_3) \equiv L_i \hat{\mathbf{e}}_i$, the commutation relation presents

$$[L_1, L_2] = L_1 L_2 - L_2 L_1 = i\hbar L_3 \qquad (5.21.1)$$

$$[L_2, L_3] = L_2 L_3 - L_3 L_2 = i\hbar L_1 \qquad (5.21.2)$$

$$[L_3, L_1] = L_3 L_1 - L_1 L_3 = i\hbar L_2 \qquad (5.21.3)$$

or in index notation

$$[L_i, L_j] = L_i L_j - L_j L_i = \sum_{n=1}^{3} i\hbar \in_{ijn} L_n \qquad (5.21.4)$$

Similar commutation relations are also valid for spin angular momentum components and the total angular momentum components as follows:

$$[S_i, S_j] = S_i S_j - S_j S_i = \sum_{n=1}^{3} i\hbar \in_{ijn} S_n \qquad (5.21.5)$$

$$[J_i, J_j] = J_i J_j - J_j J_i = \sum_{n=1}^{3} i\hbar \in_{ijn} J_n \qquad (5.21.6)$$

Although neither of the two components of the orbital angular momentum commutes, all components individually commute to the square angular momentum. Mathematically, this statement will be

$$[L^2, L_1] = [L^2, L_2] = [L^2, L_3] = 0 \qquad (5.21.7)$$

Now let's pick L_3 which commutes with L^2. That means they both have common eigenstates and eigenvalues. As they do not commute with L_1 and L_2, these two operators will not have well-defined eigenvalues. It means an eigenstate has well-defined magnitude L^2, with a well-defined L_3 component but all open possibilities for L_1 and L_2. This is possible only when we assume a right circular cone of a specific height, ensuring well-defined nature of L_3 and L^2 but all possible values for L_1 and L_2 on the plane aligned to the base of the cone. Please refer to Fig. 5.7a, where both positive and negative axes of L_3 are shown with two inverted cones as one set. Also please note from Eq. 5.17.7 that for a specific quantum number l, the L^2 value is well defined but from Eq. 5.17.8 the L_3 values are unique in the range from $-l$ to $+l$. For example, if $l = 2$, $L^2 = 6\hbar$, and $L_3 = -2\hbar$, $-\hbar$, 0, $+\hbar$, $+2\hbar$. Now, for each case of $L^2 - L_3$ combination, a total of three sets of cones will resolve to show the uncertainty of L_1 and L_2 distributed over the base of the respective cones, as is shown in Fig. 5.7b.

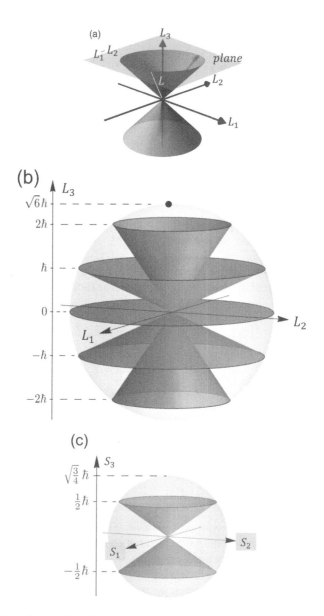

FIGURE 5.7 a) Schematics of the cone showing OAM States, b) azimuthal and magnetic quantum number for OAM, and c) same for spin angular momentum.

Quantum Mechanics for Engineers

It is also true for spin angular momentum and total angular momentum as follows

$$\left[S^2, S_1\right] = \left[S^2, S_2\right] = \left[S^2, S_3\right] = 0 \tag{5.21.8}$$

$$\left[J^2, J_1\right] = \left[J^2, J_2\right] = \left[J^2, J_3\right] = 0 \tag{5.21.9}$$

For spin, let's test with azimuthal spin quantum number $s = 1/2$. Then the spin magnetic quantum numbers will be $m_s = -\frac{1}{2}, 0, +\frac{1}{2}$. From Eq. 5.17.12 The square spin quantum number $S^2 = \frac{3}{4}\hbar$. The $S_3 = -\frac{1}{2}\hbar, 0$ and $+\frac{1}{2}\hbar$. Following similar concept of right-angle cones, when S^2 and S_3 are well defined, SAM for a fermion, with quantum number one half, there will be two sets of cones that represents the uncertainty of S_1 and S_2,. Please refer to Fig. 5.7c. Spinors will not have any well-defined space but will have three S_1 and S_2 planes at $S_3 = -\frac{1}{2}\hbar, 0$ and $+\frac{1}{2}\hbar$.

Please note that similar commutation relation for TAM, OAM, and SAM is also applicable for photons in electrodynamics and phonons in elastic and acoustic waves discussed in Chapter 6. In AM relation for elastic and acoustic waves, the Plank constant is not present. Also, the SAM is purely for spin-1 boson type particles in electrodynamics and elastodynamics (please refer section 6.7.3.2 in Chapter 6).

5.2.4.3 Convention to Express the Angular Momentum

Irrespective of spin or orbital or total angular momentum, due to the uncertainty principle described above, it is necessary to express the angular momenta using two numbers, one is the square angular momentum and the other is the third component of the angular momentum. These TAM, OAM, and SAM will be expressed by the eigenstates using respective azimuthal quantum number (AQN) and magnetic quantum number (MQN) as follows:

$$|AQN\ MQN\rangle \tag{5.22}$$

i.e., $|j\ m_j\rangle$ for TAM, $|l\ m_l\rangle$ for OAM, and $|s\ m_s\rangle$ for SAM. The eigenstate Ψ used in earlier sections (especially in section 5.2.2, Eqs. 5.11.13 through 5.11.24) is now equivalent to the bra-ket notation described using two numbers. Such a Ψ or $|AQN\ MQN\rangle$ eigenstate is synonymous with the eigenfunction $Y_{m_l}^l$ introduced for the L_3 operator of OAM in Eq. 5.17.10. When it is obvious that the angular momentum is expressed for a specific type, TAM or OAM or SAM, then only m is used to express the magnetic quantum number, eliminating their subscripts, satisfying respective square angular momentum and x_3-axis angular momentum in Eqs. 5.17.2 through 5.17.13.

Any eigenstate of angular momentum (TAM, OAM or SAM) can be expressed by Eq. 5.22. In this section, to be generic for all angular momenta (TAM, OAM or SAM), angular momentum is expressed as $\mathbf{A} \equiv (A_1, A_2, A_3) \equiv A_i\hat{\mathbf{e}}_i$. Generic AQN and MQN are used to express the eigenstate as $|a\ m\rangle$. The following statements are true for all three angular momenta once the x_3 axis is defined. The x_1 and x_2 components for the angular moment operator are A_1 and A_2. Except for zero eigenstate $|0\ 0\rangle$, if A_1 or A_2 is operated on the respective square angular momentum operator A^2, the operation does not affect the square angular momentum at all. This statement can be mathematically expressed as follows. As per the definition of azimuthal and magnetic quantum number

$$A_3 = m\hbar; \; A^2 = a(a+1)\hbar^2 \tag{5.23.1}$$

Taking the commuter operator for A^2 and A_1, and applying Eqs. 5.18 and 5.21.7 through 5.21.9

$$\left[A^2, A_1\right]|a\; m\rangle = \left(A^2 A_1 - A_1 A^2\right)|a\; m\rangle = 0; \; A^2\left(A_3|a\; m\rangle\right) = 0 \tag{5.23.2}$$

Similarly,

$$\left[A^2, A_2\right]|a\; m\rangle = \left(A^2 A_2 - A_2 A^2\right)|a\; m\rangle = 0; \; A^2\left(A_2|a\; m\rangle\right) = 0 \tag{5.23.3}$$

Hence, except for zero eigenstate $|0\; 0\rangle$, A_1 and A_2 do not affect the respective square angular momentum. This is also clear from Fig. 5.7a–c. As per the section 5.2.4.2, due to the uncertainty principle, the A_3 component of the operator is affected by A_1 and A_2 as they do not commute. All possible linear combinations of A_1 and A_2 will also not commute with A_3. To keep the analysis, separate for A_1 and A_2, it is always useful to introduce a complex operator (see sections 5.2.2 and 5.2.3) using both the terms; the following two operators were introduced [3].

$$A^+ = A_1 + iA_2; \; A^- = A_1 - iA_2 \tag{5.23.4}$$

Next, if the new complex angular momentum operators are commuted with the A_3 operator, using Eqs. (5.21.1) through (5.21.9) as needed, the commutation will be

$$\left[A^+, A_3\right] = \hbar A^+; \; \left[A^-, A_3\right] = -\hbar A^- \tag{5.23.5}$$

Considering the commutator for the positive operator only

$$\left[A^+, A_3\right]|a\; m\rangle = \left(A^+ A_3 - A_3 A^+\right)|a\; m\rangle = \hbar A^+|a\; m\rangle \tag{5.23.6}$$

$$\left[A^+, A_3\right]|a\; m\rangle = \left(\hbar A^+ + A_3 A^+\right)|a\; m\rangle \tag{5.23.7}$$

$$\left[A^+, A_3\right]|a\; m\rangle = A^+(\hbar + A_3)|a\; m\rangle \tag{5.23.8}$$

Substituting, Eq. 5.23.1,

$$\left[A^+, A_3\right]|a\; m\rangle = A_3\left(A^+|a\; m\rangle\right) = \hbar(m+1)A^+|a\; m\rangle = A_3\left(A^+|a\; m\rangle\right) \tag{5.23.9}$$

Eq. 5.23.9 signifies that if the complex operator is operated on the current eigenstate and produces a new eigenstate $A^+|a\; m\rangle$, then the new state will have an x_3 magnetic quantum number equal to $(m+1)$ not (m). It immediately implies that every time the A_3 operator is operated with the complex operator A^+, the new $A^+|a\; m\rangle$ eigenstate increase by one quantum number and goes up the ladder to $|a\; m+1\rangle$. A

Quantum Mechanics for Engineers 175

similar state was described in section 5.2.2. Although not shown, similar is true for the complex operator A^-, which will result in quantum numbers going down the ladder in a negative direction from $|a\ m\rangle$ eigenstate to $|a\ m-1\rangle$. All possible such values of quantum states will create a ladder. The new raised or lowered eigenstates can be written in terms of their quantum numbers as follows:

Generic form:

$$A^+|a\ m\rangle = \hbar\sqrt{a(a+1)-m(m+1)}\ |a\ m+1\rangle \qquad (5.23.10)$$

$$A^-|a\ m\rangle = \hbar\sqrt{a(a+1)-m(m-1)}\ |a\ m-1\rangle \qquad (5.23.11)$$

OAM:

$$L^+|l\ m\rangle = \hbar\sqrt{l(l+1)-m(m+1)}\ |l\ m+1\rangle \qquad (5.23.12)$$

$$L^-|l\ m\rangle = \hbar\sqrt{l(l+1)-m(m-1)}\ |l\ m-1\rangle \qquad (5.23.13)$$

SAM:

$$S^+|s\ m\rangle = \hbar\sqrt{s(s+1)-m(m+1)}\ |s\ m+1\rangle \qquad (5.23.14)$$

$$S^-|s\ m\rangle = \hbar\sqrt{s(s+1)-m(m-1)}\ |s\ m-1\rangle \qquad (5.23.15)$$

TAM:

$$J^+|j\ m\rangle = \hbar\sqrt{j(j+1)-m(m+1)}\ |j\ m+1\rangle \qquad (5.23.16)$$

$$J^-|j\ m\rangle = \hbar\sqrt{j(j+1)-m(m-1)}\ |j\ m-1\rangle \qquad (5.23.17)$$

Further details on the ladder operation are omitted in this section. Pictographic views of these ladders are not shown here but can be found elsewhere [2–4, 10]. For a quantum analog application of acoustic phenomena, such ladders need to be created and are discussed in the next chapters.

Specific to spin angular momentum, it can be said that the S^+ and S^- operators change the S_3 eigenstate to the next higher or lower eigenstate, while they raise and lower the magnetic quantum number by one count, respectively. A similar statement is valid for OAM and TAM.

5.2.5 Pauli's Matrix

Please recollect from the section 5.2.3.2 that spin unit vectors $\hat{\mathbf{s}}_1$ and $\hat{\mathbf{s}}_2$ are noted respectively $\left(\frac{1}{2}\hbar\right)$ and $\left(-\frac{1}{2}\hbar\right)$. For $S_3(\uparrow)$ and $S_3(\downarrow)$ they will respectively be

$$\hat{\mathbf{s}}_1 = 1,\ \hat{\mathbf{s}}_2 = 0;\ \hat{\mathbf{s}}_1 = 0,\ \hat{\mathbf{s}}_2 = 1 \qquad (5.24.1)$$

or

$$\mathbf{s}^\uparrow = \left\{ \begin{array}{c} \hat{\mathbf{s}}_1 \\ \hat{\mathbf{s}}_2 \end{array} \right\}^\uparrow = \left\{ \begin{array}{c} 1 \\ 0 \end{array} \right\} \text{ and } \mathbf{s}^\downarrow = \left\{ \begin{array}{c} \hat{\mathbf{s}}_1 \\ \hat{\mathbf{s}}_2 \end{array} \right\}^\downarrow = \left\{ \begin{array}{c} 0 \\ 1 \end{array} \right\} \tag{5.24.2}$$

$\mathbf{s}^\uparrow$ and $\mathbf{s}^\downarrow$ *are the two eigenspinors.* Hence, using these spinors, except for the zero state $|0\ 0\rangle$ all the other states for the S_3 spin angular momentum and S^2 square angular momentum in a matrix form can be written as

$$S_3 = \frac{\hbar}{2} \begin{bmatrix} 1 & 0 \\ 0 & -1 \end{bmatrix} \quad S^2 = \frac{3\hbar}{4} \begin{bmatrix} 1 & 0 \\ 0 & 1 \end{bmatrix} \tag{5.24.3}$$

or

$$S_3 = \frac{\hbar}{2} \begin{bmatrix} \hat{\mathbf{s}}_1^\uparrow & \hat{\mathbf{s}}_1^\downarrow \\ \hat{\mathbf{s}}_2^\uparrow & -\hat{\mathbf{s}}_2^\downarrow \end{bmatrix} \quad S^2 = \frac{3\hbar}{4} \begin{bmatrix} \hat{\mathbf{s}}_1^\uparrow & \hat{\mathbf{s}}_1^\downarrow \\ \hat{\mathbf{s}}_2^\uparrow & \hat{\mathbf{s}}_2^\downarrow \end{bmatrix} \tag{5.24.4}$$

or

$$S_3 = \frac{\hbar}{2} \begin{bmatrix} \mathbf{s}^\uparrow & \mathbf{s}^\downarrow \end{bmatrix} \quad S^2 = \frac{3\hbar}{4} \begin{bmatrix} \mathbf{s}^\uparrow & \mathbf{s}^\downarrow \end{bmatrix} \tag{5.24.5}$$

Next let's test what the $S^+|s\ m\rangle$ and $S^-|s\ m\rangle$ eigenstate does to these spin matrices. For a generic eigenspinor with spin azimuthal quantum number $s = \frac{1}{2}$, magnetic quantum numbers will be $m = -\frac{1}{2}, 0, +\frac{1}{2}$. Considering spinor eigenstate $|\frac{1}{2}\ \frac{1}{2}\rangle$ and Eq. 5.23.14,

$$S^+ \left| \frac{1}{2}\ \frac{1}{2} \right\rangle = \hbar \sqrt{\frac{1}{2}\left(\frac{1}{2}+1\right) - \frac{1}{2}\left(\frac{1}{2}+1\right)} \left| \frac{1}{2}\ \frac{3}{2} \right\rangle = 0 \tag{5.24.6}$$

$$S^- \left| \frac{1}{2}\ \frac{1}{2} \right\rangle = \hbar \sqrt{\frac{1}{2}\left(\frac{1}{2}+1\right) - \frac{1}{2}\left(\frac{1}{2}-1\right)} \left| \frac{1}{2}\ -\frac{1}{2} \right\rangle = \hbar \left| \frac{1}{2}\ -\frac{1}{2} \right\rangle \tag{5.24.7}$$

Hence, it can be easily shown how the complex operators changes the spin eigenstates of the spinors as follows.

$$S^-\mathbf{s}^\uparrow = \hbar\ \mathbf{s}^\downarrow; \ S^+\mathbf{s}^\uparrow = 0; \ S^-\mathbf{s}^\downarrow = 0; \ S^+\mathbf{s}^\downarrow = \hbar\ \mathbf{s}^\uparrow \tag{5.24.8}$$

Spin up eigenstate changes to spin down and spin down eigenstate changes to spin up, only if operated with S^- and S^+ operators, respectively. Next let's assume a spinor with combination of two spin states, one spin up and another spin down.

$$s = \mathbf{s}^\uparrow + \mathbf{s}^\downarrow \tag{5.24.9}$$

Quantum Mechanics for Engineers

Complex operators operated on this combined state, using Eq. 5.24.8, will give

$$S^+\left(\mathbf{s}^\uparrow + \mathbf{s}^\downarrow\right) = \mathbf{0} + \hbar\, \mathbf{s}^\uparrow;\ S^-\left(\mathbf{s}^\uparrow + \mathbf{s}^\downarrow\right) = \hbar\, \mathbf{s}^\downarrow + \mathbf{0} \tag{5.24.10}$$

Note that the $\mathbf{0}$ is a vector and thus left bold. Utilizing unit vector notation in Eq. 5.24.2, the complex operators will visualize as

$$S^+ = \hbar\left\{\begin{matrix} 0 \\ 0 \end{matrix}\right\} + \hbar\left\{\begin{matrix} \hat{s}_1 \\ \hat{s}_2 \end{matrix}\right\}^\uparrow = \hbar\begin{bmatrix} 0 & 1 \\ 0 & 0 \end{bmatrix};\ S^- = \hbar\left\{\begin{matrix} \hat{s}_1 \\ \hat{s}_2 \end{matrix}\right\}^\downarrow + \hbar\left\{\begin{matrix} 0 \\ 0 \end{matrix}\right\} = \hbar\begin{bmatrix} 0 & 0 \\ 1 & 0 \end{bmatrix} \tag{5.24.11}$$

Now, using basic algebra with Eq. 5.16.5, $S^+ = S_1 + iS_2$ and $S^- = S_1 - iS_2$, it is possible to solve the S_1 and S_2 operators. After solution we get

$$S_1 = \frac{1}{2}\left(S^+ + S^-\right) = \frac{\hbar}{2}\begin{bmatrix} 0 & 1 \\ 1 & 0 \end{bmatrix} \tag{5.24.12}$$

$$S_2 = \frac{1}{2i}\left(S^+ - S^-\right) = \frac{\hbar}{2}\begin{bmatrix} 0 & -i \\ i & 0 \end{bmatrix} \tag{5.24.13}$$

Reiterating,

$$S_3 = \frac{\hbar}{2}\begin{bmatrix} 1 & 0 \\ 0 & -1 \end{bmatrix} \tag{5.24.14}$$

$$S^2 = \frac{3\hbar}{4}\begin{bmatrix} 1 & 0 \\ 0 & 1 \end{bmatrix} \tag{5.24.15}$$

Separating the $\dfrac{\hbar}{2}$ as a common factor, Pauli's matrices are noted as

$$\sigma_{x1} = \begin{bmatrix} 0 & 1 \\ 1 & 0 \end{bmatrix}\ \sigma_{x2} = \begin{bmatrix} 0 & -i \\ i & 0 \end{bmatrix}\ \sigma_{x3} = \begin{bmatrix} 1 & 0 \\ 0 & -1 \end{bmatrix} \tag{5.24.16}$$

The first column represents the 'spin up' $\mathbf{s}^\uparrow$ state of the spinor, and the second column represents the 'spin down' $\mathbf{s}^\downarrow$ state of the spinor. These matrices are Hermitian, with symmetric real part and antisymmetric imaginary part.

If it is required to measure the SAM, S_1 or S_2 with well-defined values, then the only possible outcomes would be $\left(\frac{1}{2}\hbar\right)$ or $\left(-\frac{1}{2}\hbar\right)$ as the eigenvalues. Then it is possible to find the corresponding spinors that cause these two states as follows. Consider the S_1 operator first. Let's assume any two arbitrary spinors χ^{1+} and χ^{1-}. Following the operator, eigenvalue and eigenstate properties are

$$S_1\,\chi^{1+} = \frac{1}{2}\hbar\,\chi^{1+} \text{ and } S_1\,\chi^{1-} = -\frac{1}{2}\hbar\,\chi^{1-} \tag{5.24.17}$$

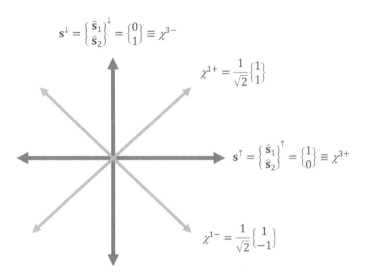

FIGURE 5.8 The spinors with their respective component plotted using orthogonal coordinate system. An overall factor of −1 has no physical effect on a quantum state, so, the negated spinors represent the same four physical states. A normalized spinor has unit length on this diagram, but the diagram cannot show spinors whose components are complex.

$$\chi^{1+} = \left\{ \begin{array}{c} \hat{c}_1 \\ \hat{c}_2 \end{array} \right\}^{\uparrow} \chi^{1-} = \left\{ \begin{array}{c} \hat{c}_1 \\ \hat{c}_2 \end{array} \right\}^{\downarrow} \qquad (5.24.18)$$

where the spinor states are represented by two arbitrary numbers assuming they are components of the eigenstates written in Eq. 5.24.2. Solving Eqs. 5.24.17 and 5.24.18, we get

$$\chi^{1+} = \frac{1}{\sqrt{2}} \left\{ \begin{array}{c} 1 \\ 1 \end{array} \right\}; \chi^{1-} = \frac{1}{\sqrt{2}} \left\{ \begin{array}{c} 1 \\ -1 \end{array} \right\} \qquad (5.24.19)$$

Similar eigenspinors with well-defined value for S_2 operator can also be found following the above process. If a similar attempt is made for the well-defined value of S_3 operator, we get the spinors described in Eq. 5.24.2. Figure 5.8 shows the orthogonal spinors states of S_3 and S_1 SAM.

5.3 SOLUTION OF SCHRÖDINGER EQUATION IN PERIODIC POTENTIAL

The eigenvalue solution of the Schrödinger equation is presented in section 5.1.7. In these solutions, it was considered that the potential function $\mathcal{U}$ had no role to play. However, in crystals, the periodic molecular orientation will contribute to the overall solution of the Schrödinger equation. In materials, molecules are arranged in specific patterns. These patterns are periodic, and thus their unit block can be identified as a unit cell. Based on the geometric repetition of the unit cells, crystals are divided into seven fundamental classes: a) simple cubic, b) tetragonal, c) orthorhombic, d) rhombohedral, e) monoclinic, f) triclinic, and g) hexagonal. These crystal orientations are shown in Fig. 5.9a. The solution of the Schrödinger equation in complex period media will demand more attention and understanding, which is beyond the scope of this book. However acoustic and elastic wave propagation in periodic media demands fundamental understanding of the wave solution in periodic media, which was first described by Felix Bloch in 1928 [13] while describing the motion of an electron in crystalline solids. A similar mathematical concept was described by George William Hill in 1877 [14], Gaston Floquet in 1883 [15], and Alexander Lyapunov in 1892 [16], multiple times before Bloch and the discovery of quantum mechanics. The fundamental solution, and the pattern of the wave function used by Bloch, is extremely valuable for solving the wave propagation in periodic solids and fluid media. To understand the basics, let's first assume a linear periodic arrangement of molecule in a crystal as shown in Fig. 5.9b.

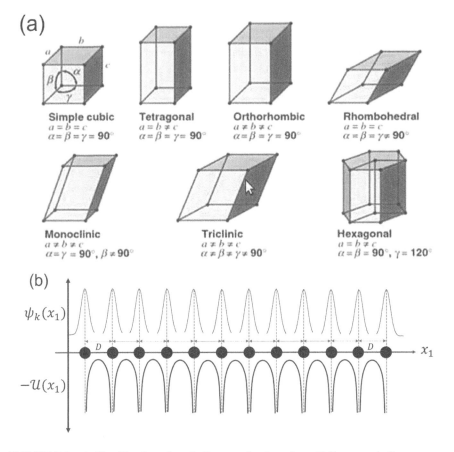

FIGURE 5.9 a) Classification of periodic crystal orientations, b) linear periodic arrangement of molecules in a crystal and Bloch solution in a periodic potential.

180   Metamaterials in Topological Acoustics

5.3.1 BLOCH SOLUTION

In this section, first a one-dimensional periodic system along x_1 is considered. The periodic potential, which is real, can be expressed using a Fourier series and can be written as

$$\mathcal{U}(x_1) = \sum_G U_G e^{iG_n x_1} = \sum_n U_n e^{i\frac{2\pi n}{D} x_1} \tag{5.25.1}$$

where $G_n = \dfrac{2\pi n}{D}$ is the reciprocal lattice vector and D is the periodicity of the molecules. $n = -\infty$ to $+\infty$ takes only the integer values, and $\sum_G$ indicates the sum over all possible values of reciprocal wave vectors G_n. The Fourier coefficients can be found by applying the orthogonality condition of the eigenfunction $e^{iG_n x_1}$. Hence, the Fourier coefficients are

$$U_G = \frac{1}{D} \int_{-D/2}^{D/2} \mathcal{U}(x_1) e^{-iG_n x_1} dx_1 = \frac{1}{D} \int_{-D/2}^{D/2} \mathcal{U}(x_1) e^{-i\frac{2\pi n}{D} x_1} dx_1 \tag{5.25.2}$$

Referring Eq. 5.9.6, the time-independent one-dimensional solution of the Schrödinger equation can be reiterated as

$$\Psi(x_j, t) = \psi(x_1) e^{-iEt/\hbar} \tag{5.25.3}$$

According to the solution presented in Table 5.1, considering only the spatial part of the wave function, the wave function can be written as the summation of the eigenfunctions like a plane wave expansion. The following solution is written in a bit different form to be self-explanatory.

$$\Psi(x_1) = \psi(x_1) = \sum_{n=0}^{\infty} A_n\, e^{ip_{1n} x_1/\hbar} = \sum_{n=0}^{\infty} A_n\, e^{ik_n x_1} \tag{5.25.4}$$

$$\Psi(x_1) = \sum_k A(k) e^{ikx_1} \tag{5.25.5}$$

where according to Table 5.1, $k_n = \dfrac{p_{1n}}{\hbar}$ are the eigenvalues of the wave numbers. Next, the Fourier coefficients in Eq. 5.25.5, which are the function of the wave number k, can be found as follows:

$$A(k) = \frac{1}{D} \int_{-D/2}^{D/2} \Psi(x_1) e^{-ikx_1} dx_1 \tag{5.25.6}$$

As the $\mathcal{U}(x_1)$ potential is periodic and has the same value at every periodic boundary (Fig. 5.9b), the wave potential $\Psi(x_1)$ must be equal to the potential at every periodic boundary. Hence, the $\Psi(x_1 + D) = \Psi(x_1)$ boundary condition must be satisfied.

Quantum Mechanics for Engineers

This signifies that for such a boundary condition, the plane wave can only take discrete values.

$$k = \frac{2\pi n}{D} \text{ with } n = -\infty\ldots,-3,\ -4,-2,-1,\ 0,\ +1,+2,+3,\ldots,\ \infty \qquad (5.25.7)$$

As of now the pattern in the solution is recognized, but the actual solution is yet to be derived using Schrödinger equation Eq. 5.9.5. Schrödinger equations in 3D and then 1D are.

$$\left[-\frac{\hbar^2}{2m}\left(\frac{\partial^2}{\partial x_1^2} + \frac{\partial^2}{\partial x_2^2} + \frac{\partial^2}{\partial x_3^2} \right) + \mathcal{U} \right]\Psi = i\hbar\frac{\partial}{\partial t}\Psi \qquad (5.25.8)$$

$$\left[-\frac{\hbar^2}{2m}\left(\frac{\partial^2}{\partial x_1^2} \right) + \mathcal{U} \right]\Psi = i\hbar\frac{\partial}{\partial t}\Psi \qquad (5.25.9)$$

Substituting the energy operator and making the equation time independent will get

$$-\frac{\hbar^2}{2m}\left(\frac{\partial^2}{\partial x_1^2}\Psi(x_1) \right) + \mathcal{U}(x_1)\Psi(x_1) = E\Psi(x_1) \qquad (5.25.10)$$

Substituting the periodic potential function in Eq. 5.25.1 and the wave function in Eq. 5.25.5, we get

$$-\frac{\hbar^2}{2m}\left(\frac{\partial^2}{\partial x_1^2}\sum_k \mathcal{A}(k)e^{ikx_1} \right) + \sum_G U_G e^{iG_n x_1}\sum_k \mathcal{A}(k)e^{ikx_1} = E\sum_k \mathcal{A}(k)e^{ikx_1} \qquad (5.25.11)$$

$$-\frac{\hbar^2}{2m}\left(\frac{\partial^2}{\partial x_1^2}\sum_k \mathcal{A}(k)e^{ikx_1} \right) + \sum_G\sum_k U_G \mathcal{A}(k)e^{i(k+G_n)x_1} = E\sum_k \mathcal{A}(k)e^{ikx_1} \qquad (5.25.12)$$

Here the objective is to solve the coefficients $\mathcal{A}(k)$ assumed in the wave function. Assuming, $\bar{k} = k + G_n$ and $k = \bar{k} - G_n$, the second term of the left-hand side of equation 5.25.12 can be rewritten as

$$\sum_G\sum_k U_G \mathcal{A}(k)e^{i(k+G_n)x_1} = \sum_G\sum_{\bar{k}} U_G \mathcal{A}(\bar{k} - G_n)e^{i(\bar{k})x_1} \qquad (5.25.13)$$

As there is no obvious distinction between k and the $\bar{k}$ except the integer eigenvalues over a long infinite range, the second term and the equation will be

$$\sum_G\sum_k U_G \mathcal{A}(k - G_n)e^{ikx_1} \qquad (5.25.14)$$

$$\frac{\hbar^2 k^2}{2m}\left(\sum_k \mathcal{A}(k)e^{ikx_1} \right) + \sum_G\sum_k U_G \mathcal{A}(k - G_n)e^{ikx_1} = E\sum_k \mathcal{A}(k)e^{ikx_1} \qquad (5.25.15)$$

$$\left[\frac{\hbar^2 k^2}{2m}\sum_k A(k) + \sum_G \sum_k U_G A(k - G_n)\right]e^{ikx_1} = E\sum_k A(k)e^{ikx_1} \tag{5.25.16}$$

Taking all the terms inside summation,

$$\sum_k \left[\left[\frac{\hbar^2 k^2}{2m} - E\right]A(k) + \sum_G U_G A(k - G_n)\right]e^{ikx_1} = 0 \tag{5.25.17}$$

As the eigenfunction $e^{ikx_1} \neq 0$,

$$\sum_k \left[\left[\frac{\hbar^2 k^2}{2m} - E\right]A(k) + \sum_G U_G A(k - G_n)\right] = 0 \tag{5.25.18}$$

With careful observation it can be deduced that the function $A(k)$ is coupled with the function $A(k - G_n)$, which also includes $A(k + G_n)$, as n ranges between $-\infty$ to $+\infty$. It is also apparent that the value of k does not depend on the value of $k + G_n$ or $k - G_n$ unless they are one or multiples of n reciprocal lattice vectors apart in both directions. Hence, it can be said that if k is the characteristic wave number of the system, then $k + G_n$ is also the characteristic wave number where $n = -\infty \ldots, -3, -4, -2, -1, 0, +1, +2, +3, \ldots, \infty$. Hence, the above logic with Eq. 5.25.5 can be written in mathematical form as follows:

$$\Psi(x_1) = \sum_k A(k)e^{ikx_1} = \sum_G A(k + G_n)e^{i(k+G_n)x_1} \tag{5.25.19}$$

For every wave number k

$$\Psi(x_1) = \left[\sum_G A(k + G_n)e^{i(G_n)x_1}\right]e^{ikx_1} = \psi_k(x_1)e^{ikx_1} \tag{5.25.20}$$

where $k + G_n$ and $k - G_n$ are used synonymously.

Further, in periodic media the boundary condition holds $\psi_k(x_1) = \psi_k(x_1 + D)$ and can be proved as follows:

$$\psi_k(x_1 + D) = \sum_G A\left(k + \frac{2\pi n}{D}\right)e^{i\left(\frac{2\pi n}{D}\right)(x_1 + D)} = \sum_G A\left(k + \frac{2\pi n}{D}\right)e^{i\left(\frac{2\pi n}{D}\right)x_1}e^{i(2\pi n)}$$

$$= \sum_G A\left(k + \frac{2\pi n}{D}\right)e^{i\left(\frac{2\pi n}{D}\right)x_1} = \psi_k(x_1) \tag{5.25.21}$$

where $e^{i(2\pi n)} = 1$ for all values of n. Hence, wave function in periodic media in a generalized form can be expressed as

$$\Psi(x_1) = \sum_G A(k + G_n)e^{i(k+G_n)x_1} = \sum_n A_n e^{i(k+G_n)x_1} = \left[\sum_n A_n e^{i(G_n)x_1}\right]e^{i(k)x_1} \tag{5.25.22}$$

Quantum Mechanics for Engineers

Comparing Eqs. 5.25.22 and 5.25.20,

$$\psi_k(x_1) = \left[\sum_n A_n e^{i(G_n)x_1}\right]; \ \Psi(x_1 + D) = \Psi(x_1) = \psi_k(x_1)e^{i(k)x_1} \ \forall \ k \qquad (5.25.23)$$

where the $\forall$ symbol means "for all the values of". Substituting the wave function in periodic media for every wave number k, the Schrödinger equation results in an eigenvalue problem as follows:

$$\left[\frac{\hbar^2 k^2}{2m} - E\right]A(k) + \sum_G U_G A(k + G_n) = 0 \qquad (5.25.24)$$

The eigenvalue problem can be written in a matrix form to visualize how the infinite set of reciprocal wave numbers G_n contributes to the solution. Again, recollecting $G_n = \frac{2\pi n}{D}$, we simply write G_{-1} to imply $n = -1$, and so on for all values of n. Eq. 5.25.24 reads

$$\left[\frac{\hbar^2 k^2}{2m} - E\right]A(k + G_0) + \sum_G U_G A(k + G_n) = 0 \qquad (5.25.25)$$

The vector of unknown coefficients with such n reciprocal wave vectors will read

$$\begin{bmatrix} A(k + G_{-n}) \\ \vdots \\ A(k + G_2) \\ A(k + G_{-1}) \\ A(k) \\ A(k + G_{+1}) \\ A(k + G_2) \\ \vdots \\ A(k + G_n) \end{bmatrix} \qquad (5.25.26)$$

Now Eq. 5.25.25 will read as an eigenvalue problem as follows, where the $E(k)$ represents the eigenvalues of the system at a specific wave number value k. The wave number domain for the k vector can be divided over a large range, between zero and infinity. The $E(k)$ eigenvalues then can be solved for each wave number k. There could be multiple energy states at a specific k value. Together $E(k)$ will provide an energy dispersion diagram for the system governed by the Schrödinger equation when the periodic potential and its Fourier coefficients are known. After the next section, it will be apparent that the wave number k between $-\dfrac{\pi}{D}$ and $+\dfrac{\pi}{D}$ will suffice for this solution.

$$
\begin{bmatrix}
\dfrac{\hbar^2(k+G_{-n})^2}{2m} & 0 & 0 & 0 & 0 & 0 & 0 \\
0 & \ddots & 0 & 0 & 0 & 0 & 0 \\
0 & 0 & \dfrac{\hbar^2(k+G_{-1})^2}{2m} & 0 & 0 & 0 & 0 \\
0 & 0 & 0 & \dfrac{\hbar^2 k^2}{2m} & 0 & 0 & 0 \\
0 & 0 & 0 & 0 & \dfrac{\hbar^2(k+G_1)^2}{2m} & 0 & 0 \\
0 & 0 & 0 & 0 & 0 & \ddots & 0 \\
0 & 0 & 0 & 0 & 0 & 0 & \dfrac{\hbar^2(k+G_n)^2}{2m}
\end{bmatrix}
$$

$$
+
\begin{bmatrix}
\cdots & \cdots & \cdots & \cdots & \cdots & \cdots & \cdots \\
\cdots & 0 & U_{G_{-1}} & U_{G_{-2}} & U_{G_{-3}} & U_{G_{-4}} & \cdots \\
\cdots & U_{G_1} & 0 & U_{G_{-1}} & U_{G_{-2}} & U_{G_{-3}} & \cdots \\
\cdots & U_{G_2} & U_{G_1} & 0 & U_{G_{-1}} & U_{G_{-2}} & \cdots \\
\cdots & U_{G_3} & U_{G_2} & U_{G_1} & 0 & U_{G_{-1}} & \cdots \\
\cdots & U_{G_4} & U_{G_3} & U_{G_2} & U_{G_1} & 0 & \cdots \\
\cdots & \cdots & \cdots & \cdots & \cdots & \cdots & \cdots
\end{bmatrix}
\begin{bmatrix}
A(k+G_{-n}) \\ \vdots \\ A(k+G_2) \\ A(k+G_{-1}) \\ A(k) \\ A(k+G_{+1}) \\ A(k+G_2) \\ \vdots \\ A(k+G_n)
\end{bmatrix}
= E(k)
\begin{bmatrix}
A(k+G_{-n}) \\ \vdots \\ A(k+G_2) \\ A(k+G_{-1}) \\ A(k) \\ A(k+G_{+1}) \\ A(k+G_2) \\ \vdots \\ A(k+G_n)
\end{bmatrix}
$$

$$(5.25.27)$$

$$
\left[H_p + H_u\right] A(k+G_n) = E(k) A(k+G_n) \tag{5.25.28}
$$

H_p is the Hamiltonian of the system generated from the kinetic energy of the system, and the H_u is the Hamiltonian generated from the potential energy composed of Fourier coefficients of the periodic potential function of the system. Together H_p and H_u give the total Hamiltonian of the system as it is described in Eqs. 5.11.6 and 5.11.7. Here H is the total Hamiltonian operator in a matrix form. Please keep in mind that an operator could also be a matrix, not just a differential operator in a differential equation. In quantum mechanics most of the operators are written in matrix form. The above equation is equivalent to Eq. 5.8.7 or $H\Psi = E\Psi$, where the vector is the amplitude coefficients of the wave function Ψ.

5.3.2 Reduced-Order Solution and Brillouin Zone

It is evident from the above solution that all possible wave numbers with periodicity G_n are also the solution of the system. This means that if they have the same phase, all possible wave solutions with different frequencies or wave numbers (when velocity is constant) are also the solution of the system. Figure 5.10a shows two waves that have different frequency but have the same phase. According to the above equation, they are both viable solutions of the system. This instigate the idea of finding the most fundamental wave number (k) solution that is unique to the equation. It can be recognized from the above equations that wave numbers (k) between 0 and $\frac{2\pi}{D}$ mapped to a full circle of 2π are the only unique wave numbers. However, to accommodate wave propagation in both directions, the full circle of 2π could also be mapped between $-\frac{\pi}{D}$ and $\frac{\pi}{D}$. It is thus customary to solve the eigenvalue problem for k only within the band of $-\frac{\pi}{D}$ and $\frac{\pi}{D}$. All other solutions are the repetition of the same circular mapping. From the infinite range of wave numbers along both positive and negative axis, only the $-\frac{\pi}{D}$ to $+\frac{\pi}{D}$ range is selected. It can be seen as the range is reduced from the infinite range of k. Solutions for energy eigenvalues for all possible wave numbers k are presented only within this reducer zone and thus called reduced-order solution. This reduced zone is called the first Brillouin zone, proposed by Léon Brillouin in 1930 [17, 18]. It is known that the wave number domain is the reciprocal (Fourier transform) of the space domain. Hence, it is also

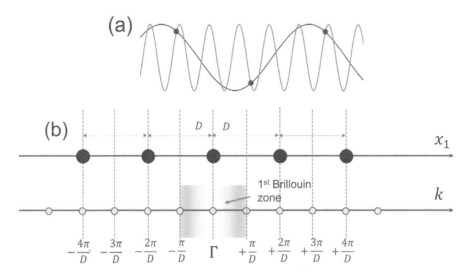

FIGURE 5.10 a) Repetition of the wave number solution, two different wave numbers are considered as the solutions of the system having the same phase. All such wave numbers are the solution of the system, b) periodic system and its respective Brillouin zone.

true that the periodicity of the space domain, which is D, will also result in a periodicity in the reciprocal wave vector domain. According to the fundamental rule of the Fourier transform, the domain of interest is mapped over a full circle. Thus, here if the D is mapped over a circle of 2π radians, the reciprocal wave number should be mapped over $\frac{2\pi}{D}$. As indicated before, to accommodate the wave propagation vectors along both the directions, the $\frac{2\pi}{D}$ range is mapped between $-\frac{\pi}{D}$ to $+\frac{\pi}{D}$. Figure 5.10b shows the Brillouin zone for a mono atomic periodic system in Fig. 5.9b. The central point is called the Γ point where $k = 0$. Traditionally, the first interval $+\frac{\pi}{D}$ is called K point in reciprocal space, and the $-\frac{\pi}{D}$ is called the $-K$ point in the reciprocal space. It is also necessary to recognize that the reciprocal space $-K$ to $+K$ creates a full closed loop of an energy state. *In reciprocal space it is possible to travel along a closed loop.*

Similarly, if the crystal is doubly periodic in two dimensions, then the reciprocal wave vectors from each direction should be considered. The first Brillouin zone for such two-dimensional system will become a 2D square or rectangle, for equi-periodic ($D = D_1 = D_2$) or non-equal ($D_1 \neq D_2$) periodic system, respectively. Similarly, it can be imagined that for a different order and different dimension of periodicity, the system will give unique patterned spaces for the Brillouin zones. Brillouin zones are further discussed and used under the acoustic and elastic waves in periodic media in the later chapters as needed.

Bloch potentials substituted in the Schrödinger equation create an eigenvalue problem as described in Eq. 5.25.26. The actual solution of the quantum energy bands as a function of k-space, and their dispersion is not the topic of this book. Thus, dispersion solution of the Schrödinger equation is not explicitly discussed, but a solution as a possible example of the energy states due to periodicity is presented in Fig. 5.11. However, not to confuse here, energy bands, or rather the equivalent frequency bands as a function of k-space for acoustic and elastic waves, are indeed the topic of this book and are covered in the next chapter.

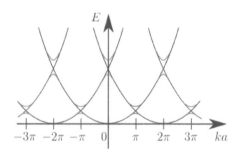

FIGURE 5.11 Sample energy band structure after solving the eigenvalue problem in Eq. 5.25.28.

Quantum Mechanics for Engineers

5.3.3 Finding Energy Bands Using the Perturbation Method

5.3.3.1 Modified Hamiltonian in Periodic Media

In a periodic system made of atoms, the total energy of the system (Hamiltonian of the system) is the integration of multiple energy states and their interactions. The Hamiltonian consists of

- kinetic energies due to the motion of the atomic particles e.g., nuclei and electrons
- potential energy due to the nuclear forces, interactions between protons and neutrons
- potential energy from the interaction of multiple electrons
- magnetic energy due to the orbital motion and spin of the atomic particles

When the Hamiltonian is written using all possible energy terms, the solution of the Schrödinger equation becomes a formidable task. Thus, generally the solutions are carried out by introducing several approximations. To exploit or to use the knowledge of the band structure for topological behavior, it is required to know the full dispersion behavior of the energy bands as a function of wave number ($E - \mathbf{k}$ relationship). To obtain the relationship in a periodic crystal, the potential function $\mathcal{U}$ must be explicitly described. However, this is not always feasible. On the contrary, in a system of period crystals, the electrons and holes are mostly populated or located within a certain energy band, within a fraction of an electron Volt (eV). Hence, when the wave function and its carrier energies are known at an extreme point of an energy band, it is possible to apply a perturbation approach to find the energies at the other points in the Brillouin zone. Some of those extreme points could be the Brillouin boundaries (called K points) or the center of the Brillouin zone (called Γ point). This may help find the $E - \mathbf{k}$ relationship without explicitly solving the Schrödinger equation. Similarly for acoustics or elastic waves, a perturbation approach will help find the $\omega - \mathbf{k}$ relationship. **k.p**-perturbation method is one of such methods that was found to be useful. **k.p**-perturbation method [19] is discussed below.

5.3.3.2 k.p Perturbation Method

Let's start with the Schrödinger equation again. The time dependent Schrödinger equation is reiterated as

$$\left[-\frac{\hbar^2}{2m}\left(\frac{\partial^2}{\partial x_1^2} + \frac{\partial^2}{\partial x_2^2} + \frac{\partial^2}{\partial x_3^2} \right) + \mathcal{U} \right]\Psi = i\hbar \frac{\partial}{\partial t}\Psi \qquad (5.26.1)$$

And the time-independent Schrödinger is written replacing the energy term as

$$\left[-\frac{\hbar^2}{2m}\left(\frac{\partial^2}{\partial x_1^2} + \frac{\partial^2}{\partial x_2^2} + \frac{\partial^2}{\partial x_3^2} \right) + \mathcal{U} \right]\Psi = E\Psi \qquad (5.26.2)$$

The solution for the time-independent Schrödinger equation is previously derived as

$$\Psi(x_j) = \sum_{n,p,q=0}^{\infty} A_{nrq} e^{-i p_{nrq} \cdot x/\hbar} \qquad (5.26.3)$$

$$\Psi(x_j) = \sum_{n,p,q=0}^{\infty} A_{nrq} e^{-i k_{nrq} \cdot x} \quad \because \mathbf{k} = \mathbf{p}/\hbar \qquad (5.26.4)$$

The above solution represents the superposition of different wave numbers in the solution. However, in a periodic media, for every wave number $\mathbf{k}$, the above solution will have additional summations over the reciprocal lattice vectors that are inherent to the coefficient A_{nrq}. If one dimensional system is considered, then the reciprocal lattice vector would be $G_n = \dfrac{2\pi n}{D}$.

For a generic three-dimensional system, however, it is necessary to discuss the Bloch solution as a three-dimensional periodic crystal. Let's assume the periodicity of the crystal in three dimensions along x_1, x_2, x_3 are D_1, D_2, D_3 respectively. The reciprocal lattice vectors would be $G_n = \dfrac{2\pi n}{D_1}$, $G_m = \dfrac{2\pi m}{D_2}$ and $G_q = \dfrac{2\pi q}{D_3}$ where $n, m, q = -\infty$ to $+\infty$, takes only the integer values. If Σ_G indicates the sum over all possible values of reciprocal wave vectors G_n, G_m and G_q, then in a three-dimensional periodic crystal for every wave number $\mathbf{k}$, the Bloch wave function in Eq. 5.25.20 modifies to

$$\Psi(\mathbf{x}) = \left[\sum_G A(\mathbf{k}+\mathbf{G}) e^{i\mathbf{G}\cdot\mathbf{x}} \right] e^{i\mathbf{k}\cdot\mathbf{x}} = \left[\psi_{\mathbf{k}}^{nmq}(\mathbf{x}) \right] e^{i\mathbf{k}\cdot\mathbf{x}} = \Psi(\mathbf{x}+\mathbf{D}) \qquad (5.26.5)$$

where $\mathbf{x} = x_i \hat{e}_i$, $\mathbf{D} = D_i \hat{e}_i$, $\mathbf{k} = k_i \hat{e}_i$, $\mathbf{G} = G_n \hat{e}_1 + G_m \hat{e}_2 + G_q \hat{e}_3$, $\mathbf{k}.\mathbf{x} = k_i x_i$ and $\mathbf{G}.\mathbf{x} = G_n x_1 + G_m x_2 + G_q x_3$. Standard index notation is assumed. Taking the first and second derivatives of the wave function in Eq. 5.26.5 with respect to x_1, we get

$$\frac{\partial}{\partial x_1} \Psi(\mathbf{x}) = (ik_1) \left[\psi_{\mathbf{k}}^{nmq}(\mathbf{x}) \right] e^{i\mathbf{k}\cdot\mathbf{x}} + \frac{\partial \psi_{\mathbf{k}}^{nmq}(\mathbf{x})}{\partial x_1} e^{i\mathbf{k}\cdot\mathbf{x}} \qquad (5.26.6)$$

$$\frac{\partial^2}{\partial x_1^2} \Psi(\mathbf{x}) = -(k_1)^2 \left[\psi_{\mathbf{k}}^{nmq}(\mathbf{x}) \right] e^{i\mathbf{k}\cdot\mathbf{x}} + 2(ik_1) \frac{\partial \psi_{\mathbf{k}}^{nmq}(\mathbf{x})}{\partial x_1} e^{i\mathbf{k}\cdot\mathbf{x}} + \frac{\partial^2 \psi_{\mathbf{k}}^{nmq}(\mathbf{x})}{\partial x_1^2} e^{i\mathbf{k}\cdot\mathbf{x}} \qquad (5.2.6.7)$$

Similarly, with respect to x_2 and x_3,

$$\frac{\partial^2}{\partial x_2^2} \Psi(\mathbf{x}) = -(k_2)^2 \left[\psi_{\mathbf{k}}^{nmq}(\mathbf{x}) \right] e^{i\mathbf{k}\cdot\mathbf{x}} + 2(ik_2) \frac{\partial \psi_{\mathbf{k}}^{nmq}(\mathbf{x})}{\partial x_2} e^{i\mathbf{k}\cdot\mathbf{x}} + \frac{\partial^2 \psi_{\mathbf{k}}^{nmq}(\mathbf{x})}{\partial x_2^2} e^{i\mathbf{k}\cdot\mathbf{x}} \qquad (5.26.8)$$

$$\frac{\partial^2}{\partial x_3^2} \Psi(\mathbf{x}) = -(k_3)^2 \left[\psi_{\mathbf{k}}^{nmq}(\mathbf{x}) \right] e^{i\mathbf{k}\cdot\mathbf{x}} + 2(ik_3) \frac{\partial \psi_{\mathbf{k}}^{nmq}(\mathbf{x})}{\partial x_3} e^{i\mathbf{k}\cdot\mathbf{x}} + \frac{\partial^2 \psi_{\mathbf{k}}^{nmq}(\mathbf{x})}{\partial x_3^2} e^{i\mathbf{k}\cdot\mathbf{x}} \qquad (5.26.9)$$

Quantum Mechanics for Engineers

Substituting the second-order derivatives in Eq. 5.26.2 and assuming $\mathbf{k}^2 = (k_1)^2 + (k_2)^2 + (k_3)^2$, we get

$$-\frac{\hbar^2}{2m}\left[\left(\frac{\partial^2}{\partial x_1^2} + \frac{\partial^2}{\partial x_2^2} + \frac{\partial^2}{\partial x_3^2}\right)\right]\Psi(\mathbf{x})$$

$$= \left[\frac{\hbar^2 \mathbf{k}^2}{2m}\left[\psi_{\mathbf{k}}^{nmq}(\mathbf{x})\right]e^{i\mathbf{k}.\mathbf{x}} - 2i\frac{\hbar^2}{2m}\left(k_1\frac{\partial\psi_{\mathbf{k}}^{nmq}(\mathbf{x})}{\partial x_1} + k_2\frac{\partial\psi_{\mathbf{k}}^{nmq}(\mathbf{x})}{\partial x_2} + k_3\frac{\partial\psi_{\mathbf{k}}^{nmq}(\mathbf{x})}{\partial x_3}\right)e^{i\mathbf{k}.\mathbf{x}}\right.$$

$$\left.-\frac{\hbar^2}{2m}\left(\frac{\partial^2\psi_{\mathbf{k}}^{nmq}(\mathbf{x})}{\partial x_1^2} + \frac{\partial^2\psi_{\mathbf{k}}^{nmq}(\mathbf{x})}{\partial x_2^2} + \frac{\partial^2\psi_{\mathbf{k}}^{nmq}(\mathbf{x})}{\partial x_3^2}\right)e^{i\mathbf{k}.\mathbf{x}}\right]$$

$$(5.26.10)$$

Based on Eqs. 5.11.2 through 5.11.5, the momentum operators could be visualized as

$$p_1^2 = -\hbar^2\frac{\partial^2}{\partial x_1^2} \quad p_2^2 = -\hbar^2\frac{\partial^2}{\partial x_2^2} \quad p_3^2 = -\hbar^2\frac{\partial^2}{\partial x_3^2} \tag{5.26.11}$$

$$p_1 = -i\hbar\frac{\partial}{\partial x_1} \quad p_2 = -i\hbar\frac{\partial}{\partial x_2} \quad p_3 = -i\hbar\frac{\partial}{\partial x_3} \tag{5.26.12}$$

Hence, in Eq. 5.26.10 the following terms could be replaced by their respective momentum operators:

$$-\hbar^2\frac{\partial^2\psi_{\mathbf{k}}^{nmq}(\mathbf{x})}{\partial x_1^2} = p_1^2\psi_{\mathbf{k}}^{nmq}(\mathbf{x}); \; -\hbar^2\frac{\partial^2\psi_{\mathbf{k}}^{nmq}(\mathbf{x})}{\partial x_2^2} = p_2^2\psi_{\mathbf{k}}^{nmq}(\mathbf{x}); \; -\hbar^2\frac{\partial^2\psi_{\mathbf{k}}^{nmq}(\mathbf{x})}{\partial x_3^2} = p_3^2\psi_{\mathbf{k}}^{nmq}(\mathbf{x})\ldots$$

$$(5.26.13)$$

According to Eq. 5.11.5, the kinetic energy term is written as $\mathcal{K} = \dfrac{p_1^2 + p_2^2 + p_3^2}{2m} = \dfrac{p^2}{2m}$. Thus, the left-hand side of the Schrödinger equation will read

$$-\frac{\hbar^2}{2m}\left[\left(\frac{\partial^2}{\partial x_1^2} + \frac{\partial^2}{\partial x_2^2} + \frac{\partial^2}{\partial x_3^2}\right)\right]\Psi(\mathbf{x}) = \left[\frac{\hbar^2\mathbf{k}^2}{2m} + \frac{\hbar}{m}(k_1 p_1 + k_2 p_2 + k_3 p_3) + \frac{p^2}{2m}\right] \tag{5.26.14}$$

$$\left[\psi_{\mathbf{k}}^{nmq}(\mathbf{x})\right]e^{i\mathbf{k}.\mathbf{x}}\ldots$$

It is apparent that the $(k_1 p_1 + k_2 p_2 + k_3 p_3)$ term in above equation is a dot product of the wave number vector $\mathbf{k}$ and the momentum vector $\mathbf{p}$ [19–21]. Thus, the time-independent Schrödinger equation will read

$$\left[\left(\frac{\hbar^2\mathbf{k}^2}{2m} + \frac{\hbar}{m}(\mathbf{k}.\mathbf{p}) + \frac{p^2}{2m}\right) + \mathcal{U}\right]\psi_{\mathbf{k}}^{nmq}(\mathbf{x})e^{i\mathbf{k}.\mathbf{x}} = E(\mathbf{k})\psi_{\mathbf{k}}^{nmq}(\mathbf{x})e^{i\mathbf{k}.\mathbf{x}} \tag{5.26.15}$$

$$\left[\left(\frac{\hbar^2 \mathbf{k}^2}{2m} + \frac{\hbar}{m}(\mathbf{k}.\mathbf{p}) + \frac{p^2}{2m}\right) + \mathcal{U}\right]\psi_{\mathbf{k}}^{nmq}(\mathbf{x}) = E(\mathbf{k})\psi_{\mathbf{k}}^{nmq}(\mathbf{x})$$

(5.26.16)

$$e^{i\mathbf{k}.\mathbf{x}} \neq 0$$

Further, let's visualize and understand the Schrödinger equation in a periodic media using the Hamiltonian operator. As discussed before, it is clear that the left-hand side of Eq. 5.26.16 is the Hamiltonian of the system. Referring to Eqs. 5.11.7 and 5.26.16, it is apparent that there some extra terms appeared in the Hamiltonian due to the periodicity. Let's express them separately as follows

$$[H(0)]\psi_0^{nmq}(\mathbf{x}) = E(0)\psi_0^{nmq}(\mathbf{x}) \ @ \ \mathbf{k} = 0$$

(5.26.17)

$$[H(0) + H(\mathbf{k})]\psi_{\mathbf{k}}^{nmq}(\mathbf{x}) = \left(E(\mathbf{k}) - \frac{\hbar^2 \mathbf{k}^2}{2m}\right)\psi_{\mathbf{k}}^{nmq}(\mathbf{x})$$

(5.26.18)

$$[H(0) + H(\mathbf{k})]\psi_{\mathbf{k}}^{nmq}(\mathbf{x}) = E'(\mathbf{k})\psi_{\mathbf{k}}^{nmq}(\mathbf{x})$$

(5.26.19)

$$[H(0) + W(\mathbf{k})]\psi_{\mathbf{k}}^{nmq}(\mathbf{x}) = E(\mathbf{k})\psi_{\mathbf{k}}^{nmq}(\mathbf{x})$$

(5.26.20)

where, referring to Eq. 5.11.7 and the above equation, we can argue and conclude that

- $H(0) = \frac{p^2}{2m} + \mathcal{U}$ is the Hamiltonian of the system at $\mathbf{k} = 0$, for which the energy eigenvalue $E(0)$ is known. Thus the $H(\mathbf{k}) = \frac{\hbar}{m}(\mathbf{k}.\mathbf{p})$ is the perturbation to the Hamiltonian $H(0)$ for determining $\psi_{\mathbf{k}}^{nmq}(\mathbf{x})$ and $E(\mathbf{k})$ near the vicinity of $\mathbf{k} = 0$, which concerns us further on. They are determined in terms complete set of cell periodic wave functions and the energy eigenvalues $E(0)$ at $\mathbf{k} = 0$.
- In Eq. 5.26.19, $E'(\mathbf{k}) = E(\mathbf{k}) - \frac{\hbar^2 \mathbf{k}^2}{2m}$ is a useful notation for writing the perturbed Hamiltonian $[H(0) + H(\mathbf{k})]$, which will be used later discussing the $\mathbf{k}.\mathbf{p}$ perturbation method.
- In Eq. 5.26.20 $W(\mathbf{k}) = \frac{\hbar}{m}(\mathbf{k}.\mathbf{p}) + \frac{\hbar^2 \mathbf{k}^2}{2m}$ or $W(\mathbf{k}) = H(\mathbf{k}) + \frac{\hbar^2 \mathbf{k}^2}{2m}$ signifies the wave number dependent part of the Hamiltonian.
- $E(\mathbf{k})$ is the energy state of the system at any other wave number value $\mathbf{k}$, which is unknown and could be found from the eigenvalues of the perturbed Hamiltonian $W(\mathbf{k})$, when solved. Assuming the solution of the perturbed Hamiltonian is $eig(W(\mathbf{k}))$ for a specific energy band, the energy eigenvalues could be written as

$$E(\mathbf{k}) = E(0) + eig(W(\mathbf{k}))$$

(5.26.21)

- The perturbed Hamiltonian $H(\mathbf{k})$ or $W(\mathbf{k})$ has the diagonal element zero, and thus is also called the 'reduced Hamiltonian' [19].

Quantum Mechanics for Engineers

5.3.3.2.1 *Finding Energy Dispersion for a Single Energy Band without Spin*

The **k.p** perturbation theory assumed that the function ψ_k^{nmq} comes from the periodicity of the crystals. For a fixed value of **k**, the wave function ψ_k^{nmq} consists of a complete set of orthogonal bases with all possible bands in the $E - \mathbf{k}$ dispersion plot. These bases are wave number dependent and do not change if the band changes. Let's first assume that the wave function for an arbitrary band 'b' at any wave number **k** is $\psi_{k,b}^{nmq}(\mathbf{x})$. Referring to Eq. 5.26.5 the wave function $\psi_{k,b}^{nmq}$ could be written for the band 'b' as follows where, the index b takes only the integer values $1 - \infty$:

$$\psi_{k,b}^{nmq}(\mathbf{x}) = \left[\sum_G A_b(\mathbf{k}+\mathbf{G})e^{i\mathbf{G}.\mathbf{x}} \right] e^{i\mathbf{k}.\mathbf{x}} \tag{5.27.1}$$

Hence, now at any wave number **k**, considering all the bands, the wave function $\psi_k^{nmq}(\mathbf{x})$ could be further written by summing the orthogonal complete set of bases (Eq. 5.27.1) multiplied with their band contributing coefficient c_b, respectively.

$$\psi_k^{nmq}(\mathbf{x}) = \left[\sum_b c_b \left(\sum_G A_b(\mathbf{k}+\mathbf{G})e^{i\mathbf{G}.\mathbf{x}} \right) \right] e^{i\mathbf{k}.\mathbf{x}} = \sum_b c_b \psi_{k,b}^{nmq}(\mathbf{x})e^{i\mathbf{k}.\mathbf{x}} \tag{5.27.2}$$

Hence, it is obvious that the above equation is valid both at the center of the Brillouin zone (Γ point) and at the Brillouin zone boundaries (K points). At the Γ point ($\mathbf{k} = 0$) we could write

$$\psi_k^{nmq}(\mathbf{x}) = \sum_b c_b \psi_{k,b}^{nmq}(\mathbf{x})e^{i\mathbf{k}.\mathbf{x}}; \; \psi_0^{nmq}(\mathbf{x}) = \sum_b c_b \psi_{0,b}^{nmq}(\mathbf{x}) \tag{5.27.3}$$

It is required that individually each basis must satisfy the Schrödinger equations; thus, substituting $\psi_{k,b}^{nmq}(\mathbf{x})$ for the b-th band in Eq. 5.26.16 will read

$$\left[\left(\frac{\hbar^2 \mathbf{k}^2}{2m} + \frac{\hbar}{m}(\mathbf{k}.\mathbf{p}) + \frac{p^2}{2m} \right) + \mathcal{U} \right] \psi_{k,b}^{nmq}(\mathbf{x}) = E_b(\mathbf{k})\psi_{k,b}^{nmq}(\mathbf{x}) \tag{5.27.4}$$

Let's assume the minima is located at $\mathbf{k} = 0$ (i.e., the Γ), about which the perturbation is intended. Then the above equation will again revise to

$$\left[\frac{p^2}{2m} + \mathcal{U} \right] \psi_{0,b}^{nmq}(\mathbf{x}) = E_b(0)\psi_{0,b}^{nmq}(\mathbf{x}) \tag{5.27.5.1}$$

$$[H(0)]\psi_{0,b}^{nmq}(\mathbf{x}) = E_b(0)\psi_{0,b}^{nmq}(\mathbf{x}) \tag{5.27.5.2}$$

Using Eq. 5.27.2, substituted into Eqs. 5.27.4 and 5.27.5.1 and summing over all bands, we get

$$\sum_b c_b \left[\left(\frac{\hbar^2 \mathbf{k}^2}{2m} + \frac{\hbar}{m}(\mathbf{k}.\mathbf{p}) + \frac{p^2}{2m} \right) + \mathcal{U} \right] \psi_{k,b}^{nmq}(\mathbf{x}) = \sum_b c_b \psi_{k,b}^{nmq}(\mathbf{x})E_b(\mathbf{k}) \tag{5.27.6}$$

$$\sum_b c_b \left[\frac{p^2}{2m} + \mathcal{U} \right] \psi_{0,b}^{nmq}(\mathbf{x}) = \sum_b c_b \psi_{0,b}^{nmq}(\mathbf{x}) E_b(\mathbf{0}) \qquad (5.27.7)$$

Comparing the above equations, the Schrödinger equation in periodic media would read

$$\sum_b c_b \left[E_b(\mathbf{0}) + \frac{\hbar^2 \mathbf{k}^2}{2m} + \frac{\hbar}{m}(\mathbf{k}.\mathbf{p}) \right] \psi_{0,b}^{nmq}(\mathbf{x}) = \sum_b c_b E_b(\mathbf{k}) \psi_{0,b}^{nmq}(\mathbf{x}) \qquad (5.27.8)$$

Applying the orthogonality condition of the bases (i.e., multiplying both sides by $\psi_{0,d}^{nmq}(\mathbf{x})$ (the index d takes only the integer values $1 - \infty$)) and integrating over the whole volume of a unit cell, we get

$$c_d \left[E_b(\mathbf{k}) - E_d(\mathbf{0}) - \frac{\hbar^2 \mathbf{k}^2}{2m} \right] - \sum_b c_b \frac{\hbar}{m}(\mathbf{k}.\mathbf{p}_{db}) = 0 \qquad (5.27.9)$$

where

$$\mathbf{p}_{db} = \int_{\Omega} \psi_{0,d}^{nmq}(\mathbf{x})^* \; \mathbf{p} \; \psi_{0,b}^{nmq}(\mathbf{x}) d\Omega \qquad (5.27.10)$$

Please note that when the band index $b = d$ (i.e, when they refer to the same band), the coefficients survive; otherwise, when $b \neq d$ they are orthogonal and the integral vanishes. Eq. 5.27.9 gives the energy eigenvalues $E_b(\mathbf{k})$ when $E_d(\mathbf{0})$ and $\mathbf{p}_{db}$ are known or found the degree of accuracy increases if more bands are considered.

Next, let's assume a situation where the most contribution comes from a single band and contribution coefficients of other bands are significantly weaker. Let's reiterate Eq. 5.27.3 in the following form, where a single band with index $b = N$ is assumed to contribute most and other bands with variable index d assumed to contribute less. Together we can split the contribution and reiterate the Eq. 5.27.3 as

$$\psi_0^{nmq}(\mathbf{x}) = c_N \psi_{0,N}^{nmq}(\mathbf{x}) + \sum_d c_d \psi_{0,d}^{nmq}(\mathbf{x}) \qquad (5.27.11)$$

It is assumed that the wave function $\psi_{0,b}^{nmq}(\mathbf{x})$ creates a complete set of orthogonal bases, and thus the square of the magnitude of the contribution coefficients (c_b) should be equal to 1 (i.e., $\sum_b |c_b|^2 = 1$). Hence, as per Eq. 5.27.11, one can write

$$|c_N|^2 + \sum_d |c_d|^2 = 1 \qquad (5.27.12)$$

If the contribution by several bands represented by the index d is far less than the contribution by the band N, then it is safe to assume $c_N \gg \sum_d c_d$ or $c_N \approx 1$. With this light and stationary perturbation theory, Eq. 5.27.9 will read

$$E_N(\mathbf{k}) \approx E_N(\mathbf{0}) + \frac{\hbar^2 \mathbf{k}^2}{2m} + \frac{\hbar}{m}(\mathbf{k}.\mathbf{p}_{NN}) \qquad (5.27.13)$$

Quantum Mechanics for Engineers 193

where all the c_d coefficients are ignored in the N-th equation. Next, without ignoring any contribution, Eq. 5.27.9 at $\mathbf{k} = 0$ will further give

$$c_d \approx c_N \frac{\hbar}{m} \left(\frac{1}{E_N(0) - E_d(0)} \right) (\mathbf{k}.\mathbf{p}_{dN}) \qquad (5.27.14)$$

Here, not to confuse, index b and d are synonymous, as no specific index was specified.

Substituting Eq. 5.27.14 back to Eq. 5.27.9, the second-order approximation with $\mathbf{k}.\mathbf{p}$ perturbation will read

$$E_N(\mathbf{k}) \approx E_N(0) + \frac{\hbar^2 \mathbf{k}^2}{2m} + \frac{\hbar}{m}(\mathbf{k}.\mathbf{p}_{NN}) + \sum_{d \neq N} \left(\frac{\hbar}{m} \right)^2 \left(\frac{|\mathbf{k}.\mathbf{p}_{dN}|^2}{E_N(0) - E_d(0)} \right) \qquad (5.27.15)$$

Please note that at $\mathbf{k} = 0$, $\mathbf{p}_{NN} = 0$; thus Eq. 5.27.15 result in a parabolic equation as follows:

$$E_N(\mathbf{k}) \approx E_N(0) + \frac{\hbar^2 \mathbf{k}^2}{2m} \left[1 + 2 \sum_{d \neq N} \frac{\hbar^2}{m} \left(\frac{|\hat{\mathbf{e}}.\mathbf{p}_{dN}|^2}{E_N(0) - E_d(0)} \right) \right] \qquad (5.27.16)$$

where $\mathbf{k}^2 = (k_1)^2 + (k_2)^2 + (k_3)^2$, $\mathbf{k} = k_i \hat{e}_i$, and $\hat{\mathbf{e}}$, represents the unit vector along the direction of the wave vector $\mathbf{k}$. Comparing Eq. 5.26.18 with second-order approximation, we can say for the N-th band

$$eig(W_N(\mathbf{k})) = \frac{\hbar^2 \mathbf{k}^2}{2m} \left[1 + 2 \sum_{d \neq N} \frac{\hbar^2}{m} \left(\frac{|\hat{\mathbf{e}}.\mathbf{p}_{dN}|^2}{E_N(0) - E_d(0)} \right) \right] \qquad (5.27.17)$$

5.3.3.2.2 *Finding Energy Dispersion for Four Energy Bands without Spin*

This section would be more relevant in Chapter 6 in relation to phononics [20]. However, from the quantum mechanics perspective, spin-free four energy bands are represented here to find the energy of the bands using $\mathbf{k}.\mathbf{p}$ perturbation. When only one band is considered, it is assumed that $c_N \gg \sum_d c_d$. However, it is not appropriate for this four-band case. Let's assume that the band-dependent wave functions $\psi_{\mathbf{k},b}^{nmq}(\mathbf{x})$ for these four bands (1, 2, 3, and 4) are summed over by multiplying their respective contribution coefficient to obtain the total wave function $\psi_{\mathbf{k}}^{nmq}(\mathbf{x})$. Hence, the total wave function will visualize as

$$\psi_{\mathbf{k}}^{nmq}(\mathbf{x}) = \left[c_1(\mathbf{k})\psi_{\mathbf{k},1}^{nmq}(\mathbf{x}) + c_2(\mathbf{k})\psi_{\mathbf{k},2}^{nmq}(\mathbf{x}) + c_3(\mathbf{k})\psi_{\mathbf{k},3}^{nmq}(\mathbf{x}) + c_4(\mathbf{k})\psi_{\mathbf{k},4}^{nmq}(\mathbf{x}) \right] e^{i\mathbf{k}.\mathbf{x}} \ldots \qquad (5.27.18)$$

Referring to Eq. 5.26.19, $E'(\mathbf{k}) = E(\mathbf{k}) - \dfrac{\hbar^2 \mathbf{k}^2}{2m}$ is a useful notation to express a perturbed Hamiltonian more easily. Hence, following a similar procedure (applying orthogonality of the Bloch eigenfunctions) as in the previous section and exploiting

Eq. 5.27.9 for four bands, the following four linear homogeneous equations could be written as follows:

$$c_1(\mathbf{k})\left[E'(\mathbf{k})-E_1(\mathbf{0})\right]-\frac{h}{m}\mathbf{k}.\left[c_2(\mathbf{k})\mathbf{p}_{12}+c_3(\mathbf{k})\mathbf{p}_{13}+c_4(\mathbf{k})\mathbf{p}_{14}\right]=0 \qquad (5.27.19)$$

$$c_2(\mathbf{k})\left[E'(\mathbf{k})-E_2(\mathbf{0})\right]-\frac{h}{m}\mathbf{k}.\left[c_1(\mathbf{k})\mathbf{p}_{21}+c_3(\mathbf{k})\mathbf{p}_{23}+c_4(\mathbf{k})\mathbf{p}_{24}\right]=0 \qquad (5.27.20)$$

$$c_3(\mathbf{k})\left[E'(\mathbf{k})-E_3(\mathbf{0})\right]-\frac{h}{m}\mathbf{k}.\left[c_1(\mathbf{k})\mathbf{p}_{31}+c_2(\mathbf{k})\mathbf{p}_{32}+c_4(\mathbf{k})\mathbf{p}_{34}\right]=0 \qquad (5.27.21)$$

$$c_4(\mathbf{k})\left[E'(\mathbf{k})-E_4(\mathbf{0})\right]-\frac{h}{m}\mathbf{k}.\left[c_1(\mathbf{k})\mathbf{p}_{41}+c_2(\mathbf{k})\mathbf{p}_{42}+c_3(\mathbf{k})\mathbf{p}_{43}\right]=0 \qquad (5.27.22)$$

where $E_1(\mathbf{0})$, $E_2(\mathbf{0})$, $E_3(\mathbf{0})$, and $E_4(\mathbf{0})$ are the energy eigenvalues of the four bands at $\mathbf{k}=\mathbf{0}$, respectively, and are known. It can be expressed as $H(\mathbf{0})\psi_0^{nmq}(\mathbf{x})=E(\mathbf{0})\psi_0^{nmq}(\mathbf{x})$. The above equations can be arranged in a matrix form as follows

$$\begin{bmatrix} \left[E'(\mathbf{k})-E_1(\mathbf{0})\right] & -\frac{h}{m}(\mathbf{k}.\mathbf{p}_{12}) & -\frac{h}{m}(\mathbf{k}.\mathbf{p}_{13}) & -\frac{h}{m}(\mathbf{k}.\mathbf{p}_{14}) \\ -\frac{h}{m}(\mathbf{k}.\mathbf{p}_{21}) & \left[E'(\mathbf{k})-E_2(\mathbf{0})\right] & -\frac{h}{m}(\mathbf{k}.\mathbf{p}_{23}) & -\frac{h}{m}(\mathbf{k}.\mathbf{p}_{24}) \\ -\frac{h}{m}(\mathbf{k}.\mathbf{p}_{31}) & -\frac{h}{m}(\mathbf{k}.\mathbf{p}_{32}) & \left[E'(\mathbf{k})-E_3(\mathbf{0})\right] & -\frac{h}{m}(\mathbf{k}.\mathbf{p}_{34}) \\ -\frac{h}{m}(\mathbf{k}.\mathbf{p}_{41}) & -\frac{h}{m}(\mathbf{k}.\mathbf{p}_{42}) & -\frac{h}{m}(\mathbf{k}.\mathbf{p}_{43}) & \left[E'(\mathbf{k})-E_4(\mathbf{0})\right] \end{bmatrix}\begin{Bmatrix} c_1(\mathbf{k}) \\ c_2(\mathbf{k}) \\ c_3(\mathbf{k}) \\ c_4(\mathbf{k}) \end{Bmatrix}=\begin{Bmatrix} 0 \\ 0 \\ 0 \\ 0 \end{Bmatrix}\cdots$$

$$(5.27.23)$$

The momentum vector $\mathbf{p}_{db}$ is expressed in Eq. 5.27.10. Separating the eigenvalue matrix at $\mathbf{k}=\mathbf{0}$, the Eq. 5.27.23 can further be written as

$$\begin{bmatrix} \left[E'(\mathbf{k})\right] & -\frac{h}{m}(\mathbf{k}.\mathbf{p}_{12}) & -\frac{h}{m}(\mathbf{k}.\mathbf{p}_{13}) & -\frac{h}{m}(\mathbf{k}.\mathbf{p}_{14}) \\ -\frac{h}{m}(\mathbf{k}.\mathbf{p}_{21}) & \left[E'(\mathbf{k})\right] & -\frac{h}{m}(\mathbf{k}.\mathbf{p}_{23}) & -\frac{h}{m}(\mathbf{k}.\mathbf{p}_{24}) \\ -\frac{h}{m}(\mathbf{k}.\mathbf{p}_{31}) & -\frac{h}{m}(\mathbf{k}.\mathbf{p}_{32}) & \left[E'(\mathbf{k})\right] & -\frac{h}{m}(\mathbf{k}.\mathbf{p}_{34}) \\ -\frac{h}{m}(\mathbf{k}.\mathbf{p}_{41}) & -\frac{h}{m}(\mathbf{k}.\mathbf{p}_{42}) & -\frac{h}{m}(\mathbf{k}.\mathbf{p}_{43}) & \left[E'(\mathbf{k})\right] \end{bmatrix}\begin{Bmatrix} c_1(\mathbf{k}) \\ c_2(\mathbf{k}) \\ c_3(\mathbf{k}) \\ c_4(\mathbf{k}) \end{Bmatrix}$$

$$=\begin{bmatrix} E_1(\mathbf{0}) & 0 & 0 & 0 \\ 0 & E_2(\mathbf{0}) & 0 & 0 \\ 0 & 0 & E_3(\mathbf{0}) & 0 \\ 0 & 0 & 0 & E_4(\mathbf{0}) \end{bmatrix}\begin{Bmatrix} c_1(\mathbf{k}) \\ c_2(\mathbf{k}) \\ c_3(\mathbf{k}) \\ c_4(\mathbf{k}) \end{Bmatrix}$$

$$(5.27.24)$$

Quantum Mechanics for Engineers

and further

$$
\begin{bmatrix}
0 & -\dfrac{h}{m}\left(\mathbf{k}.\mathbf{p}_{12}\right) & -\dfrac{h}{m}\left(\mathbf{k}.\mathbf{p}_{13}\right) & -\dfrac{h}{m}\left(\mathbf{k}.\mathbf{p}_{14}\right) \\
-\dfrac{h}{m}\left(\mathbf{k}.\mathbf{p}_{21}\right) & 0 & -\dfrac{h}{m}\left(\mathbf{k}.\mathbf{p}_{23}\right) & -\dfrac{h}{m}\left(\mathbf{k}.\mathbf{p}_{24}\right) \\
-\dfrac{h}{m}\left(\mathbf{k}.\mathbf{p}_{31}\right) & -\dfrac{h}{m}\left(\mathbf{k}.\mathbf{p}_{32}\right) & 0 & -\dfrac{h}{m}\left(\mathbf{k}.\mathbf{p}_{34}\right) \\
-\dfrac{h}{m}\left(\mathbf{k}.\mathbf{p}_{41}\right) & -\dfrac{h}{m}\left(\mathbf{k}.\mathbf{p}_{42}\right) & -\dfrac{h}{m}\left(\mathbf{k}.\mathbf{p}_{43}\right) & 0
\end{bmatrix}
\begin{Bmatrix}
c_1(\mathbf{k}) \\
c_2(\mathbf{k}) \\
c_3(\mathbf{k}) \\
c_4(\mathbf{k})
\end{Bmatrix}
$$

$$
+\begin{bmatrix}
\left[E'(\mathbf{k})\right] & 0 & 0 & 0 \\
0 & \left[E'(\mathbf{k})\right] & 0 & 0 \\
0 & 0 & \left[E'(\mathbf{k})\right] & 0 \\
0 & 0 & 0 & \left[E'(\mathbf{k})\right]
\end{bmatrix}
\begin{Bmatrix}
c_1(\mathbf{k}) \\
c_2(\mathbf{k}) \\
c_3(\mathbf{k}) \\
c_4(\mathbf{k})
\end{Bmatrix}
$$

$$
=\begin{bmatrix}
E_1(\mathbf{0}) & 0 & 0 & 0 \\
0 & E_2(\mathbf{0}) & 0 & 0 \\
0 & 0 & E_3(\mathbf{0}) & 0 \\
0 & 0 & 0 & E_4(\mathbf{0})
\end{bmatrix}
\begin{Bmatrix}
c_1(\mathbf{k}) \\
c_2(\mathbf{k}) \\
c_3(\mathbf{k}) \\
c_4(\mathbf{k})
\end{Bmatrix}\cdots
$$

$$\tag{5.27.25}$$

Comparing the above matrix form with the Eqs 5.26.17 through 5.26.20, one can write

$$
\left[H(\mathbf{k})+E'(\mathbf{k})\right]\psi_k = H(\mathbf{0})\psi_k \tag{5.27.26}
$$

Here, $H(\mathbf{k})$ is the reduced or perturbed Hamiltonian specifically found from the $\mathbf{k}.\mathbf{p}$ perturbation approach. $H(\mathbf{0})$ is the Hamiltonian of the system at $\mathbf{k}=\mathbf{0}$, and $\psi_k = \{c_1(\mathbf{k})\ c_2(\mathbf{k})\ c_3(\mathbf{k})\ c_4(\mathbf{k})\}^T$. Solving the eigenvalue problem of the perturbed Hamiltonian, energy eigenvalues at other $\mathbf{k}$ points can be calculated and the Bloch eigenfunction $\psi_{\mathbf{k}}^{nmq}(\mathbf{x})$ can be explicitly found. Once the Bloch eigenfunctions or the wave functions are found, they can be used to calculate the phase of the wave modes along a specific path in the Brillouin zone. This will be discussed further when introducing the geometric phase in section 5.5.

5.4 RELATIVISTIC PARTICLES WITH SPIN ZERO AND SPIN HALF

5.4.1 KLEIN-GORDON EQUATION FOR SPIN 0 RELATIVISTIC PARTICLES

The Dirac equation and its understanding starts fundamentally from the energy equation for the relativistic particles expressed in Eq. 5.4.6, reiterated here by multiplying the wave function Ψ on either side of the equation as

$$
E^2\Psi = \left[\left(pc\right)^2 + \left(mc^2\right)^2\right]\Psi \tag{5.28.1}
$$

The necessary quantum operators are discussed in section 5.2 and would be required here to deduce the concept of the Dirac equation. The momentum operator, square momentum operator, energy operator, and Hamiltonian operator are reiterated as

$$p = -i\hbar\nabla;\ p^2 = -\hbar^2\nabla^2;\ E = -\frac{\hbar}{i}\frac{\partial}{\partial t} = i\hbar\frac{\partial}{\partial t};\ H = -\frac{\hbar^2}{2m}\nabla^2 \qquad (5.28.2)$$

Substituting the energy operator and the square momentum operator, Eq. 5.28.1 would read

$$-\hbar^2\frac{\partial^2}{\partial t^2}\Psi = c^2\left[-\hbar^2\nabla^2 + m^2c^2\right]\Psi \qquad (5.28.3)$$

Further simplifying the above equation, Eq. 5.28.3 can be read as a wave equation with a second-order time derivate. This is unlike the Schrödinger wave equation, where it has only the first-order time derivate in the equation. The wave equation for relativistic particle (i.e., for an electron) with obvious second-order time derivative is called the Klein-Gordon equation. It was first derived by physicists Oskar Klein and Walter Gordon, in 1926 [22].

$$\left[\nabla^2 - \frac{m^2c^2}{\hbar^2}\right]\Psi = \frac{1}{c^2}\frac{\partial^2}{\partial t^2}\Psi \qquad (5.28.4)$$

$$\left[\left(\frac{\partial^2}{\partial x_1^2} + \frac{\partial^2}{\partial x_2^2} + \frac{\partial^2}{\partial x_3^2}\right) - \frac{m^2c^2}{\hbar^2}\right]\Psi = \frac{1}{c^2}\frac{\partial^2}{\partial t^2}\Psi \qquad (5.28.5)$$

The eigenfunction of the equation could be expressed as

$$\Psi = e^{-i(Et - \mathbf{p}.\mathbf{x})};\ \Psi = e^{-i(Et - \mathbf{k}.\mathbf{x})} \qquad (5.28.6)$$

where wave number and momentum are synonymous in quantum mechanics.

5.4.2 ENERGY SQUARE ROOT PARITY

Revising Eqs. 5.4.5 and 5.28.1, one can argue that energy must be squared to get the Klein-Gordon equation (Eq. 5.28.4). Thus, it could also be argued, why not start from the energy equation as follows?

$$E\Psi = \pm\sqrt{\left[(pc)^2 + (mc^2)^2\right]}\Psi \qquad (5.28.7)$$

They are mathematically equivalent. Let's rederive the wave equation for relativistic particle like an electron from Eq. 5.28.7. Substituting Eq. 5.28.2, one gets

$$i\hbar\frac{\partial}{\partial t}\Psi = \pm c\sqrt{\left[-\hbar^2\nabla^2 + m^2c^2\right]}\Psi \qquad (5.28.8)$$

Quantum Mechanics for Engineers 197

$$i\frac{\partial}{\partial t}\Psi = \pm c\sqrt{\left[-\nabla^2 + \frac{m^2 c^2}{\hbar^2}\right]}\Psi \tag{5.28.9}$$

$$\pm\sqrt{\left[\nabla^2 - \frac{m^2 c^2}{\hbar^2}\right]} = \frac{1}{c}\frac{\partial}{\partial t}\Psi \tag{5.28.10}$$

$$\pm\sqrt{\left[\left(\frac{\partial^2}{\partial x_1^2} + \frac{\partial^2}{\partial x_2^2} + \frac{\partial^2}{\partial x_3^2}\right) - \frac{m^2 c^2}{\hbar^2}\right]}\Psi = \frac{1}{c}\frac{\partial}{\partial t}\Psi \tag{5.28.11}$$

Are Eqs. 5.28.5 and 5.28.11 the same or different? It is apparent that the energy takes negative values in Eq. 5.28.11, which will result in a negative probability density function as a possible solution. However, a negative probability density function is impossible when negative energy is not much of a problem. This very question intrigued Paul Dirac in 1928 [23]. An alternative unambiguous relativistic wave equation was necessary.

5.4.3 Dirac Equation for Spin Half Particles

In mathematics any squared quantity can be written as its self-multiplication. Hence, Dirac proposed Eq. 5.28.5 to be rewritten as follows. Including the second-order time derivate term with Laplace operator the Eq. 5.28.5 will visualize

$$\left[\sqrt{\left[\left(\frac{\partial^2}{\partial x_1^2} + \frac{\partial^2}{\partial x_2^2} + \frac{\partial^2}{\partial x_3^2}\right) - \frac{1}{c^2}\frac{\partial^2}{\partial t^2}\right]}\right]^2 \Psi = \frac{m^2 c^2}{\hbar^2}\Psi \tag{5.28.12}$$

Next, Dirac assumed $\sqrt{A^2 + B^2} = aA + bB$, which is conventionally not possible. This step was an ingenious stroke by Dirac. Considering mathematics under special circumstances, the square root on the left-hand side would read

$$\sqrt{\frac{\partial^2}{\partial x_1^2} + \frac{\partial^2}{\partial x_2^2} + \frac{\partial^2}{\partial x_3^2} - \frac{1}{c^2}\frac{\partial^2}{\partial t^2}} = i\left(\left(\Gamma^1\frac{\partial}{\partial x_1} + \Gamma^2\frac{\partial}{\partial x_2} + \Gamma^3\frac{\partial}{\partial x_3}\right) + \Gamma^0\frac{\partial}{\partial t}\right) \tag{5.28.13}$$

where Γ^0 and Γ^i are the arbitrary constants. The new first order wave equation would read

$$i\left(\left(\Gamma^1\frac{\partial}{\partial x_1} + \Gamma^2\frac{\partial}{\partial x_2} + \Gamma^3\frac{\partial}{\partial x_3}\right) + \Gamma^0\frac{\partial}{\partial t}\right)\Psi = \frac{mc}{\hbar}\Psi \tag{5.28.14}$$

In natural units with a metric signature where $\hbar$ and c are unity the above equation, will revise to

$$\left(i\gamma^0\frac{\partial}{\partial t} + i\gamma^1\frac{\partial}{\partial x_1} + i\gamma^2\frac{\partial}{\partial x_2} + i\gamma^3\frac{\partial}{\partial x_3} - m\right)\Psi = 0 \tag{5.28.15}$$

$$\left(i\gamma^0 \frac{\partial}{\partial t} + i\boldsymbol{\gamma} \cdot \nabla - m\right)\Psi = 0; \ (i\gamma^\mu \partial_\mu - m)\Psi = 0 \tag{5.28.16}$$

where in index notation $\boldsymbol{\gamma} = \gamma^\nu \hat{e}_\nu$ and $\nabla = \frac{\partial}{\partial x_\nu} \hat{e}_\nu$, with ν takes index values 1, 2, and 3.

Also, $\partial_\mu = \left(\frac{\partial}{\partial t}, \frac{\partial}{\partial x_1}, \frac{\partial}{\partial x_2}, \frac{\partial}{\partial x_3}\right)$ and $\gamma^\mu = \left\{\gamma^0, \gamma^1, \gamma^2, \gamma^3\right\}$ are used in the above equation.

So far, what γ^μ would read, is not discussed. Derivation had no hurdles without the formal introduction of γ^μ except a small glitch. First, we should argue that Eq. 5.28.13 is not right and cannot be used as it is in Eq. 5.28.16. Yes, it's partially correct in mathematics but not completely true. The transition from Eqs. 5.28.12 to 5.28.13 is not valid if Γ^μ are just the numbers. But the expressions are true if Γ^μ or the γ^μ are the matrices. This means an eigenstate is split into two new eigenstates. Hence, it is necessary to find the γ^μ matrices. But why, suddenly, one would need a matrix for a relativistic particle like an electron unless electrons have additional degrees of freedom? Immediately, all was kind of falling into place because by 1928 Pauli had expressed the spin matrices for electrons as written in Eq. 5.24.16. Dirac recognized that the wave equation expressed by Eq. 5.28.16 is for relativistic particles with half spins. It's like spin 0 state is ripped into two overlapped counter spin states of +1/2 and −1/2.

The only thing left is to find the γ^μ matrices. To solve the problem, the following steps were taken. The Dirac equation and its conjugate were multiplied, and it is imposed that it should consummate to the Klein-Gordon equation, because both must be true.

$$\Psi^*\left(-i\gamma^0 \frac{\partial}{\partial t} - i\boldsymbol{\gamma} \cdot \nabla + m\right)\left(i\gamma^0 \frac{\partial}{\partial t} + i\boldsymbol{\gamma} \cdot \nabla - m\right)\Psi = 0 \tag{5.28.17.1}$$

$$\Psi^*\left(\left(\gamma^0\right)^2 \frac{\partial^2}{\partial t^2} + \left(\gamma^1\right)^2 \frac{\partial^2}{\partial x_1^2} + \left(\gamma^2\right)^2 \frac{\partial^2}{\partial x_2^2} + \left(\gamma^3\right)^2 \frac{\partial^2}{\partial x_3^2} - m^2\right)\Psi = 0 \tag{5.28.17.2}$$

Ignoring the linear terms for momenta and comparing the Klein-Gordon equation with the Laplace operator, the following conditions must be true:

$$\left(\gamma^0\right)^2 = 1, \ \left(\gamma^\nu\right)^2 = -1 \ \gamma^\mu\gamma^\xi + \gamma^\xi\gamma^\mu = 0 \ for \ \forall \ \mu \neq \xi \tag{5.28.18}$$

with $\nu = 1, 2, 3$, $\mu, \xi = 0, 1, 2, 3$, which presents an anti-commutation relation. Referring to Pauli's matrices in Eq. 5.24.16, the possible solutions that satisfy this relation are

$$\gamma^0 = \begin{pmatrix} \mathbf{I} & \mathbf{0} \\ \mathbf{0} & -\mathbf{I} \end{pmatrix} \gamma^\nu = \begin{pmatrix} \mathbf{0} & \sigma^\nu \\ -\sigma^\nu & \mathbf{0} \end{pmatrix} \tag{5.28.19}$$

Pauli's matrices (refer to section 5.2.5) are reiterated herein with the required equivalent notation.

$$\sigma^1 = \sigma_{x1} = \begin{pmatrix} 0 & 1 \\ 1 & 0 \end{pmatrix} \sigma^2 = \sigma_{x2} = \begin{pmatrix} 0 & -i \\ i & 0 \end{pmatrix} \sigma^3 = \sigma_{x3} = \begin{pmatrix} 1 & 0 \\ 0 & -1 \end{pmatrix} \tag{5.28.20}$$

Quantum Mechanics for Engineers 199

So, the gamma matrices could be read in full as follows:

$$\gamma^0 = \begin{pmatrix} 1 & 0 & 0 & 0 \\ 0 & 1 & 0 & 0 \\ 0 & 0 & -1 & 0 \\ 0 & 0 & 0 & -1 \end{pmatrix} \quad \gamma^1 = \begin{pmatrix} 0 & 0 & 0 & 1 \\ 0 & 0 & 1 & 0 \\ 0 & -1 & 0 & 0 \\ -1 & 0 & 0 & 0 \end{pmatrix}$$

$$\gamma^2 = \begin{pmatrix} 0 & 0 & 0 & -i \\ 0 & 0 & i & 0 \\ 0 & i & 0 & 0 \\ -i & 0 & 0 & 0 \end{pmatrix} \quad \gamma^3 = \begin{pmatrix} 0 & 0 & 1 & 0 \\ 0 & 0 & 0 & -1 \\ -1 & 0 & 0 & 0 \\ 0 & 1 & 0 & 0 \end{pmatrix}$$

(5.28.21)

5.4.4 HAMILTONIAN FOR SPIN ½ FERMIONS AND SPINORS

Substituting the gamma matrices from Eq. 5.28.21 into Eq. 5.28.16, the Dirac equation will explicitly be written as

$$\begin{pmatrix} i\dfrac{\partial}{\partial t} & 0 & i\dfrac{\partial}{\partial x_3} & i\dfrac{\partial}{\partial x_1} + \dfrac{\partial}{\partial x_2} \\ 0 & i\dfrac{\partial}{\partial t} & i\dfrac{\partial}{\partial x_1} - \dfrac{\partial}{\partial x_2} & -i\dfrac{\partial}{\partial x_3} \\ -i\dfrac{\partial}{\partial x_3} & -i\dfrac{\partial}{\partial x_1} - \dfrac{\partial}{\partial x_2} & -i\dfrac{\partial}{\partial t} & 0 \\ -i\dfrac{\partial}{\partial x_1} + \dfrac{\partial}{\partial x_2} & i\dfrac{\partial}{\partial x_3} & 0 & -i\dfrac{\partial}{\partial t} \end{pmatrix} \begin{pmatrix} \Psi^1 \\ \Psi^2 \\ \Psi^3 \\ \Psi^4 \end{pmatrix} = m\mathbf{I} \begin{pmatrix} \Psi^1 \\ \Psi^2 \\ \Psi^3 \\ \Psi^4 \end{pmatrix}$$

(5.28.22)

Eq. 5.28.22 can further be written in the following form:

$$\mathbf{H}_{rs}\Psi = \mathbf{E}_{rs}\Psi$$

(5.28.23)

where $\mathbf{H}_{rs}$ is the Hamiltonian of the relativistic particles with spin ½ or in other words the Hamiltonian for spin ½ fermions in relativistic quantum field theory. Eigenfunctions or the wave functions for fermions would be simply the product of the plane wave ($e^{-i\mathbf{p}\cdot\mathbf{x}}$), representing the wave particle duality and a spinor. The wave function thus would read

$$\Psi\left(x^{\mu}\right) = u\left(p^{\mu}\right) e^{-i\mathbf{p}\cdot\mathbf{x}}$$

(5.28.24)

Substituting the wave function in Eq. 5.28.22, the spinors could be solved using the following equation

$$\left(\mathbf{H}_{rs} - m\right) u(p) = 0$$

(5.28.25)

For a particle at rest where $p = 0$ (*which is also synonymous to the wavenumber* $k = 0$ *at the center of a Brillouin zone in a periodic media*), the equation will take the following form

$$\left(i\gamma^0 \frac{\partial}{\partial t} - m \right) \psi = \left(\gamma^0 \mathbf{E}_{rs} - m \right) \psi = 0; \; Eu = \begin{pmatrix} m\mathbf{I} & 0 \\ 0 & -m\mathbf{I} \end{pmatrix} u \qquad (5.28.26)$$

and the four eigenspinors would be solve as

$$u^1 = \begin{pmatrix} 1 \\ 0 \\ 0 \\ 0 \end{pmatrix} u^2 = \begin{pmatrix} 0 \\ 1 \\ 0 \\ 0 \end{pmatrix} u^3 = \begin{pmatrix} 0 \\ 0 \\ 1 \\ 0 \end{pmatrix} u^4 = \begin{pmatrix} 0 \\ 0 \\ 0 \\ 1 \end{pmatrix} \qquad (5.28.27)$$

Thus, the spinor wave functions could be written as

$$\Psi^1 = e^{-i\mathbf{p}\cdot\mathbf{x}} u^1; \; \Psi^2 = e^{-i\mathbf{p}\cdot\mathbf{x}} u^2; \; \Psi^3 = e^{+i\mathbf{p}\cdot\mathbf{x}} u^3; \; \Psi^4 = e^{+i\mathbf{p}\cdot\mathbf{x}} u^4 \qquad (5.28.28)$$

Interestingly, the spinors describe four states with four eigenspinors, and these are not the states in t, x_1, x_2, and x_3 space. However, earlier while describing Pauli's matrix in section 5.2.5, only two spinor eigenstates were described. But the Dirac equation describes four eigenstates of a spinor. The question arises, if the two eigenspinor states are for the electrons, then what are these two additional spinor states that are automatically appearing from the Dirac solution? The answer was found by Dirac himself, who described that these two additional spinors are coming from the negative energy states (i.e., two spinor states for $E = m$ and two spinor states for $E = -m$). They describe two different spin states (spin up $\uparrow$ and spin down $\downarrow$) for each energy state (e.g., positive $+$ and negative $-$). It was concluded that, besides electrons, there are two more particles of positive charge $(+e)$ that are represented by the Dirac solution. Later they were named positron [24]. Richard Feynman in 1949 [24] described these particles as negative charge electrons propagating backward in time or positive charged positrons propagating forward in time. This way it can also be said that the positrons propagating backward in time are manifested to us as a negative charged electron propagating forward in time.

Next, for a relativistic particle that is not at rest – that is, the momentum $p \neq 0$ or equivalently $k \neq 0$ (i.e., a solution at another wave number value beyond the Γ point in the Brillouin zone) – the Dirac equation would read

$$(\mathbf{H}_{rs} - m)(u_T \quad u_B) = \begin{pmatrix} E - m & -\boldsymbol{\sigma}\cdot\mathbf{p} \\ \boldsymbol{\sigma}\cdot\mathbf{p} & -E - m \end{pmatrix} \begin{pmatrix} u_T \\ u_B \end{pmatrix} = 0 \qquad (5.28.29)$$

where u_T and u_B are the 1×2 vectors describing the top and the bottom parts of the spinor. When $p \neq 0$, it was found that the solutions for u_T and u_B are coupled.

$$u_T = \frac{\boldsymbol{\sigma}\cdot\mathbf{p}}{E - m} u_B; \; u_B = \frac{\boldsymbol{\sigma}\cdot\mathbf{p}}{E + m} u_T \qquad (5.28.30)$$

Quantum Mechanics for Engineers

First assuming two states of the spinor $u_T = \{1 \quad 0\}$ and $\{0 \quad 1\}$ respectively, u_B can be found as follows:

$$u_T = \left\{ \begin{array}{c} 1 \\ 0 \end{array} \right\}; u_B = \left\{ \begin{array}{c} \dfrac{p_3}{(E+m)} \\[2mm] \dfrac{(p_1+ip_2)}{(E+m)} \end{array} \right\} \tag{5.28.31}$$

$$u_T = \left\{ \begin{array}{c} 0 \\ 1 \end{array} \right\}; u_B = \left\{ \begin{array}{c} \dfrac{(p_1-ip_2)}{(E+m)} \\[2mm] \dfrac{-p_3}{(-E+m)} \end{array} \right\} \tag{5.28.32}$$

Alternatively, with two states of the spinor $u_B = \{1 \quad 0\}$ and $\{0 \quad 1\}$ respectively, u_T can be found as follows:

$$u_B = \left\{ \begin{array}{c} 1 \\ 0 \end{array} \right\}; u_T = \left\{ \begin{array}{c} \dfrac{-p_3}{(-E+m)} \\[2mm] \dfrac{(-p_1-ip_2)}{(-E+m)} \end{array} \right\} \tag{5.28.33}$$

$$u_B = \left\{ \begin{array}{c} 0 \\ 1 \end{array} \right\}; u_B = \left\{ \begin{array}{c} \dfrac{(-p_1+ip_2)}{(-E+m)} \\[2mm] \dfrac{p_3}{(-E+m)} \end{array} \right\} \tag{5.28.34}$$

Hence, the four wave functions (as in Eq. 5.28.28) with spinors with four spin states with energy $E = \pm \sqrt{(p)^2 + (m)^2}$ could be written as follows:

For negatively charged fermions – electrons

$$\Psi^1 = \left\{ \begin{array}{c} 1 \\ 0 \\ \dfrac{p_3}{(E+m)} \\[2mm] \dfrac{(p_1+ip_2)}{(E+m)} \end{array} \right\} e^{-ip.x}; \Psi^2 = \left\{ \begin{array}{c} 0 \\ 1 \\ \dfrac{(p_1-ip_2)}{(E+m)} \\[2mm] \dfrac{-p_3}{(-E+m)} \end{array} \right\} e^{-ip.x}; E = \sqrt{(p)^2 + (m)^2} \tag{5.28.35}$$

As in Eq. 5.24.2, the eigenvectors for spin-up and spin-down fermions (i.e., electrons) are, respectively,

$$\mathbf{s}_f{}^{\uparrow} = \left\{ \begin{array}{c} \hat{s}_1 \\ \hat{s}_2 \\ \hat{s}_3 \\ \hat{s}_4 \end{array} \right\}^{\uparrow} = \left\{ \begin{array}{c} 1 \\ 0 \\ \dfrac{p_3}{(E+m)} \\ \dfrac{(p_1+ip_2)}{(E+m)} \end{array} \right\}; \quad \mathbf{s}_f{}^{\downarrow} = \left\{ \begin{array}{c} \hat{s}_1 \\ \hat{s}_2 \\ \hat{s}_3 \\ \hat{s}_4 \end{array} \right\}^{\downarrow} = \left\{ \begin{array}{c} 0 \\ 1 \\ \dfrac{(p_1-ip_2)}{(E+m)} \\ \dfrac{-p_3}{(-E+m)} \end{array} \right\} \qquad (5.28.36)$$

For positively charged antifermions – positrons

$$\Psi^3 = \left\{ \begin{array}{c} \dfrac{-p_3}{(-E+m)} \\ \dfrac{(-p_1-ip_2)}{(-E+m)} \\ 1 \\ 0 \end{array} \right\} e^{+ip.x}; \quad \Psi^4 = \left\{ \begin{array}{c} \dfrac{(-p_1+ip_2)}{(-E+m)} \\ \dfrac{p_3}{(-E+m)} \\ 0 \\ 1 \end{array} \right\} e^{+ip.x}; \quad E = -\sqrt{(p)^2+(m)^2} \qquad (5.28.37)$$

As in Eq. 5.24.2, the eigenvectors for spin-up and spin-down antifermions (i.e., positrons) are, respectively,

$$\mathbf{s}_{af}{}^{\uparrow} = \left\{ \begin{array}{c} \hat{s}_1 \\ \hat{s}_2 \\ \hat{s}_3 \\ \hat{s}_4 \end{array} \right\}^{\uparrow} = \left\{ \begin{array}{c} \dfrac{-p_3}{(-E+m)} \\ \dfrac{(-p_1-ip_2)}{(-E+m)} \\ 1 \\ 0 \end{array} \right\}; \quad \mathbf{s}_{af}{}^{\downarrow} = \left\{ \begin{array}{c} \hat{s}_1 \\ \hat{s}_2 \\ \hat{s}_3 \\ \hat{s}_4 \end{array} \right\}^{\downarrow} = \left\{ \begin{array}{c} \dfrac{(-p_1+ip_2)}{(-E+m)} \\ \dfrac{p_3}{(-E+m)} \\ 0 \\ 1 \end{array} \right\} \qquad (5.28.38)$$

5.5 DIRAC CONES AND DIRAC-LIKE CONES

5.5.1 DIRAC CONES

Dirac cones [25] are specifically described in condensed matter physics, where the behavior of statistical quasi-particles is predominant. Although the Dirac cones are named after Paul Dirac for his ingenious contribution to quantum mechanics for describing fermions and antifermions, there was no immediate transition from the Dirac equation to the understanding of Dirac cones, which was later described through spin [26]. Dirac cones are the features in band structures for a material. They were discovered independently in condensed matter physics. It was found that in some materials, in the reciprocal space (i.e., k-space) the valance band and the conduction band tend to merge with local linear dispersion. At a specific point, the energy state is equal for those two bands; however, particles tend to avoid these junctures. The point is a zero dimensional (i.e., the energy is equal only at a point

Quantum Mechanics for Engineers 203

in the k-space, and the energy values are different at all the other wave numbers or k values). Near this unique point, due to localized linear dispersion in three dimensions, the dispersion behavior looks like a cone connected with an inverted cone at this point. Due to the formation of these cones, near this unique point the electrical conduction is synonymous with the flow of charges made of massless relativistic particles. Section 5.4.3 explains how the behavior of a relativistic massless fermion can be described using the Dirac equation. Hence, the flow of massless quasi particles near this unique point can be solved theoretically utilizing the Dirac equation. Thus, these unique points are called the Dirac points, and the cones formed at these points are called Dirac cones, as shown in Fig. 5.12a. The surface of the cones is formed by different energy values at different wave numbers.

When two or more energy states or wave modes intersect, cross, or overlap with each other, the bands are said to be degenerated. This is also called the band degeneracy. Thus, it is obvious that at Dirac points the bands are degenerated. If two bands are degenerated, they are called doubly degenerated bands. If three bands are degenerated, then they are called triply degenerated bands and so on.

Materials like graphene and other photonic metamaterials exhibit this Dirac cone behavior, and many such natural or metamaterials are thus called Dirac matters [25]. Dirac cone behaviors were also reported for acoustic waves in acoustic metamaterials, which will be discussed in the next chapter. The Dirac cones almost all the time are common at the boundary of the Brillouin zone. For a very specific symmetry type in Dirac matter (e.g., C_3 and C_6 symmetry), Dirac cones at the Brillouin boundary are achieved with certainty. Additionally at the Dirac cone the massless particles (fermions) at the Fermi level led to various quantum Hall effect, which will be discussed in section 5.8.

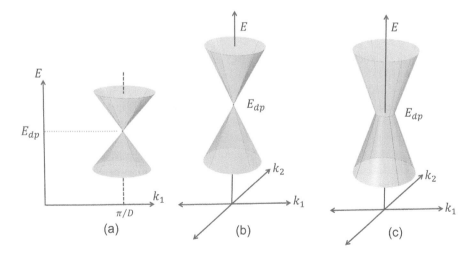

FIGURE 5.12 E_{dp} is the energy level at the Dirac points: a) Dirac cone at the Brillouin boundary $k = \pi/D$, where, D is the periodicity in the crystal; b) Dirac-like cone (Hermitian) at the Γ point $k = 0$; and c) Dirac-like cone (non Hermitian), Spawning ring at exceptional point with two connected cones at the Dirac point energy level.

5.5.2 Dirac-Like Cones

Unlike Dirac cones at the Brillouin boundary, at the center of the Brillouin zone (i.e., at $k = 0$) or at any other k values, Dirac cones are not always common but may form accidentally due to material types, geometry, symmetry, and their unique configuration [19]. Degeneracies at Dirac-like cones [27, 28] can be tuned by modifying the structural and geometrical parameters. Dirac-like cones at $k = 0$ points can be defined in to two groups: Hermitian and non-Hermitian. Hermitian Dirac-like cones have similar energy levels for two or more bands only at a point at $k = 0$ but do not extend beyond the Dirac point. However, non-Hermitian Dirac-like cones may extend from $k = 0$ point to non-zero k values. The non-Hermitian Dirac-like cones may form a spawning ring (an exceptional point) that can be visualized as two cones fused into each other. Figure 5.12b and c show examples of Dirac-like cones that are Hermitian and non-Hermitian. At a Hermitian Dirac-like cone [29], the Hamiltonian of the Dirac equation at the Dirac point must be Hermitian matrix, whereas it is not necessary for the Hamiltonian to be Hermitian in case of an exceptional point. These will be further discussed in the next chapter with acoustics context.

5.6 INTRODUCTION TO TOPOLOGY AND THE GEOMETRIC PHASE

Waves primarily have four canonical characteristics:

- frequency (ω), which is equivalent to energy (E)in quantum mechanics
- wave number (k), which is equivalent to momentum (p) in quantum mechanics
- amplitude (A), which is equivalent to the probability $|\Psi|^2$ of finding a particle
- phase (φ), which is equivalent to phase (φ) is quantum mechanics

Momentum energy, wave number energy ($E(\mathbf{p})$ or $E(\mathbf{k})$), and wave number frequency ($\omega(\mathbf{k})$) domain dispersion behaviors of waves have been explored in solid state physics and acoustics in the recent years [30, 31]. Many exotic wave behaviors were extracted from the wave dispersion curves as discussed in Chapter 6. These behaviors are called spectral behaviors [32]. Spectral properties like passband and stopband due to Bragg scattering [33], loss of transmission, or loss of propagation of energy due to local resonance [34], beam splitting, negative refraction [35, 36], zero-angle refraction, obtaining zero index material [37] etc., were discovered and reported in the past two decades. So far, all the exotic properties are limited to the exploration of the first two mutually independent canonical characteristics of wave phenomena, such as wave frequency and wave number.

Only recently, researchers have focused their attention on finding exotic wave properties by exploring the other two canonical characteristics. They are amplitude and phase. Wave amplitude could be real or complex. Complex amplitude has an inherent phase associated with it, which differs from the phase that engineers are familiar with. Phase is divided into two parts. One is the dynamic phase (φ_D), which is well recognized, and engineers describe it for most dynamical systems. Second is

Quantum Mechanics for Engineers 205

the geometric phase (φ_g), which is uncommon to engineers. However, recently it was found that they are relevant for acoustics, elastic oscillators, vibration, and elastic waves. Traditional classical mechanics is restricted to the dynamic phase only. The geometric phase however, manifested in numerous dynamic and acoustic experiments [38, 39], but explicitly they were not understood unless the topological notion was introduced. In its most general form, a complex amplitude of wave could be expressed by $A = \mathbb{A}e^{i\varphi_g}$ where the complex amplitude acquires a geometric phase (φ_g). Connection of this geometric phase φ_g with the spectral properties – e.g., $E(\mathbf{p})$ or $E(\mathbf{k})$ or $\omega(\mathbf{k})$ – discussed above inherently brought the concept of topology into the study of photonics and phononics. Topology offers a unique understanding of the phenomenon associated with solid state physics and acoustic and mechanical systems that acquire geometric phase which are explored herein.

5.6.1 WHAT IS TOPOLOGY?

The word 'topology' is an ancient word from Greek or rather originated from Albanian 'Trokë', which means a place or a location or earth surface. The contemporary meaning of topology is more mathematical than physical. In mathematics and geometry, if an object transforms through a deformation but is able to retain its properties, then the start and end objects are called topologically connected as if they are the same place or Trokë. The most well-known example of topology is depicted using a donut and a coffee mug that are topologically connected. These objects share few similar properties. A hole in the middle of a donut is topologically equivalent to a hollow handle in a coffee mug. If the properties of a geometric object are protected during continuous geometric deformation, then the properties are called the topology protected properties. If a phenomenon prevails in a donut, then it must prevail in a mug too (Fig. 5.13a). They are called 'topologically protected' phenomenon. Topological properties are the 'topological invariants' attached to the topology, even when the manifested geometries are continuously deformed, such as twisted, bent, stretched, etc.

In modern times, during the 18th century, while solving the famous seven-bridge problem [40], Leonhard Euler borrowed an idea from Grottfried Leibniz and introduced the concept of geometrically similar problems related to topology. However, the term topology was formally introduced by the German mathematician Johann Benedict Listing in the 19th century for solving mathematically similar problems. Later, at the end of the 19th century, Henri Poincaré formally introduced the concept of homotopy and homology to describe topology in a more mathematically tractable form [41]. In Poincaré's depiction, he used many tools from abstract algebra. Later, in the 20th century, the study of topology [41] was mathematically studied by Andrey Kolmogorov, Vladimir Arnold, and Jürgen Kurt Moser. Their study on topology gave birth to the famous Kolmogorov-Arnold-Moser (KAM) theory of topology [42] that primarily describes the invariable quasi-periodic motion of Hamiltonian dynamical system under perturbation. KAM theory is applicable to any n-dimensional tori. Thus, the theory is applicable to a three-dimensional torus, which is described as a product of two circles. A donut shape is called a torus.

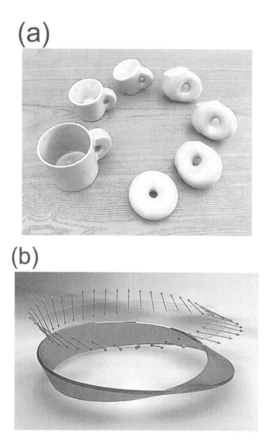

FIGURE 5.13 a) A ceramic model by Keenan Crane and Henry Segerman depicts that a donut can transform into a coffee mug with homeomorphic transformation or continuous deformation, b) Mobius strip showing the change in polarization of light due to topological behavior. *Photo courtesy of Peter Banzer/Max Planck Institute for the Science of Light.*

Towards the end of 20th century topology became more popular in quantum mechanics. It was also popular in solid state physics and in acoustics with the advent of the quantum Hall effect (QHE). Topology started to make its way into quantum mechanics through Dirac's theoretical formulation of magnetic monopoles (e.g., Dirac cone) [43] and the Aharonov-Bohm effect [44].

Topology is our point of interest in this book because it offers exotic understanding of a few unique phenomenon that are associated with electromagnetic wave, electron waves (wave-particle duality), acoustic waves, elastic waves, and mechanical systems, which are otherwise not possible to understand using conventional classical physics. Some non-conventional topological structures like helicoidal twists also create unique topological phenomena. If such one helicoidal twist is joined by closing the loop, then the single surface helicoid becomes a Möbius strip (Fig. 5.13b). It means that if a cyclic geometrical transformation occurs on the surface of a Möbius strip, then this transformation is unique and topologically protected by the geometry.

Quantum Mechanics for Engineers

5.6.2 What Is the Geometric Phase?

Contemporary to the mathematical description of topology, the geometric phase was discovered independently during the 20th century through several experiments with the flow of electrons, electromagnetic waves, or light in optics. They were explainable neither by Newtonian-Lagrangian classical mechanics nor by the contemporary quantum mechanics. It was found that while explaining the geometric phases, the topological notion emerged, and the topological understanding was found to be useful in quantum mechanics. Later the geometric phase and its topological understanding was adopted for generalized wave physics. Thus, sometimes the geometric phase is also called the topological phase.

The geometric phase was first discovered in 1956 by Shivaramakrishnan Pancharatnam, born in Calcutta, India in 1934. Advised by Nobel laureate C.V. Raman, Pancharatnam discovered the geometric phase while passing a polarized light through a series of polarizers (or several crystals). He concluded [45] that the polarization vector of an electromagnetic wave contributes to the reflected and refracted waves through the polarizers. Polarization adds to the degrees of freedom of the electromagnetic wave. He further concluded that during the propagation of an electromagnetic wave through the polarizers, the evolution of the polarization vector must be considered in addition to the typical dynamic phase, which is expected to be $e^{i\mathbf{k}\cdot\mathbf{x}-\omega t}$ where $\mathbf{k}$ is the wave number and ω is the frequency. If an electromagnetic plane wave of an angular frequency ω propagates in the x_3-direction with magnitudes $\mathcal{E}_1$ and $\mathcal{E}_2$ along x_1 and x_2 directions, respectively, and passes through a set of polarizers, the electric field can be described as

$$\mathbf{E}(\mathbf{x},t) = \left(\mathcal{E}_1\hat{\mathbf{e}}_1 + \mathcal{E}_2 e^{i\delta}\hat{\mathbf{e}}_2\right)e^{i(k_3 x_3 - \omega t)}e^{i\varphi_g} = \left(\mathcal{E}_1\hat{\mathbf{e}}_1 + \mathcal{E}_2 e^{i\delta}\hat{\mathbf{e}}_2\right)e^{i\varphi_D}e^{i\varphi_g} \tag{5.29.1}$$

where δ is the phase difference between the two components of the fields $\mathcal{E}_1$ and $\mathcal{E}_2$. The dynamic phase would be $\varphi_D = (k_3 x_3 - \omega t)$. The geometric phase φ_g is the additional phase acquired due to polarization. The dynamic phase φ_D and its evolution during wave propagation is governed by the Maxwell's wave equation, described in Chapter 4. However, if the polarization of the wave changes during this propagation, in addition to the dynamic phase the wave will acquire an extra contribution to its phase called the geometric phase or the topological phase. This phase is called Pancharatnam's phase. Later, such phenomenon were found in quantum mechanics, specifically in a periodic crystal and in relation to the quantum Hall effect. The Aharonov-Bohm effect, demonstrated in 1959 [44], is also one such effect, where the wave function of a charged particle passing through a solenoid manifested a phase shift due to enclosed magnetic field. In 1983, 27 years after Pancharatnam's discovery, Michael Victor Berry [46] mathematically described the geometric phase within the quantum mechanics framework. Berry generalized the concept of the geometric phase, and thus it is called the Berry phase. In certain conditions, the Berry phase manifested the Pancharatnam phase in optics, and thus sometime the geometric phase is also called the Pancharatnam-Berry phase [47].

5.6.3 How Are They Connected?

Topology and the geometric phase are related by a cause-and-effect relationship. Topology is the reason. Topology contributes to the physical manifestation of the geometric phase (or topological phase) in the wave function. Although many researchers have connected the understanding of topology and geometric phase in recent times, historically many such phenomena in classical mechanics resulted from the geometric phase but were not recognized as a function of topology. The geometric phase and topology offer deep connections among many unrelated classical phenomena because they were never tied together into one framework.

The famous Foucault's pendulum [48] in the Pantheon in Paris was established in 1851 to prove Earth's rotation. A full cycle of the pendulum took approximately 32 hours along the latitude 48°52'. The plane of swing of the pendulum was slowly shifted due to the rotation of the Earth. However, returning to the original state took longer (32 hours) than the period of Earth's rotation (24 hours). That means that after 24 hours, the plane of swing was not in line with the original plane of the swing when the pendulum started at 0 hour. Let's assume that the plane of swing at a particular instant can be represented by a polarization vector at that instant. This means that the polarization vector at 0 hour and 24 hours were not the same and had an angular difference, say α. This shift is called the phase shift due to the parallel transport of a vector along a curved surface or geodesic along the latitude 48°52'. Originally, it was recognized that physically there must be some form of exchange of momentum between the Earth's rotation and the motion of the pendulum. Later it was recognized that mathematically this was the result of the geometric phase and could be explained by the parallel transport of a vector.

It was mentioned in section 5.6.2 that polarization adds an additional degree of freedom. This additional degree of freedom is the primary cause for developing the geometric phase. Polarization can be recognized as a vector field. When a vector is transported on a curved surface and trying to keep the same direction of the polarization, the vector is said to be parallelly transported. (Please refer to Appendix in Chapter 6.) *Mathematically, the covariant derivative of a vector field along a path on a surface parametrized using parameter space is zero, then the vector is said to be parallelly transported.* Following the similar concept, the vector representing the swing plane of the pendulum is thus parallelly transported due to the Earth's motion and acquires a geometric phase after 24 hours. It takes additional 8 hours to diminish that acquired phase and report the same plane of swing after 32 hours. It is thus necessary to understand the parallel transport of a vector or an operator along a curve surface parameterized using few geometric parameters. Readers are recommended to first understand the concept of parallel transport that is depicted in the Appendix of Chapter 6. An understanding of parallel transport is assumed in further discussions. Assuming that parallel transport phenomena are known, polarity vectors (e.g., spin is considered polarity, with spin up and spin down) that are inherent to wave function would be affected by the path along which the wave travels and would cause the acquisition of the geometric phase. If the topologies of two surfaces are similar, then the transport along a closed curve on the topologically similar surfaces could cause similar acquisition of the geometric phase. Topology and the geometric phase are connected by a cause-effect relationship. In two independent spaces that are topologically different but sharing a boundary, wave function may exhibit different behavior in the bulk media but may cause localization of a specific phenomenon at the boundary. Topology thus creates a bulk-boundary distinction. Wave behavior at the boundary may not be the same as the behavior manifested in the bulk media. Here symmetry of the media would play a significant role. Symmetries are discussed in section 5.7.

Multiple times the term topology of a 'space' is used. Please note that it is not always necessary to restrict our realization of topology to a physical space. The topological space could also be described as reciprocal space. In section 5.3, the reciprocal space is introduced. In basic terms, the Fourier domain of any space is called the reciprocal space. Thus, wave number or momentum space are the reciprocal domains of the position space. Frequency is the reciprocal space of time, and so on. The inverse of any physical and temporal dimension is called the reciprocal space. In section 5.3, solutions of wave dispersions ($E(\mathbf{p})$ or $E(\mathbf{k})$ or $\omega(\mathbf{k})$) are presented in reciprocal space. If the wave is propagated in a periodic crystal, then the dispersion relation is found over the reciprocal space (wave number ($\mathbf{k}$)) within the first Brillouin zone. Considering a periodic media, it is shown that the solution of the eigenenergies or eigenfrequencies are repeated after $k_i = \pi/D_i \hat{e}_i$. Solution of the wave dispersion is relevant only within the first Brillouin zone. Thus, it can be visualized that the reciprocal space $\mathbf{k}$ is self-repeating. Beyond π/D_i, space is considered bounded within $k_i \in \left(-\dfrac{\pi}{D_i}, +\dfrac{\pi}{D_i}\right)$. Virtually connecting the two points at $k_i = -\dfrac{\pi}{D_i}$ and $k_i = +\dfrac{\pi}{D_i}$, a circular loop could be obtained in $\mathbf{k}$-space. Hence, in one-dimensional and two-dimensional periodic crystals, the Brillouin zone essentially looks like a circle and a torus, as shown in Fig. 5.14. Next if a vector field is transported along this loop, will it really repeat itself? Or will it cause acquisition of the phase that we discussed above? This will be further discussed in the following sections.

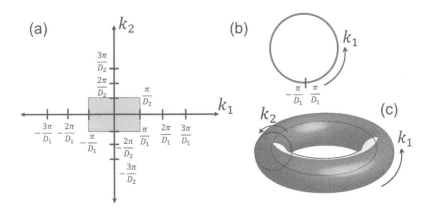

FIGURE 5.14 a) 2D representation of Brillouin zone in 2D periodic crystal, b) in one-dimensional periodic crystal, the Brillouin zone takes the shape of a circle in wave vector domain, and c) in two-dimensional periodic crystal, the Brillouin zone takes the shape of a torus in wave vector domain.

5.6.4 THE BERRY PHASE

If a quantum system is described at an instant by a Hamiltonian operator (H), then the system must have few eigenstates (E, Ψ) at that instant. If the Hamiltonian changes, the eignestate also changes. If the system as a function of its parameter space (e.g., momentum or other geometrical or physical properties) slowly evolves near a close proximity of an eignestate (E, Ψ), then it is also obvious that the new eigenstates (E', Ψ') must be the solution of the continuously evolving Hamiltonian (H'), respectively at every instant. Now, if the evolving Hamiltonian (H') is evolved through a cyclic process and comes back to its original Hamiltonian (H), then it is also intuitive that the system will regain the same eigenstate (E, Ψ) that was found before the cyclic evolution. But according to Berry in 1983 [46], if a system with a specific Hamiltonian evolves through a cyclic adiabatic change in a parameter space, say $\mathbf{R} \equiv (\mu, \gamma, \beta, \alpha ...., t)$ with arbitrary but relevant parameters, and recovers the original Hamiltonian (i.e., $H(\mathbf{R}(t = T)) = H(\mathbf{R}(t = 0))$ after a period T, then in addition to the dynamic phase of the eigenstates, the wave function will acquire an additional phase called the geometric phase. Evolution of the system happened over a closed path $\mathbf{R}(t)$ in parameter space, and Berry called the closed loop a circuit C. Mathematically assuming T being very large, Berry expressed the wave function after acquiring the geometric phase using the following three simple equations:

Following the Schrödinger (section 5.1.5) equation for a system with parameterized Hamiltonian we can write

$$H\big(\mathbf{R}(t)\big)\Psi = i\hbar \frac{\partial \Psi}{\partial t} \qquad (5.29.2)$$

Hence, the N-th eigenstate could be written as (refer to Eq. 5.8.7)

$$H\big(\mathbf{R}(t)\big)\Psi_N(\mathbf{R}) = E_N(\mathbf{R})\Psi_N(\mathbf{R}) \qquad (5.29.3)$$

where the eigenvalues and eigenfunctions become the function of the parameter space. Berry argued that the above equation does not give any relationship between the energies $E_N(\mathbf{R})$ and the phases of the eigenwave function $\Psi_N(\mathbf{R})$ at different $\mathbf{R}$. Adiabatically, the system prepared for $\Psi_N(\mathbf{R}(t = 0))$ will evolve to a state $\Psi_N(\mathbf{R}(t = T))$ obeying the Hamiltonian $H(\mathbf{R}(t))$. But $H(\mathbf{R}(t = T)) = H(\mathbf{R}(t = 0))$. Thus, any arbitrary phase could be placed in the expression for the wave function, if the eigenstate wave function $\Psi_N(\mathbf{R})$ is single-valued in a parameter space that includes a path C for the evolution. The wave function will read

$$\Psi(t) = \exp\left\{\frac{-i}{\hbar} \int_0^\tau E_N\big(R(\tau)\big)d\tau\right\} \exp\big(i\varphi_g(t)\big)\Psi_N(\mathbf{R}(t)) \qquad (5.29.4)$$

where $\varphi_g(t)$ is the acquired geometric phase due the evolution. The first exponential term is nothing but the dynamic phase, which is integrable over time. The geometric phase, however, is not integrable and not single-valued over the evolution (i.e., $\varphi_g(t = T) \neq \varphi_g(t = 0)$). Under any circumstances, the N-th eigenwave function of the

Quantum Mechanics for Engineers

N-th energy band in Eq. 5.29.4 with energy E_N must satisfy the Schrödinger equation. Thus, substituting the wave function into the Schrödinger equation (Eq. 5.29.2) $\varphi_g(t)$ could be found as follows

$$\dot{\varphi}_g(t) = i \left\langle \Psi_N\left(\mathbf{R}(t)\right) \left| \frac{\partial}{\partial \mathbf{R}} \Psi_N\left(\mathbf{R}(t)\right) \right\rangle . \dot{\mathbf{R}}(t) \right. \tag{5.29.5}$$

where $\langle .\,|\,. \rangle$ indicates the inner product with normalization. The inner product is nothing but dot product for vectors and product integral for functions. The total change in phase around the loop $\mathcal{C}$ could be written as

$$\Psi(t=T) = \exp\left\{ \frac{-i}{\hbar} \int_0^\tau E_N\left(R(\tau)\right) d\tau \right\} \exp\left(i\varphi_g(\mathcal{C})\right) \Psi(t=0) \tag{5.29.6}$$

With explicit geometric phase change

$$\varphi_g(\mathcal{C}) = i \oint_C \left\langle \Psi_N(\mathbf{R}) \left| \frac{\partial}{\partial \mathbf{R}} \Psi_N(\mathbf{R}) \right\rangle . d\mathbf{R} \right. \tag{5.29.7}$$

$\varphi_g(\mathcal{C})$ is a circuit integral over a path $\mathcal{C}$. The vector valued function $i\left\langle \Psi_N(\mathbf{R}) \left| \frac{\partial}{\partial \mathbf{R}} \Psi_N(\mathbf{R}) \right\rangle \right.$ is called the Berry connection or Berry vector valued function, useful to describe the Berry curvature.

So far, no specific path is specified for this derivation, except the evolution along the path should be slow and adiabatic in nature. Thus, the Berry phase is path independent. The $\varphi_g(\mathcal{C})$ is always a real valued quantity. The most important 'concept of Berry's phase is the existence of a continuous parameter space in which the state of the system can travel on a closed path' (e.g., $\mathcal{C}$) [49].

5.6.4.1 The Berry Phase in Bloch Media

If a medium is periodic and the wave function is expressed as a Bloch wave function, then the medium is called a Bloch medium. At the end of section 5.6.3, it is discussed that for a periodic solid, the band structure found in reciprocal space inherently carries a parameter space. The Bloch solution of the energy bands in presented in wave number ($\mathbf{k}$) domain. Wave number domain is nothing but the momentum domain in quantum mechanics. In a periodic solid, the wave number is a conserved quantity, and the wave function is expressed as $\psi_{\mathbf{k}}^{nmq}(\mathbf{x})$ (refer to section 5.3.2) where n, m, and q, the integer index, governs the number of Bloch wave vectors included in the solution. The number of Bloch wave vectors considered in the solution governs the number of eigenvalues or the energy bands found from the solution. However, to avoid confusion, energy bands should be numbered differently and sequentially. Let's assign the energy bands from the Bloch solution with number N. N is a function of the Bloch wave vector numbers n, m, and q without the loss of generality. Assigning N to the N-th energy band, the respective Bloch eigenwave function could simply read $\psi_{\mathbf{k}}^N(\mathbf{x})$.

Now, if the system is moved along a closed loop, adiabatically, (*i.e., by applying small perturbation that can vary the wave number vector* (**k**) *on a path that closes itself*) then the Bloch eigenwave function $\psi_{\mathbf{k}}^{N}(\mathbf{x})$ must incur the Berry phase. The Brillouin zone in a 2D crystal takes the form of a torus, as shown in Fig. 5.14. If the wave number is varied along the wave vector space and reaches the Brillouin boundary, then the space closes itself, giving a closed loop of parameter space along which the Hamiltonian has evolved. This situation could give rise to the geometric phase in the wave function. Ref [30] described the Berry phase in an electronic system in more detail. The Berry phase has three key properties.

- The Berry phase is gauge invariant. Thus φ_g^{N} is a gauge invariant quantity. An overall phase factor can be multiplied with any gauge that is parameter dependent and creates an opportunity to generate a loop along the parameter space. The Berry phase does not change due to the phase factor up to integer multiple of 2π. This will be possible only when the wave function is single valued over a loop in the parameter space. Hence, the Berry phase as a physical quantity should explicitly be observable from experiments.
- The Berry phase belongs to the general class of geometrical phase discussed above. As written in Eq. 5.29.7, the Berry phase is a line integral in the parameter space closing a loop along a path C. As the changes are adiabatic and slow, the rate of change of the parameter along the loop is immaterial. The Berry phase can be expressed in terms of local geometrical quantities in the parameter space. Physically, the Berry phase can also be presented as an integral of a field, called the Berry curvature, which is already present in Eq. 5.29.7 in a slightly different form.
- The Berry phase has close analogies to the gauge field theories and differential geometry.

In section 5.6.4, a parameter-dependent Hamiltonian is described and is fundamental to find the Berry phase. In a periodic, solid crystal, the motion of an electron can be described by the Hamiltonian as

$$H = \left[\frac{1}{2m}(\boldsymbol{p})^2 + \mathcal{U}(\mathbf{x}) \right] \tag{5.29.8}$$

where $\mathcal{U}(\mathbf{x}+\mathbf{D}) = \mathcal{U}(\mathbf{x})$ is the periodic potential and $\mathbf{D} = D_i \hat{e}_i$ is the periodicity vector of the crystal. Applying the Bloch theorem (refer to section 5.3.1) the eigenwave function for the above Hamiltonian can be written as $\psi_{\mathbf{k}}^{N}(\mathbf{x}+\mathbf{D}) = \left[\psi_{\mathbf{k}}^{N}(\mathbf{x}) \right] e^{i\mathbf{k}.\mathbf{x}}$ (refer to Eq. 5.26.5) where N is the N-th Bloch eigenwave function in 3D periodic crystal with Bloch wave vectors G_n, G_m, and G_q in three respective directions. It can be noted that the Hamiltonian in Eq. 5.29.8 is not a function of wave number $\mathbf{k}$. But the boundary condition or the Bloch eigenwave functions are the functions of $\mathbf{k}$. To exploit the closed-loop behavior of a wave vector as a parameter, it is necessary to make the Hamiltonian a function of $\mathbf{k}$ and parameterize as follows:

$$H(\mathbf{k}) = e^{-i\mathbf{k}.\mathbf{x}}.H.e^{i\mathbf{k}.\mathbf{x}} = \left[\frac{1}{2m}(\boldsymbol{p}+\mathbf{k}\hbar)^2 + \mathcal{U}(\mathbf{x}) \right] \tag{5.29.9}$$

Quantum Mechanics for Engineers

As the Hamiltonian is transformed, the eigenfunction must be transformed. With the boundary condition, the new transformed cell periodic part of the Bloch eigenwave function would read as

$$u_{\mathbf{k}}^N(\mathbf{x}) = \left[\psi_{\mathbf{k}}^N(\mathbf{x})\right] e^{-i\mathbf{k}.\mathbf{x}}; \; u_{\mathbf{k}}^N(\mathbf{x}+\mathbf{D}) = u_{\mathbf{k}}^N(\mathbf{x}) \qquad (5.29.10)$$

The boundary condition ensures that all the eigenstates for the Hamiltonian $H(\mathbf{k})$ belongs to the Hilbert space. Now, as the Hamiltonian is parameterized with wave vector $\mathbf{k}$, the respective basis function can simply be described as $u_N(\mathbf{k})$, which is equilavent to $u_{\mathbf{k}}^N(\mathbf{x})$ in Eq. 5.29.10, but explicitly a function of the parameter space $\mathbf{k}$. Next the Brillouin zone of wave vector space $\mathbf{k}$ could be considered as the parameter space for the Hamiltonian $H(\mathbf{k})$. Thus, any path along the parameter space $\mathbf{k}$ closing the loop will result in a Berry phase in the eigenwave function $u_N(\mathbf{k})$. As in Eq. 5.29.7, the Berry phase of the N-th energy band would now read

$$\varphi_g^{\;N} = i\oint_C \langle u_N(\mathbf{k})| \nabla_k \; u_N(\mathbf{k})\rangle .d\mathbf{k} \qquad (5.29.11)$$

where $\nabla_k = \dfrac{\partial}{\partial k_i}\hat{e}_i$ is the gradient vector with derivatives with respect to the wave vectors. Wave vector space in quantum mechanics is equivalent to the momentum space as $= \mathbf{p}/\hbar$. Thus, if a Hamiltonian $H(\mathbf{k})$ is forced to move in wave vector space $\mathbf{k}$, then it means that the Hamiltonian $H(\mathbf{p})$ is forced to move in the momentum space $\mathbf{p}$ along a closed path C. *Closed path C in momentum space can be generated either by applying an external magnetic field ($\mathbf{B}$), or by applying an electric field ($\mathbf{E}$) to the system* [50]. Such a closed loop effect can be possible only in 2D and 3D crystals. *Application of the external fields essentially introduces the broken symmetry* (refer to section 5.7) of the crystal, which contributes to the geometric phase. The Berry phase acquired due to the magneto-oscillatory effect was demonstrated in HgTe [51] and graphene [30].

Please note that the expression of the Berry phase has the Berry connection. The curl of the Berry connection is called the Berry curvature and can be written as follows in vector form:

$$\mathcal{R}_N(\mathbf{k}) = \nabla_k \times \langle u_N(\mathbf{k})| \nabla_k \; u_N(\mathbf{k})\rangle \qquad (5.29.12)$$

The Berry curvature is the property of the band structure and specific to the N-th band. It is immediately apparent that to describe the Berry curvature, one would not need to have the integral along a closed path C in parameter space. The Berry curvature still could be found for a specific energy band without the integral. *In many materials the Berry curvature carries a finite value where time-reversal or space inversion symmetry is naturally broken.* This indicates that active generation of closed path C in momentum space by applying either an external *magnetic field ($\mathbf{B}$) or electric field ($\mathbf{E}$)* to the system *is not necessary for calculating the Berry curvature.*

5.6.4.2 The Chern Number at Degenerated Band Structure

The meaning of the Berry phase and its relation to the Berry connection and Berry curvature is mathematically described. The Chern number warrants some discussion on the effect of the Berry phase when there are doubly and triply degenerated states. What happens when there are two or three energy bands that meet at an energy point, like the situations discussed in section 5.5 on Dirac cones and Dirac-like cones? Graphene is essentially a hexagonal structure and has high symmetry points at K and K' of the first Brillouin boundary (Fig. 5.12a). At this point, at a certain energy level two energy bands meet with their respective linear dispersion. A Dirac cone is formed at the high symmetry points.

Adiabatic approximation by traversing the Hamiltonian along a closed path of parameter space is restricted only to the N-th band. But when the two bands (N-th and $N-1$-th or N-th and $N+1$-th bands) meet at a single energy level, understanding of Berry connections and Berry curvature becomes complex. The Berry curvature becomes singular when two energy states ($E_N(k)$ and $E_{N-1}(k)$ or $E_{N+1}(k)$) are equal. Let's introduce another parameter called the Berry tensor, related to the Berry curvature vector, using a permutation symbol or Levi-Civita antisymmetric tensor discussed in Chapter 2 and 3. The Berry tensor is described for all the bands present in a band structure and can be written as $\mathbb{R}_{ij}^N(\mathbf{k}) = \epsilon_{ijm}(\mathcal{R}_N(\mathbf{k}))_m$ where summation is assumed over the m index. Summation of all the Berry tensors over all the energy bands in parameter space $\mathbf{k}$ must vanish. That is, $\sum_N \mathbb{R}_{ij}^N(\mathbf{k}) = 0$ for all $\mathbf{k}$. At the points of degeneracy, the Berry tensor is singular. It's like the Berry curvature generated by a monopole, which acts like a source or a sink for the Berry curvature flux [48]. As is done for an electric charge or a magnetic monopole discussed in Chapter 4, an integral of the total Berry curvature flux going out or sinking in could provide the total charge of the Dirac monopole. Thus, a spherical integral of the Berry curvature over a unit sphere contains a monopole. Number of monopoles is a scalar quantity. This is synonymous to an electric charge and can be assumed as a Berry charge. This is the Berry phase on the sphere. Thus, the Berry curvature integrated over a closed manifold will result in a quantized value to 2π, which will be equal to the number of monopoles inside the manifold. This number of monopoles or Berry charge is called the Chern number (C_n). It was found that the Berry phase at the Dirac cone near the Brillouin zone boundary is π, because there are two monopoles, whereas Berry phase at the Dirac-like cone with triple degeneracy formed at the center of the Brillouin zone is zero. The following conclusions are true for the Berry phase and Berry curvature.

- Active application of external fields breaks the symmetry, causes a closed path along a parameter space (e.g., momentum space and cause manifestation of the Berry phase).
- Without applying any external field to a crystal, if the time-reversal symmetry or space inversion symmetry is naturally broken, then the eigenfunction also incurs the Berry phase.
  - At the Brillouin boundary, or at the doubly degenerated Dirac cone (explained by Dirac equation of fermions) the Berry phase is Pi (π).
  - At the center of the Brillouin zone with triply degenerated Dirac-like cone, the Berry phase is zero (0).

Quantum Mechanics for Engineers

- The Berry phase is associated with a closed path C in a parameter space, but the Berry curvature is a local quantity of the band and provides the geometric properties of the parameter space.

5.6.5 THE ZAK PHASE

While describing the Berry phase and applying it to a periodic crystal, J. Zak in 1988 [49] derived the mathematical reality of the Berry phase that may incur in a one-dimensional crystal. In short, the Berry phase in a 1D periodic crystal is called the Zak phase. When the linear chain has no symmetry, the Berry phase can assume any value. However, when inversion symmetry is present, Berry's phase becomes quantized, and it can assume only the values 0 and π as mentioned in the previous section. Here are the derivation and arguments presented as discussed in Ref [49].

The Bloch wave function in a one-dimensional periodic crystal will have only one integer parameter in the reciprocal space (Bloch wave vector), and the Bloch wave function will read $\psi_{\mathbf{k}}^n(\mathbf{x})$. Generalized wave function for the N-th energy band will simply be $\psi_{\mathbf{k}}^N(\mathbf{x})$. The Brillouin zone is a circle in 1D periodic crystal with the reciprocal space $\mathbf{k}$ self-repeating at $k_1 = -\dfrac{\pi}{D_1}$ and $k_1 = +\dfrac{\pi}{D_1}$. In two dimensions, however, the Brillouin zone is a torus (Fig. 5.14). Further, in this section, for a 1D system k_1 is read as k without the loss of the generality. Zak started deriving the formula for the Berry phase for an electron in a 1D periodic potential $\mathcal{U}(x_1)$ in presence of an externally applied time-dependent vector potential $A(t)$. The reason for this introduction of a time-dependent vector potential was to make the Hamiltonian change in time as Berry first introduced (refer to Eq. 5.29.2). It is assumed that the function $A(t)$ changes adiabatically or slowly over time. This means that the frequencies of Fourier coefficients of the time-dependent function are much smaller than the frequency of the energy bang gaps. According to Eq. 5.9.14, the Schrödinger equation in a 1D periodic solid with external applied field can be written as

$$\left[\frac{1}{2m}(i\hbar\nabla + q\mathbf{A})^2 + \mathcal{U}(x_1)\right]\Psi = i\hbar\frac{\partial}{\partial t}\Psi \qquad (5.29.13)$$

Solution of the equation for the $N - $ th energy band can be written in the following form when the time-dependent vector potential is applied. Energy becomes a function of time.

$$\left[\frac{1}{2m}(i\hbar\nabla + q\mathbf{A}(t))^2 + \mathcal{U}(x_1)\right]\Psi_N(x_1,t) = E_N(t)\Psi_N(x_1,t) \qquad (5.29.14)$$

where $\mathcal{U}(x_1 + D_1) = \mathcal{U}(x_1)$ is the periodic potential. Applying the Bloch solution as depicted in Eq. 5.25.20, the time-dependent eigenwave function would read

$$\Psi_N(x_1,t) = \psi_{k(t)}^N(x_1)e^{ikx_1} \qquad (5.29.15)$$

Substituting Eq. 5.29.15 into Eq. 5.29.14, when $e^{ikx_1} \neq 0$ would read

$$\left[\frac{1}{2m}\left(i\hbar\nabla + q\mathbf{A}(t)\right)^2 + \mathcal{U}(x_1)\right]\psi_{k(t)}^N(x_1) = E_N(k(t))\psi_{k(t)}^N(x_1) \qquad (5.29.16)$$

Due to the presence of time-dependent vector potential, the Bloch wave number k will become a function time and $k(t) = k - \frac{q}{\hbar}A(t)$. Now it is also necessary to transform the Hamiltonian into the parameter space as is done in Eq. 5.29.9. Transforming the Hamiltonian and Bloch wave functions, the above governing equation with Eq. 5.29.10 would read

$$\left[\frac{1}{2m}\left(i\hbar\nabla + \hbar\mathbf{k} + q\mathbf{A}(t)\right)^2 + \mathcal{U}(x_1)\right]u_{k(t)}^N(x_1) = E_N(k(t))u_{k(t)}^N(x_1) \qquad (5.29.17)$$

where $u_{\mathbf{k}}^N(\mathbf{x}) = \left[\psi_{\mathbf{k}}^N(\mathbf{x})\right]e^{-i\mathbf{k}.\mathbf{x}}$ with boundary condition $u_{\mathbf{k}}^N(\mathbf{x}+\mathbf{D}) = u_{\mathbf{k}}^N(\mathbf{x})$. Next, according to Eq. 5.29.4, the Bloch wave function would incur a Berry phase and would read

$$\Psi_N(x_1,t) = \exp\left\{\frac{-i}{\hbar}\int_0^\tau E_N\left(k(\tau)\right)d\tau\right\}\exp\left(i\varphi_g^N(t)\right)u_{k(t)}^N(x_1) \qquad (5.29.18)$$

where $\varphi_g^N(t)$ is a time-dependent Berry phase for the N-th energy band. Similarly, as is described for the Berry phase, substituting Eq. 5.29.17 into the Schrödinger equation in (5.29.13), the time derivative of the Berry phase and Berry phase of the N-th energy band would read, respectively,

$$\dot{\varphi}_g^N(t) = i\int_0^{D_1}\left\{u_{k(t)}^N(x_1)^*\frac{\partial}{\partial t}u_{k(t)}^N(x_1)\right\}dx_1 \qquad (5.29.19.1)$$

$$\varphi_g^N = \frac{2\pi}{D_1}\int_{-\pi/D_1}^{\pi/D_1}\left[i\int_0^{D_1}\left\{u_{k(t)}^N(x_1)^*\frac{\partial}{\partial k}u_{k(t)}^N(x_1)\right\}dx_1\right]dk \qquad (5.29.19.2)$$

Next, it was argued that as the Berry phase in gauge invariant, a time-independent gauge would not change the Berry phase. Originally the Berry phase was described for a time-varying Hamiltonian with a periodic cycle. But in Zak's Hamiltonian, this condition is not satisfied, and hence it is required that vector potential $A(t)$ change by $\frac{2\pi}{D_1}$. This will add change to the wave vector k, as apparent from Eq. 5.29.17. But according to the Bloch theorem, such change induces a cyclic change of the wave vector and satisfies the closed-loop traversing of the Hamiltonian Eq. 5.29.17 in k parameter space. Alternately, $u_{k(t)}^N = u_{k+\frac{2\pi}{D_1}}^N = u_{k+\frac{2\pi}{D_1}}^N e^{-i2\pi x_1/D_1}$ would compensate the change in the Hamiltonian. This is perfect to satisfy Eq. 5.29.19. Zak concluded that the Bloch problem is therefore a generalization of the Berry phase to a noncyclic

Quantum Mechanics for Engineers   217

change of the Hamiltonian, up to a gauge transformation, and the Berry phase is gauge invariant. To recognize Zak's milestone contribution, the Berry phase in 1D periodic media is called the Zak phase.

5.7 UNDERSTANDING SYMMETRY AND INVARIANCE

Symmetries in nature provide unique opportunities to exploit wave behavior with exotic properties in metamaterials. Nature presents many geometrical symmetries, as in the following examples.

1. Reflection symmetry, where an object is symmetric about its central axis and both halves are a mirror reflection of the other; the man body is an example of reflection symmetry.
2. Rotational symmetry, where an object remains unaltered if the object is rotated about an axis; ancient triskelion and the structure of a piezo protein on the cell membrane are two examples of rotational symmetry.
3. Translational symmetry, where an object remains the same if each and every point in the object is translated along a path at a same rate; many frieze patterns in ancient art, including Mayan art, are example of translational symmetry.
4. Helical symmetry, where the object is recovered and looks the same when the object is translated and rotated simultaneously in space; the double helix DNA structure is an example of helical symmetry about its screw axis.
5. Scale symmetry, where the shape remains indistinguishable if an object is zoomed in or zoomed out (i.e., if the spatial and temporal scale is varied); the Koch curve, fractal curves or fractal architecture, snowflakes etc., are a few examples of scale symmetry.

Symmetry in physics, however, generally means invariance. According to physical laws, a few parameters of geometry or an object must be invariant to be symmetric of specific type. An invariant could be observable in the form of a scalar or a vector or a tensor that remains unaltered under different specific transformations. An invariant could be specific to a specific transformation. A quantity could be invariant under a specific transformation, but if the transformation method changes, the invariant may change. In quantum mechanics, quantum parameters such as momentum and energy (with continuous symmetry) respectively, are invariant under spatial translation and time translation. That means momentum is conserved in space and energy is conserved over time. A crystal has translational symmetry where the electron density is periodic. Electron density is an invariant, which makes the crystal a periodic media, and a Bloch solution can be used.

Similarly, a topological space has an invariant. The topological invariant never changes under homeomorphism (refer to topology in section 5.5.1). The earliest descriptions of these invariants are unknowingly described in physics through Gauss's law (Eq. 4.7) and Ampere's law (Eq. 4.12). In electrodynamics (refer to Chapter 4) the line or surface integrals described in Gauss's law and Ampere's law remain invariant. They remain invariant under continuous deformation of the surface. Thus, from topological perspective they are equivalent, and the integrals are topological invariants.

Symmetry can be explored in both spatial and temporal domains. Spatial symmetry is defined as $f_i(x_j) = f_i(-x_j)$, which means that the geometric parameters will result in the same values if the space is inverted. Similarly, for temporal properties, the system would result in the same function when the time is reversed, $f_i(x_j,t) = f_i(x_j,-t)$. The first one is called 'space inversion symmetry' and the second one is called 'time inversion symmetry'. *When any of this symmetry is broken, a geometric phase arises.* These symmetries can be broken in parameter space by applying external field (e.g., a magnetic field or an electric field). In some natural materials the symmetries are broken without applying any external fields. When space inversion symmetry is broken, it presents a topological boundary carrying different properties. Bulk-boundary distinction is prominent in such cases. Topologically a bulk medium, if separated by an interfacial layer that causes space inversion symmetry, will have a unique topological edge state. The edge state happens to support backscattering immune unidirectional wave propagation [52].

5.7.1 Geometric Symmetries

Symmetric operation [53] is a process by which an object is transformed but retains its shape and size, or rather looks the same after the operation. There are different symmetric operations with different symmetric elements: a plane, an axis, a line, or a point about which the symmetric operations are performed. For example, reflection symmetry requires a plane, but rotational symmetry requires an axis of rotation. Geometric symmetry is relevant because it is of interest to create metamaterials and periodic systems that can have a non-conventional topology by breaking the symmetry. To break the symmetry, it is necessary to understand how different symmetries could be analyzed. While composing periodic media, it is necessary to utilize certain symmetry operations in 2D or 3D to find the fundamental symmetry elements. Further, to understand if the metamaterial inherits any special inversion symmetry, or to intentionally create the metamaterial posing specific symmetry, or to intentionally create the metamaterial with broken symmetry, the following tabulation of symmetric elements would be helpful. These symmetric elements are classified into the following designations:

- E symmetry is called 'identity operation', which means there are no symmetry elementsEvery metamaterial will have this symmetry.
- C_n symmetry is a rotational symmetry element with n number of axes. There is a primary axis that is out of plane, encompassing $360°$ of a circular plane of the object. If n is 2, then the object has a plane perpendicular to the primary axis and divides it into two $180°$ segments of the object. Next, if the object is rotated about the primary axis by $180°$, then the object is indistinguishable, thus symmetric. Similarly, for any n, if the object is rotated by $360°/n$ or its even multiples, the object will be indistinguishable and thus symmetric.

- σ-symmetry is a reflection symmetry element. In 3D, an object may have three reflection planes about which the mirror images could be taken. If a mirror plane contains the primary axis of the object described above, it will be called vertical plane or σ_v. There could be multiple vertical planes of symmetry in an object. A mirror plane perpendicular to the primary axis will result is horizontal reflection symmetry, and thus the element is called σ_h. Between two C_2 axes a vertical reflection plane is called dihedral mirror plane and the element is called σ_d.
- i-symmetry is an inversion symmetry element. An object may have a point or a plane about which the object can be inverted and an inverted image could be recovered. For example, the human face poses an inversion symmetry. To ensure inversion symmetry, all points on the object must be projected on the opposite direction from the plane of inversion using the same distances; the respective points are located on the object such that the position vector of a point on the object would transform from $+x_j$ to $-x_j$.
- S_n symmetry is an improper rotation symmetry followed by reflection. A special object, if directly reflected about a plane, may not exhibit symmetry, but if it is rotated about an axis by some angle before the reflection, then the object may exhibit symmetry. Hence, following the similar definition presented for C_n symmetry, if the object is rotated by an angle $360°/n$ about a primary axis, but followed by a reflection about a plane perpendicular to the primary axis, then the object will be said to have an improper rotational symmetry.

Figure 5.15 shows molecular structure having different symmetries are self-explanatory.

Additional nomenclature of symmetries must be described herein to be comprehensive. They are parity symmetry and chiral symmetry.

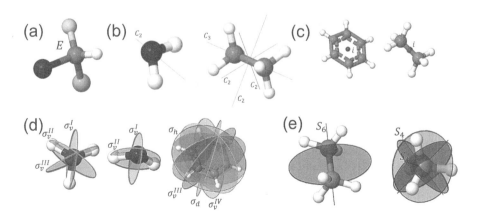

FIGURE 5.15 Symmetry elements: a) e symmetry, b) C_n symmetry, c) σ-symmetry, d) i-symmetry, and e) S_n symmetry (images created via symmetry https://symotter.org/).

Parity symmetry ($\mathcal{P}$-symmetry) is a geometrical symmetry equivalent to reflection symmetry. Spatial coordinates are flipped to describe parity symmetry. To describe electron waves, fermions, and bosons in quantum mechanics, parity symmetry is useful to describe an invariant phenomenon when the space is inverted about a point or an axis. Next, when a reflection of an object is taken about a mirror plane, and the reflected object is not identical to the original image, then the object is called 'Chiral'. For example, two palms in human hands are chiral. Thus, in short, an object is said to be chiral when it is not superimposable on its mirror image. Two chiral objects that are related are called enantiomers [53]. If an object or a medium has C_1 axis symmetry but does not have any other symmetry elements, then the object is said to be asymmetric. Objects that have one or more proper rotation axes (C_n, $n > 1$) but no improper rotation axes (S_n) are said to be dissymmetric. All asymmetric and dissymmetric objects are effectively chiral objects [53]. In other words, the absence of S_n is the necessary and sufficient condition for chirality.

Here it is necessary to mention another nomenclature of parity symmetry or $\mathcal{P}$-symmetry. That is space inversion symmetry (SIS). A topological space is inverted about an axis, and the new space may have different topology with a bulk-boundary distinction at the inversion axis. A system is said to have SIS when the elements of the objects are inverted about an axis.

5.7.2 Time-Reversal Symmetries or $\mathcal{T}$-symmetry

As the name suggests, if time-dependent phenomena or variables remain unaltered when the time becomes negative (time reversal $\equiv -t$), the system is said to demonstrate time-reversal symmetry ($\mathcal{T}$-symmetry). In classical mechanics many physical parameters, variables, and properties have time-reversal symmetry. For example, the position of a particle (x_j), acceleration ($\ddot{x}_j$), force ($\mathbf{F}$), energy (E), density (ρ), electric field ($\mathbf{E}(x_j)$), electric potential (V), electric displacement, and polarization density ($\mathbf{D}(x_j) = \epsilon \mathbf{E}(x_j) + \mathbf{P}(x_j)$) all have time-reversal symmetry but are called even $\mathcal{T}$-symmetry. However, time, velocity ($\dot{x}_j$), momentum (p_i), angular momentum ($\mathbf{L}$) magnetic field ($\mathbf{H}(x_j)$), magnetic flux density ($\mathbf{B}(x_j)$), electromagnetic potential ($\mathbf{A}(x_j)$), and electric current density ($\mathbf{J}$) are a few example of variables and fields that become negative when the time is reversed. These are called the odd variables in terms of $\mathcal{T}$-Symmetry. Although this is quite understandable for classical mechanics, in quantum mechanics the phenomena are not that straightforward. The fundamentals of quantum mechanics are based on the uncertainty principle (Eq. 5.20.4) and nonzero commutator relation (Eq. 5.18). It is indicated above that the position vector has even $\mathcal{T}$-symmetry but momentum has odd $\mathcal{T}$-symmetry (i.e., momentum becomes negative when time is reversed). When a commutator operation is taken under time-reversal conditions, $\mathbf{x}$ remains $\mathbf{x}$ but momentum $\mathbf{p}$ becomes $-\mathbf{p}$, and thus the commutator of $[\mathbf{x}, -\mathbf{p}] \neq i\hbar$. So, time reversal is not the same as parity symmetry, in which both the position $\mathbf{x}$ and the momentum $\mathbf{p}$ are negative. Thus to avoid this challenge, Eugene Wigner in 1931 [54] proposed a theorem for how symmetries could be represented on the Hilbert space of states. Without going into actual Wigner work on all different symmetries, time-reversal symmetry is explained here. This is

Quantum Mechanics for Engineers

presented considering a one-dimensional Schrödinger equation without any potentials or time-independent Hamiltonian. Referring to Eq. 5.8.3, but replacing t with $-t$ would be

$$\left(-\frac{\hbar^2}{2m}\frac{\partial^2}{\partial x^2}\right)\Psi(x,t) = -i\hbar\frac{\partial}{\partial t}\Psi(x,t) \text{ or } H(x)\Psi(x,t) = -i\hbar\frac{\partial}{\partial t}\Psi(x,t) \qquad (5.30.1)$$

But the above equation may not result in the same wave function Ψ and may not remain invariant under T-symmetry. However, it must be symmetric. Thus, let's assume an operator T that reverses the time and when operated on the wave function, Ψ will become

$$T\left[\Psi(x,t)\right] = \Psi'(x,t') = \Psi'(x,-t) \qquad (5.30.2)$$

but must satisfy the Schrödinger equation in the following form to be invariant.

$$H(x)\Psi'(x,t') = i\hbar\frac{\partial}{\partial t'}\Psi'(x,t') \qquad (5.30.3)$$

Substituting Eq. 5.30.2 in Eq. 5.30.3, we get

$$H(x)T\left[\Psi(x,t)\right] = i\hbar\frac{\partial}{\partial t'}T\left[\Psi(x,t)\right] \qquad (5.30.4)$$

Operating the inverse of the T operator on both sides will be

$$T^{-1}H(x)T\left[\Psi(x,t)\right] = T^{-1}\left(-i\hbar\frac{\partial}{\partial t}\right)T\left[\Psi(x,t)\right] \qquad (5.30.5)$$

As the Hamiltonian is time independent, the Hamiltonian and the time-reversal operator must commute. Thus, it can be said that

$$T^{-1}H(x)T = T^{-1}TH = H \qquad (5.30.6)$$

Eq. 5.30.5 will revise to

$$H(x)\Psi(x,t) = T^{-1}(-i)Th\frac{\partial}{\partial t}\left[\Psi(x,t)\right] \qquad (5.30.7)$$

To recover the original Schrödinger equation, the following is necessary.

$$T^{-1}(-i)T = i \text{ or } -iT = Ti \qquad (5.30.8)$$

The equation recovered is the same equation written in Eq. 5.30.3:

$$H(x)\Psi(x,t) = i\hbar\frac{\partial}{\partial t}\left[\Psi(x,t)\right] \qquad (5.30.9)$$

It seems the operator is taking the complex conjugate of any complex variable. To investigate further, T could be expressed as a product of two operators, U and K, where U is any unitary matrix and K is complex conjugate operator. Hence, $T = UK$. Let's assume a wave function Ψ as a linear combination of two wave functions (ψ and ϕ) with their respective contribution factors a and b. Applying $T = UK$ operator on Ψ will be

$$T[\Psi] = T[a\psi + b\phi] = UK[a\psi + b\phi] \tag{5.30.10}$$

$$\therefore T[\Psi] = U[a^* K\psi + b^* K\phi] \tag{5.30.11}$$

$$\therefore T[\Psi] = a^*[UK\psi] + b^*[UK\phi] = a^*[T\psi] + b^*[T\phi] \tag{5.30.12}$$

The property above is called the anti-linear property of the T operator. Additionally, the T operator has an anti-unitary property, which was also proved by Wigner in 1932 [55]. Let's use a wave function and find what the unitary operator could be. A one-dimensional wave function, as written in Eq. 5.5.5, is reiterated and the T operator is operated on it as follows:

$$\Psi = Ae^{\frac{i}{\hbar}(px-Et)}; \; T\Psi = UK\left[e^{\frac{i}{\hbar}(px-Et)}\right] = Ue^{\frac{-i}{\hbar}(px-Et)} \tag{5.30.13}$$

Please note that the conjugate operator K transformed the phase into a complex conjugate. The U could be 1 or any other phase factor $e^{i\varphi_g}$. Although the T is operated on the wave function, the probability of finding the particle should be calculated by taking the square magnitude as $|T\Psi(x,t)|^2 = 1$. Not to cancel the phase factor for demonstration of time reversal symmetry let's assume $U = 1$. Hence, with Eq. 5.30.2, Eq. 5.30.13 will read as time-reversed wave function with no change in energy but the momentum in the opposite direction.

$$T[\Psi(x,t)] = \Psi'(x,-t) = e^{\frac{-i}{\hbar}(px-Et)} = e^{\frac{i}{\hbar}(-px+Et)} \tag{5.30.14}$$

The conclusion is that time-reversal symmetry creates momentum in the opposite direction. While doing so, it does not change the eigensolution of the energy states. Negative momentum implies that the wave propagates in the opposite direction. However, there might be some situations where the unitary operator is not unity, and the phase factor does not cancel out automatically. Then the system is said to acquire a geometric phase φ_g, and the time-reversal symmetry will be broken. Geometric phase could also be time dependent. That means that with time reversal, momentum is not necessarily negative. This may occur in systems with non-conventional topology. Hence, it is obvious that under a symmetry-breaking condition in a non-conventional topology, the system will inevitably acquire a geometric phase as discussed in section 5.5.2. Similar situations were found not only in photonics but also in phononics while exploiting acoustic waves [38, 39, 56, 57]. The time-reversal symmetry could be broken by active means (adding energy from outside) or symmetry breaking

Quantum Mechanics for Engineers

by specific design of a metamaterial. Active methods in quantum mechanics and solid-state physics may include application of external magnetic or electric field. In acoustics such external perturbation is very difficult, but time-varying stiffness or active control of the flow of an acoustic medium could serve this purpose of active breaking of time-reversal symmetry.

5.8 CONNECTING SYMMETRY BREAKING AND GEOMETRIC PHASE

The primary objective of these sections is to facilitate the understanding of non-conventional topology that can break the symmetry of a system and may result in geometric phase in wave propagation. Along the same line, starting from section 5.5.1, the understandings [52] are arranged sequentially.

- Topology is described by a topological parameter and is an invariant.
- Topology and geometric phase are related by cause-and-effect relationship.
- If the topology of a system changes, the wave propagating in that system may acquire geometrical phase.
- If a system inherently possesses some form of symmetry, breaking of symmetry is equivalent to the change in topology or creating a non-conventional topology.
- Thus, breaking symmetry could also create geometric phase leading to non-conventional topology, or vice versa.
- Non-conventional topology could result in a unidirectional backscattering-immune edge state at the boundary of two topological states. This creates a bulk-boundary distinction.
- Breaking symmetries like time-reversal symmetry, parity symmetry, or chiral symmetry, results in geometric phase.
- Broken symmetry is synonymous with non-reciprocal wave propagation.
- Symmetry breaking can be achieved from local resonance of internal structural components.
- Internal resonance plays a key role in coupling and wave dispersion of the system that may lead to symmetry breaking.
- Symmetry breaking can be achieved purely by geometric design from internal structural characteristics, called intrinsic breaking of symmetry.
- Symmetry breaking can be achieved by external application of additional fields, causing spatio-temporal change in structural characteristics, called extrinsic breaking of the symmetry. Symmetry could be broken having external magnetic, electric field, or controlled chiral flow.
- Multiple symmetries could be broken simultaneously (e.g., time inversion symmetry, space inversion symmetry, chiral symmetry, etc.).
- Space-inversion symmetry and any other form of breaking symmetry could contribute to the geometric phase.
- The geometric phase could be explained by its topological interpretation using parallel transport of a vector field along a path in energy and

momentum space or frequency and wave number space or on some curved manifold, where momentum or wave number space are periodic due to the Bloch solution.
- The geometric phase is always considered as a part of the wave amplitude that changes as a function of wave vector, and thus the geometric amplitude can be written as $\mathcal{A}(\mathbf{k}) = \mathcal{A}_0 e^{i\varphi_g(\mathbf{k})}$.
- The amplitude of the wave function is isomorphic to the evolution of a field of vectors that are parallel to each other but tangential to a manifold (e.g., Möbius strip). The manifold could be energy or frequency space of the Bloch wave vector space that repeats at the Brillouin zone boundary.

5.9 QUANTUM HALL EFFECTS

Edwin Hall in 1879 [58] performed a unique experiment in which a conductor was subjected to a magnetic field. It was observed that the conductor gave out a voltage difference along the orthogonal direction to the flow of the current in the conductor. The voltage developed in the conductor in the orthogonal direction is called the Hall Voltage (V_H). The flow of the current is nothing but the flow of charges carriers (e.g., electrons, positrons, or ions). When the magnetic field is applied on the current, the charged particles experience the Lorentz force. The Lorentz force bends the direction of the flow of the charged particle and some charged particles accumulate at the boundary of the conductor. When the two perpendicular ends of the conductor are connected, the accumulated charges tend to flow along the new line and give out the Hall current, and thus the electric potential or Hall voltage is detected. Hall effects can be found not only in conductors, such as metals, but the phenomenon equally relevant for semiconductors.

In this book, in line with the topological behavior of materials and quantum mechanics, interest lies with three new phenomena called the quantum trio [59]. They are quantum Hall effect (QHE), quantum valley Hall effect (QVHE), and quantum spin Hall effect (QSHE). The quantum trio effect in relation to the acoustics is discussed in Chapter 7. However, the basic concepts in relation to photonics are briefly discussed below.

Discovery of the Berezinskii-Kosterlitz-Thouless (BKT) transition changed previous concepts of Landau by explaining the phase transition in a two-dimensional model, which cannot be explained by symmetry breaking theory [60, 61]. With this, the integer QHE emerged [62]. The QHE is observed in a 2D electron system with low temperatures and strong magnetic field (an external field as a key ingredient for symmetry breaking). Here, the bulk state behaves as a featureless insulator, which cannot be characterized by local order parameters. But a robust edge-state in the boundary under external deformation exists. Such behavior intuitively contradicts understanding of Anderson localization, [63] where the edge state is supposed to be affected by impurities. Thus, the existence of truly distinct phases of matter is established by QHE. Here, the bulk is characterized by a topological invariant. This is because the bulk-order parameters fail to describe the new phases. The quantized Hall conductance is an invariant that does not change with the arbitrary phase appearing in the eigenfunctions. The first analogue QHE in photonics was proposed

Quantum Mechanics for Engineers

and demonstrated in 2008 by Haldane [64]. A gyromagnetic material subjected to a magnetic field was used to break the time-reversal symmetry for a 2D photonic crystal. Later Haldane also introduced periodic magnetic flux substituting an external applied field to concur the breaking of time-reversal symmetry [64]. Such an effect was named the quantum anomalous Hall effect (QAHE) [64]. Both effects can be expressed in terms of the same topological invariant. Further similar QHE studies were extended to replace the periodic magnetic flux by a ferromagnetic insulator with strong spin-orbit coupling [65–67]. This discovery led to one of the most ground-breaking phenomena in the field of condensed matter physics. The topological insulator was discovered. Topological insulators were also found to be the result of the quantum spin Hall effect (QSHE) [68–72]. Please refer to these effects in Chapter 6 and Chapter 7 for more details in the context of acoustic and elastic waves in periodic metamaterial.

5.10 SUMMARY

Quantum mechanical understanding is presented from an engineering perspective to explore different metamaterials. Topological phenomena and their relation to the geometric phase is presented through rigorous discussions and easy-to-follow mathematical steps. Starting from wave particle duality, wave functions and wave dispersions are explained. Further, the Schrödinger equation, the Dirac equation, and the Klein-Gordon equation are discussed in simple mathematical terms that demonstrate spin mediated wave phenomena. Throughout the text quantum methods are explained in context of acoustic and elastic waves in metamaterials. Symmetry and symmetry breaking while acquiring geometric phase for topological phenomena is discussed through easy-to-follow mathematical equations. The mathematical steps in relation to quantum mechanics are explained in this chapter and are directly used in the following chapters on acoustic and elastic wave for quantum analogue applications.

REFERENCES

1. Planck, M., *Ueber das Gesetz der Energieverteilung im Normalspectrum.* Annalen Der Physik, 1901, **309**(3): p. 553–563.
2. Miller, D.A.B., *Quantum Mechanics for Scientists and Engineers.* 2008, Cambridge, England: Cambridge University Press.
3. Fromhold, A.T., *Quantum Mechanics for Applied Physics and Engineering.* 1991: Courier Dover Publications.
4. Schroeder, D.V., *Notes on Quantum Mechanics.* 2022, Weber State University.
5. Weinberger, P., *Revisiting Louis de Broglie's famous 1924 paper in the* Philosophical Magazine. Philosophical Magazine Letters, 2006, **86**(7): p. 405–410.
6. Einstein, A., *Zur Elektrodynamik bewegter Korper (On the electrodynamics of moving bodies)* Annalen Der Physik, 1905, **17**(891), p. 891–921.
7. Kandrup, H.E., *Reduced Hamiltonian descriptions.* Physical Review D, 1994, **50**(4): p. 2425–2430.
8. Neuenschwander, D.E., *Tensor Calculus for Physics.* 2015, Baltimore, MD: Johns Hopkins University Press.

9. Coddens, G., *From Spinors to Quantum Mechanics*. 2015, London: Imperial College Press.
10. Cohen-Tannoudji, C., Diu, B., Laloe, F., *Quantum Mechanics, Volume 3: Fermions, Bosons, Photons, Correlations, and Entanglement*. 2019, Hoboken, New Jersey: Wiley.
11. Blyth, T.S., Robertson, E.F., *Groups, Rings and Fields: Algebra Through Practice*. 1985, Cambridge: Cambridge University Press.
12. McCurdy, S., Venkatraman, R., *Quantitative stability for the Heisenberg–Pauli–Weyl inequality*. Nonlinear Analysis, 2021, **202**: p. 112147.
13. Bloch, F., *Über die Quantenmechanik der Elektronen in kristallgittern*. Z. Physik, 1928, **20**: p. 555–600.
14. Magnus, W., Winkler, S., *Hill's Equation*. 2004, Mineola, NY: Courier Dover.
15. Floquet, G., *Sur les équations différentielles linéaires à coefficients périodiques*. Annales De lnales équations différent, 1883, **12**: p. 47–88.
16. Lyapunov, A.M., *The General Problem of the Stability of Motion*, ed. A.T. Translated by Fuller. 1992, London: Taylor and Francis.
17. Banerjee, B., *Waves in Periodic Media*, in *An Introduction to Metamaterials and Waves in Composites*. 2011, Boca Raton, FL: CRC Press.
18. Brillouin, L., *Wave Propagation in Periodic Structures*. 1946, New York: Dover Publication.
19. Mei, J., Wu, Y., Chan, C.T., Zhang, Z.-Q., *First-principles study of Dirac and Dirac-like cones in phononic and photonic crystals*. Physical Review B, 2012, **86**: p. 035141-1-7.
20. Cao, H., Mei, J., *Dirac-Like Cones at Low-Symmetry Points in Phononic Crystals*, in *ASME International Mechanical Engineering Congress and Exposition*. 2014. American Society of Mechanical Engineers.
21. Vasileska, D.J.A.S.U., Tech. Rep, *Tutorial for Semi-Empirical Band-Structure Calculation*.
22. Gordon, W., *Der Comptoneffekt nach der Schrödingerschen Theorie*. 1026, Berlin, Germany: Springer.
23. Dirac, P.A.M., *The quantum theory of the electron*. Proceedings of the Royal Society A, 1928, **117**(778): p. 610–624.
24. Feynman, R., *The theory of positron*. Physical Review, 1949, **76**(6): p. 749–759.
25. Montambaux, G., *Artificial graphenes: Dirac matter beyond condensed matter Graphènes artificiels: Matière de Dirac au-delà de la Matière condensée*. Comptes Rendus Physique, 2018, **19**(5): p. 285–305.
26. Yves, S., Berthelot, T., Fink, M., Lerosey, G., Lemoult, F., *Measuring dirac cones in a subwavelength metamaterial*. Physical Review Letters, 2018, **121**: p. 267601.
27. Indaleeb, M.M., Banerjee, S., *Simultaneous dirac-like cones at two energy States in tunable phononic crystals: An analytical and numerical study*. Crystals, 2021, **11**(12): p. 1528.
28. Indaleeb, M.M., et al., *Deaf band-based prediction of Dirac cone in acoustic metamaterials*. Journal of Applied Physics, 2020, **127**(6): p. 064903.
29. Indaleeb, M.M., Ahmed, H., Banerjee, S., *Acoustic computing: At tunable pseudo-spin-1 Hermitian Dirac-like cone*. The Journal of the Acoustical Society of America, 2022, **152**(3): p. 1449–1462.
30. Xiao, D., Chang, M.-C., Niu, Q., *Berry phase effects on electronic properties*. Reviews of Modern Physics, 2010, **82**(3): p. 1959.
31. Swinteck, N., et al., *Bulk Elastic Waves With Unidirectional Backscattering-Immune Topological States in a Time-Dependent Superlattice*. 2015, **118**(6): p. 063103.
32. Hasan, M.A., et al., *Spectral analysis of amplitudes and phases of elastic waves: Application to topological elasticity*. Journal of the Acoustical Society of America, 2019. **146**(1): p. 748–766.

Quantum Mechanics for Engineers

33. Goffaux, C., Vigneron, J., *Theoretical study of a tunable phononic band gap system.* Physical Review B, 2001, **64**(7): p. 075118.
34. Kushwaha, M.S., *Stop-Bands for Periodic Metallic Rods: Sculptures That Can Filter the Noise.* 1997. **70**(24): p. 3218–3220.
35. Ahmed, H., et al. *Investigation of wave trapping and attenuation phenomenon for a high symmetry interlocking micro-structure composite metamaterial,* in *Smart Structures and NDE for Energy Systems and Industry 4.0.* 2019. SPIE.
36. Ahmed, H., et al., *Multifunction acoustic modulation by a multi-mode acoustic metamaterial architecture.* Journal of Physics Communications, 2018. **2**(11): p. 115001.
37. Hajian, H., Ozbay, E., Caglayan, H., Enhanced *Transmission and beaming via a zero-index photonic crystal.* Journal of Applied Physics Letter, 2016. **109**(3): p. 031105.
38. Boulanger, J., et al., Observation of a non-Adiabatic geometric phase for elastic waves. Annals of Physics, 2012. **327**(3): p. 952–958.
39. Wang, S., Ma, G., Chan, C.T., Topological transport of sound mediated by spin-redirection geometric phase. Science Advances, 2018. **4**(2): p. eaaq1475.
40. Euler, L., *Solutio problematis ad geometriam situs pertinentis.* Comment. Acad. Sci. U. Petrop, 1736, **8**: p. 128–40.
41. Hatcher, A., *Algebraic Topology.* 2002, Cambridge: Cambridge University Press.
42. Percival, I.C., *A variational principle for invariant tori of fixed frequency.* Journal of Physics A: Mathematical and General, 1979, **12**(3): p. L57–L60.
43. Atiyah, M.F., Hitchin, N., *The Geometry and Dynamics of Magnetic Monopoles.* 1988, Princeton, NJ: Princeton University Press.
44. Aharonov, Y., Bohm, D., *Significance of electromagnetic potentials in quantum theory.* Physical Review, 1959, **115**(3): p. 485–491.
45. Pancharatnam, S., *Generalized theory of interference, and its applications.* Proceedings of Indian Academy of Sciences A, 1956, **44**: p. 247–262.
46. Berry, M.V., *Quantal phase factors accompanying adiabatic changes.* Proceedings of the Royal Society of London. A. Mathematical Physical Sciences, 1984. **392**(1802): p. 45–57.
47. Rechtsman, M.C., et al., *Topological creation and destruction of edge states in photonic graphene.* Physical Review Letters, 2013. **111**(10): p. 103901.
48. Ma, G., Xiao, M., Chan, C.T., *Topological phases in acoustic and mechanical systems.* Nature Reviews Physics, 2019, **1**(4): p. 281–294.
49. Zak, J., *Berry's phase for energy bands in solids.* Physical Review Letters, 1989, **62**(23): p. 2747.
50. Cohen, E., Larocque, H., Bouchard, F., Nejadsattari, F., Gefen, Y., Karimi, E., *Geometric phase from Aharonov–Bohm to Pancharatnam–Berry and beyond.* Nature Reviews Physics, 2019, **1**: p. 437–449.
51. Gudina, S.V., Bogoliubskii, A.S., Klepikova, A.S., Neverov, V.N., Turutkin, K.V., Podgornykh, S.M., Shelushinina, N.G., Yakunin, M.V., Mikhailov, N.N., Dvoretsky, S.A., *Anomalous phase shift of magneto-oscillations in HgTe quantum well with inverted energy spectrum.* Journal of Magnetism and Magnetic Materials, 2021, **524**: p. 167655.
52. Deymier, P., Runge, K., *Sound Topology, Duality Coherence and Wave-Mixing: An Introduction to the Emerging New Science of Sound. Solid-State Science.* 2017, Cham, Switzerland: Springer.
53. McWeeny, R., *Symmetry: An Introduction to Group Theory and Its Applications.* 2002, Garden City, New York: Dover Publications.
54. Wigner, E.P., *Gruppentheorie und ihre Anwendung auf die Quanten mechanik der Atomspektren.* 1931, Braunschweig, Germany: Friedrich Vieweg und Sohn.
55. Wigner, E.P., *The Collected Works of Eugene Paul Wigner, Volume 5, Part A, Nuclear Energy.* 1992, Berlin, Germany: Springer.
56. Xiao, M., et al., Geometric phase and band inversion in periodic acoustic systems. Nature Physics, 2015, **11**(3): p. 240–244.

57. Wang, S., et al., *Spin-Orbit Interactions of Transverse Sound*. 2021. **12**(1): p. 1–9.
58. Hall, E., *On a new action of the magnet on electric currents*. American Journal of Mathematics, 1879, **2**(3), p. 287–292.
59. Kuo, X., Fan, Y., Wang, K.L., *Review of Quantum Hall trio*. Journal of Physics and Chemistry of Solids, 2019, **128**: p. 2–23.
60. Kosterlitz, J.M. Thouless, D.J., *Ordering, metastability and phase transitions in two-dimensional systems*. Journal of Physics C: Solid State Physics, 1973, **6**(7): p. 1181–1203.
61. Berezinskii, V.L., *Translation: Destruction of long-range order in one-dimensional and two-dimensional systems having a continuous symmetry group I. Classical systems*. Soviet Physics—JETP, 1971, **32**(3): p. 610–616.
62. Klitzing, K.V., Dorda, G., Pepper, M., *New method for high-accuracy determination of the fine-structure constant based on Quantized Hall resistance*. Physical Review Letters, 1980, **45**: p. 494.
63. Fang, A., et al., *Anomalous Anderson Localization Behaviors in Disordered Pseudospin Systems*. 2017. **114**(16): p. 4087–4092.
64. Raghu, S., Haldane, F.D.M., *Analogs of quantum-Hall-effect edge States in photonic crystals*. Physical Review A, 2008, **78**(3): p. 033834.
65. Qiao, Z., Yang, S.A., Feng, W., Tse, W.-K., Ding, J., Yao, Y., Wang, J., Niu, Q., *Quantum anomalous Hall effect in graphene from Rashba and exchange effects*. Physical Review B, 2010, **82**: p. 161414(R).
66. Liu, C.-X., Qi, X.-L., Dai, X., Fang, Z., Zhang, S.-C., *Quantum anomalous Hall effect in Hg1−yMnyTe quantum Wells*. Physical Review Letters, 2008, **101**: p. 146802.
67. Onoda, M., Nagaosa, N., *Quantized anomalous Hall effect in two-dimensional ferro-magnets: Quantum Hall effect in metals*. Physical Review Letters, 2003, **90**: p. 206601.
68. Kane, C.L., Mele, E.J., *Z2 topological order and the quantum spin Hall effect*. Physical Review Letters, 2005, **95**: p. 146802.
69. Bernevig, B.A., Zhang, S.-C., *Quantum spin Hall effect*. Physical Review Letters, 2006, **96**: p. 106802.
70. Wu, C., *Orbital analogue of the quantum anomalous Hall effect in p-band systems*. Physical Review Letters, 2008, **101**: p. 186807.
71. Chang, C.Z., Zhang, J., Feng, X., Shen, J., Zhang, Z., Guo, M., Li, K., Ou, Y., Wei, P., Wang, L.-L., Ji, Z.-Q., Feng, Y., Ji, S., Chen, X., Jia, J., Dai, X., Fang, Z., Zhang, S.-C., He, K., Wang, Y., Lu, L., Ma, X.-C., Xu, Q.-K., *Experimental observation of the quantum anomalous Hall effect in a magnetic topological insulator*. Science, 2013, **340**(6129): p. 167–170.
72. Jotzu, G., Messer, M., Desbuquois, R., Lebrat, M., Uehlinger, T., Greif, D., Esslinger, T., *Experimental realization of the topological Haldane model with ultracold fermions*. Nature Reviews Physics, 2014, **515**: p. 237–240.

6 Waves in Periodic Media
Quantum Analogous Application of Acoustics and Elastic Waves

6.1 PERIODIC MEDIA FOR ACOUSTIC AND ELASTIC WAVES

6.1.1 INTRODUCTION

Periodic media could be one dimensional (1D), two dimensional (2D) or three dimensional (3D). Periodicities are created artificially along the respective directions for the metamaterials. Periodic media consist of periodically arranged geometric configurations of more than one material type. Hence, the media must have a host medium that creates a unit cell called matrix. There could be one or multiple constituent materials with material properties different from the constituent matrix (creating acoustic impedance mismatch) but housed inside the unit cell. Together they create a new unit cell with host and constituent materials. Once the unit cell is created, the unit cell could be repeated along any specific dimension to create a metamaterial with geometrical periodicity. For example, Fig. 6.1a shows a base material, a constituent material of an arbitrary shape, a unit cell, and a 2D periodic metamaterial structure. Fig. 6.1b shows an example of a split ring metamaterial where the host matrix is epoxy and the central resonator with rings is made of steel [1–3]. Wave dispersion, wave scattering, and wave attenuation along with several other unique wave phenomena, completely depend on the material properties of the host matrix and the constituent material properties of the inclusions. Different geometric patterns may have different effects on wave propagation. In this chapter several such examples will be presented, followed by the derivation of the governing wave equations in periodic elastic media with concepts borrowed from Chapter 3. Analytical and numerical solution of the equations followed by interpretation of the wave dispersion in the respective media are also discussed in this chapter.

6.1.2 PERIODICITY AND SYMMETRY

Periodicity is classified as 1D, 2D, and 3D. A single-unit cell with its constituent materials could be periodic but may not be symmetric. The material presented in Fig. 6.1 a is not symmetric, but Fig. 6.1b is symmetric. Readers are encouraged to refer to section 5.6.1 in Chapter 5 to understand the classification of different symmetries in metamaterials. Fig. 6.1b shows reflection symmetry about two planes. This way, any material can be created with inherent symmetry or asymmetry. It is not necessary for a unit cell to have an inherent symmetry (Fig. 6.1a) in the periodic

DOI: 10.1201/9781003225751-6

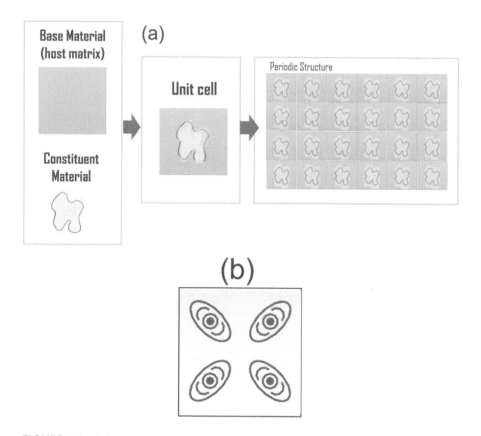

FIGURE 6.1 a) Construction of a periodic metamaterial, b) a split ring metamaterial made of steel inclusion in epoxy matrix.

media, but if the symmetry exists between multiple unit cells (Fig. 6.1b), it is customary to consider all elements that are symmetric within one unit cell, until the linear periodicity is ensured in all intended directions. After such linear periodicity, the periodic media may exhibit symmetry between two or more unit cells. Periodic media could have more than two planes of periodicity. For example, a metamaterial made of hexagonal structure, such as graphene, can have periodicity in three directions on a 2D plane. Figure 6.2 shows an example of a hexagonal metamaterial [4] that has periodicity between two planes and along three planes as indicated. The material is made of epoxy host matrix and plexiglass or poly (methyl methacrylate) a.k.a. PMMA inclusions.

Exploiting various symmetries, different metamaterials could be created. The good news is that readers can create their own geometry that has never been studied before. In fact, only a few metamaterial structures in their simplest form have been studied so far. Thus, creating and investigating topological waves in new metamaterials is virtually an ocean. The depth of this ocean has not been fully explored. As a seedling attempt, this book provides fundamental knowledge and understanding to explore them in the right direction, with the right tools and the right physics.

Waves in Periodic Media

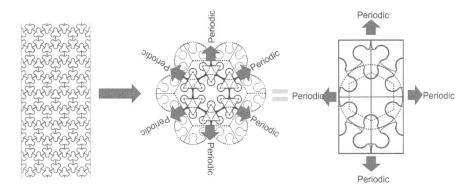

FIGURE 6.2 Hexagonal C_3 and C_6 symmetry with additional reflection symmetry about two orthogonal planes.

6.1.3 Brillouin Zones in Periodic Media

While solving wave propagation in periodic media it is necessary to understand the Brillouin zone. Understanding the Brillouin zone is directly connected to the Bloch solution of wave propagation in a periodic media. Please refer to section 5.3.1 'Bloch Solution' and section 5.3.2 'Reduced-Order Solution and Brillouin Zone' in Chapter 5. Unlike in quantum mechanics, where quantum wave function is the primary variable in Schrödinger equation, acoustic and elastic waves may use acoustic pressure, velocity, or displacements as their primary variables of interest. First, it is essential to find the dispersion behavior in the media before the primary variables are solved and understood. Starting from the governing equation, if the solution patterns of the unknown variables are substituted into the equation and then boundary conditions are applied, the governing equation gives an eigenvalue problem. The eigensolution gives the dispersion behavior.

To achieve the dispersion behavior, first a solution for the primary variables should be assumed. Like the solution method presented in section 5.3.1, the primary variables for acoustics and elastic waves are also expressed assuming the superposition of all the Bloch wave modes. Wave number domain, which is a reciprocal space (reciprocal of spatial domain), is used in the solution for the primary variables. An arbitrary n-th Bloch wave vector k_n can be expressed as $(k + G_n)$, where k is the wave number at which the frequency solution is required. $G_n = \dfrac{2\pi n}{D}$ is the reciprocal lattice vector, where D is the periodicity of the unit cell, and $n = -\infty\ to + \infty$, takes only the integer values. It is evident from the above expression that all possible wave numbers with periodicity G_n are also the solution of the system. This means that as long as the waves have the same phase, all possible wave solutions with different wave numbers are also the solution of the system at a specific frequency. Likewise, all possible frequencies at a specific wave number are also the solution of the system. A situation is expressed in Fig. 5.10a, where two waves with different wavelengths have the same phase. Thus, they both are viable solutions for a periodic acoustic system. This gives unlimited infinite solutions in both positive and negative directions with the integer n.

It is impossible to comprehend infinite solutions and use them wisely to express a wave field. This resulted in the idea of finding the most fundamental wave number (k) solution that is unique to an equation. The question should be asked, Is there any unique fundamental solution of the wave number (k) at a specific frequency that should be understood first? The answer is yes. That is where the reduced-order solution or the Brillouin zone is conceptualized. If out of all possible numbers of $n = -\infty,...,-3, -4,-2,-1, 0, +1,+2,+3,...., \infty$, zero is chosen, the solution is purely the wave number k. When $n = 1$ is chosen, it can be recognized that the wave numbers (k) between 0 and $\frac{2\pi}{D}$ mapped to a full circle of 2π are the only unique wave numbers. The following n values, if considered in the solution, will just repeat the solutions. To accommodate wave propagating in both directions, the full circle of 2π could also be mapped between $-\frac{\pi}{D}$ and $\frac{\pi}{D}$. Hence, it is customary to solve the eigenvalue problem for k only within the reciprocal space of $-\frac{\pi}{D}$ and $\frac{\pi}{D}$. This makes all the other solutions repetition of the same circular mapping. From the infinite range of wave numbers along both positive and negative axes, only the $-\frac{\pi}{D}$ to $+\frac{\pi}{D}$ range is selected. For 1D periodic media, this zone, $-\frac{\pi}{D}$ to $+\frac{\pi}{D}$, is called the first Brillouin zone [5]. Similarly, for 2D periodic media a reciprocal square zone with sides $\frac{2\pi}{D}$ centered at zero would be called the first Brillouin zone in 2D. As the integer value of n increases, the system results in higher-order Brillouin zones as indicated in Fig. 6.3, which shows the Brillouin zones for 2D square and hexagonal periodic systems. It is required to find the solutions within the first Brillouin zone hereafter. The center of

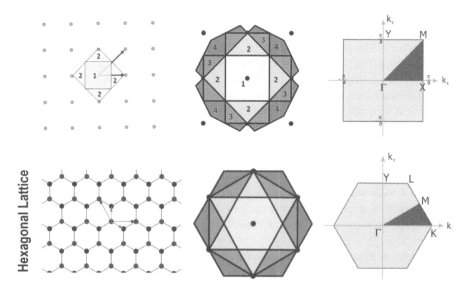

FIGURE 6.3 Brillouin zone for a square and a hexagonal unit cell.

Waves in Periodic Media

the Brillouin zone is called the gamma point, or simply the Γ-point. The right end of the first Brillouin zone is called point X. The corner of the Brillouin zone is called point M. As the Brillouin zone indicates the reciprocal space, they indicate the wave vector directions. $\Gamma - X$ indicates the wave propagation along the x_1 axis in periodic media. $X - M$ indicates wave propagation along the x_2 axis in periodic media. $\Gamma - M$ or $M - \Gamma$ indicates wave propagation along the diagonal direction at an angle of $45°$ or $30°$ with respect to the x_1 axis in square and hexagonal lattice, respectively. This way the wave propagation along any direction between zero and $360°$ could be explored. To understand the global wave characteristics in a periodic media, typically the Bloch waves mode along $\Gamma - X$, $X - M$ and $M - \Gamma$ directions are considered.

6.2 ACOUSTIC WAVES IN PERIODIC MEDIA

6.2.1 GOVERNING DIFFERENTIAL EQUATIONS

Wave propagation equations or governing partial differential equations of motion in any elastic media, described in Chapter 2 Appendix Eq. A.2.1.1 through A2.1.3, could be recalled as

$$\frac{\partial \sigma_{11}}{\partial x_1} + \frac{\partial \sigma_{12}}{\partial x_2} + \frac{\partial \sigma_{13}}{\partial x_3} + f_1 = \rho \frac{\partial^2 u_1}{\partial t^2} \tag{6.1.1}$$

$$\frac{\partial \sigma_{21}}{\partial x_1} + \frac{\partial \sigma_{22}}{\partial x_2} + \frac{\partial \sigma_{23}}{\partial x_3} + f_2 = \rho \frac{\partial^2 u_2}{\partial t^2} \tag{6.1.2}$$

$$\frac{\partial \sigma_{31}}{\partial x_1} + \frac{\partial \sigma_{32}}{\partial x_2} + \frac{\partial \sigma_{33}}{\partial x_3} + f_3 = \rho \frac{\partial^2 u_3}{\partial t^2} \tag{6.1.3}$$

And, recalling the wave propagation in fluid media discussed in Chapter 3, reiterating Eq. 3.1.1 through 3.1.3 can be written as

$$\frac{\partial \sigma_{11}}{\partial x_1} + f_1 = \rho \frac{\partial^2 u_1}{\partial t^2} \quad or \quad -\frac{\partial \mathrm{p}}{\partial x_1} + f_1 = \rho \frac{\partial^2 u_1}{\partial t^2} \tag{6.2.1}$$

$$\frac{\partial \sigma_{22}}{\partial x_2} + f_2 = \rho \frac{\partial^2 u_2}{\partial t^2} \quad or \quad -\frac{\partial \mathrm{p}}{\partial x_2} + f_2 = \rho \frac{\partial^2 u_2}{\partial t^2} \tag{6.2.2}$$

$$\frac{\partial \sigma_{33}}{\partial x_3} + f_3 = \rho \frac{\partial^2 u_3}{\partial t^2} \quad or \quad -\frac{\partial \mathrm{p}}{\partial x_3} + f_3 = \rho \frac{\partial^2 u_3}{\partial t^2} \tag{6.2.3}$$

Further referring to Chapter 3, substituting the stress displacement relations in Eq. 3.98.2 and performing the spatial derivatives, the governing differential equation in an isotropic media will be visualized as Eq. 3.107:

$$(\lambda + \mu)\underline{\nabla}(\nabla.\mathbf{u}) + \mu\nabla^2 u + \underline{\nabla}\lambda(\nabla.\mathbf{u}) + 2(\nabla\mu.\underline{\nabla})\mathbf{u} + \underline{\nabla}\mu \times \underline{\nabla} \times \mathbf{u} + \mathbf{f} = \rho(\mathbf{x})\ddot{\mathbf{u}} \tag{6.3}$$

which could be further written in three equations as follows:

$$\frac{\partial}{\partial x_1}(\lambda \nabla \cdot \mathbf{u}) + \nabla \cdot \left(\mu \left(\nabla u_1 + \frac{\partial \mathbf{u}}{\partial x_1} \right) \right) = \rho \frac{\partial^2 u_1}{\partial t^2} \tag{6.4.1}$$

$$\frac{\partial}{\partial x_2}(\lambda \nabla \cdot \mathbf{u}) + \nabla \cdot \left(\mu \left(\nabla u_2 + \frac{\partial \mathbf{u}}{\partial x_2} \right) \right) = \rho \frac{\partial^2 u_2}{\partial t^2} \tag{6.4.2}$$

$$\frac{\partial}{\partial x_3}(\lambda \nabla \cdot \mathbf{u}) + \nabla \cdot \left(\mu \left(\nabla u_3 + \frac{\partial \mathbf{u}}{\partial x_3} \right) \right) = \rho \frac{\partial^2 u_3}{\partial t^2} \tag{6.4.3}$$

All these different forms of the same equations are helpful to visualize and start the derivation according to the problem as needed. The above equations are written for a 3D general body. Thus, these equations are good for 3D metamaterials with 3D periodic geometry. Understanding such a system is a bit complicated. However, to start with the basics, let's discuss a few scenarios that are helpful to illuminate wave behaviors in 2D structures with infinite third dimension. This means that the phononic crystals or metamaterials are made of 2D geometric pattern on an $x_1 - x_2$ plane, but the structural features, including the host matrix, are extended along the third dimension (i.e., along x_3 axis) to infinity or more than three wavelengths for a specific frequency of interest. We can decouple the above equations in Eq. 6.1.1 through 6.1.3 into the following categories (refer to Fig. 6.4)

a. One-dimensional polarization: $u_3 \neq 0$ but $u_1 = u_2 = 0$ and $\frac{\partial}{\partial x_1} = \frac{\partial}{\partial x_2} = 0$

b. Out-of-plane polarized plane wave: $u_3 \neq 0$, but $u_1 = u_2 = 0$ and $\frac{\partial}{\partial x_3} = 0$,

c. In-plane polarized plane wave: $u_3 = 0$, $\frac{\partial}{\partial x_3} = 0$ but $u_1 \neq 0$; $u_2 \neq 0$

d. No out-of-plane variation: $\frac{\partial}{\partial x_3} = 0$ but $u_1 \neq 0$; $u_2 \neq 0$; $u_3 \neq 0$

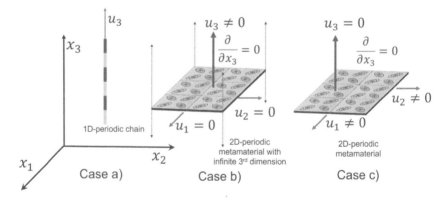

FIGURE 6.4 Different periodic media and mathematical consideration for governing wave equations.

Waves in Periodic Media

For case a) in any generic material, the governing equations will read only as follows. (Refer to Eq. 6.1.1 through 6.1.3.)

$$\frac{\partial \sigma_{33}}{\partial x_3} + f_3 = \rho \frac{\partial^2 u_3}{\partial t^2} \qquad (6.5.1)$$

This is synonymous with a one-dimension wave along the direction x_3 where periodicity is also along the direction x_3. Here, x_3 could be x_1 or x_2 as defined.

For case b) in any generic material, the governing equation will read only

$$\frac{\partial \sigma_{31}}{\partial x_1} + \frac{\partial \sigma_{32}}{\partial x_2} + f_3 = \rho \frac{\partial^2 u_3}{\partial t^2} \qquad (6.5.2)$$

For case c) in any generic material, the governing equations will read

$$\frac{\partial \sigma_{11}}{\partial x_1} + \frac{\partial \sigma_{12}}{\partial x_2} + f_1 = \rho \frac{\partial^2 u_1}{\partial t^2} \qquad (6.6.1)$$

$$\frac{\partial \sigma_{21}}{\partial x_1} + \frac{\partial \sigma_{22}}{\partial x_2} + f_2 = \rho \frac{\partial^2 u_2}{\partial t^2} \qquad (6.6.2)$$

For case d) in any generic material, the governing equations will read

$$\frac{\partial \sigma_{11}}{\partial x_1} + \frac{\partial \sigma_{12}}{\partial x_2} + f_1 = \rho \frac{\partial^2 u_1}{\partial t^2} \qquad (6.7.1)$$

$$\frac{\partial \sigma_{21}}{\partial x_1} + \frac{\partial \sigma_{22}}{\partial x_2} + f_2 = \rho \frac{\partial^2 u_2}{\partial t^2} \qquad (6.7.2)$$

$$\frac{\partial \sigma_{31}}{\partial x_1} + \frac{\partial \sigma_{32}}{\partial x_2} + f_3 = \rho \frac{\partial^2 u_3}{\partial t^2} \qquad (6.7.3)$$

6.2.2 Bloch Solution for Acoustic Waves

While solving the Schrödinger equation with a periodic potential, the Bloch solution is discussed in section 5.3.1 in Chapter 5. As is shown that the wave equations are equivalent, acoustic wave equations must have similar solutions. Here Bloch solutions for cases a), b), and c) are presented in the following sections.

6.2.2.1 One-Dimension Waves in a Continuous Periodic Chain

This situation is expressed under case a), and the governing equation is presented in Eq. 6.5.1. From the constitutive equation for any generic material (refer to Eq. 2.96 in Chapter 2) the normal stress could be expressed as

$$\sigma_{33} = \mathbb{C}_{31}e_{11} + \mathbb{C}_{32}e_{22} + \mathbb{C}_{33}e_{33} \qquad (6.8.1)$$

assuming $\mathbb{C}_{34} = \mathbb{C}_{35} = \mathbb{C}_{36} = 0$ where the effect of shear strain on a normal stress is ignored. In fact, this is the case for most metamaterials within the acoustic limit. Strains in a material due to acoustic and elastic waves are much smaller, and thus stress-stain relationship is assumed to be linear. Substituting Eq. 6.8.1 into Eq. 6.5.1 and using strain-displacement relation as in Eq. A.2.2.1 through A.2.2.3 from Chapter 2, the governing equation will be

$$\frac{\partial}{\partial x_3}\left[\mathbb{C}_{31}\frac{\partial u_1}{\partial x_1} + \mathbb{C}_{32}\frac{\partial u_2}{\partial x_2} + \mathbb{C}_{33}\frac{\partial u_3}{\partial x_3}\right] = \rho\frac{\partial^2 u_3}{\partial t^2} \tag{6.8.2}$$

where the body force f_3 is assumed to be zero for simplicity. Applying the condition for case a) $\dfrac{\partial}{\partial x_1} = \dfrac{\partial}{\partial x_2} = 0$, further the equation will be

$$\frac{\partial}{\partial x_3}\left[\mathbb{C}_{33}\frac{\partial u_3}{\partial x_3}\right] = \rho\frac{\partial^2 u_3}{\partial t^2} \tag{6.8.3}$$

$$\therefore \mathbb{C}_{33}(x_3)\frac{\partial^2 u_3}{\partial x_3^2} + \frac{\partial u_3}{\partial x_3}\cdot\frac{\partial \mathbb{C}_{33}}{\partial x_3} = \rho(x_3)\frac{\partial^2 u_3}{\partial t^2} \tag{6.8.4}$$

where the material coefficients in metamaterials are the functions of space.

Following a similar solution pattern as is used in Eq. 5.25.19 in Chapter 5 for solving Schrödinger equation, the Bloch solution for the displacement u_3 could be assumed as follows:

$$u_3 = \sum_G A(k+G_n)e^{i(k+G_n)x_3}e^{-i\omega t} \tag{6.8.5}$$

More explicitly, with unknown coefficients, the Bloch solution of the primary variable can be written as

$$u_3 = \sum_{j=-n}^{+n} A_j e^{i\left(k+\frac{2\pi j}{D_3}\right)x_3}e^{-i\omega t} \tag{6.8.6}$$

where A_j is the j-th coefficient of the j-th Bloch mode and D_3 is the special periodicity of the one-dimensional metamaterial. Bloch modes are supposed to be extended to infinity; however, for a numerically tractable solution, the series is truncated and kept between $-n$ to $+n$, within which it is assumed that the solution is converged. k is the wave number along the x_3 direction and could be expressed as k_3, to be explicit. However, for a one-dimension wave, the wave number along the direction of wave propagation is implied, and thus k would suffice. Substituting Eq. 6.8.6 into Eq. 6.8.4 the governing equation with the Bloch modes in a step-by-step derivation process can be written as

$$\frac{\partial u_3}{\partial x_3} = \sum_{j=-n}^{+n} iA_j\left(k+\frac{2\pi j}{D_3}\right)e^{i\left(k+\frac{2\pi j}{D_3}\right)x_3}e^{-i\omega t} \tag{6.8.7}$$

Waves in Periodic Media

$$\frac{\partial^2 u_3}{\partial t^2} = \sum_{j=-n}^{+n} -\omega^2 A_j e^{i\left(k+\frac{2\pi j}{D_3}\right)x_3} e^{-i\omega t} \tag{6.8.8}$$

$$\frac{\partial}{\partial x_3}\left[\mathbb{C}_{33}(x_3)\sum_{j=-n}^{+n} iA_j\left(k+\frac{2\pi j}{D_3}\right)e^{i\left(k+\frac{2\pi j}{D_3}\right)x_3}\right] = \rho(x_3)\left[\sum_{j=-n}^{+n} -\omega^2 A_j e^{i\left(k+\frac{2\pi j}{D_3}\right)x_3}\right] \tag{6.8.9}$$

$$\mathbb{C}_{33}(x_3)\sum_{j=-n}^{+n} i^2 A_j\left(k+\frac{2\pi j}{D_3}\right)^2 e^{i\left(k+\frac{2\pi j}{D_3}\right)x_3} + \sum_{j=-n}^{+n} iA_j\left(k+\frac{2\pi j}{D_3}\right)e^{i\left(k+\frac{2\pi j}{D_3}\right)x_3}\cdot\frac{\partial\mathbb{C}_{33}(x_3)}{\partial x_3}$$

$$= \rho(x_3)\sum_{j=-n}^{+n} -\omega^2 A_j e^{i\left(k+\frac{2\pi j}{D_3}\right)x_3} \tag{6.8.10}$$

where the above equation is for monochromatic wave with frequency ω and it is assumed that $e^{-i\omega t} \neq 0$. Further, in periodic media, it is assumed that the material properties change periodically. According to the Fourier series, any periodic function could be expressed as a summation of Fourier coefficients multiplied with their orthogonal bases. Thus, exploiting Fourier series expressions, $\mathbb{C}_{33}(x_3)$ and $\rho(x_3)$, the periodic functions would read

$$\mathbb{C}_{33}(x_3) = \sum_{p=-n}^{+n} C_p e^{i\left(\frac{2\pi p}{D_3}\right)x_3} \quad \text{and} \quad \rho(x_3) = \sum_{p=-n}^{+n} \rho_p e^{i\left(\frac{2\pi p}{D_3}\right)x_3} \tag{6.8.11}$$

where p is an arbitrary coefficient taking values between $-n$ to $+n$. Further, substituting Eq. 6.8.11 into Eq. 6.8.9,

$$\frac{\partial}{\partial x_3}\left[\sum_{p=-n}^{+n} C_p e^{i\left(\frac{2\pi p}{D_3}\right)x_3}\sum_{j=-n}^{+n} iA_j\left(k+\frac{2\pi j}{D_3}\right)e^{i\left(k+\frac{2\pi j}{D_3}\right)x_3}\right]$$

$$= \sum_{p=-n}^{+n} \rho_p e^{i\left(\frac{2\pi p}{D_3}\right)x_3}\left[\sum_{j=-n}^{+n} -\omega^2 A_j e^{i\left(k+\frac{2\pi j}{D_3}\right)x_3}\right] \tag{6.8.12}$$

Assuming Bloch basis vectors or reciprocal lattice vectors as $G_j = \dfrac{2\pi j}{D_3}$, and performing the following mathematical juggleries $G_j = \dfrac{2\pi j}{D_3}$; $G_p = \dfrac{2\pi p}{D_3}$; $G_j + G_p = \tilde{G}_m$; $G_p = \tilde{G}_m - G_j$, and index $p = m - j$, the governing equation will further read as an eigenvalue problem as follows:

$$\left[\sum_m\sum_j C_{m-j}A_j\left(k+G_j\right)\left(k+\tilde{G}_m\right)e^{i\tilde{G}_m x_3}\right]e^{ikx_3} = \left[\omega^2\sum_m\sum_j \rho_{m-j}A_j\left(e^{i\tilde{G}_m x_3}\right)\right]e^{ikx_3} \tag{6.8.13}$$

or

$$\left[\sum_m \sum_j C_{m-j} A_j \left(k+G_j\right)\left(k+\tilde{G}_m\right) e^{i\tilde{G}_m x_3} - \omega^2 \sum_m \sum_j \rho_{m-j} A_j e^{i\tilde{G}_m x_3}\right] = 0 \qquad (6.8.14)$$

Multiplying Eq. 6.8.14 by $e^{-i\frac{2\pi p}{D_3} x_3} \equiv e^{-iG_p x_3}$ and integrating over a period of D_3, the equation will visualize as

$$\frac{1}{D_3} \int_{-D_3/2}^{D_3/2} \sum_m \sum_j C_{m-j} A_j \left(k+G_j\right)\left(k+G_p\right) e^{i\tilde{G}_m x_3} e^{-iG_p x_3} \, dx_3$$

$$= \omega^2 \frac{1}{D_3} \int_{-D_3/2}^{D_3/2} \sum_m \sum_j \rho_{m-j} A_j e^{i\left(\tilde{G}_m - G_p\right) x_3} \, dx_3 \qquad (6.8.15)$$

This approach is equivalent to applying the orthogonality condition of the reciprocal lattice vectors. When the wave modes and their indices are equal, the integral will result, a term multiplied with their respective amplitude coefficients. As $\frac{1}{D_3} \int e^{i\left(\tilde{G}_m - G_p\right) x_3} \, dx_3 = \delta_{mp}$ is a Kronecker delta that takes value 1 when $m = p$, the eigenvalue problem could be written as

$$\sum_p \sum_j C_{p-j} A_j \left(k+G_j\right)\left(k+G_p\right) = \omega^2 \sum_p \sum_j \rho_{p-j} A_j \qquad (6.8.16)$$

In the above equation, for each value of wave number k, the system would result in multiple frequency values representing the wave modes in the system. The energy solution in Eq. 5.25.28 in Chapter 5 is a function of wave number (or momentum) (i.e., $E(k)$ represents the eigenvalue solutions). Similarly, here for acoustics and elastic waves, the frequency $\omega(k)$ is a function of wave number and represents the eigenvalues of the system made of periodic material properties. As an example, expanding the eigenvalue problem in Eq. 6.8.16 in a matrix form for Bloch modes $n = -1, 0, +1$ would visualize as

$$\begin{bmatrix} C_{-1+1}\left(k+G_{-1}\right)\left(k+G_{-1}\right) & C_{-1-0}\left(k+G_0\right)\left(k+G_{-1}\right) & C_{-1-1}\left(k+G_{+1}\right)\left(k+G_{-1}\right) \\ C_{0+1}\left(k+G_{-1}\right)\left(k+G_0\right) & C_{0-0}\left(k+G_0\right)\left(k+G_0\right) & C_{0+1}\left(k+G_{+1}\right)\left(k+G_0\right) \\ C_{+1+1}\left(k+G_{-1}\right)\left(k+G_{+1}\right) & C_{+1-0}\left(k+G_0\right)\left(k+G_{+1}\right) & C_{+1-1}\left(k+G_{+1}\right)\left(k+G_{+1}\right) \end{bmatrix}$$

$$-\omega^2 \begin{bmatrix} \rho_{-1+1} & \rho_{-1-0} & \rho_{-1-1} \\ \rho_{0+1} & \rho_{0-0} & \rho_{0-1} \\ \rho_{+1+1} & \rho_{+1-0} & \rho_{+1-1} \end{bmatrix} \begin{Bmatrix} A_{-1} \\ A_0 \\ A_{+1} \end{Bmatrix} = \begin{Bmatrix} 0 \\ 0 \\ 0 \end{Bmatrix} \qquad (6.8.17)$$

Waves in Periodic Media

This means that Eq. 6.8.17 will give the frequencies for three wave modes at each value of wave number k. It is discussed in section 6.1.3 that the wave number domain is discretized within the first Brillouin zone, which is $-\dfrac{\pi}{D_3}$ to $+\dfrac{\pi}{D_3}$. Eigenvalues are obtained for each value of k within the Brillouin zone and will result in three frequency values, or ω values, at each k. Hence, if any integer number N is used as an index in the Bloch mode expansion, the matrix in Eq. 6.8.17 will take the size of $(2N+1)\times(2N+1)$. This also means that at each wave number k the eigenvalue problem will result in $(2N+1)$ frequency values. Many such solutions for a 1D metamaterial system could be found in existing literature [6]. Thus, explicit analysis of this simple system is omitted here. However, solutions for the more complex systems for case b) and case c) are presented in the following sections.

6.2.2.2 Plane Waves with Out-of-Plane Polarity

This situation is mathematically expressed for case b) in section 6.2.1. Here, out-of-plane polarity is expressed by the displacement of the media along the x_3 direction u_3, when the wave propagation is described on the $x_1 - x_2$ plane (refer to Fig. 6.4). Based on the definition of out-of-plane polarity, the displacement along the x_3 direction can be written as a superposition of the Bloch wave modes with periodicity along the x_1 and x_2 direction as follows:

$$u_3 = \sum_{G} u_3(G) e^{i k \cdot x} e^{-i\omega t} = \sum_{G_m}\sum_{G_n} A_{mn} e^{i(k+G)\cdot x} e^{-i\omega t} \tag{6.8.18}$$

$$\text{or } u_3 = \sum_{G_m}\sum_{G_n} A_{mn} e^{i(k_1 x_1 + k_2 x_2)} e^{i(G_m x_1 + G_n x_2)} e^{-i\omega t} \tag{6.8.19}$$

Next, it is necessary to substitute the stress-strain relation in the governing equation. Referring to the discussion in section 2.14.3 and Eq. 2.100.3, the shear stress equation could be written as $\sigma_{31} = \mathbb{C}_{66}(\mathbf{x})(2e_{31})$. As the displacement along the direction x_1 $(u_1 = 0)$ is zero, substituting strain-displacement relation the stress is expressed as $\sigma_{31} = \mathbb{C}_{66}(\mathbf{x})\dfrac{\partial u_3}{\partial x_1}$. In orthotropic, transversely isotropic, and isotropic material, the coefficient $\mathbb{C}_{66}$ is the shear modulus and can be expressed by the space-dependent second Lamé constant μ (i.e., $\mathbb{C}_{66}(\mathbf{x}) = \mu$). Please note that the spatial dependency of the shear modulus is implied. Substituting the stress-strain relation and the strain-displacement relation in case b), the governing equation in Eq. 6.5.2 will further read

$$\frac{\partial}{\partial x_1}\left(\mu \frac{\partial u_3}{\partial x_1}\right) + \frac{\partial}{\partial x_2}\left(\mu \frac{\partial u_3}{\partial x_2}\right) = \rho \frac{\partial^2 u_3}{\partial t^2} \tag{6.8.20}$$

Please note that the metamaterial is periodic, and thus the periodic material properties could be expressed with their respective Fourier coefficients

240   Metamaterials in Topological Acoustics

as expressed in Eq. 6.8.11. Thus, the shear-modulus μ and density ρ could be expressed as

$$\mu = \sum_{\bar{G}_p}\sum_{\bar{G}_q}\mu_{pq}e^{i\left[(\bar{G}_p)x_1+(\bar{G}_q)x_2\right]} = \sum_{\bar{G}_p}\sum_{\bar{G}_q}\mu_{pq}e^{i\left[\bar{G}.x\right]} \tag{6.8.21}$$

$$\rho = \sum_{\bar{G}_p}\sum_{\bar{G}_q}\rho_{pq}e^{i\left[(\bar{G}_p)x_1+(\bar{G}_q)x_2\right]} = \sum_{\bar{G}_p}\sum_{\bar{G}_q}\rho_{pq}e^{i\left[\bar{G}.x\right]} \tag{6.8.22}$$

where $\bar{\mathbf{G}} = \bar{G}_p\hat{\mathbf{e}}_1 + \bar{G}_q\hat{\mathbf{e}}_2$ is the Bloch wave vector related to the material properties and hats are added to distinguish them from the Bloch wave vectors related to the displacement function, which is $\mathbf{G} = G_m\hat{\mathbf{e}}_1 + G_n\hat{\mathbf{e}}_2$. Substituting Eq. 6.8.19, Eq. 6.8.21, and Eq. 6.8.22 into Eq. 6.8.20, the term-by-term expressions would be

$$\frac{\partial u_3}{\partial x_1} = \sum_{G_m}\sum_{G_n}iA_{mn}\left(k_1 + G_m\right)e^{i(\mathbf{k}+\mathbf{G}).\mathbf{x}}e^{-i\omega t} \tag{6.8.23}$$

$$\frac{\partial u_3}{\partial x_2} = \sum_{G_m}\sum_{G_n}iA_{mn}\left(k_2 + G_n\right)e^{i(\mathbf{k}+\mathbf{G}).\mathbf{x}}e^{-i\omega t} \tag{6.8.24}$$

$$\frac{\partial}{\partial x_1}\left(\mu\frac{\partial u_3}{\partial x_1}\right)$$
$$= \sum\sum\sum\sum i^2\mu_{pq}A_{mn}\left(k_1+G_m\right)\left(k_1+G_m+\bar{G}_p\right)e^{i\left[\left(k_1+G_m+\bar{G}_p\right)x_1+\left(k_2+G_n+\bar{G}_q\right)x_2\right]} \tag{6.8.25}$$

$$\frac{\partial}{\partial x_2}\left(\mu\frac{\partial u_3}{\partial x_2}\right)$$
$$= \sum\sum\sum\sum i^2\mu_{pq}A_{mn}\left(k_2+G_n\right)\left(k_2+G_n+\bar{G}_q\right)e^{i\left[\left(k_1+G_m+\bar{G}_p\right)x_1+\left(k_2+G_n+\bar{G}_q\right)x_2\right]} \tag{6.8.26}$$

where summation over G_m, G_n, G_p, and G_q Bloch lattice vectors is implied in Eq. 6.8.25 and 6.8.26. To further simply the summation and visualization, the following mathematical juggleries are used. It is assumed that $\left(G_m + \bar{G}_p\right)\hat{\mathbf{e}}_1 = \tilde{G}_\ell\hat{\mathbf{e}}_1$ and $\left(G_n + \bar{G}_q\right)\hat{\mathbf{e}}_2 = \tilde{G}_h\hat{\mathbf{e}}_2$ are replaced with new indices ℓ and h. Hence $\mathbf{G}+\bar{\mathbf{G}} = \tilde{\mathbf{G}}$ or $\bar{\mathbf{G}} = \tilde{\mathbf{G}} - \mathbf{G}$, $\bar{G}_p = \left(\tilde{G}_\ell - G_m\right)$ and $\bar{G}_q = \left(\tilde{G}_h - G_n\right)$. Based on these mathematical treatments, the Fourier coefficients for μ and density ρ could be written as

$$\mu_{pq} = \mu_{\left(\tilde{G}_\ell-G_m\right)\left(\tilde{G}_h-G_n\right)} = \mu_{\left(\tilde{\mathbf{G}}-\mathbf{G}\right)}; \; \rho_{pq} = \rho_{\left(\tilde{G}_\ell-G_m\right)\left(\tilde{G}_h-G_n\right)} = \rho_{\left(\tilde{\mathbf{G}}-\mathbf{G}\right)} \tag{6.8.27}$$

Waves in Periodic Media   **241**

Substituting μ_{pq} and ρ_{pq} into Eqs. 6.8.25 and 6.8.26, the governing equation would read

$$\sum\sum\sum\sum i^2\left[\mu_{(\tilde{G}-G)}A_{mn}\left[\left(k_1+G_m\right)\left(k_1+\tilde{G}_t\right)+\left(k_2+G_n\right)\left(k_2+\tilde{G}_h\right)\right]\right]e^{i(\tilde{G})\cdot x}e^{ik\cdot x}$$

$$=-\omega^2\sum\sum\sum\sum\rho_{(\tilde{G}-G)}A_{mn}e^{i(\tilde{G})\cdot x}e^{ik\cdot x} \qquad (6.8.28)$$

As $e^{ik\cdot x}\neq 0$, the governing equation would now read

$$\sum_{\tilde{G}}\left[\mu_{(\tilde{G}-G)}\left[\left(k_1+G_m\right)\left(k_1+\tilde{G}_t\right)+\left(k_2+G_n\right)\left(k_2+\tilde{G}_h\right)\right]-\omega^2\rho_{(\tilde{G}-G)}\right]A_{mn}e^{i(\tilde{G})\cdot x}=0$$

$$(6.8.29)$$

$$\text{or}\ \sum_{\tilde{G}}\left[\mu_{(\tilde{G}-G)}\left[(\mathbf{k}+\mathbf{G})\cdot\left(\mathbf{k}+\tilde{\mathbf{G}}\right)\right]-\omega^2\rho_{(\tilde{G}-G)}\right]e^{i(\tilde{G})\cdot x}A_{\mathbf{G}}=0 \qquad (6.8.30)$$

Multiplying Eq. 6.8.30 by $e^{-iG\cdot x}\equiv e^{-i(G_1x_1+G_2x_2)}$ and integrating over a plane with periodicity D_1 and D_2 along x_1 and x_2 direction would result in the final eigenvalue problem.

Applying the orthogonality condition of the reciprocal lattice vectors, the equation will be

$$\frac{1}{D_1D_2}\iint e^{-iG\cdot x}\sum_{\tilde{G}}\left[\mu_{(\tilde{G}-G)}\left[(\mathbf{k}+\mathbf{G})\cdot\left(\mathbf{k}+\tilde{\mathbf{G}}\right)\right]-\omega^2\rho_{(\tilde{G}-G)}\right]e^{i(\tilde{G})\cdot x}A_{\mathbf{G}}=0 \quad (6.8.31)$$

$$\sum_{\tilde{G}}\left[\mu_{(\tilde{G}-G)}\left[(\mathbf{k}+\mathbf{G})\cdot\left(\mathbf{k}+\tilde{\mathbf{G}}\right)\right]-\omega^2\rho_{(\tilde{G}-G)}\right]e^{i(\tilde{G}-G)\cdot x}A_{\mathbf{G}}=0 \qquad (6.8.32)$$

The above equation is a matrix and depends on the number of Bloch wave vectors used in the solution. If any integer number N is used as an index in the Bloch mode expansion, the matrix in Eq. 6.8.32 will realize a size of $(2N+1)^2\times(2N+1)^2$. For example, $N=1$ will generate two sets of indices. One set is for $m=-1,0+1$ and another set is for $n=-1,0,+1$. Thus, the coefficients A_{mn} in Eq. 6.8.29 or $A_{\mathbf{G}}$ will result in a vector of $(2.1+1)^2=9$ elements as $A_{\mathbf{G}}\equiv A_{mn}=\{A_{-1-1}\ \ A_{-10}\ \ A_{-1+1}\ \ A_{0-1}\ \ A_{00}\ \ A_{0+1}\ \ A_{+1-1}\ \ A_{+10}\ \ A_{+1+1}\}^T$.

Further, there are two scenarios, when $\overline{\mathbf{G}}=0$ and $\overline{\mathbf{G}}\neq 0$ (i.e., $\left(\tilde{\mathbf{G}}=\mathbf{G}\right)$ and $\left(\tilde{\mathbf{G}}\neq\mathbf{G}\right)$) will result in the coefficients of the matrix along the diagonal terms and at the off-diagonal terms, respectively.

Please observe Eq. 6.8.32 for a moment, and it can be realized that when the wave numbers are discretized, most of the terms in the matrix could be found using the input wave number values and the Bloch reciprocal lattice vector indices (e.g., ℓ, h, m and n). But $\mu_{(\tilde{G}-G)}$ and $\rho_{(\tilde{G}-G)}$ are Fourier coefficients and are not known until a specific metamaterial or an arrangement of phononic crystals is explicitly described.

242   Metamaterials in Topological Acoustics

With specific design of a metamaterial the Fourier coefficients could be found. For a few specific simple geometries, the Fourier coefficients are found in the Appendix section of this chapter. Let's reiterate that modified Bloch wave vector $\bar{\mathbf{G}} = \tilde{\mathbf{G}} - \mathbf{G}$, & $\bar{G}_p = \left(\tilde{G}_\ell - G_m \right)$ and $\bar{G}_q = \left(\tilde{G}_h - G_n \right)$. Referring to the appendix, the Fourier coefficients for the shear modulus and density would be

- For circular phononic crystal in a host matrix with filling fraction $f = \dfrac{\pi r_0^{\,2}}{a^2}$ (refer to Fig. A.6.1)
  - Second Lamé constant

$$\mu_{\left(\tilde{\mathbf{G}} - \mathbf{G} \right)} = \mu_{\left(\bar{\mathbf{G}} \right)} = \mu_{pq} = \mu_I f + \left(1 - f \right) \mu_{II} \ \text{ when } \tilde{\mathbf{G}} = \mathbf{G} \text{ or } \bar{\mathbf{G}} = 0$$

$$(6.8.33)$$

$$\mu_{\left(\tilde{\mathbf{G}} - \mathbf{G} \right)} = \mu_{\left(\bar{\mathbf{G}} \right)} = \mu_{pq} = \left(\mu_I - \mu_{II} \right) \frac{2 f J_1 \left(G_{pq} r_0 \right)}{G_{pq} r_0} \ \text{ when } \tilde{\mathbf{G}} \neq \mathbf{G} \text{ or } \bar{\mathbf{G}} \neq 0$$

$$(6.8.34)$$

  - Density

$$\rho_{\left(\tilde{\mathbf{G}} - \mathbf{G} \right)} = \rho_{\left(\bar{\mathbf{G}} \right)} = \rho_{pq} = \rho_I f + \left(1 - f \right) \rho_{II} \ \text{ when } \tilde{\mathbf{G}} = \mathbf{G} \text{ or } \bar{\mathbf{G}} = 0$$

$$(6.8.35)$$

$$\rho_{\left(\tilde{\mathbf{G}} - \mathbf{G} \right)} = \rho_{\left(\bar{\mathbf{G}} \right)} = \rho_{pq} = \left(\rho_I - \rho_{II} \right) \frac{2 f J_1 \left(G_{pq} r_0 \right)}{G_{pq} r_0} \ \text{ when } \tilde{\mathbf{G}} \neq \mathbf{G} \text{ or } \bar{\mathbf{G}} \neq 0$$

$$(6.8.36)$$

- For a square or rectangular phononic crystal in a host matrix with filling fraction $f = \dfrac{b_1 b_2}{a_1 a_2}$ (refer to Fig. A.6.2)

  - 2nd Lamé constant

$$\mu_{\left(\tilde{\mathbf{G}} - \mathbf{G} \right)} = \mu_{\left(\bar{\mathbf{G}} \right)} = \mu_{pq} = \mu_I f + \left(1 - f \right) \mu_{II} \ \text{ when } \tilde{\mathbf{G}} = \mathbf{G} \text{ or } \bar{\mathbf{G}} = 0$$

$$(6.8.37)$$

$$\mu_{\left(\tilde{\mathbf{G}} - \mathbf{G} \right)} = \mu_{\left(\bar{\mathbf{G}} \right)} = \mu_{pq}$$

$$= \left(\mu_I - \mu_{II} \right) f \left[\frac{\sin\left(G_p \dfrac{b_1}{2} \right)}{G_p \left(\dfrac{b_1}{2} \right)} \right] \left[\frac{\sin\left(G_q \dfrac{b_2}{2} \right)}{G_q \left(\dfrac{b_2}{2} \right)} \right] \ \text{ when } \tilde{\mathbf{G}} \neq \mathbf{G} \text{ or } \bar{\mathbf{G}} \neq 0$$

$$(6.8.38)$$

  - Density

$$\rho_{\left(\tilde{\mathbf{G}} - \mathbf{G} \right)} = \rho_{\left(\bar{\mathbf{G}} \right)} = \rho_{pq} = \rho_I f + \left(1 - f \right) \rho_{II} \ \text{ when } \tilde{\mathbf{G}} = \mathbf{G} \text{ or } \bar{\mathbf{G}} = 0$$

$$(6.8.39)$$

Waves in Periodic Media

$$p_{(\tilde{G}-G)} = p_{(\overline{G})} = p_{pq}$$

$$= (\rho_I - \rho_{II}) f \left[\frac{\sin\left(G_p \dfrac{b_1}{2}\right)}{G_p\left(\dfrac{b_1}{2}\right)} \right] \left[\frac{\sin\left(G_q \dfrac{b_2}{2}\right)}{G_q\left(\dfrac{b_2}{2}\right)} \right] \text{ when } \tilde{G} \neq G \text{ or } \overline{G} \neq 0$$

$$(6.8.40)$$

Substituting the Fourier coefficients for respective properties of different types of phononic crystals into Eq. 6.8.32, the eigenvalues could be found and the dispersion relation $\omega(\mathbf{k})$ would be understood.

6.2.2.3 Plane Waves with In-Plane Polarity

This situation is mathematically expressed for case c) in section 6.2.1 (Eqs. 6.6.1 and 6.6.2). Here in-plane polarity is expressed by the displacement of the media along the x_1 direction as u_1 and along the x_2 direction as u_2, when the wave propagation is described on the $x_1 - x_2$ plane (refer to Fig. 6.4c). Based on the definition of in-plane polarity, the displacements u_1 and u_2 can be written as a superposition of the Bloch wave modes with periodicity along the x_1 and x_2 direction as follows:

$$u_1 = \sum_{\mathbf{G}} u_1(\mathbf{G}) e^{i\mathbf{k}\cdot\mathbf{x}} e^{-i\omega t}$$

$$= \sum_{G_m}\sum_{G_n} A_{mn} e^{i(\mathbf{k}+\mathbf{G})\cdot\mathbf{x}} e^{-i\omega t} = \sum_{G_m}\sum_{G_n} A_{mn} e^{i(k_1 x_1 + k_2 x_2)} e^{i(G_m x_1 + G_n x_2)}$$

$$(6.9.1)$$

$$u_2 = \sum_{\mathbf{G}} u_2(\mathbf{G}) e^{i\mathbf{k}\cdot\mathbf{x}} e^{-i\omega t}$$

$$= \sum_{G_m}\sum_{G_n} A_{mn} e^{i(\mathbf{k}+\mathbf{G})\cdot\mathbf{x}} e^{-i\omega t} = \sum_{G_m}\sum_{G_n} B_{mn} e^{i(k_1 x_1 + k_2 x_2)} e^{i(G_m x_1 + G_n x_2)}$$

$$(6.9.2)$$

Please note that, unlike Eq. 6.8.19, there are two sets of unknown coefficients, A_{mn} and B_{mn}. Two different sets of coefficients for two displacements are considered. Next, the stress-strain relationship to be used (refer to Eq. 2.100.2 in Chapter 2) is

$$\begin{Bmatrix} \sigma_{11} \\ \sigma_{22} \\ \sigma_{33} \\ \sigma_{23} \\ \sigma_{31} \\ \sigma_{12} \end{Bmatrix} = \begin{bmatrix} \mathbb{C}_{11}(\mathbf{x}) & \mathbb{C}_{12}(\mathbf{x}) & \mathbb{C}_{13}(\mathbf{x}) & 0 & 0 & 0 \\ \mathbb{C}_{21}(\mathbf{x}) & \mathbb{C}_{22}(\mathbf{x}) & \mathbb{C}_{23}(\mathbf{x}) & 0 & 0 & 0 \\ \mathbb{C}_{31}(\mathbf{x}) & \mathbb{C}_{32}(\mathbf{x}) & \mathbb{C}_{33}(\mathbf{x}) & 0 & 0 & 0 \\ 0 & 0 & 0 & \mathbb{C}_{44}(\mathbf{x}) & 0 & 0 \\ 0 & 0 & 0 & 0 & \mathbb{C}_{55}(\mathbf{x}) & 0 \\ 0 & 0 & 0 & 0 & 0 & \mathbb{C}_{66}(\mathbf{x}) \end{bmatrix} \begin{Bmatrix} e_{11} \\ e_{22} \\ e_{33} \\ 2e_{23} \\ 2e_{31} \\ 2e_{12} \end{Bmatrix}$$

$$(6.9.3)$$

Please note that σ_{12} is at the sixth row and the material coefficient used to express shear stress is $\mathbb{C}_{66}(\mathbf{x})$. Implying the spatial dependency for ease of writing the equations, the stress-strain equation is rewritten:

$$
\begin{Bmatrix} \sigma_{11} \\ \sigma_{22} \\ \sigma_{33} \\ \sigma_{23} \\ \sigma_{31} \\ \sigma_{12} \end{Bmatrix} = \begin{bmatrix} C_{11} & C_{12} & C_{13} & 0 & 0 & 0 \\ C_{21} & C_{22} & C_{23} & 0 & 0 & 0 \\ C_{31} & C_{32} & C_{33} & 0 & 0 & 0 \\ 0 & 0 & 0 & C_{44} & 0 & 0 \\ 0 & 0 & 0 & 0 & C_{55} & 0 \\ 0 & 0 & 0 & 0 & 0 & C_{66} \end{bmatrix} \begin{Bmatrix} e_{11} \\ e_{22} \\ e_{33} \\ 2e_{23} \\ 2e_{31} \\ 2e_{12} \end{Bmatrix}
\tag{6.9.4}
$$

For the case c) $\dfrac{\partial u_3}{\partial x_3} = 0$ explicitly, the three stresses in Eqs. 6.6.1 and 6.6.2 can be expressed as

$$
\sigma_{11} = C_{11}e_{11} + C_{12}e_{22}; \quad \sigma_{22} = C_{21}e_{11} + C_{22}e_{22}; \quad \sigma_{12} = 2C_{66}e_{12}
\tag{6.9.5}
$$

Stresses with displacement gradients can be written as

$$
\sigma_{11} = C_{11}\frac{\partial u_1}{\partial x_1} + C_{12}\frac{\partial u_2}{\partial x_2}
\tag{6.9.6.1}
$$

$$
\sigma_{22} = C_{21}\frac{\partial u_1}{\partial x_1} + C_{22}\frac{\partial u_2}{\partial x_2}
\tag{6.9.6.2}
$$

$$
\sigma_{12} = C_{66}\left(\frac{\partial u_1}{\partial x_2} + \frac{\partial u_2}{\partial x_1}\right)
\tag{6.9.6.3}
$$

Performing the derivatives and substituting Eqs. 6.9.1 and 6.9.2 into Eqs. 6.9.6.1 through 6.9.6.3, we get

$$
\sigma_{11} = \left[C_{11}\sum_{G_m}\sum_{G_n} i(k_1 + G_m)A_{mn} + C_{12}\sum_{G_m}\sum_{G_n} i(k_2 + G_n)B_{mn} \right] e^{i(\mathbf{k}+\mathbf{G})\cdot\mathbf{x}}
\tag{6.9.7.1}
$$

$$
\sigma_{22} = \left[C_{21}\sum_{G_m}\sum_{G_n} i(k_1 + G_m)A_{mn} + C_{22}\sum_{G_m}\sum_{G_n} i(k_2 + G_n)B_{mn} \right] e^{i(\mathbf{k}+\mathbf{G})\cdot\mathbf{x}}
\tag{6.9.7.2}
$$

$$
\sigma_{12} = C_{66}\left[\sum_{G_m}\sum_{G_n} i(k_2 + G_n)A_{mn} + \sum_{G_m}\sum_{G_n} i(k_1 + G_m)B_{mn} \right] e^{i(\mathbf{k}+\mathbf{G})\cdot\mathbf{x}}
\tag{6.9.7.3}
$$

Waves in Periodic Media 245

Metamaterial is periodic and thus, as in the previous section, the material coefficients C_{ij} (any coefficient form the matrix in Eq. 6.9.4) can be written in terms of Fourier coefficients as follows:

$$C_{ij} = \sum_{\bar{G}_p} \sum_{\bar{G}_q} C_{ij}(\mathbf{G}) e^{i\mathbf{G}\cdot\mathbf{x}} = \sum_p \sum_q C_{ij}^{pq} e^{i\left(\bar{G}_p x_1 + \bar{G}_q x_2\right)} \tag{6.9.8}$$

Derivatives of the material coefficients with respect to any ℓ-th direction will be

$$\frac{\partial C_{ij}}{\partial x_\ell} = (-1)^{\ell+1} \sum_p \sum_q i C_{ij}^{pq} \frac{2\pi\left(p(2-\ell)\right)+q(1-\ell)}{D_\ell} e^{i\bar{G}\cdot\mathbf{x}} \tag{6.9.9}$$

Next, substituting the Fourier representation of the material constants into Eqs. 6.9.7.1 through 6.9.7.3, we get

$$\sigma_{11} = \sum_{\bar{G}_p} \sum_{\bar{G}_q} \sum_{G_m} \sum_{G_n} \left[i C_{11}^{pq} (k_1 + G_m) A_{mn} + i C_{12}^{pq} (k_2 + G_n) B_{mn} \right] e^{i\left(k+G+\bar{G}\right)\cdot\mathbf{x}} \tag{6.9.10.1}$$

$$\sigma_{22} = \sum_{\bar{G}_p} \sum_{\bar{G}_q} \sum_{G_m} \sum_{G_n} \left[i C_{21}^{pq} (k_1 + G_m) A_{mn} + i C_{22}^{pq} (k_2 + G_n) B_{mn} \right] e^{i\left(k+G+\bar{G}\right)\cdot\mathbf{x}} \tag{6.9.10.2}$$

$$\sigma_{12} = \sum_{G_p} \sum_{G_q} \sum_{G_m} \sum_{G_n} \left[i C_{66}^{pq} (k_2 + G_n) A_{mn} + (k_1 + G_m) B_{mn} \right] e^{i\left(k+G+\bar{G}\right)\cdot\mathbf{x}} \tag{6.9.10.3}$$

As in the previous section, $\bar{\mathbf{G}}$ is specifically used to represent the Bloch vectors that originated from the material coefficients and can be expressed as $\bar{\mathbf{G}} = \bar{G}_p \hat{\mathbf{e}}_1 + \bar{G}_q \hat{\mathbf{e}}_2$, while recollecting $\mathbf{G} = G_m \hat{\mathbf{e}}_1 + G_n \hat{\mathbf{e}}_2$. Next, assuming $\mathbf{G} + \bar{\mathbf{G}} = \tilde{\mathbf{G}}$ or $\bar{\mathbf{G}} = \tilde{\mathbf{G}} - \mathbf{G}$, $\bar{G}_p = \left(\tilde{G}_\ell - G_m\right)$ and $\bar{G}_q = \left(\tilde{G}_h - G_n\right)$ and performing the derivative of the stresses, we get

$$\frac{\partial \sigma_{11}}{\partial x_1} = \sum_{\left(\tilde{G}-G\right)} \sum_{G} i^2 \left[C_{11}^{pq} (k_1 + G_m)(k_1 + G_m + \bar{G}_p) A_{mn} \right.$$
$$\left. + C_{12}^{pq} (k_2 + G_n)(k_1 + G_m + \bar{G}_p) B_{mn} \right] e^{i\left(k+\tilde{G}\right)\cdot\mathbf{x}} \cdots \tag{6.9.11}$$

or

$$\frac{\partial \sigma_{11}}{\partial x_1} = \sum_{\left(\tilde{G}-G\right)} \sum_{G} i^2 \left[C_{11}^{pq} (k_1 + G_m)(k_1 + \tilde{G}_\ell) A_{mn} + C_{12}^{pq} (k_2 + G_n)(k_1 + \tilde{G}_\ell) B_{mn} \right] e^{i\left(k+\tilde{G}\right)\cdot\mathbf{x}} \tag{6.9.12}$$

or

$$\frac{\partial \sigma_{11}}{\partial x_1} = \sum_{(\tilde{G}-G)} \sum_{G} i^2 \left[C_{11}^{\tilde{G}-G}\left(k_1 + G_m\right)\left(k_1 + \tilde{G}_\ell\right) A_{mn} + C_{12}^{\tilde{G}-G}\left(k_2 + G_n\right)\left(k_1 + \tilde{G}_\ell\right) B_{mn} \right] e^{i\left(k+\tilde{G}\right)\cdot x}$$

(6.9.13)

Similarly,

$$\frac{\partial \sigma_{22}}{\partial x_2} = \sum_{(\tilde{G}-G)} \sum_{G} i^2 \left[C_{21}^{\tilde{G}-G}\left(k_1 + G_m\right)\left(k_2 + \tilde{G}_h\right) A_{mn} + C_{22}^{\tilde{G}-G}\left(k_2 + G_n\right)\left(k_2 + \tilde{G}_h\right) B_{mn} \right] e^{i\left(k+\tilde{G}\right)\cdot x}$$

(6.9.14)

and

$$\frac{\partial \sigma_{12}}{\partial x_2} = i^2 \sum_{\tilde{G}-G} \sum_{G} C_{66}^{\tilde{G}-G} \left[\left(k_2 + G_n\right)\left(k_2 + \tilde{G}_h\right) A_{mn} + \left(k_1 + G_m\right)\left(k_2 + \tilde{G}_h\right) B_{mn} \right] e^{i\left(k+\tilde{G}\right)\cdot x}$$

(6.9.15)

$$\frac{\partial \sigma_{12}}{\partial x_1} = i^2 \sum_{\tilde{G}-G} \sum_{G} C_{66}^{\tilde{G}-G} \left[\left(k_2 + G_n\right)\left(k_1 + \tilde{G}_\ell\right) A_{mn} + \left(k_1 + G_m\right)\left(k_1 + \tilde{G}_\ell\right) B_{mn} \right] e^{i\left(k+\tilde{G}\right)\cdot x}$$

(6.9.16)

Next plugging into Eqs. 6.6.1 and 6.6.2, we get
$$\frac{\partial \sigma_{11}}{\partial x_1} + \frac{\partial \sigma_{12}}{\partial x_2} = \rho \frac{\partial^2 u_1}{\partial t^2} \text{ equivalent to}$$

$$\sum_{\tilde{G}-G} \sum_{G} \left[\left[C_{11}^{\tilde{G}-G}\left(k_1 + G_m\right)\left(k_1 + \tilde{G}_\ell\right) + C_{66}^{\tilde{G}-G}\left(k_2 + G_n\right)\left(k_2 + \tilde{G}_h\right) \right] A_{mn} \right.$$

$$\left. + \left[C_{12}^{\tilde{G}-G}\left(k_2 + G_n\right)\left(k_1 + \tilde{G}_\ell\right) + C_{66}^{\tilde{G}-G}\left(k_1 + G_m\right)\left(k_2 + \tilde{G}_h\right) \right] B_{mn} \right] e^{i\left(k+\tilde{G}\right)\cdot x}$$

$$= \omega^2 \sum_{\tilde{G}-G} \sum_{G} \rho_{\left(\tilde{G}-G\right)} A_{mn} e^{i\left(k+\tilde{G}\right)\cdot x}$$

(6.9.17)

$$\sum_{\tilde{G}-G} \sum_{G} \left[\left[C_{21}^{\tilde{G}-G}\left(k_1 + G_m\right)\left(k_2 + \tilde{G}_h\right) + C_{66}^{\tilde{G}-G}\left(k_2 + G_n\right)\left(k_1 + \tilde{G}_\ell\right) \right] A_{mn} \right.$$

$$\left. + \left[C_{22}^{\tilde{G}-G}\left(k_2 + G_n\right)\left(k_2 + \tilde{G}_h\right) + C_{66}^{\tilde{G}-G}\left(k_1 + G_m\right)\left(k_1 + \tilde{G}_\ell\right) \right] B_{mn} \right] e^{i\left(k+\tilde{G}\right)\cdot x}$$

$$= \omega^2 \sum_{\tilde{G}-G} \sum_{G} \rho_{\left(\tilde{G}-G\right)} B_{mn} e^{i\left(k+\tilde{G}\right)\cdot x}$$

(6.9.18)

Waves in Periodic Media

Similar approaches presented in the previous section are performed on both the equations above. Multiplying Eqs. 6.9.17 and 6.9.18 by $e^{-i\mathbf{G}\cdot\mathbf{x}} \equiv e^{-i(G_1 x_1 + G_2 x_2)}$ and integrating over a plane with periodicity D_1 and D_2 along x_1 and x_2 direction, would result in the final eigenvalue problem. Please note $e^{i(\tilde{\mathbf{G}}-\mathbf{G})\cdot\mathbf{x}} = e^{i(\bar{\mathbf{G}})\cdot\mathbf{x}}$.

Applying orthogonality condition of the reciprocal lattice vectors the equation will be

$$\frac{1}{D_1 D_2} \iint e^{-i\mathbf{G}\cdot\mathbf{x}} \left[Eq.\,(6.9.17) \right] = 0$$

$$\frac{1}{D_1 D_2} \iint e^{-i\mathbf{G}\cdot\mathbf{x}} \left[Eq.\,(6.9.18) \right] = 0$$

(6.9.19)

The integral will result in matrices in a following form:

$$\begin{bmatrix} S_{11}^{\tilde{\mathbf{G}}-\mathbf{G}} & S_{12}^{\tilde{\mathbf{G}}-\mathbf{G}} \\ S_{21}^{\tilde{\mathbf{G}}-\mathbf{G}} & S_{22}^{\tilde{\mathbf{G}}-\mathbf{G}} \end{bmatrix} \begin{Bmatrix} A_{\mathbf{G}} \\ B_{\mathbf{G}} \end{Bmatrix} = \omega^2 \begin{bmatrix} \rho_{(\tilde{\mathbf{G}}-\mathbf{G})} & \mathbf{0} \\ \mathbf{0} & \rho_{(\tilde{\mathbf{G}}-\mathbf{G})} \end{bmatrix} \begin{Bmatrix} A_{\mathbf{G}} \\ B_{\mathbf{G}} \end{Bmatrix}$$

(6.9.20)

where the components of the elements would be

$$S_{11}^{\tilde{\mathbf{G}}-\mathbf{G}} = \left[C_{11}^{\tilde{\mathbf{G}}-\mathbf{G}} \left(k_1 + G_m\right)\left(k_1 + \tilde{G}_\ell\right) + C_{66}^{\tilde{\mathbf{G}}-\mathbf{G}} \left(k_2 + G_n\right)\left(k_2 + \tilde{G}_h\right) \right]$$ (6.9.21)

$$S_{12}^{\tilde{\mathbf{G}}-\mathbf{G}} = \left[C_{12}^{\tilde{\mathbf{G}}-\mathbf{G}} \left(k_2 + G_n\right)\left(k_1 + \tilde{G}_\ell\right) + C_{66}^{\tilde{\mathbf{G}}-\mathbf{G}} \left(k_1 + G_m\right)\left(k_2 + \tilde{G}_h\right) \right]$$ (6.9.22)

$$S_{21}^{\tilde{\mathbf{G}}-\mathbf{G}} = \left[C_{21}^{\tilde{\mathbf{G}}-\mathbf{G}} \left(k_1 + G_m\right)\left(k_2 + \tilde{G}_h\right) + C_{66}^{\tilde{\mathbf{G}}-\mathbf{G}} \left(k_2 + G_n\right)\left(k_1 + \tilde{G}_\ell\right) \right]$$ (6.9.23)

$$S_{22}^{\tilde{\mathbf{G}}-\mathbf{G}} = \left[C_{22}^{\tilde{\mathbf{G}}-\mathbf{G}} \left(k_2 + G_n\right)\left(k_2 + \tilde{G}_h\right) + C_{66}^{\tilde{\mathbf{G}}-\mathbf{G}} \left(k_1 + G_m\right)\left(k_1 + \tilde{G}_\ell\right) \right]$$ (6.9.24)

Here, to expand the matrix and visualize how to solve the eigenvalue problem, first the number of Bloch wave vectors should be assumed. Let's assume $m = -1, 0, +1$ and $n = -1, 0, +1$. As A_{mn} is a two-indices coefficient, it will generate nine coefficients. Similarly, B_{mn} will result in another nine coefficients. Thus, the size of the matrices in Eq. 6.9.20 would be 18×18 for $m = -1, 0, +1$ and $n = -1, 0, +1$. For any arbitrary Bloch wave number N, the size of the matrices would be $2(2N+1)^2 \times 2(2N+1)^2$.

For an isotropic media $C_{11} = C_{22} = C_{33} = \lambda + 2\mu$, $C_{66} = \mu, C_{12} = (C_{11} - 2C_{66}) = C_{21} = C_{23} = C_{32}$, placing the coefficients at their respective place the stress-strain relation would be

$$\begin{Bmatrix} \sigma_{11} \\ \sigma_{22} \\ \sigma_{33} \\ \sigma_{23} \\ \sigma_{31} \\ \sigma_{12} \end{Bmatrix} = \begin{bmatrix} C_{11} & (C_{11} - 2C_{66}) & (C_{11} - 2C_{66}) & 0 & 0 & 0 \\ (C_{11} - 2C_{66}) & C_{11} & (C_{11} - 2C_{66}) & 0 & 0 & 0 \\ (C_{11} - 2C_{66}) & (C_{11} - 2C_{66}) & C_{11} & 0 & 0 & 0 \\ 0 & 0 & 0 & C_{66} & 0 & 0 \\ 0 & 0 & 0 & 0 & C_{66} & 0 \\ 0 & 0 & 0 & 0 & 0 & C_{66} \end{bmatrix} \begin{Bmatrix} e_{11} \\ e_{22} \\ e_{33} \\ 2e_{23} \\ 2e_{31} \\ 2e_{12} \end{Bmatrix}$$

(6.9.25)

248   Metamaterials in Topological Acoustics

where λ and μ are two implied spatial functions of Lamé constants.

With the specific design of a metamaterial, the Fourier coefficients of the respective material coefficients in Eq. 6.9.19 can be found. Referring to Appendix, the Fourier coefficients for any coefficients C_{ij} and density ρ would be

- For circular phononic crystal in a host matrix with filling fraction $f = \dfrac{\pi r_0^2}{a^2}$ (refer to Fig. A.6.1)
  - C_{ij} constant

$$C_{ij}^{\tilde{G}-G} = C_{ij}^{\bar{G}} = C_{ij}^{pq} = \left(C_{ij}\right)_I f + \left(1-f\right)\left(C_{ij}\right)_{II} \text{ when } \tilde{G} = G \text{ or } \bar{G} = 0 \qquad (6.9.26)$$

$$C_{ij}^{\tilde{G}-G} = C_{ij}^{\bar{G}} = C_{ij}^{pq} = \left(\left(C_{ij}\right)_I - \left(C_{ij}\right)_{II}\right) \frac{2fJ_1\left(G_{pq}r_0\right)}{G_{pq}r_0} \text{ when } \tilde{G} \neq G \text{ or } \bar{G} \neq 0$$
$$(6.9.27)$$

- For density refer to Eq. 6.8.35 and 6.8.36.
- For a square or rectangular phononic crystal in a host matrix with filling fraction $f = \dfrac{b_1 b_2}{a_1 a_2}$ (refer to Fig. A.6.2)
  - For second Lamé constant

$$C_{ij}^{\tilde{G}-G} = C_{ij}^{\bar{G}} = C_{ij}^{pq} = \left(C_{ij}\right)_I f + \left(1-f\right)\left(C_{ij}\right)_{II} \text{ when } \tilde{G} = G \text{ or } \bar{G} = 0$$
$$(6.9.28)$$

$$C_{ij}^{\tilde{G}-G} = C_{ij}^{\bar{G}} = C_{ij}^{pq}$$

$$= \left(\left(C_{ij}\right)_I - \left(C_{ij}\right)_{II}\right) f \left[\frac{\sin\left(G_p \dfrac{b_1}{2}\right)}{G_p\left(\dfrac{b_1}{2}\right)}\right]\left[\frac{\sin\left(G_q \dfrac{b_2}{2}\right)}{G_q\left(\dfrac{b_2}{2}\right)}\right] \text{ when } \tilde{G} \neq G \text{ or } \bar{G} \neq 0$$
$$(6.9.29)$$

- For density refer to Eq. 6.8.39 and 6.8.40.

Substituting the Fourier coefficients for the respective properties of different types of phononic crystals into Eq. 6.9.20, the eigenvalues can be found and the dispersion relation $\omega(\mathbf{k})$ will be understood. However, solving the eigenvalue problem from Eq. 6.9.20 is not simple. It requires some understanding, and an extensive computer code should be written.

6.2.2.4 Computer Code in MATLAB to Find Wave Dispersion

Here a sample MATLAB code is presented for circular phononic crystal. The code has many comments and is self-explanatory.

Waves in Periodic Media

```matlab
%% %%%%%%%%%%%%%%%%%%%%%%%%%%%%%%%%%%%%%%%%%%%%%%%%%%%%%%%%%%%%%%%%%%%%%%%
%  A comprehensive code for wave dispersion  in phononic crystals (PnCs)  %
%  arranged in a period fashion. This code will calculate the 2D band     %
%  diagram using plane-wave expansion method for three dimension          %
%  Coupled u1 and u2 displacement components for X-Y polarized formulation %
%  And decoupled  u3 displacement component  for Z-polarized formulation  %
%  input XYPol=1;ZPol=0 for XY and ZPol=1;XYPol=0 for Z-polarized wave.    %
%  The code is written based on the derivation presented in two articles  %
%  published by our team in 2019 and 2020                                 %
%
%  Ref. in Physical Review B and Journal of Applied Physics.
%
%%%%%%%%%%%%%%%%%%%%%%%%%%%%%%%%%%%%%%%%%%%%%%%%%%%%%%%%%%%%%%%%%%%%%%%%%%%
%% NOTE: There are Amn and Bmn displacement coefficients in XY modes       %
%            but Amn           displacement coefficients only in Z modes   %
%%%%%%%%%%%%%%%%%%%%%%%%%%%%%%%%%%%%%%%%%%%%%%%%%%%%%%%%%%%%%%%%%%%%%%%%%%%
%% CODE WRITTEN by %% Sourav Banerjee
%
%  2017-2022         Mustahseen Indaleeb & Hossain Ahmed
%
%       i-MAPS Laboratory, Mechanical Engineering,
%
%                       University of South Carolina
%
%%%%%%%%%%%%%%%%%%%%%%%%%%%%%%%%%%%%%%%%%%%%%%%%%%%%%%%%%%%%%%%%%%%%%%%%%%%
clear;
clc;
XYPol=0;
ZPol=1;
%% Lattice constant and dimensions of phononic crystals
lat_const=.0254;                   % 1 inch = 0.0254 m
b=.0086868;                        % 0.342 inch = 0.0086868 m
                                   % b=dimension of the
                                   square PnCS.
                                   % b=diameter of Circular
                                   PnCs
fill_fraction = (pi*(b/2)^2)/(lat_const)^2; % Filling Fraction
```

```
%% Number of Block wave vector to be included in the analysis
bloch_size=1;                          % iteration
Num_mn=(2*bloch_size+1);
NumCoeff=(Num_mn)^2;
theta=0;           % direction of wave propagation
Num_kinterX=40;  % number of intervals within 0 - pi/D  or k1
Num_kinterY=Num_kinterX;  % number of intervals within 0
- pi/D  or k2
Num_kinterXY=50;  % number of intervals along diagonal within
0 - pi/D - ML

% Max normalized k1 or k2 will be 0.5

del_k1=0.5/Num_kinterX;
del_k2=0.5/Num_kinterY;
del_k12=0.5/Num_kinterXY;

NumPassbands=9; % identify how many pass bands are required <
Num_Coeff

%% Phononic crystal (A material) density and longitudinal
sound of speed
% rho_A = 1760; % kg/m^3
rho_A = 2790; % kr/m^3
% vel_A = 1284; % m/sec
%% Phononic crystal (A material) elastic modulus and Poisson's
ratio
% E_A = 2.9e9;     % Pa
% nu_A = 0.4;
% %Calculation of constitituve matrix
% C11_A = (E_A/(1-nu_A^2));
% C12_A = (E_A*nu_A)/(1-nu_A^2);
% C21_A = (E_A*nu_A)/(1-nu_A^2);
% C22_A = E_A/(1-nu_A^2);
% C66_A = E_A/(2*(1+nu_A));
%
% C_A = [C11_A C12_A 0; C21_A C22_A 0; 0 0 C66_A];
% %  C11 C12  0
% %  C21 C22  0
% %   0   0   C66
% %Stiffness matrix of inclusion ends here
%% Phononic crystal (A material) constitituve matrix directly
given
C66_A=27.333;        % mu
C11_A=111.439;
C12_A=C11_A-2*C66_A; % Lambda-2*mu
C22_A=C11_A;
C21_A=C12_A;

C_A = [C11_A C12_A 0; C21_A C22_A 0; 0 0 C66_A];
%   C11 C12  0
```

Waves in Periodic Media

```
%   C21  C22   0
%    0    0   C66
%Stiffness matrix of inclusion ends here
%% Matrix (B Material) density and longitudinal sound of
   speed
% rho_B=1.2929;    % density of the base medium
rho_B=1389;        % Density of the base medium
% vel_B=343;       % wave speed in the base medium
% %% Phononic crystal (B material) elastic modulus and
Poisson's ratio
% E_A = 2.9e9;     % Pa
% nu_air = 0.2854; % Poisson's ratio of base material
%% Phononic crystal (B material) constitituve matrix directly
given
C66_B=2.067;
C11_B=8.891;
C12_B=C11_B-2*C66_B;
C22_B=C11_B;
C21_B=C12_B;

C_B = [C11_B C12_B 0; C21_B C22_B 0; 0 0 C66_B];
%   C11  C12   0
%   C21  C22   0
%    0    0   C66
%Stiffness matrix of inclusion ends here
%% Reduced stiffness and density

ratio_lam=C11_A/C11_B;
ratio_mu=C66_A/C66_B;
ratio_den=rho_A/rho_B;
% see derivation
del_lam=(ratio_lam-1)/
(fill_fraction*ratio_lam+1-fill_fraction);
del_mu=(ratio_mu-1)/(fill_fraction*ratio_mu+1-fill_fraction);
del_rho=(ratio_den-1)/
(fill_fraction*ratio_den+1-fill_fraction);

C66_avg=fill_fraction*C66_A+(1-fill_fraction)*C66_B;   %
Cavg=CA*f+CB*(1-f)
C11_avg=fill_fraction*C11_A+(1-fill_fraction)*C11_B;   %
Cavg=CA*f+CB*(1-f)
C12_avg=fill_fraction*C12_A+(1-fill_fraction)*C12_B;   %
Cavg=CA*f+CB*(1-f)
C11_norm=C11_avg/C66_avg;
C22_norm=C11_norm;
C12_norm=C12_avg/C66_avg;

rho_avg=fill_fraction*rho_A+(1-fill_fraction)*rho_B;
                                        % Required when G
                                          not eq Gbar
del_C11norm=(C11_A-C11_B)/C66_avg;      % (CA-CB)/C66
```

```
del_C12norm=(C12_A-C12_B)/C66_avg;                  % (CA-CB)/C66
del_C66norm=del_mu;

ShearVel_avg=sqrt(C66_avg/rho_avg);                 % To multiply
                                                      norm. frequency


%% Bloch wave vector G = G1*e1+G2*e2
% method for C, C++, Fortran, Python
k=1;
for i =1:NumCoeff
    for j =1:NumCoeff
        G1(k)=i;
        G2(k)=j;
        k=k+1;
    end
end
% alternate Method Matlab, Python
mn=(-bloch_size:bloch_size);  % create a vector [-1 0 +1] if
bloch_size=1;
% Next create a dummy vector to create G2
dummyVec=(-floor((bloch_size*2+1)^2/2):floor((bloch_
size*2+1)^2/2));
G1=repmat(mn,1,bloch_size*2+1);% [ -1  0 +1 -1 0 +1 -1  0
+1];
G2=round(dummyVec/(bloch_size*2+1)); % [ -1 -1 -1  0 0  0 +1
+1 +1];


% At the end we will get G1 and G2 vectors
% G1 for x direction, G2 for y direction

%% CALCULATION START

%% Loop for the wave vectors k1 and k2 based on direction
%   There are three directions that we consider
%   L-X X-M M-L  pls Note L is inverted gamma.

Omega=cell(3,1);     % preallocate frequecy matrix for 3 directions
% [S]{Amn;Bmn}=w^2[M]{Amn;Bmn}
% [inv[M]*[S]-w^2[I]]*{Amn;Bmn}=0
% Create Identity matrix
iden=ones(1,length(2*NumCoeff));
I=diag(iden);
%*************************************************************************
% LOOP 1 : FOR DIRECTION
for direction=1:3       % for three different dierctions
%*************************************************************************
    if (direction==1)
    K1=del_k1:del_k1:0.5; % a vector of normalized wave
numbers along X L-X
```

Waves in Periodic Media

253

```
    K2=zeros(1,length(K1));
    Kplot=K1;                  % A vector for plotting
    elseif (direction==2)
    K2=del_k2:del_k2:0.5; % a vector of normalized wave
numbers along X L-X
    K1=0.5*ones(1,length(K2));
    Kplot=K2;
    elseif (direction==3)
    % a vector of normalized wave numbers along X L-X
    K12=del_k12:del_k12:(sqrt(0.5^2+0.5^2));
    K1=K12.*cosd(45);
    K2=K12.*sind(45);
    Kplot=K1;
    end
    Num_K=length(K1);

% LOOP 2 : for wave vector
    % preallocate freq. matrix for all wave numbers
    OmegaK=zeros(Num_K,NumPassbands);
    for k=1:Num_K
    %%%%%%%%%%%%%%%%%%%%%%%%%%%%%%%%%%%%%%%%%%%%%%%%%%%%%%%%%%%%%%%%%

    % Condition imposed for XY-polarized solution or
      Z-polarized solution
    %+++++++++++++++++++++++++++++++++++++++++++++++++++++++++++++
XY-polarized
    if (XYPol==1 && ZPol==0)
            % Define stiffness matrix S and mass matrix M
            S=size(2*NumCoeff,2*NumCoeff);  % prepolulate S
            M=size(2*NumCoeff,2*NumCoeff);  % prepolulate M
            % LOOP 3 & 4 : to get G and G bar See derivation
            % Create S and M matrixes
                for i=1:NumCoeff
                  for j=1:NumCoeff
            iC=(i-1)*2; % This is because we have Amn and
Bmn coefficients
            jC=(j-1)*2; % This is because we have Amn and
Bmn coefficients
                            % Thus each G we will result 2x2
                              matrix
                            % Please note in derivation
                              there are 2 eq. 2 coeffs.
                      if (i==j) % This is for (Gbar-G) see
                              derivation
                        S(iC+1,jC+1)=(C11_norm*((K1(k)+G1(i))^2)
                        )+(K2(k)+G2(i))^2; % C11_norm=C11_avg/
                        C66_avg
                        S(iC+1,jC+2)=(C12_norm+1)*(K2(k)+G2(i))*
(K1(k)+G1(i));     % C12_norm=C12_avg/C66_avg
                        S(iC+2,jC+1)=(C12_norm+1)*(K1(k)+G1(i))*
(K2(k)+G2(i));     % C12_norm=C12_avg/C66_avg
```

```matlab
                    S(iC+2,jC+2)=(C22_norm*((K2(k)+G2(i))^2)
)+(K1(k)+G1(i))^2; % C22_norm=C22_avg/C66_avg
                    M(iC+1,jC+1)=1.0; % normalized density
                    M(iC+2,jC+2)=1.0;
                    M(iC+1,jC+2)=0.0;
                    M(iC+2,jC+1)=0.0;

                else
                    Gr_mag=sqrt(4*pi*fill_fraction*((G1(i)
                    -G1(j))^2+(G2(i)-
                                G2(j))^2)); % to get
                                F(Gbar-G)
                    % G1=2*pi*m/a ; G2=2*pi*n/a
                    % G=sqrt((G1^2+G2^2));
                    % G*r=sqrt(r^2*(G1^2+G2^2)); f=pi*r^/
                    a^2
                    % Gr_mag^2=4*pi*f*(G1^2+G2^2)

                    %Getting the Fourier Coefficient
                     F(Gbar-G)
                    F=2*fill_fraction*(1/
                    Gr_mag)*besselj(1,Gr_mag);

                    % bloch phases in 4 unique - see
                      derivation
                    % (k1+G1m)*(k1+G1o)  m and o for
                      direction 1
                    phase1=(K1(k)+G1(i))*(K1(k)+G1(j));
                    % (k2+G2n)*(k2+G2l)  n and l for
                      direction 2
                    phase2=(K2(k)+G2(i))*(K2(k)+G2(j));
                    % (k1+G1m)*(k2+G2l)  m and l for
                      direction 1 and 2
                    phase3=(K1(k)+G1(i))*(K2(k)+G2(j));
                    % (k2+G2n)*(k1+G1o)  n and o for
                      direction 2 and 1
                    phase4=(K2(k)+G2(i))*(K1(k)+G1(j));

                    S(iC+1,jC+1)=F*(del_C11norm*phase1+
                    del_C66norm*phase2);
                    S(iC+1,jC+2)=F*(del_C12norm*phase3+
                    del_C66norm*phase4);
                    S(iC+2,jC+1)=F*(del_C12norm*phase4+
                    del_C66norm*phase3);
                    S(iC+2,jC+2)=F*(del_C11norm*phase2+
                    del_C66norm*phase1);

                    M(iC+1,jC+1)=F*del_rho;
                    M(iC+2,jC+2)=F*del_rho;
                    M(iC+1,jC+2)=0.0;
                    M(iC+2,jC+1)=0.0;
```

Waves in Periodic Media   255

```
        end   % If loop ends
      end   % G_bar loop ends
    end   % G loop ends

  % Now For a specific dircetion and for a specific
    wave number
  % The S and M matrices are ready

  % Next eigenvalue problem
  %%% calculate eigenvalues and frequency
    A=S*(M^-1);              % This is   Mgg * inv(Ngg)
    Kushawa et.al
    EigenVal=eig(A);

    NonZeroOmegaSqnorm=find(EigenVal~=0);
    OmegaSqnorm=sort(EigenVal(NonZeroOmegaSqnorm));
    % Factor is Cs^2*(2pi/a)^2
    OmegaSq=OmegaSqnorm*((ShearVel_avg*2*pi/
    lat_const)^2);


    % Square root of omegaSq are the frequency omega
    % Store frequencies for all wave number
    OmegaK(k,:)=[real(sqrt(OmegaSq(1:NumPassbands)))];

    %+++++++++++++++++++++++++++++++++++++++++++++++++++
Z-Polarized
    elseif (XYPol==0 && ZPol==1)
        % Define stiffness matrix S and mass matrix M
        S=size(NumCoeff,NumCoeff);  % prepolulate S
        M=size(NumCoeff,NumCoeff);  % prepolulate M
        % LOOP 3 & 4 : To Get G And G Bar - See
derivation
        % Create S and M matrixes
            for i=1:NumCoeff
              for j=1:NumCoeff

                if (i==j) % This is for (Gbar-G) see
                derivation
                  S(i,j)=(K1(k)+G1(i))^2+(K2(k)+G
                  2(i))^2;
                  % C66_norm=C66_avg/C66_avg = 1 % unlike
                    line 198
                  M(i,j)=1.0; % normalized density
                else
    Gr_mag=sqrt(4*pi*fill_fraction*((G1(i)-G1(j))^2+(G2(i)-
                            G2(j))^2)); % to get
                            F(Gbar-G)
                  % G1=2*pi*m/a ; G2=2*pi*n/a
                  % G=sqrt((G1^2+G2^2));
```

Metamaterials in Topological Acoustics

```matlab
        % G*r=sqrt(r^2*(G1^2+G2^2)); f=pi*r^/
          a^2
        % Gr_mag^2=4*pi*f*(G1^2+G2^2)

        %Getting the Fourier Coefficient
          F(Gbar-G)
        F=2*fill_fraction*(1/
        Gr_mag)*besselj(1,Gr_mag);

        % bloch phases in 4 unique - see
          derivation
        % (k1+G1m)*(k1+G1o)   m and o for
          direction 1
        phase1=(K1(k)+G1(i))*(K1(k)+G1(j));
        % (k2+G2n)*(k2+G2l)   n and l for
          direction 2
        phase2=(K2(k)+G2(i))*(K2(k)+G2(j));

        S(i,j)=F*(del_C66norm*phase1+
        del_C66norm*phase2);
        M(i,j)=F*del_rho;


      end  % If loop ends
     end   % G_bar loop ends
    end    % G loop ends

  % Now For a specific dircetion and for a specific
    wave number
  % The S and M matrices are ready

  % Next eigenvalue problem
  %%% calculate eigenvalues and frequency
   A=S*(M^-1);              % This is   Mgg * inv(Ngg)
                                        Kushawa et.al
   EigenVal=eig(A);

   NonZeroOmegaSqnorm=find(EigenVal~=0);
   OmegaSqnorm=sort(EigenVal(NonZeroOmegaSqnorm));
   % factor is Cs^2*(2pi/a)^2
   OmegaSq=OmegaSqnorm*((ShearVel_avg*2*pi/
   lat_const)^2);


   % Square root of omegaSq are the frequency
     omega
   OmegaK(k,:)=[real(sqrt(OmegaSq(1:NumPassbands)))];
   % Store frequencies for all wave number

end % End if Loop that Triggered XY Polarization or Z
Polarization
```

Waves in Periodic Media   257

```
%%%%%%%%%%%%%%%%%%%%%%%%%%%%%%%%%%%%%%%%%%%%%%%%%%%%%%%%%%%%%%%%%%%%%%%%%%
end              % Wavenumber loop ends S and M will reset to
Empty Matrix

    Omega[7]=OmegaK;      % Storing the Freqiency Matrix in a
cell


    %% Plotting the band diagram - for each direction

    figure=figure(1);

    % Direction 1
    if direction==1 % Ploting the eigenvalues for L-X
    direction
     axis1 = subplot(1,3,1);
     for pb=1:NumPassbands
       GammaXX=plot(Kplot,Omega[7](:,pb),'o','MarkerFaceColor',
       'k');hold on
     end
     position1=get(axis1,'Position');
     set(axis1,'YLim',[0,max(Omega{direction}
     (:,NumPassbands))]);
     set(axis1,'XLim',[0,0.5]);
     set(axis1,'nextplot','add');
     set(axis1,'xtick',[]);
     set(axis1,'xticklabel',{});
     ylabel('Frequency (kHz)');
     yaxislim=max(Omega{direction}(:,NumPassbands));

    % Direction 2
    elseif direction==2 % Ploting the eigenvalues for X-M
    direction

     axis2= subplot(1,3,2);
     for pb=1:NumPassbands
       XM2=plot(Kplot,Omega{direction}(:,pb),'sq','MarkerFaceC
       olor','r'); hold on
     end
     position2=get(axis2,'Position');
     set(axis2,'nextplot','add');
     position2(1)=position1(1)+position1(3);%% Width
     position2(4)=position1(4); %%
     set(axis2,'position',position2);
     set(axis2,'YLim',[0,yaxislim]);
     set(axis2,'XLim',[0,.5]);
     set(axis2,'ytick',[]);
     set(axis2,'xticklabel',{});

    % Direction 3
    else %% Ploting the eigenvalues for M-L direction
```

258   Metamaterials in Topological Acoustics

```
axis3=  subplot(1,3,3);
for pb = 1:NumPassbands
   MGamma=plot(flip(Kplot),Omega{direction}(:,pb),'o','Mar
kerFaceColor','b'); hold on
   end
set(axis3,'nextplot','add');
position3=get(axis3,'Position');
position3(1)=position2(1)+position2(3);
position3(4)=position1(4);
set(axis3,'position',position3);
set(axis3,'YLim',[0,yaxislim]);
set(axis3,'XLim',[0,.5]);


%% Labelling - reduced wave vectors
text(-1,-.05*max(OmegaK(:,NumPassbands)),'\Gamma');
text(-.5,-.05*max(OmegaK(:,NumPassbands)),'X');
text(0,-.05*max(OmegaK(:,NumPassbands)),'M');
text(.5,-.05*max(OmegaK(:,NumPassbands)),'\Gamma');
text(-.3,-.1*max(OmegaK(:,NumPassbands)),'Reduced Wave
Vector','HorizontalAlignment','center');
   text(-.3,1.075*max(OmegaK(:,NumPassbands)),'Dispersion
Relation ','HorizontalAlignment','Center');

   end
end                   % Direction loop ends

%% PROGRAM END
```

A few notes on the code: Material *I* is material *a*; Material *II* is material *b*.
Additionally,

$$C_{66avg} = C_{66}^I f + (1-f) C_{66}^{II} \tag{6.10.1}$$

$$C_{11avg} = C_{11}^I f + (1-f) C_{11}^{II} \tag{6.10.2}$$

$$C_{ij\ norm} = \frac{C_{ijavg}}{C_{66avg}} \tag{6.10.3}$$

$$\Delta C_{ij\ norm} = \frac{C_{ij}^I - C_{ij}^{II}}{C_{66avg}} \tag{6.10.4}$$

Average shear wave velocity: $c_s = \sqrt{\dfrac{C_{66avg}}{\rho_{avg}}}$ $\qquad$ (6.10.5)

6.2.2.5 Dispersion Curves

After running the code, one could find the dispersion curves for x_3 polarized wave
and x_1, x_2 polarized waves as shown in Fig. 6.5a and 6.5b, respectively. In this case,
an inch-by-inch unit call of air matrix (Material *II* in derivation and Material *b* in

Waves in Periodic Media   259

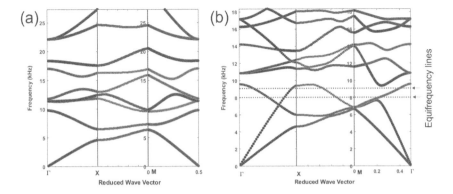

FIGURE 6.5 Dispersion behavior in a metamaterial made of circular phononic crystals in air media: a) in a vertically polarized plane wave (x_3 or Z polarized), b) in plane polarized plane wave (X-Y polarized wave).

the code above) is considered where a 0.342-inch diameter phononic crystal made of polyvinyl chloride (PVC) (Material *I* in derivation and Material *a* in the code above) is placed inside the unit cell. Then the unit cell is repeated in both directions as depicted in Fig. 6.1a.

6.3 BLOCH WAVE VECTORS FOR OTHER LATTICE STRUCTURES

In the above derivations and computer code, square arrangement of phononic crystals are considered. However, mathematical derivation is not restricted to the square arrangement. A unit cell can also be called a lattice of the metamaterial. There are four common metamaterial structures that are conceptualized for their various applications. Commonly they have the following lattice structures (refer to Chapter 1 and Fig. 6.6a-d):

- Square lattice (Fig. 6.6a)
- Hexagonal lattice or graphene lattice (Fig. 6.6b)
- Kagomé lattice (Fig. 6.6c)
- Triangular lattice (Fig. 6.6d)

To avoid complexity, a standard notation *a* is used to describe the length of each side of the lattices. For a square lattice the periodicity $D_1 = D_2 = a$ is depicted in Appendix. Please refer to the derivation above where two-unit vectors are used throughout the derivation. They are $\hat{e}_1$ and $\hat{e}_2$, respectively. For a square lattice, they are the unit vector along the x_1 and x_2 direction respectively as shown in Fig. 6.6a. However, in hexagonal graphene-type lattice, they may not be oriented along the x_1 and x_2 directions. Let's introduce two new unit vectors $\hat{i}$ and $\hat{j}$ along x_1 and x_2 directions, respectively. They will help to orient the $\hat{e}_1$ and $\hat{e}_2$ unit vectors at any possible direction in the lattice. The $\hat{e}_1$ and $\hat{e}_2$ directions are shown in Fig. 6.6 for the respective lattices. Let's call $\hat{e}_1$ and $\hat{e}_2$ the lattice basis vectors or direct lattice vectors. Table 6.1 shows the lattice basis vectors and their corresponding reciprocal Bloch basis vectors $\mathbf{G} = \mathbf{G}_1\hat{e}_1 + \mathbf{G}_2\hat{e}_2$ for different lattice structures. *a* is the length of

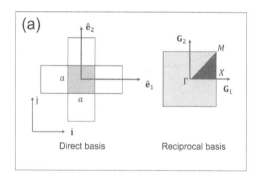

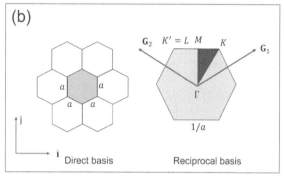

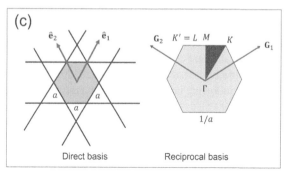

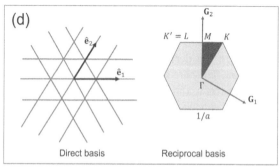

FIGURE 6.6 Representation of direct basis vectors and respective reciprocal basis vector: a) square lattice, b) hexagonal graphene type lattice, c) Kagomé lattice, d) triangular lattice.

Waves in Periodic Media 261

TABLE 6.1

Translation Vectors for Four Basic Lattices

Topology of the Lattice	Basis Vectors	Reciprocal Basis Vectors
Square lattice	$\hat{\mathbf{e}}_1 = a\hat{\mathbf{i}}; \hat{\mathbf{e}}_2 = a\hat{\mathbf{j}}$	$\mathbf{G}_m = \dfrac{2\pi m}{a}\hat{\mathbf{i}};\ \mathbf{G}_n = \dfrac{2\pi n}{a}\hat{\mathbf{j}}$
Hexagonal graphene-type lattice	$\hat{\mathbf{e}}_1 = \sqrt{3}a\left(\dfrac{1}{2}\hat{\mathbf{i}} + \dfrac{\sqrt{3}}{2}\hat{\mathbf{j}}\right)$	$\mathbf{G}_m = \dfrac{2\pi m}{\sqrt{3}a}\left(\hat{\mathbf{i}} + \dfrac{1}{\sqrt{3}}\hat{\mathbf{j}}\right)$
	$\hat{\mathbf{e}}_2 = \sqrt{3}a\left(-\dfrac{1}{2}\hat{\mathbf{i}} + \dfrac{\sqrt{3}}{2}\hat{\mathbf{j}}\right)$	$\mathbf{G}_n = \dfrac{2\pi n}{\sqrt{3}a}\left(-\hat{\mathbf{i}} + \dfrac{1}{\sqrt{3}}\hat{\mathbf{j}}\right)$
Kagomé lattice	$\hat{\mathbf{e}}_1 = 2a\left(\dfrac{1}{2}\hat{\mathbf{i}} + \dfrac{\sqrt{3}}{2}\hat{\mathbf{j}}\right)$	$\mathbf{G}_m = \dfrac{\pi m}{a}\left(\hat{\mathbf{i}} + \dfrac{1}{\sqrt{3}}\hat{\mathbf{j}}\right)$
	$\hat{\mathbf{e}}_2 = 2a\left(-\dfrac{1}{2}\hat{\mathbf{i}} + \dfrac{\sqrt{3}}{2}\hat{\mathbf{j}}\right)$	$\mathbf{G}_n = \dfrac{\pi n}{a}\left(-\hat{\mathbf{i}} + \dfrac{1}{\sqrt{3}}\hat{\mathbf{j}}\right)$
Triangular lattice	$\hat{\mathbf{e}}_1 = a\hat{\mathbf{i}}$	$\mathbf{G}_m = \dfrac{2\pi m}{a}\left(\hat{\mathbf{i}} - \dfrac{1}{\sqrt{3}}\hat{\mathbf{j}}\right)$
	$\hat{\mathbf{e}}_2 = a\left(\dfrac{1}{2}\hat{\mathbf{i}} + \dfrac{\sqrt{3}}{2}\hat{\mathbf{j}}\right)$	$\mathbf{G}_n = \dfrac{2\pi n}{a}\left(\dfrac{2}{\sqrt{3}}\hat{\mathbf{j}}\right)$

each side of the lattices. Please note that the reciprocal basis vectors $\mathbf{G}_1$ and $\mathbf{G}_2$, or equivalently the Bloch basic vectors $\mathbf{G}_m$ and $\mathbf{G}_n$, are indicated as vectors, not scalar. This does not change the understanding of the derivation.

The irreducible first Brillouin zone points of the four lattices are presented in Table 6.2.

Substituting appropriate Bloch waves for respective lattices in a metamaterial, the eigenvalue problem can be constructed and the eigenvalue solution will generate dispersion curves for the respective metamaterials.

6.3.1 GENERALIZED BLOCH WAVE SOLUTION

Metamaterials are created with multiple material constituents. It is not necessary that only two materials are used to construct a metamaterial. The constituent's geometry could be arbitrary, as is shown in Fig. 6.1. The derivation using plane-wave expansion method in section 6.2 is presented for a general material, but the material property matrix is created only for regular geometry like circular or square inclusion in the metamaterials. When inclusion geometry is arbitrary, then it would be difficult to implement. Hence, the following two subsections present alternative methods for finding the Bloch wave solutions in metamaterials.

6.3.1.1 Fast-Plane Wave Expansion Method

Governing equations from the fourth and remaining case – case d) – is considered here. However, the method will be equally applicable for the other cases. In

TABLE 6.2

Irreducible First Brillouin Zone

Topology of the Lattice	Cartesian Basis	Reciprocal Basis
Square lattice	$\Gamma = (0,0)$	$\Gamma = (0,0)$
	$X = \dfrac{1}{a}\left(\dfrac{1}{2},0\right)$	$X = \left(\dfrac{1}{2},0\right)$
	$M = \dfrac{1}{a}\left(\dfrac{1}{2},\dfrac{1}{2}\right)$	$M = \left(\dfrac{1}{2},\dfrac{1}{2}\right)$
Hexagonal graphene-type lattice	$\Gamma = (0,0)$	$\Gamma = (0,0)$
	$K = \dfrac{1}{a}\left(\dfrac{1}{3\sqrt{3}},\dfrac{1}{3}\right)$	$K = \left(\dfrac{2}{3},\dfrac{1}{3}\right)$
	$M = \dfrac{1}{a}\left(0,\dfrac{1}{3}\right)$	$M = \left(\dfrac{1}{2},\dfrac{1}{2}\right)$
Kagomé lattice	$\Gamma = (0,0)$	$\Gamma = (0,0)$
	$K = \dfrac{1}{a}\left(\dfrac{1}{3\sqrt{3}},\dfrac{1}{3}\right)$	$K = \left(\dfrac{2}{3},\dfrac{1}{3}\right)$
	$M = \dfrac{1}{a}\left(0,\dfrac{1}{3}\right)$	$M = \left(\dfrac{1}{2},\dfrac{1}{2}\right)$
Triangular lattice	$\Gamma = (0,0)$	$\Gamma = (0,0)$
	$K = \dfrac{1}{a}\left(\dfrac{1}{3},\dfrac{1}{\sqrt{3}}\right)$	$K = \left(\dfrac{1}{3},\dfrac{2}{3}\right)$
	$M = \dfrac{1}{a}\left(0,\dfrac{1}{\sqrt{3}}\right)$	$M = \left(0,\dfrac{1}{2}\right)$

this method, the Fourier series expansion of material constants at arbitrary spatial locations is obtained to quantify the spatially varying metamaterial properties. In Eq. A.6.2 in Appendix, $\alpha(\mathbf{x})$ is used to represent the material parameter as a function of spatial location. Let's assume that it represents the two Lamé constants and the density as it is used in section 6.2. The Fourier series expansion of material constant $\alpha(\mathbf{x})$ is also derived in Eq. A.6.29.4 for a square unit cell with a square or rectangular constituent material (Material *II* inside Material *I*). In the next step to discretize a geometry of any arbitrary shape, the unit cell is divided in to $N \times N$ square pixels, as shown in Fig. 6.7 [8]. Assuming that P_L is a set of pixel elements filled with Material I inside a host matrix made of Material II in a unit cell, the Fourier coefficients of the material constant $\alpha_0\left(\overline{\mathbf{G}}\right)$ at the center pixel ($P_0 \in P_L$) are defined as

$$\alpha_0\left(\overline{\mathbf{G}}\right) = \alpha_{pq}^0 = \alpha_I f + \left(1-f\right)\alpha_{II} \ \text{ when } \tilde{\mathbf{G}} = \mathbf{G} \text{ or } \overline{\mathbf{G}} = \mathbf{0} \qquad (6.11.1)$$

$$\alpha_0\left(\overline{\mathbf{G}}\right) = \alpha_{pq}^0 = \left(\alpha_I - \alpha_{II}\right)F\left(\overline{\mathbf{G}}\right) \ \text{ when } \tilde{\mathbf{G}} \neq \mathbf{G} \text{ or } \overline{\mathbf{G}} \neq \mathbf{0} \qquad (6.11.2)$$

Waves in Periodic Media

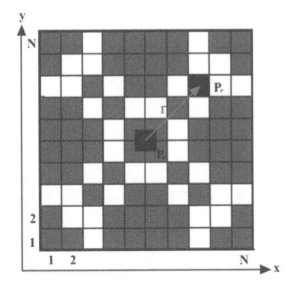

FIGURE 6.7 Discretization of a unit cell with any constituent arbitrary geometry into multiple square inclusions (white pixels are the host matrix, Material II and inclusion Material I are in gray). This figure is reprinted from Ref [8] with permission.

For a square pixel at the center, the structural factor $F(\bar{G})$ is

$$F(\bar{G}) = f \left[\frac{\sin\left(G_p \frac{b}{2}\right)}{G_p \left(\frac{b}{2}\right)} \right] \left[\frac{\sin\left(G_q \frac{b}{2}\right)}{G_q \left(\frac{b}{2}\right)} \right] \qquad (6.11.3)$$

where b is the dimension of the square pixel at the center. If the unit cell of dimension a is divided into N pixels, then $b = \frac{a}{N}$ and the structural factor will read

$$F(\bar{G}) = f \left[\frac{\sin\left(G_p \frac{a}{2N}\right)}{G_p \left(\frac{a}{2N}\right)} \right] \left[\frac{\sin\left(G_q \frac{a}{2N}\right)}{G_q \left(\frac{a}{2N}\right)} \right] = f \operatorname{sinc}\left(G_p \frac{a}{2N}\right) \operatorname{sinc}\left(G_q \frac{a}{2N}\right) \qquad (6.11.4)$$

Please note that in mathematics sinc $x = \sin x/x$. Now, according to the location of the respective pixel (at a distance $\mathbf{r}$ from the origin) in the unit cell, the Fourier coefficients of any pixel ($P_r \in P_L$) filled with Material *II* can be calculated as

$$\alpha_r(\bar{G}) = \alpha_0(\bar{G}) e^{i\bar{G}\cdot\mathbf{r}} \qquad (6.11.5)$$

Superimposing the effects of all pixels filled with Material *II* would read

$$\alpha\left(\bar{\mathbf{G}}\right)= \sum_{P_r \in P_L} \alpha_r\left(\bar{\mathbf{G}}\right) = \sum_r \alpha_r\left(\bar{\mathbf{G}}\right)\delta(r) = \alpha_0\left(\bar{\mathbf{G}}\right)\sum_r e^{i\bar{\mathbf{G}}\cdot\mathbf{r}}\delta(\mathbf{r}) \qquad (6.11.6)$$

where $\delta(\mathbf{r})$ is defined as [8].

$$\delta(\mathbf{r})=\left[\delta(r_1),\ \delta(r_2),\ \cdots,\ \delta(r_k),\ \cdots,\ \delta(r_L)\right]=\left\{\begin{array}{ll} 1 & P_r \in P_0 \\ 0 & other \end{array}\right\} \qquad (6.11.7)$$

where r_k represents the center location of the k-th pixel and L is the total number of pixels in the unit cell. Assuming $e\left(\bar{\mathbf{G}}\right)=e^{i\bar{\mathbf{G}}\cdot\mathbf{r}}$, a vector made of all the exponentials of the dot product of the Bloch wave vector and the location of the pixels (r_k) the Fourier coefficients could be written as

$$\alpha\left(\bar{\mathbf{G}}\right)= \alpha_0\left(\bar{\mathbf{G}}\right)\ e\left(\bar{\mathbf{G}}\right)\cdot\delta(\mathbf{r}) \qquad (6.11.8)$$

Bloch theorem for the unknown functions like displacement would read

$$u_1 = \sum_G u_1(\mathbf{G})e^{i\mathbf{k}\cdot x}e^{-i\omega t} = \sum_{G_m}\sum_{G_n}A_{mn}e^{i(k_1 x_1+k_2 x_2)}e^{i(G_m x_1+G_n x_2)} \qquad (6.11.9)$$

$$u_2 = \sum_G u_2(\mathbf{G})e^{i\mathbf{k}\cdot x}e^{-i\omega t} = \sum_{G_m}\sum_{G_n}B_{mn}e^{i(k_1 x_1+k_2 x_2)}e^{i(G_m x_1+G_n x_2)} \qquad (6.11.10)$$

$$u_2 = \sum_G u_2(\mathbf{G})e^{i\mathbf{k}\cdot x}e^{-i\omega t} = \sum_{G_m}\sum_{G_n}C_{mn}e^{i(k_1 x_1+k_2 x_2)}e^{i(G_m x_1+G_n x_2)} \qquad (6.11.11)$$

Substituting the displacement functions and the Fourier coefficients for the material constants, as is demonstrated in section 6.2.2.2 and 6.2.2.3, skipping few derivation steps, the governing equations for case d) would turn into an eigenvalue problem and would read

$$-\omega^2\sum_G \rho\left(\bar{\mathbf{G}}\right)e^{i\bar{\mathbf{G}}\cdot\mathbf{r}}e^{-i\omega t}\sum_{G_0}e^{i(G_0+k)\cdot\mathbf{r}}\mathbf{u}_k^{x_1}(\mathbf{G_0})=\left(\left(2\sum_{\bar{G}}\mu\left(\bar{\mathbf{G}}\right)e^{i\bar{\mathbf{G}}\cdot\mathbf{r}}+\sum_{\bar{G}}\lambda\left(\bar{\mathbf{G}}\right)e^{i\bar{\mathbf{G}}\cdot\mathbf{r}}\right)\right.$$

$$\left(e^{-i\omega t}\sum_{G_0}e^{i(G_0+k)\cdot\mathbf{r}}\mathbf{u}_k^{x_1}(\mathbf{G_0})\right)_{x_1} + \sum_{\bar{G}}\lambda\left(\bar{\mathbf{G}}\right)e^{i\bar{\mathbf{G}}\cdot\mathbf{r}}\left(e^{-i\omega t}\sum_{G_0}e^{i(G_0+k)\cdot\mathbf{r}}\mathbf{u}_k^{x_2}(\mathbf{G_0})\right)_{x_2}\right)_{x_1}$$

$$+\left(\sum_{\bar{G}}\mu\left(\bar{\mathbf{G}}\right)e^{i\bar{\mathbf{G}}\cdot\mathbf{r}}\left(\left(e^{-i\omega t}\sum_{G_0}e^{i(G_0+k)\cdot\mathbf{r}}\mathbf{u}_k^{x_1}(\mathbf{G_0})\right)_{x_2}+\left(e^{-i\omega t}\sum_{G_0}e^{i(G_0+k)\cdot\mathbf{r}}\mathbf{u}_k^{x_2}(\mathbf{G_0})\right)_{x_1}\right)\right)_{x_2}\right)$$

$$\qquad (6.11.12)$$

Waves in Periodic Media

$$-\omega^2 \sum_{\bar{G}} \rho(\bar{G}) e^{i\bar{G}\cdot r} e^{-i\omega t} \sum_{G_0} e^{i(G_0+k)\cdot r} \mathbf{u}_k^{x_2}(G_0)$$

$$= \left(\left(2\sum_{\bar{G}} \mu(\bar{G}) e^{i\bar{G}\cdot r} + \sum_{\bar{G}} \lambda(\bar{G}) e^{i\bar{G}\cdot r}\right)\left(e^{-i\omega t} \sum_{G_0} e^{i(G_0+k)\cdot r} \mathbf{u}_k^{x_2}(G_0)\right)\right)_{x_2}$$

$$+ \sum_{\bar{G}} \lambda(\bar{G}) e^{i\bar{G}\cdot r}\left(e^{-i\omega t} \sum_{G_0} e^{i(G_0+k)\cdot r} \mathbf{u}_k^{x_2}(G_0)\right)_{x_1}\Bigg)_{x_2} + \left(\sum_{\bar{G}} \mu(\bar{G}) e^{i\bar{G}\cdot r}\right.$$

$$\left.\left(\left(e^{-i\omega t} \sum_{G_0} e^{i(G_0+k)\cdot r} \mathbf{u}_k^{x_2}(G_0)\right)_{x_1} + \left(e^{-i\omega t} \sum_{G_0} e^{i(G_0+k)\cdot r} \mathbf{u}_k^{x_1}(G_0)\right)_{x_2}\right)\right)_{x_1}$$

$$\tag{6.11.13}$$

$$-\omega^2 \sum_{\bar{G}} \rho(\bar{G}) e^{i\bar{G}\cdot r} e^{-i\omega t} \sum_{G_0} e^{i(G_0+k)\cdot r} \mathbf{u}_k^{x_3}(G_0)$$

$$= \left(\sum_{\bar{G}} \mu(\bar{G}) e^{i\bar{G}\cdot r}\left(e^{-i\omega t} \sum_{G_0} e^{i(G_0+k)\cdot r} \mathbf{u}_k^{x_3}(G_0)\right)_{x_1}\right)_{x_1}$$

$$\tag{6.11.14}$$

$$+ \left(\sum_{\bar{G}} \mu(\bar{G}) e^{i\bar{G}\cdot r}\left(e^{-i\omega t} \sum_{G_0} e^{i(G_0+k)\cdot r} \mathbf{u}_k^{x_3}(G_0)\right)_{x_2}\right)_{x_2}$$

In Eqs. 6.11.12, 6.11.13, and 6.11.14, the jump discontinuities are present due to the discretized pixels. It is mentioned in reference [8] that the α (material parameters ρ, λ and μ) and $\dfrac{\partial u_i}{\partial x_j}$ are piecewise continuous but interstitially smooth and bounded periodic functions, in which the concurrent jump discontinuities (or the jump discontinuity points) exist at the boundaries of different materials. However, the dot product function, $h_n = \alpha \cdot \dfrac{\partial u_i}{\partial x_j}$, is continuous along the material boundaries.

By eliminating the contact discontinuity points while solving for h_n, a continuous function can be obtained across these boundaries. This is due to the Laurent inverse theorem [9]. The Fourier coefficients for h_k will read

$$h_n = \sum_{m=-\infty}^{m=+\infty} [f]_{n-m}^{-1} \alpha_m \tag{6.11.15}$$

where α_m is the Fourier coefficient of α. The Fourier series expansion of the function $h(\mathbf{x})$ can then be written as

$$h(x) = \sum_{n=-\infty}^{n=+\infty} h_n e^{inx} = \sum_{n=-\infty}^{n=+\infty}\sum_{m=-\infty}^{m=+\infty} [f]_{n-m}^{-1} \alpha_m e^{inx} \tag{6.11.16}$$

For practical applications and without loss of generality, Eq. 6.11.16 can be written as

$$h_n^{(M)} = \sum_{m=-M}^{m=M} [f]_{n-m}^{-1} g_m \tag{6.11.17}$$

$$h_n^{(M)} = \sum_{n=-M}^{n=M} h_n^{(M)} e^{inx} \tag{6.11.18}$$

Substituting Eq. 6.11.8 and Eq. 6.11.1 into Eqs. 6.11.12 through 6.11.14, the eigenvalue problem will be

$$-\omega^2 \sum_{\tilde{G}}\sum_{G_0} \left([\rho_0]_{\tilde{G}-G_0} [e]_{\tilde{G}-G_0} \cdot \delta(r) \right)^{-1} \mathbf{u}_k^{x_1}(G_0) e^{i\tilde{G}\cdot r} e^{i(G_0+k)\cdot r}$$

$$= \left(\left(\sum_{\tilde{G}}\sum_{G_0} \left(2[\mu_0]_{\tilde{G}-G_0} + [\lambda_0]_{\tilde{G}-G_0} \right) [e]_{\tilde{G}-G_0} \cdot \delta(r) \right)^{-1} \mathbf{u}_k^{x_1}(G_0) e^{i\tilde{G}\cdot r} e^{i(G_0+k)\cdot r} \right)_{x_1}$$

$$+ \left(\sum_{\tilde{G}}\sum_{G_0} \left([\lambda_0]_{\tilde{G}-G_0} [e]_{\tilde{G}-G_0} \cdot \delta(r) \right)^{-1} \mathbf{u}_k^{x_2}(G_0) e^{i\tilde{G}\cdot r} e^{i(G_0+k)\cdot r} \right)_{x_2} \Bigg)_{x_1}$$

$$+ \left(\left(\sum_{\tilde{G}}\sum_{G_0} \left([\mu_0]_{\tilde{G}-G_0} [e]_{\tilde{G}-G_0} \cdot \delta(r) \right)^{-1} \mathbf{u}_k^{x_1}(G_0) e^{i\tilde{G}\cdot r} e^{i(G_0+k)\cdot r} \right)_{x_2} \right.$$

$$+ \left. \left(\sum_{\tilde{G}}\sum_{G_0} \left([\mu_0]_{\tilde{G}-G_0} [e]_{\tilde{G}-G_0} \cdot \delta(r) \right)^{-1} \mathbf{u}_k^{x_2}(G_0) e^{i\tilde{G}\cdot r} e^{i(G_0+k)\cdot r} \right)_{x_1} \right)_{x_2}$$

$$\tag{6.11.19}$$

Waves in Periodic Media

$$-\omega^2 \sum_{\tilde{G}} \sum_{G_0} \left([\rho_0]_{\tilde{G}-G_0} [e]_{\tilde{G}-G_0} \cdot \delta(r) \right)^{-1} \mathbf{u}_k^{x_2}(\mathbf{G_0}) e^{i\tilde{G}\cdot r} e^{i(G_0+k)\cdot r}$$

$$= \left(\left(\sum_{\tilde{G}} \sum_{G_0} \left(\left(2[\mu_0]_{\tilde{G}-G_0} + [\lambda_0]_{\tilde{G}-G_0} \right) [e]_{\tilde{G}-G_0} \cdot \delta(r) \right)^{-1} \right. \right.$$

$$\left. \mathbf{u}_k^{x_2}(\mathbf{G_0}) e^{i\tilde{G}\cdot r} e^{i(G_0+k)\cdot r} \right)_{x_2}$$

$$+ \left(\sum_{\tilde{G}} \sum_{G_0} \left([\lambda_0]_{\tilde{G}-G_0} [e]_{\tilde{G}-G_0} \cdot \delta(r) \right)^{-1} \mathbf{u}_k^{x_1}(\mathbf{G_0}) e^{i\tilde{G}\cdot r} e^{i(G_0+k)\cdot r} \right)_{x_1} \right)_{x_2}$$

$$+ \left(\left(\sum_{\tilde{G}} \sum_{G_0} \left([\mu_0]_{\tilde{G}-G_0} [e]_{\tilde{G}-G_0} \cdot \delta(r) \right)^{-1} \mathbf{u}_k^{x_2}(\mathbf{G_0}) e^{i\tilde{G}\cdot r} e^{i(G_0+k)\cdot r} \right)_{x_1} \right)$$

$$+ \left(\sum_{\tilde{G}} \sum_{G_0} \left([\mu_0]_{\tilde{G}-G_0} [e]_{\tilde{G}-G_0} \cdot \delta(r) \right)^{-1} \mathbf{u}_k^{x_1}(\mathbf{G_0}) e^{i\tilde{G}\cdot r} e^{i(G_0+k)\cdot r} \right)_{x_2} \right)_{x_1}$$

$$(6.11.20)$$

$$-\omega^2 \sum_{\tilde{G}} \sum_{G_0} \left([\rho_0]_{\tilde{G}-G_0} [e]_{\tilde{G}-G_0} \cdot \delta(r) \right)^{-1} \mathbf{u}_k^{x_3}(\mathbf{G_0}) e^{i\tilde{G}\cdot r} e^{i(G_0+k)\cdot r}$$

$$= \left(\left(\sum_{\tilde{G}} \sum_{G_0} \left([\mu_0]_{\tilde{G}-G_0} [e]_{\tilde{G}-G_0} \cdot \delta(\mathbf{r}) \right)^{-1} \mathbf{u}_k^{x_3}(\mathbf{G_0}) e^{i\tilde{G}\cdot r} e^{i(G_0+k)\cdot r} \right)_{x_1} \right)_{x_1}$$

$$+ \left(\left(\sum_{\tilde{G}} \sum_{G_0} \left([\mu_0]_{\tilde{G}-G_0} [e]_{\tilde{G}-G_0} \cdot \delta(\mathbf{r}) \right)^{-1} \mathbf{u}_k^{x_3}(\mathbf{G_0}) e^{i\tilde{G}\cdot r} e^{i(G_0+k)\cdot r} \right)_{x_2} \right)_{x_2}$$

$$(6.11.21)$$

Further the eigenvalue problem with this method of improved fast plane wave expansion method would read

$$\omega^2 \sum_{\mathbf{G_0}} \left[\frac{1}{\rho_0\left(\tilde{\mathbf{G}}-\mathbf{G_0}\right) e\left(\tilde{\mathbf{G}}-\mathbf{G_0}\right)} \cdot \frac{1}{\delta(r)} \right]^{-1} u_{\mathbf{k+G}}^{x_1}$$

$$= \sum_{\mathbf{G_0}} \left[\left[\frac{1}{\lambda_0\left(\tilde{\mathbf{G}}-\mathbf{G_0}\right) e\left(\tilde{\mathbf{G}}-\mathbf{G_0}\right)} \cdot \frac{1}{\delta(r)} \right]^{-1} (k+\mathbf{G_0})_{x_1} \left(k+\tilde{\mathbf{G}}\right)_{x_1} \right.$$

$$+ \left[\frac{1}{\mu_0\left(\tilde{\mathbf{G}}-\mathbf{G_0}\right) e\left(\tilde{\mathbf{G}}-\mathbf{G_0}\right)} \cdot \frac{1}{\delta(r)} \right]^{-1} \qquad (6.11.22)$$

$$\left. \left((k+\mathbf{G_0})_{x_2} \left(k+\tilde{\mathbf{G}}\right)_{x_2} + 2(k+\mathbf{G_0})_{x_1} \left(k+\tilde{\mathbf{G}}\right)_{x_1} \right) \right] u_{\mathbf{k+G}}^{x_1}$$

$$+ \sum_{\mathbf{G_0}} \left[\left[\frac{1}{\lambda_0\left(\tilde{\mathbf{G}}-\mathbf{G_0}\right) e\left(\tilde{\mathbf{G}}-\mathbf{G_0}\right)} \cdot \frac{1}{\delta(r)} \right]^{-1} (k+\mathbf{G_0})_{x_2} \left(k+\tilde{\mathbf{G}}\right)_{x_1} \right.$$

$$+ \left. \left[\frac{1}{\mu_0\left(\tilde{\mathbf{G}}-\mathbf{G_0}\right) e\left(\tilde{\mathbf{G}}-\mathbf{G_0}\right)} \cdot \frac{1}{\delta(r)} \right]^{-1} (k+\mathbf{G_0})_{x_1} \left(k+\tilde{\mathbf{G}}\right)_{x_2} \right] u_{\mathbf{k+G}}^{x_2}$$

$$\omega^2 \sum_{\mathbf{G_0}} \left[\frac{1}{\rho_0\left(\tilde{\mathbf{G}}-\mathbf{G_0}\right) e\left(\tilde{\mathbf{G}}-\mathbf{G_0}\right)} \cdot \frac{1}{\delta(r)} \right]^{-1} u_{\mathbf{k+G}}^{x_2}$$

$$= \sum_{\mathbf{G_0}} \left[\left[\frac{1}{\lambda_0\left(\tilde{\mathbf{G}}-\mathbf{G_0}\right) e\left(\tilde{\mathbf{G}}-\mathbf{G_0}\right)} \cdot \frac{1}{\delta(r)} \right]^{-1} (k+\mathbf{G_0})_{x_2} \left(k+\tilde{\mathbf{G}}\right)_{x_2} \right.$$

$$+ \left[\frac{1}{\mu_0\left(\tilde{\mathbf{G}}-\mathbf{G_0}\right) e\left(\tilde{\mathbf{G}}-\mathbf{G_0}\right)} \cdot \frac{1}{\delta(r)} \right]^{-1} \qquad (6.11.23)$$

$$\left. \left((k+\mathbf{G_0})_{x_1} \left(k+\tilde{\mathbf{G}}\right)_{x_1} + 2(k+\mathbf{G_0})_{x_2} \left(k+\tilde{\mathbf{G}}\right)_{x_2} \right) \right] u_{\mathbf{k+G}}^{x_2}$$

$$+ \sum_{\mathbf{G_0}} \left[\left[\frac{1}{\lambda_0\left(\tilde{\mathbf{G}}-\mathbf{G_0}\right) e\left(\tilde{\mathbf{G}}-\mathbf{G_0}\right)} \cdot \frac{1}{\delta(r)} \right]^{-1} (k+\mathbf{G_0})_{x_1} \left(k+\tilde{\mathbf{G}}\right)_{x_2} \right.$$

$$+ \left. \left[\frac{1}{\mu_0\left(\tilde{\mathbf{G}}-\mathbf{G_0}\right) e\left(\tilde{\mathbf{G}}-\mathbf{G_0}\right)} \cdot \frac{1}{\delta(r)} \right]^{-1} (k+\mathbf{G_0})_{x_2} \left(k+\tilde{\mathbf{G}}\right)_{x_1} \right] u_{\mathbf{k+G}}^{x_1}$$

Waves in Periodic Media

$$\omega^2 \sum_{\mathbf{G}_0} \left[\frac{1}{\rho_0 (\tilde{\mathbf{G}} - \mathbf{G}_0) e(\tilde{\mathbf{G}} - \mathbf{G}_0)} \cdot \frac{1}{\delta(r)} \right]^{-1} u_k^{x_3}(\mathbf{G}_0)$$

$$= \sum_{\mathbf{G}_0} \left[\frac{1}{\mu_0 (\tilde{\mathbf{G}} - \mathbf{G}_0) e(\tilde{\mathbf{G}} - \mathbf{G}_0)} \cdot \frac{1}{\delta(r)} \right]^{-1} (k + \mathbf{G}_0)_{x_1} (k + \tilde{\mathbf{G}})_{x_1} u_k^{x_3}(\mathbf{G}_0)$$

$$+ \sum_{\mathbf{G}_0} \left[\frac{1}{\mu_0 (\tilde{\mathbf{G}} - \mathbf{G}_0) e(\tilde{\mathbf{G}} - \mathbf{G}_0)} \cdot \frac{1}{\delta(r)} \right]^{-1} (k + \mathbf{G}_0)_{x_2} (k + \tilde{\mathbf{G}})_{x_2} u_k^{x_3}(\mathbf{G}_0)$$

(6.11.24)

Solving eigenvalues from Eqs. 6.11.22 through 6.11.24, Fig. 6.8 shows a representative solution. The solution presents a dispersion behavior considering an arbitrary constituent geometry discretized into multiple square pixels as fill material in a host matrix. In this case, lead square pixels are assumed in an epoxy matrix. Knowing the constituents and geometry discretization will help find the dispersion behavior. Alternatively, this problem could be set as an inverse problem where the metamaterial geometry is not known but the desired dispersion behavior is known. Interestingly, the above formulation could be easily used for topology optimization of the pixels for any specific desired band structure or features. Different optimizers including a genetic algorithm could be used for this optimization. Knowing the required band structure, specific constituent geometry could be obtained that will result in a desired dispersion behavior. In this chapter many dispersion behaviors are presented that are of interest for topological acoustics. Hence, knowing what feature is required in a dispersion curve, the metamaterial geometry could be optimized using the above formulations. This is an open field of research, and many different metamaterial constructs could be created, driven by specific topological applications. Extensive discussion on genetic algorithm-based optimization of metamaterials is omitted herein and can be found elsewhere.

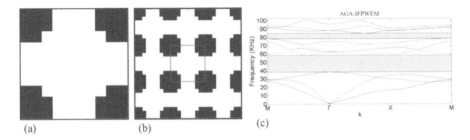

FIGURE 6.8 a) Example of a discretized unit cell with a patterned fill of Material I (lead in black) inside Material II (epoxy in white), b) repeating the unit cell, the metamaterial is constructed, c) dispersion behavior obtained from the eigenvalue solution. This figure is reprinted from Ref [8] with permission.

6.3.1.2 Finite Element Simulation Method

As long as the geometries of the constituent materials are regular, they can be divided into a reasonable number of pixels. If two material systems are used, the above in sections 6.2 and 6.3.1.1 could be readily applicable. However, there are infinite possibilities of making metamaterials with various geometries and multiple material types. This makes the problem very complex. The finite element method with a reduced-order Bloch wave solution [10, 11] could help alleviate this problem. For example, the metamaterial presented in Figs. 6.1b and 6.2 are not easy to implement with the above formulations. An FEM solution of the dispersion curve set as an eigenvalue problem could be helpful. Any finite element software could be used for solving the Bloch wave modes and presenting the dispersion curve as depicted in this chapter. However, the process is not automatic. Thus, a basic level understanding of Bloch wave vectors, wave numbers, their directions and discretization and frequency solutions must be skillfully used and understood. The dispersion curves presented in the following sections of this chapter are obtained from FEM-based COMSOL Multiphysics simulation [12].

6.4 FEATURES OF WAVE DISPERSION ($\omega - k$)

In quantum mechanics and solid-state physics, energy wave number relation or $E - k$ plot ($E(\mathbf{k})$) governs the dispersion. Similarly in acoustics, wave dispersion is governed by the frequency wave number or $\omega - k$ ($\omega(\mathbf{k})$) plot. Earlier, in Chapter 5, some dispersion behaviors are discussed. It is highly recommended to first read sections 5.3 through 5.5 before going through the following sections. Now is the best time to understand the dispersion behavior with much deeper understanding and correlate the understanding in acoustics. Interestingly, the understanding of the dispersion behaviors from $E(\mathbf{k})$ or $\omega(\mathbf{k})$ plots are not different. These understandings are universal. In this section many, subsections are created on different features but point to the previous discussion in Chapter 5. It is expected to correlate the understanding in acoustics simultaneously.

Please refer to Fig. 6.9 for the following discussion. There are many wave modes that describe wave dispersion in the media. Wave dispersion of a wave mode means

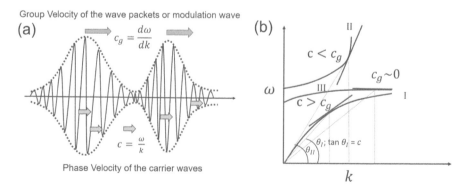

FIGURE 6.9 a) Depiction of group velocity and phase velocity, b) hypothetical dispersion curve to show different cases of dispersion.

Waves in Periodic Media

that if frequency changes, the wave number (magnitude of the wave vector) changes. That also means that if the wave number changes, the frequency also changes for that specific wave mode. The phase velocity of a wave mode is written as $c = \omega/k$, and the group velocity is equal to $c_g = d\omega/dk$. Hence, wave dispersion in a medium also means that the wave velocities of different wave modes in the medium change as the frequency of excitation changes.

Substituting the expression for angular frequency in terms of phase velocity into the expression for the group velocity, we get

$$c_g = c + k\frac{dc}{dk} \qquad (6.12.1)$$

In general, a system is said to be dispersive when its phase velocity depends on the angular frequency or on the wavelength. Thus, in a metamaterial, the angular frequencies and the wave numbers follow a dispersion relation. Figures 6.9a and 6.9b describe a few hypothetical dispersion relations through $\omega(\mathbf{k})$ plot. Three (I, II, III) different types of wave modes or dispersion curves are described.

- *Convex dispersion*: $c > c_g$ (i.e., $\dfrac{dc}{dk} < 0$): It will appear that the wave originates at the back of the wave packet (Fig. 6.9a) and travels toward the front of the wave packet. The shape of the packet along the length of the travel direction of the wave changes continuously. This situation is called the positive dispersion. Convex curvature of a dispersion curve is a result of this situation. For example, curve I in Fig. 6.9b shows that the dispersion curve has lower slope than the black dotted line at angle θ_I.

- *Zero Dispersion*: $c = c_g$ (i.e., $\dfrac{dc}{dk} = 0$): When the phase velocity is equal to the group velocity, the carrier wave will appear to be stagnant. The wave packet retains its shape and is called a zero-dispersion situation. Geometrically, this situation occurs when the slope of the dispersion curve or the frequency spectrum matches the slope of the triangles generated by the dotted lines in Fig. 6.9b.

- *Local Resonance with zero group velocity*: A flat dispersion curve (curve III in Fig. 6.9b) results in a unique situation that is also nondispersive. This is because the group velocity of the wave packet (Fig. 6.7a) becomes zero. It means that the wave packet does not move. This situation occurs when the phase velocity is directly proportional to the dc/dk. This signifies that at any specific frequency, multiple wave numbers can be present. Geometrically, this situation occurs when a straight-line dispersion curve is present due to local resonance of an embedded resonator in a metamaterial.

- *Concave dispersion*: $c < c_g$ (i.e., $\dfrac{dc}{dk} > 0$): When phase velocity is less than the group velocity, the carrier waves will fall back and will not be able to keep up with the wave packets, which move faster. This situation is called negative dispersion. Concave curvature of a dispersion curve generally indicates this situation as described by curve II in Fig. 6.9b. Here, the slope of the tangent is higher than the tan of the angle θ_{II}.

6.4.1 PHONONS

The dispersive wave modes in a dispersion relation ($\omega - k$ ($\omega(\mathbf{k})$)) plot are equivalently called phonons, or bands. Just as photons in quantum mechanics quantize the light energy, phonons are the energy quantum of lattice vibrations and the main carriers of the heat energy. In acoustic metamaterials, the wave modes carry the energy and thus they are synonymously called phonons. They play an essential role in acoustic energy propagation, energy localization, heat conduction, thermal barrier coating, heat–electricity energy conversion, and superconductivity. Photons in quantum mechanics are described as spin-free bosons (refer to Chapter 5). Electrons are described as fermions with spin (1/2) particles. Intuitively, phonons in acoustics do not have the spin degrees of freedom. Hence, phonons in metamaterials could be synonymously considered as spin-free boson particles. Despite the differences, photons, electrons, and phonons in periodic media are solved using similar mathematical techniques with the Bloch wave function expansion as described in this chapter and in Chapter 5. Hence, the topological descriptions using the geometric phase, such as Zack phase, Berry phase, Berry connection, Berry curvature, and topological invariant or topological charge like Chern number, which are defined (see Chapter 5) on the basis of Bloch wave functions, are equally applicable for phonon propagation in acoustic metamaterials.

6.4.2 EQUIFREQUENCY CONTOURS

As the terminology indicates, an equifrequency contour is a plot of all the wave vectors at a specific frequency of a specific wave mode. Such plots at different frequencies, when plotted together, will give and equifrequency contour map.

Referring to Fig. 6.5b, it is understood that for a specific wave mode, frequency (ω) is the function of the wave number (k). Two horizontal lines are identified in the figure, and they are the equifrequency lines at 8 kHz and 9 kHz, respectively. Along those two lines, a specific wave mode has different wave numbers along different directions (e.g., $\Gamma - X$, $X - M$, and $M - \Gamma$). Although only three directions are presented in Fig. 6.5b, a mode can be traced along every possible direction in 360° from the Γ point. This will give wave numbers in all possible directions in 360° for the specific wave mode at a specific frequency. A polar plot generated is called the equifrequency contour. Superposing multiple equifrequency contours at different frequencies will give an equifrequency contour map in 2D or a topological frequency map of the wave modes in 3D. Many hidden features can be understood by exploring such equifrequency contours. Figure 6.10 shows the dispersion behavior for the metamaterial presented in Fig. 6.1b [1]. Mode 9 and mode 10 were arbitrarily selected to demonstrate the concept of equifrequency contours. Figure 6.11a shows the equifrequency contours on a 2D plane, with frequencies identified on the scale between 21 kHz – 22 kHz. Fig. 6.11b alternatively shows the frequency topology map of mode 9 and mode 10 in 3D for the split ring metamaterial. Essentially, many horizontal slices of this 3D frequency map between a selected frequency band (e.g., 21 kHz – 22 kHz herein) construct the equifrequency contours. This way, after finding the dispersion behavior in a specific metamaterial, the equifrequency contours and the frequency topology map could be constructed for any number of modes that

Waves in Periodic Media

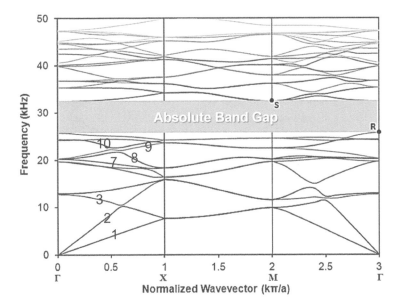

FIGURE 6.10 Dispersion behavior in a metamaterial shown in Fig. 6.1b.

are identified to be important between a selected frequency band. Interestingly in this split ring metamaterial, an absolute band gap was found between 24 kHz and 33 kHz. In this book however, the discussions on the band gaps are not covered as there are many other books and articles [13–16] that discuss this behavior.

6.4.3 Band Degeneracies

Degeneracies in dispersion plots ($\omega(\mathbf{k})$) are identified when two wave modes are merged or crossed together at any specific frequency. Degeneracies could be

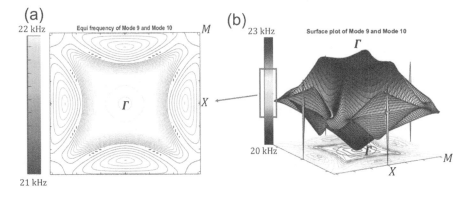

FIGURE 6.11 a) Equifrequency contours between frequency 21 kHz – 22 kHz, b) frequency topology map in 3D for modes 9 and 10.

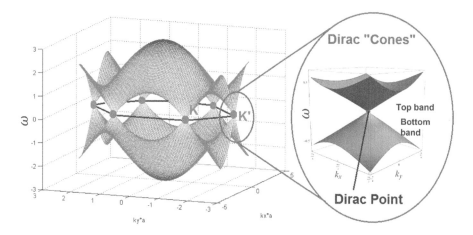

FIGURE 6.12 a) Equifrequency contour map with frequency topology in a hexagonal graphene type metamaterial, b) formation of Dirac cone at the Brillouin boundary where top band and the bottom band touch each other.

identified at different wave number values at different wave vector directions. They could be identified at the Γ or X or K or at the M points. Interestingly, Fig. 6.11b shows that, despite being a square lattice, mode 9 and mode 10 are degenerated entirely along the first Brillouin zone boundary. They do not, however, degenerate at the Γ point. Let's see other examples of band degeneracies. If a graphene-type hexagonal lattice is used in a metamaterial, it is almost inevitable that at least two bands will degenerate at the K and K' points.

Figure 6.12 shows the band degeneracies at the Brillouin zone boundary but only at the points K and K'. An additional case for band degeneracies is presented in Fig. 6.13. In this case the bands are degenerated at the Γ point. Square polyvinyl chloride (PVC) rods are embedded in an air host matrix (Fig. 6.13a, b, c) and the dispersion plot (Fig. 6.13d) is generated [17, 18]. Two possible degeneracies were identified at DRA and DRB as shown in the figure, but only DRB is explored here. It was found that three bands are degenerated at the Γ point at DRB. Unlike Fig. 6.12, where only two bands called top band and the bottom band degenerated, in Fig. 6.13 three bands, namely top band, bottom band, and a deaf band (see section 6.4.3) are degenerated. The degeneracies at the K and K' points are almost certain with the lattice geometry; however, the degeneracies at Γ point are not certain and can be found only through parameter optimization. Thus, the degeneracies at the K point are called firm degeneracies and degeneracies at the Γ point are called the accidental degeneracies.

The degeneracies at the Γ point can be lifted or shifted via parameter tuning. For example, in this case, the dimension of the square PVC crystal or the rotation of the crystal could shift or lift these degeneracies. That being said, it is obvious that as long as the lattice geometry (hexagonal graphene-type lattice) is consistent, the firm degeneracies could not be lifted. However, this statement is not entirely true. The degeneracy could be lifted at the boundary of space inversion symmetry (to understand space inversion symmetry please refer to section 5.7 in Chapter 5) of

Waves in Periodic Media

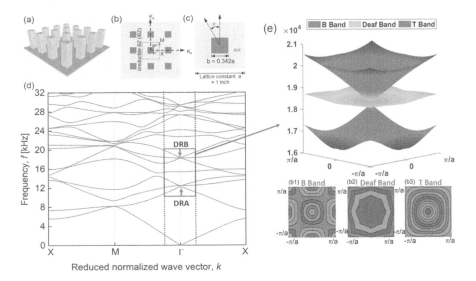

FIGURE 6.13 a) Schematic representation of the two-dimensional periodic systems of hard square rods in a square lattice. The filling fraction is 0.1169, and the square array has the same orientation as the lattice. b) Transverse cross-section of 2D square lattice in a 3 × 3 supercell with nine squares of $b = 0.342a$ (where a = lattice constant) hosted in air media, depicting the irreducible Brillouin zone (BZ) with wave-vector directions (ΓX-XM-MΓ). c) Schematic view of a single phononic crystal inside irreducible BZ composed of PVC, where the two constituents are denoted by white and blue as air and PVC rods, respectively. d) Acoustic band structure of PVC square rods with filling fractions $f = 0.1169$ embedded in an air host. The rotational angle is $\Theta = 0°$. The colored blocked regions around the Γ point denote probable Dirac region A (bottom) and Dirac region B (top). e) Equifrequency contour map with frequency topology at the Γ point with three degenerated bands.

a metamaterial in 2D. This is shown when the topological wave is discussed with the quantum valley Hall effect (QVHE). Additionally, this degeneracy could also be lifted in 3D metamaterials at the out-of-plane wave vector direction (i.e., along the k_3) and are called Weyl points [19].

With the foregoing discussion, it is necessary to mention that beyond the above cases of degeneracies at Γ point (at Brillouin zone center) or at the K and K' points (at Brillouin boundaries) or at high symmetry lines, degeneracy can happen anywhere in the Brillouin zone having any **k** wave vector. In a periodic system without structural symmetry, such degeneracies anywhere in the Brillouin zone are mostly double degeneracies and may result in a Dirac cone. These degeneracies are also called accidental degeneracies because they can be achieved by closing a band gap or tuning a wave mode at a specific wave vector **k**. Some form of engineered tuning [20] or optimization techniques [21–24] help achieve such degeneracies [25].

6.4.4 Deaf Bands

Group wave velocity of a wave mode is written as $d\omega/dk$. As long as a dispersion curve shows a change in frequency with respect to the wave numbers, the wave mode

276   Metamaterials in Topological Acoustics

will exhibit a group velocity. The group velocity could be positive or negative, when frequency increases or decreases with increasing wave numbers, respectively. But when $d\omega$ is very small, close to zero, compared to dk, the group velocity will be close to zero (i.e., $c_g \approx 0$). In this situation, the acoustic wave cannot pass through the metamaterial. Locally at a specific narrow band frequency, the wave transmission drops significantly compared to other frequencies. This phenomenon is not like the total band gap with wider frequency range (see Fig. 6.8 and Fig. 6.10). Here the physics of non-transmissibility of sound is different compared to the physics of the band gaps and local resonances. Due to the most flat band and non-transmissibility of acoustic waves, this band is called a 'deaf band'. An example of a deaf band is shown in Fig. 6.13e, where it is trapped between a top band and a bottom band. Typically, such a locally trapped flat band is called a deaf band and occurs near the Γ point. Later it is shown that this situation occurs due to an antisymmetric phonon.

6.4.5 DIRAC CONES AT K POINT

Dirac cones at the Dirac points are the band degeneracies. At a Dirac point two bands merge together and demonstrate almost linear dispersion. It may appear that the modes have crossed each other; however, the upper concave mode merged with the lower convex mode with linear dispersion behavior. These are non-accidental degeneracies and depend only on the overall geometry of the unit cell. They also do not depend on the geometric dimension of orientation of the constituent material. If two bulk media with similar hexagonal lattice is placed side by side but the $\mathcal{P}$ –symmetry or the space inversion symmetry is broken at a boundary, then at the boundary the degeneracy may be lifted. Such a situation is discussed under the quantum valley Hall effect (QVHE). Please refer section 5.5.1 to pictographically understand Dirac cones. An example of a Dirac cone at the K or at K' point is shown in Fig. 6.12. We can call the Dirac cone degeneracies at the K and K' point symmetry-protected degeneracies, and it can be found that they carry a Berry phase of π. Integrating Berry curvature further gives modal topological charge, and they may result in a nonzero Chern number. Please refer to section 5.6.4 in Chapter 5 to understand this statement better. Mathematically a Dirac cone can be expressed with an effective Hamiltonian with linear dispersion. In Chapter 5 it was discussed that to describe a Dirac cone dispersion, it is not necessary to have a complete solution of the Schrödinger equation; rather a locally applicable (locally around a band), effective Hamiltonian would suffice. For acoustic media, a similar approach would work, using the **k.p** perturbation method discussed in Chapter 5. At a Dirac cone there are two linear bands, and thus two eigenvalues are necessary. Two eigenvalues are the group wave velocities of those two modes. Hence, the effective Hamiltonian at a Dirac point can be written as a 2×2 Hamiltonian as follows

$$H(\mathbf{k}) = \omega_D \left(k_1 \sigma_{x1} + \xi k_2 \sigma_{x2} \right) \tag{6.12.2}$$

$$\sigma_{x1} = \begin{bmatrix} 0 & 1 \\ 1 & 0 \end{bmatrix} \sigma_{x2} = \begin{bmatrix} 0 & -i \\ i & 0 \end{bmatrix} \sigma_{x3} = \begin{bmatrix} 1 & 0 \\ 0 & -1 \end{bmatrix} \tag{6.12.3}$$

Waves in Periodic Media 277

where $\xi = \pm 1$ for K or K' point respectively. σ_{x_1}, σ_{x_2} and σ_{x_3} are the Pauli's matrix described in Chapter 5 Eq. 5.24.16.

6.4.6 Dirac-Like Cones

Please refer to section 5.5.2 to pictographically understand the Dirac-like cones at the Γ point. Fig. 6.13e is an example of a Dirac-like cone at the Γ point. Dirac-like cones can be Hermitian or non-Hermitian. Hermitian Dirac cones are further classified into two categories. Hermitian Dirac cones with pseudo spin 1 and pseudo spin ½. They individually create interesting wave phenomena and have unique topological effects discussed in Chapter 7. At a Dirac-like cone three bands degenerate, and one of those is a deaf band. Due to the deaf band, the degeneracy at the Γ point is called triple degeneracy. Here, a top and bottom band form a Dirac cone, and a deaf band is trapped in between. Wave behavior at such point is not quite like a Dirac cone, and thus it is called a Dirac-like cone. The top band and the bottom have a linear dispersion and can be described by an effective Hamiltonian in Eq. 6.12.2, but the presence of a deaf band prohibits them from doing so. As there are three bands at a Dirac-like cone, the effective Hamiltonian should thus be a 3×3 matrix and can be written in such a way that one of the three eigenvalues for the deaf band at the Dirac frequency represents a constant value. This necessity gives the effective Hamiltonian from the **k.p** perturbation approach as [25]

$$H(\mathbf{k}) = \begin{pmatrix} \omega_D & c_g\sqrt{k_1^2 + k_2^2} & 0 \\ c_g\sqrt{k_1^2 + k_2^2} & \omega_D & 0 \\ 0 & 0 & \omega_D \end{pmatrix} \qquad (6.12.4)$$

and the eigenvalue could be written as

$$\omega_1 = \omega_D + c_g\sqrt{k_1^2 + k_2^2}; \; \omega_2 = \omega_D - c_g\sqrt{k_1^2 + k_2^2}; \; \omega_3 = \omega_D \qquad (6.12.5)$$

Near the Dirac-point frequency ω_D, the first two frequency eigenvalues represent linear dispersion (in a conical shape). The group wave velocity c_g indicates the slope of the cone with linear dispersion at an arbitary magnitude of wave vector $|\mathbf{k}| = \sqrt{k_1^2 + k_2^2}$. However, the third eigenvalue, $\omega_3 = \omega_D$, is for the deaf band. As the Hamiltonian in Eq. 6.12.4 is not like the Dirac Hamiltonian in Eq. 6.12.2, this typical dispersion does not give any Berry phase or topological charge. Thus, the Chern number is also zero. There are no phononic Dirac monopoles, neither sink nor source. Please refer to section 5.6.4 in Chapter 5 to understand this statement better. The linear dispersions with triple degeneracy at the Γ point is induced by accidental degeneracy instead of structural symmetry. Superposition of different modal symmetries and anti-symmetries govern the existence of a Dirac-like cone [26, 27]. A Dirac-like cone at the Γ point may result in zero effective refractive index in the Dirac frequency. Some zero-index materials are realized exploiting the physics of Dirac-like cone [28].

6.4.7 Weyl Point

Weyl points are the Dirac points in 3D. They always appear in pairs. They appear along the k_3 wave vector direction at $k_3 = 0$ and $k_3 = \dfrac{\pi}{D_3}$, where D_3 is the periodicity of the metamaterial along the x_3 direction. Figure 6.14 shows an example of a 3D hexagonal lattice where the Weyl points were found [19]. Figure 6.14a shows three-dimensional periodic systems of carefully designed hexagonal lattice. Further, Figure 6.14b through e show the hexagonal unit cell and its design. Figure 6.14f shows the first Brillouin zone in 3D for the 3D geometry. Please note that as the metamaterial is 3D, an additional wave vector k_3 should be considered (compared to the derivation in section 6.2. and 6.3). With an additional k_3 wave vector, the Brillouin zone for the unit cell will take a shape of a hexagonal cylinder in the reciprocal space. Due to the k_3 wave vector in addition to K point ($k_3 = 0$), an additional Brillouin boundary in 3D is identified at $H\left(k_3 = \dfrac{\pi}{D_3}\right)$, where D_3 is the height of the unit cell in Fig. 6.14d. KH distance is the inverse of the height of each unit cell in Fig. 6.14d or e.

It is discussed in the previous section that if a hexagonal lattice is used as a unit cell in a metamaterial, then it is inevitable that at a specific frequency two bands will degenerate at the K point. This happens due to spatial symmetry. A Dirac cone would form. From the band diagram of the metamaterial, it was found that the

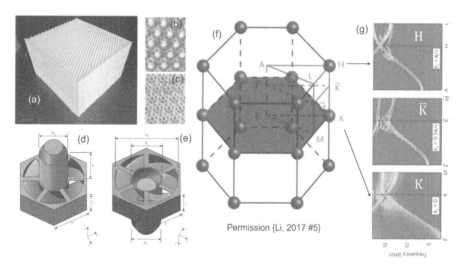

Permission {Li, 2017 #5}

FIGURE 6.14 a) Schematic representation of the three-dimensional periodic systems of charily designed hexagonal lattice, b) unit cell, c) unit cell alternate view, d) hexagonal unit cell from top view, e) hexagonal unit cell from bottom view, f) the architecture of first Brillouin zone in 3D for the 3D geometry. Due to the vertical wave vector, an additional Brillouin boundary in 3D is identified at H. KH distance is the inverse of the height of each unit cell in d) or e). g) The Dirac-cone was identified at k point and point h. However, in between the degeneracy is lifted. As an example, the lifted degeneracy is shown at $\bar{K}$ arbitrarily chosen at the mid point between K and H. This pair of Dirac cones in 3D is called Weyl point. This figure is reprinted from Ref [19] with permission.

Dirac-cone was indeed formed at the K point. However, in addition to the Dirac cone at K point, an additional degeneracy formed at the point H. These two degeneracies are identified as a pair of Dirac cones in the metamaterial. Interestingly with further investigation of the Dirac cones it was found that along the wave vector path k_3 (i.e., along KH), the Dirac cones were not present and the degeneracies were lifted. This pair of Dirac cones in 3D geometry is called the Weyl point [19]. An interesting topological acoustical phenomenon called backscattering immune unidirectional acoustic wave propagation was demonstrated in this metamaterial. To verify the existence and clearly visualize the Weyl point, the lifted degeneracy is shown at the $\bar{K}$ point, arbitrarily chosen at the mid-point between K and H. The Weyl point has its own unique topological behavior.

Please note, if they exist, Weyl points must exist in pairs. It is described that the Dirac cones in Weyl points must have opposite charges (topological invariant or topological charge) and are connected by a spectral excitation called a Fermi arc [19]. As in the previous sections using the **k.p** perturbation method discussed in Chapter 5, an effective Hamiltonian for a Weyl point could also be written explicitly. At a Weyl point there are two Dirac points. Thus, the physics near Weyl points can be captured by Hamiltonians at two-levels (at K and at H). Interestingly, the new two-level Hamiltonian will contain all the components of Pauli's matrices.

$$H(\mathbf{k}) = \omega_D \left(k_1 \begin{bmatrix} 0 & 1 \\ 1 & 0 \end{bmatrix} + k_2 \begin{bmatrix} 0 & -i \\ i & 0 \end{bmatrix} + k_3 \begin{bmatrix} 1 & 0 \\ 0 & -1 \end{bmatrix} \right) \qquad (6.12.6)$$

It is thus obvious that such a mathematical form of Weyl point is robust against Hermitian perturbations. The Dirac cones at a Weyl point are Hermitian in nature. Non-Hermitian Dirac cones are discussed in the next section.

6.4.8 Spawning Rings at Exceptional Points

Dirac points in 2D and Weyl points in 3D may turn in to an exceptional point. Some Dirac cone dispersion may get altered due to certain non-Hermitian perturbations that can deform a Dirac cone. Exceptional points are the degeneracies of non-Hermitian type and result in counterintuitive topological phenomena. Non-Hermitian systems can give rise to a special kind of spectral degeneracy called an exceptional point. A spawning ring (or an exceptional point), which can be visualized as two cones fused into each other, is an example of non-Hermitian Dirac cone.

The definition of Hermitian and non-Hermitian is no different from what is known for matrices. If a material is created or perturbed in such a way that non-Hermicity is ensured, the system may demonstrate a degeneracy that is of non-Hermitian type. Generally, any lossless medium in a closed system is defined as a non-Hermitian system. Thus, material loss or gain or open boundaries are a few examples of such perturbation that may lead a system to behave like a non-Hermitian system. As loss or gain must be ensured; the eigenvalues of a non-Hermitian system should be complex [29], composed of both real and imaginary parts. Ensuring this positive or negative damping at the degeneracy points, Dirac-like cones or the Weyl point may transform into an exceptional point. Real and complex parts of the eigenvalues would be very different. The exceptional point is nothing but a singularity in a non-Hermitian system. At the exceptional point, two or more eigenvalues and their associated eigenfunctions collapse into one eigenvalue or eigenfunction [30, 31]. When there is no loss, the Hermitian-type Dirac-like cone results in two dipole modes that are degenerated with a single antisymmetric mode at the Γ point, as discussed in section 6.4.6. But when material loss or gain is introduced, the Dirac-like cone will deform into a flat band enclosing a ring of exceptional points. This ring is called the spawning ring [32]. Inside the spawning ring, the real modes (i.e., the real parts of the eigenvalues) demonstrate degenerated flat dispersion. On the contrary, the imaginary mode (i.e., the imaginary part of the eigenvalues) splits into two branches inside the spawning ring. The modes are just the opposite outside the spawning ring at the Dirac frequency. Here, the real mode splits into two modes and shows linear dispersion, while the imaginary modes degenerate into a small flat band. This situation is graphically explained in Fig. 6.15 [33]. As in the previous sections using the **k.p** perturbation method discussed in Chapter 5, an effective Hamiltonian for a non-Hermitian Dirac-like cone (for three band degeneracies) could be written as.

$$H(\mathbf{k}) = \begin{pmatrix} \omega_D & c_g\sqrt{k_1^2 + k_2^2} & 0 \\ c_g\sqrt{k_1^2 + k_2^2} & \omega_D - i\eta_D & 0 \\ 0 & 0 & \omega_D - i\eta_D \end{pmatrix} \qquad (6.12.7)$$

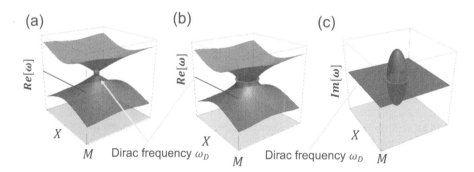

FIGURE 6.15 a) A schematic showing a Dirac-like cone at Γ point without the flat deaf band (for better clarity); no loss condition gives a Hermitian Dirac-like cone; only real eigenvalues are shown; for a Hermitian system the imaginary part of the eigenvalues is zero. b) A schematic showing a non-Hermitian Dirac-like cone at Γ point without the flat deaf band (for better clarity); here the real part of the frequency eigenvalues is shown. At the Dirac frequency ω_D the real eigenvalues are degenerated over an area on the 2D plane of the wave number space, that forms an exceptional point. c) A schematic showing a non-Hermitian Dirac-like cone at Γ point without the flat deaf band (for better clarity); here the imaginary part of the frequency eigenvalues is shown; at the Dirac frequency ω_D the imaginary eigenvalues are degenerated over an entire area on the 2D plane of the wavenumber space except inside the exceptional point enclosed within the spawning ring. This figure is reprinted from Ref [33] with permission.

Waves in Periodic Media

such that the three eigenvalues for the three modes are

$$\omega_1 = \omega_2 = \omega_D - i\frac{\eta_D}{2} + c_g\sqrt{\left(k_1^2 + k_2^2\right) - \left(\frac{\eta_D}{2c_g}\right)^2} \; ; \; \omega_3 = \omega_D - i\eta_D \quad (6.12.8)$$

where η_D is a damping factor simulating material loss introduced to have complex eigenvalues from a non-Hermitian dispersion.

6.4.9 DOUBLE DIRAC CONES AND SPINORS

Double Dirac cones are an interesting feature in dispersion band structure. Double Dirac cones are two sets of Dirac cones that are degenerated at a single frequency point at the center of the Brillouin zone (Fig. 6.16 a). These two Dirac cones (each with its own two bands) have a total of four bands degenerated, having different group velocities and dispersion. Such overlapping Dirac cones can be observed at the Γ point. These degeneracies are accidental in nature. Double Dirac cones can be obtained utilizing a fourfold accidental degeneracy and/or through folding the Dirac cones at the K and K' points in a graphene-type hexagonal lattice. Double Dirac cones are explained by phononic pseudospins. This unique feature gives a unique topological behavior and is discussed under the quantum spin Hall effect (QSHE) in Chapter 7. Unlike electrons, which are fermions with spin 1/2, phonons inherently do not have a spin but can be categorized as spin 0 Boson type particles. Like spin in quantum mechanics and photonics, phonon spins are categorized as pseudospin 1 in acoustics. This is because the band degeneracies and their characteristics are very similar but are understood differently by mimicking the pseudospin. It is extremely difficult to create a double Dirac cone state in acoustics. This is possible only if the degrees of freedom of a metamaterial unit cell are increased significantly, which may create artificial acoustic Kramer's doublets [34, 35]. They will help to obtain a pair of pseudospin states analogous to electrons.

Further, mimicking the spin degrees of freedom, double Dirac cones in acoustic media are viewed as multipolar pseudospin states where spin-like clockwise and anticlockwise displacement fields can be explained by pseudospin down and pseudospin up state, respectively. This viewpoint helps realize the QSHE [36, 37] in acoustic metamaterials. Say, for example, if a Double Dirac cone is formed accidentally at a parameter state by having increased degrees of freedom, then very likely this degeneracy could be lifted by changing the parameter or geometric configuration that gave rise to the accidental quadruple degeneracy in the first place. In some cases, this tuning or changing parameter space could go in both directions from its degenerated state. This means that if a parameter like rotational angle of a constituent element in a metamaterial is responsible to create the quadruple degeneracy at a specific angle say $\theta = \theta_D$, then the angle could be either increased or decreased (i.e., parameter space could change in both directions, i.e., $\theta < \theta_D$ or $\theta > \theta_D$) to lift the degeneracy. Angular rotation is just an example herein. Likewise, any geometric dimensional parameter responsible for a double Dirac cone could

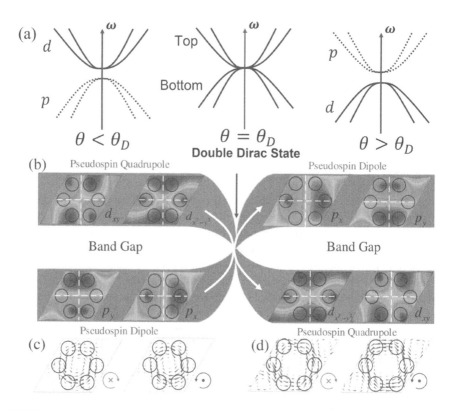

FIGURE 6.16 a) Shows a state of formation of double Dirac cone along the frequency scale (center). There are two top bands and two bottom bands degenerated to a quadruple degeneracy when an arbitrary parameter $\theta = \theta_D$ is used for tuning. Tuning is possible due to increased degrees of freedom of the phononic unit cell. Increased degrees of freedom help mimic the pseudospin states. When the parameter is lower (left) than the Dirac cone parameter, the top bands and the bottom bands separate and degeneracy is lifted while two top bands and two bottom bands are independently degenerated. When the parameter is higher (right) than the Dirac cone parameter, the top bands and the bottom bands again separate and degeneracy is lifted while two top bands and two bottom bands are independently degenerated. Interestingly, beyond the $\theta = \theta_D$ in parameter space, the nature of the modes of the top bands and bottom bands flipped, which is shown as solid and dotted lines. This is called topological mode inversion. b) Topological mode inversions undergoing the transition: p_x and p_y show pseudospin dipole modes; d_{xy} and $d_{x^2-y^2}$ show pseudospin quadrupole modes. Before and after the double Dirac cone state, the modalities are flipped. p type pseudospin state becomes the top bands and the d type states becomes bottom bands. c) Example of trivial regime. d) Example of nontrivial regime. This figure is reprinted from Ref [34] with permission.

be altered in both ways to lift the degeneracy. This is where the double Dirac cone becomes interesting. In some cases, it is found that these two sets of top bands (called d type) and two sets of bottom bands (called p type; see Fig. 6.16) have flipped mode shapes on opposite sides of the parameter value at which the quadruple degeneracy is formed. Although confusing, these d type and p type bands are synonymous to the d and p orbitals in quantum mechanics and thus similar nomenclature is still used in acoustics.

Waves in Periodic Media

Interestingly, in the case of a double Dirac cone, a pair of dipolar states (p type, i.e., bottom bands) are accompanied by a pair of quadrupolar states (d type or top bands) when $\theta < \theta_D$. The dipoles, or p type or bottom band mode shapes, are either even or odd, symmetrical about two orthogonal axes x_1 and x_2. They are called p_x and p_y, respectively. The quadrupoles, or d type or top bands mode shapes, are odd, symmetrical to axes x_1 and x_2. They are called d_{xy} and $d_{x^2-y^2}$ (see Fig. 6.16b). Next, when $\theta > \theta_D$, the pair of dipolar states flip and p type become top bands, while the pair of quadrupolar states d type become bottom bands. This is called topological mode inversion. The drastic band inversion before and after the double Dirac cone parameter will cause topologically protected edge states and reconfigurable topological one-way transmission.

Recollecting the discussion on spinors in section 5.4.4 in Chapter 5, it is said that a two-dimensional vector carrying spin eigenvalues in a three-dimensional space is called a 'spinor' [34, 35, 38, 39]. These two spin eigenstates are orthogonal and are related to the spin unit vectors as are used in Cartesian coordinate system $\hat{e}_i$. Hence, mimicking this understanding, here the p type and d type bands could also be written with their respective spin eigenstates. One pseudospin down (from p type) and one pseudospin up (from d type) together could represent a spinor. Hence at a double Dirac cone there are two spinors, each with one pseudospin down (clockwise rotational mode) and one pseudospin up (anticlockwise rotational mode). Alternatively, this could be viewed as a four-spinor state with four eigenvectors. Please refer to the Hamiltonian discussed in section 5.4.4 for four spinors. Similarly, here due to a double Dirac cone, there are four bands, and thus the effective Hamiltonian would be a 4×4 matrix near the Γ point. Referring to Eq. 5.28.22 in Chapter 5, using **k.p** perturbation method, keeping only the spatial derivative terms (i.e., $\frac{\partial}{\partial t} = 0$), equivalently from acoustic wave equation, the effective Hamiltonian for four spinor state could be written as

$$H(\mathbf{k}) = \begin{pmatrix} \Omega - \Delta_2 k^2 & \Delta_1(k_1 + ik_2) & 0 & 0 \\ \Delta_1^*(k_1 - ik_2) & -\Omega + \Delta_2 k^2 & 0 & 0 \\ 0 & 0 & \Omega - \Delta_2 k^2 & \Delta_1(k_1 - ik_2) \\ 0 & 0 & \Delta_1^*(k_1 + ik_2) & -\Omega + \Delta_2 k^2 \end{pmatrix} \qquad (6.12.9)$$

where $\Omega = (\omega_d - \omega_p)/2$, noted as the difference between the frequencies of p type dipolar and d type quadrupolar pseudospin eigenstates in the case of $\theta < \theta_D$ $\omega_d > \omega_p$ and thus $\Omega > 0$. But due to band inversion beyond the double Dirac point, $\theta > \theta_D$, $\omega_d < \omega_p$ and thus $\Omega < 0$. Δ_1 is off-diagonal first-order perturbation term (refer section 5.3.3.2) and $\Delta_2 < 0$ is derived from the diagonal elements of the second-order perturbation term.

6.4.10 TOPOLOGICAL CHARGE AND INVARIANT

Most of the discussion here could be easily referred to section 5.6.4 in Chapter 5. Please read section 5.6.4 to proceed further. To recapitulate, a spherical integral of the Berry curvature over a unit sphere that contains a monopole will provide a scalar quantity synonymous to an electric charge and can be assumed as a Berry charge, which is the Berry phase on a sphere. Thus, the Berry curvature integrated over a closed manifold will result in a quantized value of 2π, which will be equal to the number of monopoles inside the manifold. This number of monopoles or Berry charges is called the Chern number. The Chern number is a topological invariant and is called a topological charge.

It was found that the Berry phase at the Dirac cone near Brillouin zone boundary (at K) is π, because there are two monopoles. Whereas the Berry phase at the Dirac-like cone with triple degeneracy formed at the center of the Brillouin zone is zero. The following conclusions are true for the Berry phase and Berry curvature.

Active application of external fields may break the symmetry, causing a closed path along a parameter space (e.g., wave vector space) and causing manifestation of the Berry phase

- Without applying any external field to an acoustic metamaterial, if the time-reversal symmetry or space inversion symmetry is naturally broken, then the eigenfunction also incurs the Berry phase.
  - At the Brillouin boundary, or at the doubly degenerated Dirac cone (explained by a Dirac equation of fermions) the Berry phase is Pi (π). The topological charges or Chern numbers are ± 1.
  - At the center of the Brillouin zone with a triply degenerated Dirac-like cone, the Berry phase is zero (0). The topological charge, or Chern number, is zero.
  - At the center of the Brillouin zone with a quadropoly degenerated double Dirac cone, the topological charges, or Chern numbers, are $\pm\dfrac{1}{2}\big[\mathrm{sgn}(\Omega)+\mathrm{sgn}\,(\Delta_2)\big]$, where the 'sgn' function carries the number with the sign of the real number in the bracket.
- The Berry phase is associated with a closed path $\mathcal{C}$ in a parameter space, but the Berry curvature is a local quantity of the band and provides the geometric properties of the parameter space.

6.5 EXAMPLES: COUNTERINTUITIVE NON-TOPOLOGICAL WAVE PHENOMENA

In addition to the above characteristics, dispersion curves may present many other unique wave-propagation phenomena that are non-topological in nature. A few of those characteristics are reported in this section.

6.5.1 ACOUSTIC TRANSPARENCY, BEAM SPLITTING, NEGATIVE REFRACTION, AND SUPER LENSING

Instead of going though different features adopted for different metamaterials, exotic acoustical features such as acoustic transparency, ultrasonic beam focusing, acoustic

Waves in Periodic Media

band gap, and super lensing capability are demonstrated using a single metamaterial structure. Ultrasonic wave focusing, by virtue of negative refraction and simultaneous transparency of the metamaterial at sonic frequencies, is uncommon due to frequency disparity. To avoid the unwanted distortion of a wave at sonic frequencies, a metamaterial with an array of butterfly-shaped thin split ring resonators is conceptualized [1, 40] to achieve the beam focusing at ultrasonic frequency (37.3 kHz) and keep the structure transparent to the sonic frequencies (<20 kHz). The butterfly metamaterial with local ring resonators or butterfly crystals (BC) is proposed [1] to create wide band gaps (~7 kHz) at ultrasonic frequencies above 20 kHz. A unique sub-wavelength scale wave-focusing capability of the butterfly metamaterial utilizing the negative refraction phenomenon is also explored, while keeping the metamaterial block transparent to the propagating wave at lower sonic frequencies below the bandgap frequencies.

6.5.1.1 Butterfly Crystal Dispersion

A dispersion relation for the proposed unit cell in Fig. 6.11b and Fig. 6.17a is obtained by performing an eigenfrequency analysis of the structure. The geometric dimensions and the material specifications [1] are summarized in Table 6.3. One of the objectives of selecting this unique butterfly structure is the presence of local anisotropy in the

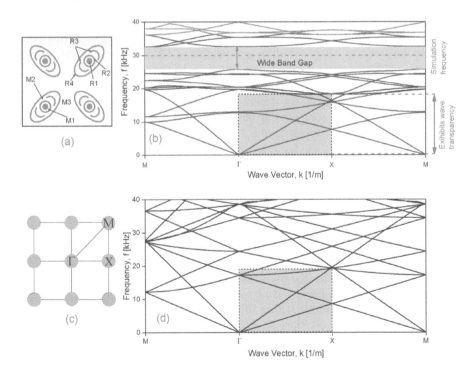

FIGURE 6.17 Dispersion relationship of the proposed butterfly design and the epoxy base material: a) proposed unit cell consisting of steel balls and elliptical steel rings in epoxy, b) dispersion relation of the butterfly design, frequency vs. normalized wave vector, c) irreducible first Brillouin zone, d) dispersion relation of the base epoxy material.

TABLE 6.3

Geometric Dimensions and Material Properties (Fig. 6.17a)

Component Name	Outer Dimension (mm)	Material
M1	Diameter – 5.387	Epoxy
M2	Major Radius – 10.2	E = 2.35 GPa
	Minor Radius – 5.08	ρ = 1110 kg/m^3
M3	2 X 2 Square	ν = 0.38
R1	Diameter – 3.591	Stainless Steel
R2, R3	Diameter – 7.183	E = 205 GPa
R4	Major Radius – 11.05	ρ = 7850 kg/m^3
	Minor Radius – 5.969	ν = 0.28

unit cell, but it has near isotropic behavior at the metamaterial scale. The dispersion band structure shown in Fig. 6.17b was computed for the MΓXM boundary of the first Brillouin zone [41] (refer to Fig. 6.17c), using the Bloch-Floquet periodic boundary condition [42].

In Fig. 6.17b, a large band gap from ~26 kHz to ~32 kHz is evident [1]. Further investigation of Fig. 6.17b reveals near linear dispersion (frequency vs. normalized wave vector) relation below ~18 kHz, irrespective of ΓX (dashed window) or XM directions. This demonstrates the near isotropic behavior that is found in classical bulk isotropic materials [43]. Applying a similar Bloch-Floquet periodic boundary condition on the base epoxy material (i.e., without the butterfly constituents), dispersion relation was computed at the MΓXM boundary of the first Brillouin zone (Fig. 6.17d). Upon comparing the dispersion relation in ΓX (dashed window) or XM direction for both the butterfly structure and the base epoxy material, it is evident that both geometries have similar dispersion behavior. Hence an infinitely repeated butterfly unit cell placed in a 2D epoxy medium can act as a single isotropic material, and the presence of these unit cells will almost be unrealized below ~18 KHz. To prove the 'acoustic transparency', a frequency domain simulation was performed. To achieve this objective, a 112 mm × 100 mm epoxy plate was modeled, using COMSOL Multiphysics V4.3. A butterfly crystal (BC) region, designed by an array of unitary cells identical to the one considered in Fig. 6.17a, is placed in the base plate. In particular, the BC arrangement consists of 22 rows and 6 columns of butterfly unit cells. A plane wave front was generated by the periodic displacement of a rectangular source with a dimension of 127 × 12.7 mm^2. A perfectly matched layer boundary condition was considered at all the boundaries of the base plate to approximate negligible wave reflection from the edges. Figure 6.18a shows the geometric configuration of the setup without an actuation of the rectangular exciter. Figure 6.18b shows the simulation outcome performed between the frequencies.) No displacement is observed either at the BC or at the epoxy and the host material on the right. This confirms that no wave is transmitted through BC within the band gap region.

Waves in Periodic Media

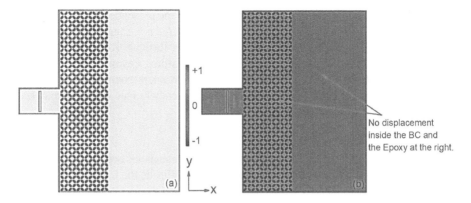

FIGURE 6.18 a) Geometric configuration without excitation, b) simulation of wave field at a frequency from the band gap frequency, 30 kHz.

Next, to demonstrate the acoustic transparency, a plane crested wave was generated at the audible frequency range in Γ-X direction (along x_1-direction). Simulation was performed between 0 and ~20 kHz frequency within the audible frequency range. Results of these two simulations are presented in Fig. 6.19 to have a visible comparison between the wave fields generated in the base material with and without the presence of BC [40]. Figure 6.19 is self-explanatory. The presence of the BC region does not affect the wave propagation at ~5 kHz, ~15 kHz and ~18 kHz. Transmission of the circular wave fronts is clearly visible. While the presence of the butterfly metamaterial region is unrealized by the incident waves at or below

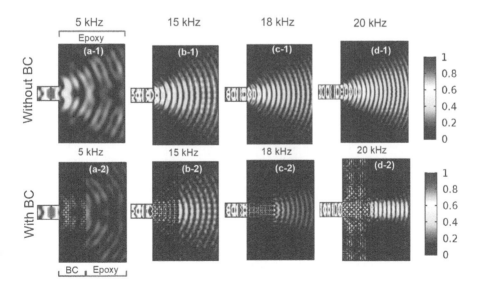

FIGURE 6.19 Normalized total displacement comparing the wave propagation at 5 kHz, 15 kHz, 18 kHz, and 20 kHz with and without BC region.

~18 kHz, the circular wave front starts to alter due to the presence BCs at or beyond the ~19 kHz (i.e., towards the end of the audible frequency range). The proposed structure acts as an *acoustically transparent* media through 90% of the audible frequency range but demonstrates metamaterial features beyond ~20 KHz. The *acoustic transparency* feature is particularly important in such cases where the effect of BC is desirable only in the ultrasonic frequency ranges when the material is undetectable at the audible ranges.

6.5.1.2 Wave Bifurcation

In this section, higher-order dispersion curves (i.e., frequencies beyond the complete band gap), are studied along the ΓX and MΓ direction inside the first Brillouin zone. The acoustic energy of elastic waves directly depends on the group velocity. It is well known that a higher group velocity of a propagating wave results in higher acoustical energy [44]. To understand the qualitative measure of the acoustic energy, the mode shapes at different points on the dispersion curves at the frequency range between ~37.085 kHz and ~37.43 kHz were analyzed. In Fig. 6.20, a portion of the dispersion curve in ΓX direction, along with the mode shapes at ~37.3 kHz, is presented. In the top section of Fig. 6.20, mode shapes at points 'a' through 'f' are shown next to the frequency vs. a normalized wave vector plot in ΓX direction. It can be noted that points 'b', 'e' and 'f' are located on three different modes around *equal frequency*

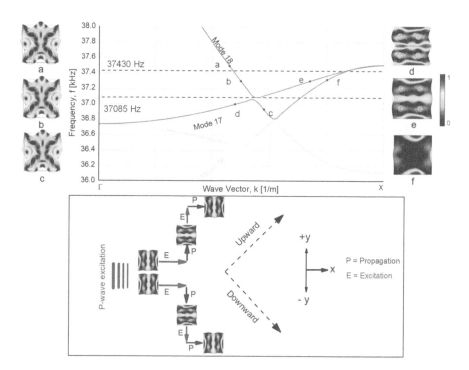

FIGURE 6.20 Analysis of mode shape (ΓX direction) and wave propagation direction showing the mechanism for wave bifurcation.

level. Since the measure of the group velocity directly depends on the slope of the dispersion curve, it can be observed that

$$\frac{dw}{dk}\bigg|_b > \frac{dw}{dk}\bigg|_f > \frac{dw}{dk}\bigg|_e \; ; \; C_g^b > C_g^f > C_g^e; \; E_b > E_f > E_b \qquad (6.13)$$

Point 'e' possesses lower resonant energy compared to the other two points. Considering this argument, the mode shape of point 'e' can be taken into consideration for a wave to be propagated in ΓX direction. The displacement patterns of this mode shape indicate that the transmitting wave needs to be propagated in orthogonal direction to ΓX. Therefore, the mode shape of point 'e' directs the propagating wave to transmit in both $+x_2$ and $-x_2$ directions locally. However, the transmitting wave from the first unit cell to its adjacent second unit is again dominated by the mode shape of point 'e'. The lower part of Fig. 6.20 (boxed part) explains this feature, where 'E' and 'P' indicate the *'excitation'* and *'propagation'* directions, respectively. Hence, the incident plane wave is bifurcated into upward ($+x_2$) and downward ($-x_2$) directions locally while preserving the global wave propagation (i.e., the wave vector) in x_1-direction. Hence, it is necessary that the transmitted wave should bifurcate inside the BC structure. This is called beam splitting or wave bifurcation. To verify the claim, a frequency domain study was performed in a frequency range of ~37.085 kHz to ~37.43 kHz. At ~37.3 kHz, wave bifurcation and wave refocusing phenomena were observed, which is shown in Fig. 6.21. In Fig. 6.21a, it can be observed that the excited plane wave initially bifurcates into two wave directions inside the BC, and afterward these two waves converge into a single point at the outside of the BC due to negative refraction discussed in the next section.

There are many other scenarios that could be found, based on group velocity value and group velocity direction that may cause wave bifurcation. Examples of wave bifurcation in a metamaterial made of PVC phononic crystal embedded in air matrix can be found in Refs [45, 46].

6.5.1.3 Negative Refraction and Wave Focusing

At the BC and epoxy, while the metamaterial is excited with a source, the propagating wave faces an interesting situation explained below. Here at the interface, incident

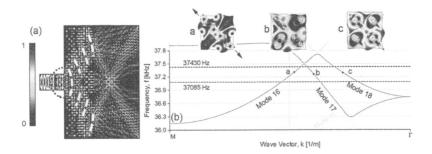

FIGURE 6.21 a) Determination of dominant mode shape in wave bifurcation, b) mode shape of points 'a', 'b' and 'c' in MΓ direction, c) identification of maximum total displacement at the focal point over the focusing frequency range.

waves impinge at MΓ direction (in both +45° and −45° direction to x_1-axis). To identify the dominant mode shapes, the portion of the dispersion curve related to MΓ direction was analyzed. At ~37.3 kHz, the mode shapes of mode 16, 17, and 18 are presented in Fig. 6.21b. Slope at point 'a' on mode 16 has the lowest value, which indicates the lowest group velocity compared to the points 'b' and 'c'. Hence, the mode shape at point 'a' is the dominant mode shape in the MΓ direction. Considering the displacement pattern of the mode shape at point 'a', it can be observed that the displacement of particles along the two diagonals has different values. While the displacement along one diagonal is zero, the displacement of another diagonal is positive. The diagonalized displacement pattern creates a 45° local wave propagation direction indicated by negative 45° black arrows in Fig. 6.21b. Therefore, any wave incident at positive 45° at BC-epoxy interface, will refract at a negative 45° direction. By symmetry, waves incident at negative 45° at the BC-Epoxy interface, transmit at positive 45° direction. As the waves come out of the BC region and transmit through the isotropic epoxy media, the wave propagation directions remain the same and eventually cross each other to form a focal region. Here, the negative wave refraction and wave focusing are demonstrated. Further, the negative wave refraction is explained in Fig. 6.22. Fig. 6.22a shows the incident and negative refraction angle of the transmitted wave at

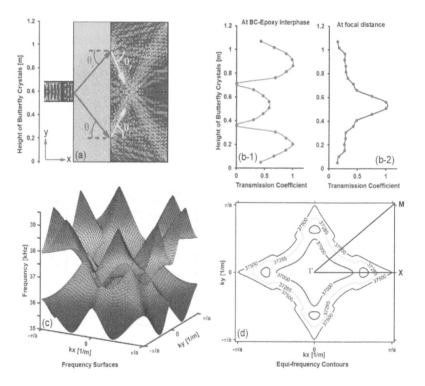

FIGURE 6.22 (COLOR): Demonstration of negative refraction phenomena: a) incident and refracted wave directions, b-1) wave transmission at the BC and epoxy interphase, b-2) wave transmission at the focal point, c) 3D frequency topology map between 35 – 39 kHz, d) equi-frequency contours for a frequency range of 37 kHz to 37.5 kHz.

Waves in Periodic Media

37.285 kHz. Figure 6.22b-1 and b-2 demonstrate the wave transmission coefficient at BC and epoxy interphase and at the focal distance, respectively. To explain, equifrequency surfaces are plotted for a frequency band of 35kHz to 39 kHz in Fig. 6.22c. From these surfaces, equifrequency contours are constructed in Fig. 6.22d for a frequency range of 37 kHz to 37.5 kHz. It can be noted that the direction of the energy flux is outward from the center of the Brillouin zone (Γ point) because the positive group velocity direction is always along the direction of increasing frequency. Hence, any line passing through Γ and M points would intersect the 37.28 kHz equifrequency contour, resulting in outward energy flux, and the normal to this intersection point determines the direction of negatively refracted waves.

6.5.1.4 Superlensing: Beyond the Diffraction Limit

The formation of acoustic focal points indicates the existence of negative refraction property of the BC metamaterial. The smallest feature that can be identified by a conventional acoustic flat lens is limited by the spatial frequency. In 1873, Ernst Abbe proposed a fundamental 'diffraction limit' for optics, which is half the wavelength of the propagating energy [47]. To overcome this diffraction limit, a structure having negative refraction property can be utilized as a super lens. To investigate the possibility of superlensing capability of the BCs, two different simulation configurations were designed. First, only one rectangular exciter of 12.7 mm interface incident 12.7 mm was excited at a frequency ~37.285 KHz. Next, two exciters (Fig. 6.23) of size 3.81 mm were

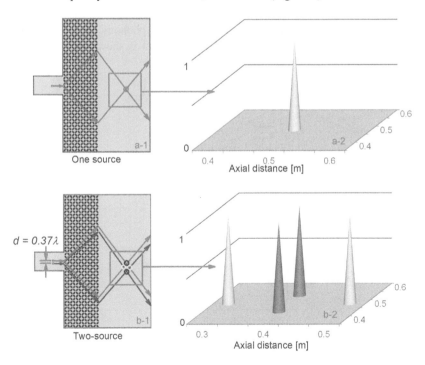

FIGURE 6.23 Demonstration of superlensing capability of butterfly structure: a) design configuration with one exciter, b) design configuration with two sources.

excited at the same frequency, ~37.285 kHz, keeping them at a distance of less than the wavelength. In epoxy, the p-wave velocity was found at 1990.78 m/s. As, wavelength, $\lambda = V_P/f$, at $f = 37.285$ kHz, the calculated wavelength was found as 53.39 mm. In the numerical simulation (Fig. 6.23), the distance d was kept at 19.75 mm which is about 0.37λ. Figs. 6.23(a-1) and 6.23(b-1) show the schematic of wave propagation direction originating from one and two source excitation systems respectively. The rectangular boxes in these two figures are further enlarged, respectively, in Figs. 6.23(a-2) and Fig. 6.23(b-2) with the displacement amplitudes at the circular points. While plotting the displacements, a binary format is used. Any displacement amplitude measured at the focal point(s) greater than 125 μm was marked as 1 or 0. In the case of one source configuration, a prominent single focal point was found with a displacement amplitude of 228.6 μm. However, for the two-source configuration, four focal points are generated as shown in the schematic diagram of Fig. 6.23(b-2). Displacement amplitudes of these four points were calculated numerically (190.5 μm, 177.8 μm, 167.1 μm and 127.0 μm). Hence, four focal points with peak value 1 were detected. Therefore, the BC can create focal points of acoustic sources that are separated by 0.37λ of the base epoxy material. This evident the super lensing capability beyond the diffraction limit is possible followed by negative refraction in a BC metamaterial.

6.5.2 Orthogonal Wave Transport at Dirac-like Cone

Please refer to Fig. 6.13, where a metamaterial with a square lattice structure having square phononic crystals is used to find a dispersion band structure. Similarly, a metamaterial with a square lattice structure having circular phononic crystals (Fig. A.6.1) is used to find another dispersion band structure and is shown in Fig. 6.24 [48].

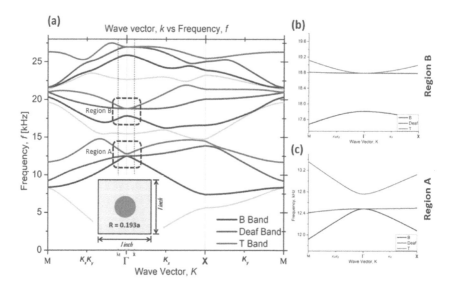

FIGURE 6.24 a) Band structure of the unit cell, showing region A and region B, two possible Dirac-like degenerated point, b) and c) zoomed in parts of region A and region B. It is to be noted that, ~M and ~X denote 10% of ΓM and ΓX respectively.

Waves in Periodic Media

Both structures have a unique feature. They both tend to have triple degeneracy and tend to generate Dirac-like cones at two different energy states, namely DRA and DRB. In both cases, one top band (T-band), one bottom band (B-band), and a 'Deaf band' (D-band) in between, as is discussed in section 6.4.6, are degenerated.

A square phononic crystal of dimension $0.342a$ and a circular phononic crystal of diameter $0.193a$ were used arbitrarily to generate the dispersion bands, keeping the filling fraction constant where a is the dimension of the square lattice. However, this arbitrary selection *was not appropriate* to generate the perfect state of a Dirac-like cone. It was found that the dimensions had to be tuned. When the diameter of the circular phononic crystals was tuned, at a specific dimension they generated the Dirac-like cones at their respective energy level. Figure 6.25 [49] shows the dimensions of the circular phononic crystals that created the perfect states of Dirac-like cones. Although tuning is possible by increasing or decreasing the diameter of the circular phononic crystals, it is not possible to tune the metamaterial structure in real time, by means of a mechanism or a process. A total replacement of the phononic crystals is necessary.

It was found that for square phononic crystals, Dirac-like cones could be perfected by adding a rotational degree of freedom and by breaking unit cell symmetry.

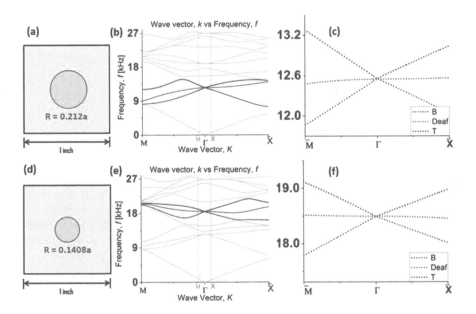

FIGURE 6.25 Accidental degeneracy at 'Region A' DRA and 'Region B' DRB: a) a unit cell for region 'A' with PnCs of radius $r = 0.212a$ in air matrix, b) dispersion relation for region 'A' after increasing the radius $r = 0.193a$ to $r = 0.212a$ where the 'T' band is lowered with respect to the 'deaf' band and having an accidental degeneracy for the frequency $f = 12.551$ kHz, c) magnified view of the 'Region A' Dirac-like point, d) a unit cell for region 'B' with PnCs of radius $r = 0.1408a$ in air matrix, e) dispersion relation after decreasing the radius $r = 0.193a$ to $r = \sim 0.1408a$ where the 'B' band moves upward with respect to the 'deaf' band and having an accidental degeneracy for the frequency $\omega = 18.512$ kHz, and f) magnified view of the 'Region B' Dirac-like point.

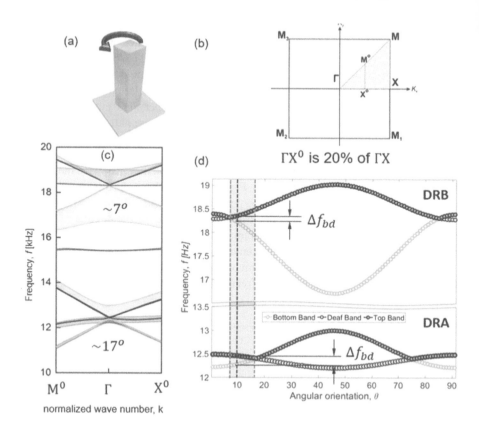

FIGURE 6.26 a) Geometric representation of square rod PnCs embedded in the air matrix and rotated counterclockwise (CCW). This angular tuning was conducted at 1° intervals over a range of 0°–90°. b) First Brillouin zone and normalized wave number segment used in c). c) superposed band structure obtained from different CCW unit cells. The dispersion relationship shows that a Dirac cone is achieved at a ~7° rotation for DRB and at a ~17° rotation for DRA. d) Modulation of the T, deaf, and B bands as a function of the rotational angle with different CCW unit cells. A red box is identified between ~7° and ~17° where the t and deaf bands are degenerated, but the B band is separated by a band gap Δf_{db}. The angle of rotation is identified where Δf_{db} is equal for DRA and DRB.

Figure 6.26 [49] shows the tuning process, and all the bands are superposed at different tuning angle to identify the perfect state of the Dirac-like cones. It was found that with counterclockwise rotation of ~7° and ~17° angles, Dirac-like cones are developed at the DRB and DRA, respectively, at frequencies ~18.38 kHz ~12.46 kHz. Next, the modes shapes and wave transmissibility are tested at near deaf band frequency at the DRA energy level. Figure 6.27 shows that the deaf band indeed caused a drastic local drop in wave transmissibility, as discussed in section 6.4.4. It also shows that the top band and the bottom band mode shapes are dipolar, with opposite polarity, whereas the deaf band is an antisymmetric bipolar mode [49]. Thus, when the wave is excited at the Dirac-like cone frequency the T-band and B-band

Waves in Periodic Media

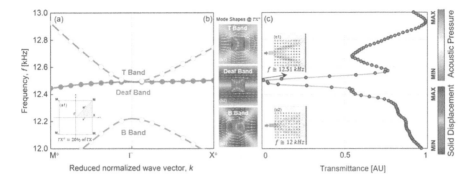

FIGURE 6.27 a) Dispersion diagram for the DRA location, showing a small band gap between doubly degenerated 'T band' and 'Deaf band' and a 'B band'. The irreducible BZ is demonstrated on the inset. b) Respective mode shapes of a unit lattice of 'T band', 'Deaf band', and 'B band' along ΓX normalized wave vector direction. Both the absolute acoustic pressure distribution field and the solid displacement field are shown in two different colormaps, with their respective displacement vectors (white-arrow vectors depict absolute acoustic pressure, and red arrows inside the square phononic crystals depict solid displacement. c) The wave transmission decreases to zero near the deaf band if transmitted along ΓX *[100]* (top). The numerically calculated acoustic pressure field and solid displacement fields are shown for a 10 × 10 phononic crystal matrix, excited at 'Deaf band' frequency and another arbitrary frequency for B band.

wave modes cancel each other out, creating an opportunity for the antisymmetric deaf-band mode to prevail. As the deaf band antisymmetric dipolar mode is purely orthogonal to the direction of the wave propagation, the wave takes a bend and is propagated orthogonally in the metamaterial structure.

Similar physics prevails if the circular phononic crystal is used in the metamaterial. It will be easier to understand the mechanism through circular geometry and hence results from a metamaterial with circular phononic crystals are explained herein. However, please note that the physics is similar and the results from a square phononic system can be found in Ref [17]. In general, the deaf bands, due to the antisymmetric acoustic modes, exhibit strong spatial localization and cannot be excited; however, due to the tuned triple degeneracy at the Dirac-like cone frequency, the bands actually maintain their antisymmetric shape to propagate orthogonally. Figure 6.28, which is self-explanatory, helps explain the scenario graphically. In Fig. 6.28a, band structure with mode shapes of the T-band, D-band, and the B-band are shown for the ΓX and MΓ directions. The T-band and B-band mode shapes are quadrupolar modes, orthogonal to each other; in contrast, the deaf band mode is a dipolar mode. Hence, at the Dirac frequency, the T-band and the B-band nullify each other, keeping the deaf band dipolar mode to dominate the propagation of waves in the material. Examining the wave transmission behavior from the acoustic pressure field at the Dirac-like cone, it can be seen that to keep the dominant dipolar mode alive, a 45° bent line must be developed to carry over the plane wave orthogonally.

Wave guiding and propagation patterns were studied inside and outside the metamaterial. Figure 6.29a shows the acoustical pressure field distribution at a frequency

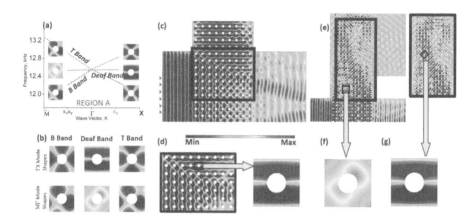

FIGURE 6.28 a) band structure with mode shapes of t, deaf and B bands for ΓX and MΓ direction, b) the total acoustic pressure mode shape in XY plane for ΓX and MΓ direction, c) the total acoustic pressure field in the metamaterial showing the bent line of transmitting the plane waves orthogonally, d) the zoomed part of c), depicting the resemblance of the acoustic pressure distribution during orthogonal transportation with the ΓX deaf band mode shape e) the acoustic pressure field for MΓ direction excitation and the plane wave transportation, f) and g) the deaf band mode shape for both MΓ and ΓX direction, showing the wave transportation pattern resemblances.

at 10 kHz, slightly below the Dirac frequency. At the Dirac frequency (~12.55 kHz) only a little acoustical energy propagates in the direction of actuation, creating a partial bandgap. But at this Dirac frequency, wave energy is transported orthogonally as shown in Fig. 6.29b. As the wave propagates inside, the orthogonal transport becomes prominent, leaving a 45° diversion line as shown in Fig. 6.28d and e. While traveling through the wave, orthogonally transported energy again takes 90°-turn towards any opening along the path. Thus, the original wave propagation direction into the C1 and C2 channels was retrieved at a distant location, through leaky plane

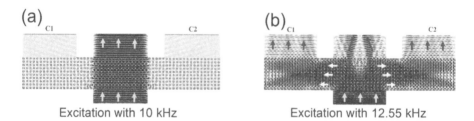

FIGURE 6.29 a) and b) the absolute acoustic pressure field for a plane wave generated in ΓX direction for frequencies of f = ~10 kHz and f = ~12.551 kHz respectively. It can be seen that while the plane wave was propagating through the metamaterial at f = ~10 kH without changing any direction, the wave is transported orthogonally in a converging–diverging pattern when propagated at the Dirac-like frequency of f = ~12.551 kHz. After reaching the end of the tunnel, the wave again turned orthogonally keeping the plane wave pattern undisturbed along C1 and C2 channels.

waves, which is also a result of negative refraction phenomena. For more details, please refer to [49]. Similar orthogonal wave transport is demonstrated with square phononic crystals and can be found in Ref [17, 18]. For experimental verification of the above phenomena refer to [17, 18, 48, 49].

6.5.3 Acoustic Computing at Dirac-like Cone

Exploiting the behavior presented above in a metamaterial with square phononic crystals, several acoustic computing units are devised. Functionality is demonstrated through virtual experiments. First, a single block of 35 × 35 elements is composed in a matrix. Two input terminals, A and B, are defined on either side of the matrix. Simultaneously, the other two orthogonal directions are considered to be the output channels with Q as shown in Fig. 6.30. Figure 6.30 also shows a typical input and output schedule for AND, OR, NAND, and NOR gates. All inputs, if not zero, are considered acoustic wave excitation input from the respective terminals. Frequency of excitation was kept at the Dirac frequencies, e.g., ~12.46 kHz, or ~18.38 kHz (Fig. 6.26). Here acoustic computing of Boolean algebra is demonstrated at ~18.38 kHz. However, similar phenomena persist at ~12.46 kHz (not shown). Please note that 9.7° rotation of the square phononic crystals results in a Dirac cone behavior, while a 45° rotation results in a bandgap (refer to Fig. 6.26). 9.7° is specifically selected between 7° and 17° based on the degeneration frequency gap depicted in Fig. 6.26. Every input is considered 1 if the wave is excited at the frequency ~18.38 kHz at the respective terminals A or B. From the output terminal Q, if the transmission coefficient is above 0.75, then the terminal output is otherwise the output is considered 0. The following logics are created to perform all the tasks that are necessary to perform by each 'gate' as depicted in Fig. 6.31.

- **AND Gate**: Unless both A and B are 'on', with positive input of wave (i.e., 1), the phononic crystal orientation will be fixed to 45° rotation. Hence, when A and B are excited independently (i.e., A = 1, B = 0, or A = 0, B = 1),

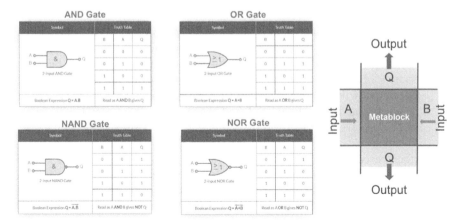

FIGURE 6.30 a) schedule input/output for different gates, b) metablock structure to test the computing logics.

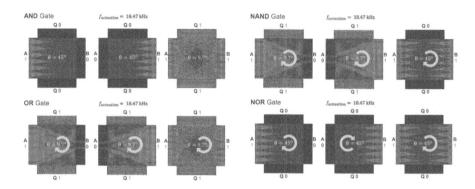

FIGURE 6.31 Functionality of AND, OR, NAND, and NOR gates.

the output terminal will read 0, resulting in A AND B operation. However, when both A and B are 'on', the phononic crystals automatically rotate to 9.7°, creating the opportunity for orthogonal transport. In such a scenario, output terminal will read 1.

- **OR Gate**: This gate is fixed at a fixed rotation of 9.7° of the phononic crystals. When A = 1 and B = 0, it will result in transmission through the output terminal result output 1. This will be the same for B = 1 and A = 0. When both A and B are on, (i.e., A = 1, B = 1), the output will still read 1 due to the orthogonal transport at the Dirac cone. From the output amplitude at the Q terminals, the input status of the A and B terminals can be deduced.
- **NAND Gate**: In this gate, the phononic crystals orientation will be 9.7° unless both A and B are 'on' with positive input of wave (i.e., 1). When A = 1 and B = 0, this gate will result in output Q = 1, and vice versa. However, when both A and B terminals excite the wave with input 1, then the system will rotate to 45°. This will result in output Q = 0. In an acoustic wave, zero inputs cannot be functionalized, and hence, A = 0, B = 0, and Q = 1 cannot be achieved.
- **NOR Gate**: This gate is fixed at a fixed rotation of 45° of the phononic crystals. When A = 1 and B = 0, it will prohibit transmission through the output terminal, which will result in output Q = 0. This will be the same for B = 1 and A = 0. When both A and B are on (i.e., A = 1, B = 1), the output will still read 0 due to the bandgap. In an acoustic wave, zero inputs cannot be functionalized, and hence, A = 0, B = 0, and Q = 1 cannot be achieved.

Further, based on this concept, a six-degree-of-freedom system is created to perform more complex computing tasks where every single component is utilized to perform Boolean algebra. A conceptual design is presented in Fig. 6.32. Six square metablock matrices (T1, T2, C1, C2, B1, and B2) are created. Each matrix will have its own control of rotation. This control will help them to rotate to a specific angle of 45° or 9.7° as needed to perform the gate operation presented in Fig. 6.31. Although multiple input terminals could be created, only one input terminal, at

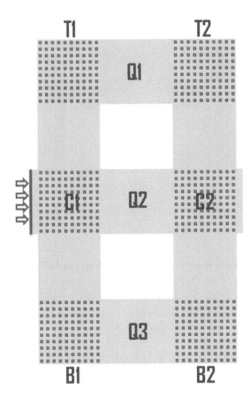

FIGURE 6.32 A six degrees of freedom system for acoustic computing.

C1, is considered for a simple demonstration. This demonstration is simulated at ~12.46 kHz at DRA. A similar outcome could be achieved at DRB frequency as demonstrated in Fig. 6.31. Keeping a constant input of wave excitation at terminal C1, outputs through the horizontal gray channels (Q1, Q2, and Q3) are investigated. Terminals are rotated independently to a achieve certain output. Figure 6.33 shows a few such scenarios (cases).

- **Case 1**: C1 is excited and kept at 0° rotation, while all the other matrix terminals are rotated to 45°. This situation shows no transport and no output through Q1 and Q3, but Q2 has an output (i.e., Q2 = 1). If C1 was also rotated to 45°, all outputs would be 0.
- **Case 2**: C1 is excited while all the matrix terminals (C1, C2, T1, T2, B1, and B2) are rotated at 9.7°. This situation shows full orthogonal transport, and outputs through Q1, Q2, and Q3 are all equal to 1.
- **Case 3**: C1 is excited. C1, C2, T1, and T2 are rotated at 9.7° while B1 and B2 are rotated at 45°. This situation resulted in Q1 = 1, Q2 = 1, and Q3 = 0.
- **Case 4**: Vice versa, C1 is excited. But C1, C2, B1, and B2 are rotated at 9.7°. T1 and T2 are rotated at 45°. This situation resulted in Q1 = 0, Q2 = 0, and Q3 = 1.

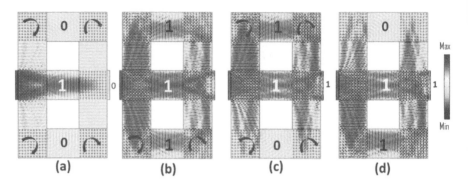

FIGURE 6.33 Acoustic computing using a multi-degree of freedom system to exploit multi-channel input and multi-channel output: a) Case 1, b) Case 2, c) Case 3, d) Case 4.

Following the above rules, many more situations could be created using the metablock structure made of phononic crystals. Other output terminals that are vertical channels in Fig. 6.32 (gray channels) could be added as additional outputs to investigate (not shown) more complex scenarios.

6.6 ACTIVE BREAKING OF TIME-REVERSAL SYMMETRY

Before starting this section, it is highly recommended for readers to read section 5.6 in Chapter 5 and the Appendix section 6.8.3 of this chapter. It was pointed out that time-reversal symmetry breaking is one method to achieve topological behavior. It is also mentioned that other symmetry-breaking scenarios, like the breaking of space inversion symmetry, are the cause of topological phenomena.

The breaking of time-reversal symmetry can be achieved by introducing active energy to the system, or it may happen naturally. Here time-reversal symmetry breaking is considered when an external field is applied. The approach is discussed for elastic waves in metamaterial. In a periodic metamaterial, if the stiffness is actively but gradually modulated, interesting topological phenomena emerge. At first glance, the phenomena may not manifest as topological. However, with careful investigation using the tools at our disposal, the topological features can be extracted. One of these tools includes parallel transport of a vector field for investigating the geometric phase, if present on a wavenumber space. Others are the Berry phase in 2D and 3D or Zak phase in 1D. They are the processes described in Chapter 5 that could be used to find the topological phenomena, if they exist.

6.6.1 TOPOLOGICAL BAND GAPS

First a one-dimension periodic metamaterial is considered with spatial modulation of stiffness. To break time-reversal symmetry, a time-dependent modulation to the stiffness is incorporated to extract topological behavior. Hence, time-dependent elastic superlattices are considered herein. At first glance this scenario seems impractical. However, discovery of an intermediate phase [50] in glasses opened a new horizon for this possible modulation. In Ref [51] it was shown how stiffness of chalcogenide glasses (Ge-Se) resulting from a photo-elastic effect could be modulated. By spatially and temporally varying the light intensity, the material stiffness could be changed if the material is arranged in a specific spatial orientation. It was shown that the longitudinal elastic property ($\lambda + 2\mu$) could be decreased by half with gradual changes when the Ge-Se chalcogenide glasses were illuminated at near bandgap laser radiation with increasing power. As the illumination power increases, the stiffness coefficient decreases. However, these changes are not permanent and are recoverable when the shining laser power is reduced. Similarly, this scenario could be achieved utilizing many other smart materials. For example, shape memory alloy [52] changes its stiffness with applied voltage. Both magneto-elastic particles [53] and elasto-magneto rheological fluid [52] change their shapes viscosity with controlled application of magnetic field, respectively. Piezoelectric material causes a change in displacement under application of voltage [54, 55]. A specific arrangement of such a piezoelectric system with reversed polarity could be exploited for spatial-temporal modulation of stiffness when the AC voltage supply is applied. Nonlinear media are known for responding with nonlinear amplitude when force is increased linearly. Hence, with specific tuning, a nonlinear media could be designed to have modulated stiffness [56]. Considering these are the practical scenarios, in the following example, a spatio-temporal modulation of stiffness is presented and topological effect is explained.

6.6.1.1 Spatio-Temporal Modulation of Material Coefficients

For simplicity a one-dimensional system is considered. Refer to Fig. 6.34. A one-dimensional periodic metamaterial is shown with its equivalent lattice structure. The internal spring coefficients are changed over time along the length of the material. The elastic modulus and density are written as a function of the space and time. Figure 6.35 pictographically shows the spatio-temporal change in modulus and/or density with a constant velocity, e.g., $\mathbf{v} = \lambda_p/T_p$. The frequencies of the temporal and spatial modulations are Ω_p and G_p, with period T_p and D_p, respectively.

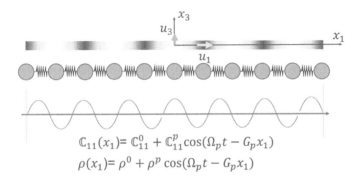

FIGURE 6.34 Schematics showing an example elastic metamaterial with spatio-temporal modulation of elastic properties.

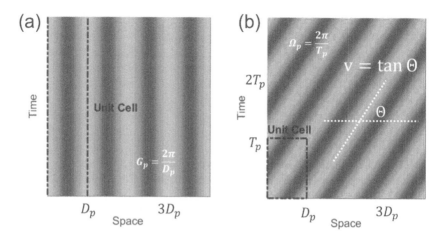

FIGURE 6.35 Spatio-temporal modulation of elastic properties: a) spatial modulation with zero temporal modulation, b) with both spatial and temporal modulation. Velocity of modulation (v) is the slope of the modulation graph which is $v = T_p/D_p$ [57].

6.6.1.2 Governing Differential Equations

The governing differential equation for elastic wave propagation in the system along two orthogonal directions could be written as follows. Equation 6.14.1 is like the equation discussed in section 6.2.2 for the longitudinal wave. Equation 6.14.2 is a flexural wave equation where the wave is propagated along x_1 but the displacement is considered along x_3. The material is periodic both in time and space. Thus, a Bloch wave vector is required both in the space and time domains as shown in Fig. 6.35 [57]. Equation 6.14.3 presents the modified primary canonical variable where Bloch wave vectors in their respective reciprocal spaces are added to the primary variable for frequency ω and wavenumber k. As this example presents only a one-dimensional problem, the k_1 wave number is replaced with k, without the loss of generality.

$$\frac{\partial}{\partial x_1}\left[\mathbb{C}_{11}(x_1,t)\frac{\partial u_1}{\partial x_1}\right] = \frac{\partial}{\partial t}\left[\rho(x_1,t)\frac{\partial u_1}{\partial t}\right] \quad (6.14.1)$$

$$\frac{\partial^2}{\partial x_1^2}\left[\mathbb{C}_{11}(x_1,t)\frac{\partial^2 u_3}{\partial x_1^2}\right] + \left(\frac{A}{I}\right)\frac{\partial}{\partial t}\left[\rho(x_1,t)\frac{\partial u_3}{\partial t}\right] = 0 \quad (6.14.2)$$

$$\omega \to \omega + \Omega_p^n = \omega + \frac{2\pi n}{T_p}; \quad k \to k + G_p^n = k + \frac{2\pi n}{D_p} \quad (6.14.3)$$

As the frequency and wave number both are Bloch periodic, the displacement field could be expressed as follows (refer to section 6.2.2 for more clarity):

$$u_1 = \sum_{G,\Omega} u_1(G,\Omega)e^{ikx}e^{-i\omega t} = e^{i(kx-\omega t)}\sum_{G_n,\Omega_n} A_n e^{i(G_p^n)\cdot x} e^{-i\Omega_p^n t} \quad (6.14.4)$$

Waves in Periodic Media

$$u_3 = \sum_{G,\Omega} u_3(G,\Omega) e^{ikx} e^{-i\omega t} = e^{i(kx-\omega t)} \sum_{G_n,\Omega_n} B_n e^{i(G_p^n)\cdot x} e^{-i\Omega_p^n t} \qquad (6.14.5)$$

Similarly, the material coefficients could be expressed with their respective Fourier coefficients as follows. Please note that the indices used in Eqs. 6.14.4 and 6.14.5 differ from the following equations.

$$\mathbb{C}_{11}(x_1) = \sum_{-\infty}^{+\infty} C_m e^{i(G_p^m)x_1} e^{-i(\Omega_p^m)t}; \ \rho(x_1) = \sum_{-\infty}^{+\infty} P_m e^{i(G_p^m)x_1} e^{-i(\Omega_p^m)t} \qquad (6.14.6)$$

6.6.1.3 Dispersion Bands with Directional Bandgaps

Substituting the displacement and material property expressions into Eq. 6.14.1 for longitudinal waves and into Eq. 6.14.2 for transverse waves, the final eigenvalue problem will be as follows (please refer to 6.2.2 where similar approach discussed in detailed and is adopted herein):

$$\sum_{n=-N}^{+N} C_{m-n}\left[\left(k+G_p^n\right)\left(k+G_p^m\right)\right]A_n = \sum_{n=-N}^{+N} P_{m-n}\left[\left(\omega+\Omega_p^n\right)\left(\omega+\Omega_p^m\right)\right]A_n \qquad (6.14.7)$$

$$\sum_{n=-N}^{+N} C_{m-n}\left[\left(k+G_p^n\right)\left(k+G_p^m\right)\right]^2 B_n = \left(\frac{A}{I}\right)\sum_{n=-N}^{+N} P_{m-n}\left[\left(\omega+\Omega_p^n\right)\left(\omega+\Omega_p^m\right)\right]B_n \qquad (6.14.8)$$

where N is any arbitrary number for this expansion that represents how many Bloch wave modes are included in the expression. Higher numbers would help extract a higher number of modes. If N Bloch modes are used, then for the above equations, each would result in $(2N+1 \times 2N+1)$ matrices. Once the eigenvalue problem is set and eigenvalues are found, they would create a dispersion curve in frequency–wave number domain. Now, it's time to explore the features from the dispersion curves. For convenience in Ref [57], two new normalized parameters are used, such as ξ_p and η_p (Eq. 6.14.9), to normalize the longitudinal material coefficient and the density.

$$\xi_p = \frac{\mathbb{C}_{11}^p}{\mathbb{C}_{11}^0}; \ \eta_p = \frac{\rho^p}{\rho^0} \qquad (6.14.9)$$

Dispersion diagrams are shown in Fig. 6.36 [57]. Longitudinal waves propagation in the material with different cases of material coefficient and density are investigated. Figure 6.36a shows the dispersion behavior for the metamaterial with a non-modulated beam when $\xi_p = 0$ and $\eta_p = 0$. Figure 6.36b shows the dispersion behavior for the metamaterial when a space-modulated metamaterial is considered and $\xi_p = 0.40$ and $\eta_p = 0$. Figure 6.36c shows the dispersion diagram when both space- and time-modulated metamaterial is created with $\xi_p = 0.40$ and $\eta_p = 0.05$. Figure 6.36d shows the dispersion diagram for the metamaterial with space- and time-modulated stiffness and density with $\xi_p = 0.40$ and $\eta_p = 0.20$. It can be seen that when $\xi_p \neq 0$ and $\eta_p \neq 0$, the mirror symmetry of the dispersion curve along ΓX and $-\Gamma X$ is broken. This scenario presents an opportunity for one-directional wave propagation.

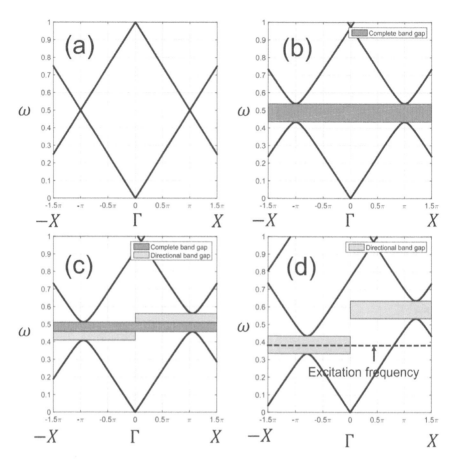

FIGURE 6.36 Dispersion diagrams for beam in longitudinal motion and harmonic modulation: a) non-modulated beam with $\xi_p = 0$ and $\eta_p = 0$, b) space-modulated only beam with $\xi_p = 0.40$ and $\eta_p = 0$, c) space- and time-modulated beam with $\xi_p = 0.40$ and $\eta_p = 0.05$, d) space – and time-modulated beam with $\xi_p = 0.40$ and $\eta_p = 0.20$. When $\xi_p \neq 0$ and $\eta_p \neq 0$ the mirror symmetry is broken causing one directional propagation. This figure is reprinted from Ref [57] with permission.

When a specific frequency near the pass band for ΓX direction is excited, it will not propagate along the $-\Gamma X$ direction due to a band gap (see Fig. 6.36d) actively created from spatio-temporal modulation of stiffness.

6.6.1.4 How Directional Band Gaps Are Topological

So far it is seen that waves will propagate along only one direction due to the symmetry breaking along the wave number space. But how this behavior is topological? Please refer to section 5.6 in Chapter 5, where it is said that to ensure a topological phenomenon the wave function must acquire a geometric phase. To be a topological phenomenon, the example should be explained through geometric phase. Figure 6.36 does not present sufficient clues on this behavior, except that the wavenumber space

Waves in Periodic Media

symmetry is broken. It was said before that the active modulation of spatio-temporal material parameter causes breaking of time-reversal symmetry. If the time-reversal symmetry is broken, then the eigenvalues at specific wave numbers in both the positive frequency axis (i.e., ω) and the negative frequency axis (i.e., $-\omega$) should not match. This is not apparent in Fig. 6.36. This symmetry is called particle ($+\omega$) hole ($-\omega$) symmetry. Both particle-hole symmetry and wave number space symmetry are broken. This can be demonstrated using another example very nicely presented in Ref [6]. Similar sinusoidal modulation of material parameters was used. The metamaterial composed in this example was very small for molecular dynamics study. Using the spectral energy density (SED) method, the eigenvalue solution was found on the entire frequency–wave number space where the frequency space includes both the $+\omega$ (particle) side and the $-\omega$ (hole) side. From this detailed study a two-dispersion band structure map was obtained when the velocities were $\mathbf{v}$ = zero and 350m/s.

These two band structures are shown in Fig. 6.37a and b. When the modulation velocity is zero the temporal modulation is absent. As expected, the dispersion map

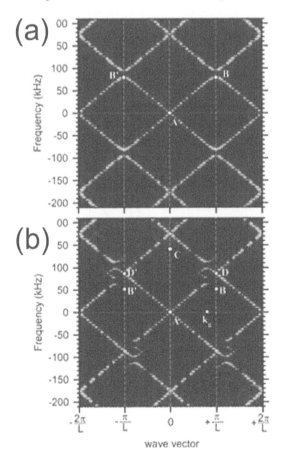

FIGURE 6.37 Band structure with modulating velocity: a) v = 0, b) v = 350 m/s. This figure is reprinted from Ref [6] with permission.

resembles Fig. 6.36a. However, when the velocity is nonzero, the dispersion map has many features. Frequency–wave number eigenvalues are non-symmetric about the Γ point along the ΓX and $-\Gamma X$ space. Interestingly there are many faint frequency–wave number maps that seem uniformly shifted and caused a band gap. Due to particle-hole symmetry breaking, the same bad gap is not present in $-\omega$ space. The shifts in these faint frequency bands are uniform and were found to be $\Omega_p^n = \dfrac{2\pi}{T_p}$

where, T_p is the temporal periodicity of the stiffness modulation in the metamaterial. The faint bands gradually diminish (Fig. 6.37b) with increasing n, meaning their intensity decreases as the temporal Bloch wave index n increases. A hybridization gap (or a directional band gap) between frequency-shifted bands (Fig. 6.37b) and bands from the time-independent stiffnesses (Fig. 6.37a) was identified. Two such band gaps were identified along the wave number axis. However, they are not symmetric about the Γ point. At wave number k_g in the ΓX space, the hybridization gap was found along the $+\omega$ frequency space (a.k.a particle space). But along the hole $(-\omega)$ frequency space such a gap does not exist. Similarly, due to symmetry breaking, such a bad gap is not present at $-k_g$ along $-\Gamma X$ space. Please note that within ΓX and $-\Gamma X$ only one k_g point occurs. This is very interesting and intriguing, helping us explain the directional band gap using geometric phase. It is argued that if the velocity of the temporal modulation $\mathbf{v}$ is changed to a negative value, then the complete band structure will flip horizontally, and the hybridization gap will form at $-k_g$ wave number on $-\Gamma X$ space. In such case, no such point will be identified on the ΓX space. Hence, whether positive or negative velocity of modulation is imposed, one and only one hybridization gap is formed either at k_g point or $-k_g$ point, respectively.

6.6.1.5 Manifold with Parallel Transport of the Wave Function

It is described in Chapter 5 that a Brillouin zone in a periodic metamaterial creates a circular loop that closes itself (@ $k = +\dfrac{\pi}{D_p}, -\dfrac{\pi}{D_p}$), represented by a circle with an angular rotation of 2π. If such a loop is self-repeating without any twist, like a Mobius strip (refer to Fig. 5.13 in Chapter 5), then a vector transported along that space should have no effect, as shown in Fig. 6.38a. The staring vector and the repeating vector will have the same polarity at any arbitrary junction point between $-\Gamma X$ and ΓX. However, this would not be the case if a band between $-\Gamma X$ and ΓX has a twist like a Mobius strip at any point between $+\dfrac{\pi}{D_p}$ and $-\dfrac{\pi}{D_p}$ along k wavenumber axis. Such a twist is identified at wave number k_g in the previous section at the point of hybridization gap (see Fig. 6.37b) when the stiffness of the metamaterial is temporally modulated. Also, it is mentioned that there is only one k_g point within $+\dfrac{\pi}{D_p}$ and $-\dfrac{\pi}{D_p}$ or along the $-\Gamma X$ and ΓX space, due to the symmetry breaking. Hence, a Mobius strip is formed with only one twist due to a spatio-temporal modulation of stiffness in the metamaterial. Next, it is required to see the parallel transport of a wave function, which is a wave displacement function for the elastic wave. Figure 6.38b shows a Mobius strip with parallel transport of an orthogonal vector. Understanding the mechanism and mathematics from section 6.8.3 in Appendix, it

Waves in Periodic Media

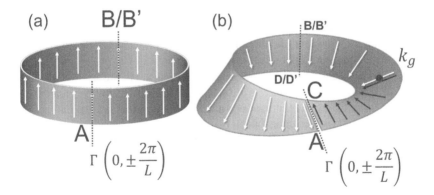

FIGURE 6.38 a) Conventional wave number space within Brillouin zone in a time-independent regular periodic metamaterial, b) illustration of a Möbius strip representing the wave number space manifold that is supporting the wave propagation in time-dependent periodic metamaterial. Time modulation of stiffness caused the generation of the manifold about the wave number k_g.

is clearly visible that the parallel transport of a vector (i.e., the polarity of the wave function) will flip at the wavenumber k_g. This will cause an opposite polarity at the Γ point. The flipping at wavenumber k_g is not apparent in the figure as the front view seems obvious due to the smooth transition of the vector. This appears by virtue of the topology of the manifold. The polarity of the wave function will reverse at k_g and wave function after completing a 2π rotation will incur a phase of π. This is the geometric phase acquired by the topology of a Mobius strip formed due to the symmetry breaking. Here at the gap, the wave displacement function transitions between the zeroth order to the first-order wave function. If the Berry phase along this path is calculated, a monopole will be found at k_g. The Berry phase value will be π. Thus, the Chern number would be 1. If the vector keeps repeating its path, after crossing the k_g wave number a second time the parallelly transported vector field will lose the acquired Berry phase and the geometric phase will be zero after 4π rotation. This causes the non-reciprocity of the wave and causes directional band gap or unidirectional topological wave propagation.

6.7 QUANTUM ANALOGOUS ELASTIC WAVES

Behaviors of elastic and acoustic wave propagation and their features are compared throughout this book. In Chapter 4, Table 4.4 presents the equivalency between the electromagnetic waves and the acoustic waves with respect to their parameters (wave functions, polarization, etc.). In Chapter 5, comparisons between elastic wave and quantum wave equations (Schrödinger equation) are presented in Table 5.1. Different quantum phenomena and analysis techniques are applied to explain electromagnetic systems and their unique behavior. This was possible only by introducing renewed but novel interpretation of the Helmholtz equation in light of the Schrödinger equation. As discussed in this book, the quantum Hall effect, topological waves [58], Dirac degeneracies [17, 18, 48, 49], Anderson localization [59], Bloch

308   Metamaterials in Topological Acoustics

oscillations [60], and supersymmetry (SUSY) [61] behaviors are synonymous with electromagnetic waves and quantum waves. In some cases, acoustic waves show similar behavior. However, the demonstration of the wave equivalency is not complete until the quantum operations that are presented in Chapter 5 are derived for the elastic waves. Out of many, a few specific quantum operations that are thought to be unique in quantum mechanics are surprisingly found to have equivalent nature in elastic waves. A few such operations are presented in this section and applied to acoustic and elastic waves under special circumstances. These operations are ladder operations (refer to section 5.2.2 in Chapter 5) and operations that lead the Klein-Gordon equation to Dirac equations (refer to section 5.4 in Chapter 5).

6.7.1 Hamiltonian and Ladder Operation for Elastic Waves

6.7.1.1 Elastic Hamiltonian

First it is necessary to express the elastic wave equations in their respective Hamiltonian. In elastic waves, both longitudinal (L) and transverse (T) wave modes are present. These wave modes are discussed in Chapter 3, using their respective scalar (ϕ) and vector (Ψ) potentials, and were called P and S waves, respectively. The displacement function **u** at a point due to propagating waves in an elastic media could be expressed using the Helmholtz decomposition, if the media is isotropic. These two modes have orthogonal polarities and are coexistent. This contrasts with the electromagnetic and acoustic waves that have either transverse or longitudinal polarizations. Coexistent polarities could lead to anomalous polarization in anisotropic media under certain conditions. A few such cases were discussed in section 3.2.2.4 in Chapter 3. Interesting polarization behaviors, if manipulated wisely, could provide new directions in information propagation. Elastic waves can enhance computing possibilities beyond what was possible using acoustic waves [17]. Utilizing diverse possibilities of polarization and their superposition, polarization-driven control of *plane waves* could be achieved. A few such scenarios are presented in oblique diffraction, anisotropic wave modes and their interactions [62], mode conversion and or trans modal coupling [63], splitting of transverse wave mode in hyper elastic neo-Hookean material [64], double zero-index material [65] etc.

Elastic wave equations are more complex due to their tensorial expressions (refer to Chapter 3), and it has been difficult to transfer them into simpler scalar equations that are quantum analogous (e.g., Schrödinger equation). It is shown that for a 1D problem classical wave equation can be mapped to the Schrödinger equation. But when the L and T modes are present, it is not easy. Under certain circumstances, and with many assumptions, quantum-inspired physics and functionalities of elastic waves are presented herein.

First, it is assumed that the isotropic material properties periodically vary in a metamaterial, but actual variation is not known for quantum analogous applications. It is assumed that the Lamé constants and density are the functions of x_1 axis only, $\mathbb{L}(x_1) = \lambda$, $\mathrm{m}(x_1) = \mu$, $\rho = \rho(x_1)$. Similar expressions were used in section 3.2.3 in Chapter 3. It is also assumed that the wave propagation is solved in an orthogonal to the spatial variation (i.e., along the x_3 direction). In-plane wave solution on a $x_1 - x_3$ plane is intended.

Waves in Periodic Media   **309**

Starting from Eq. 3.107, using Eq. 3.103 in Chapter 3, a special case for in-plane waves where, $\dfrac{\partial}{\partial x_3} = 0$ and $u_3 = 0$, the governing equation would read

$$(\lambda+\mu)\frac{\partial}{\partial x_1}(\nabla.\mathbf{u})+\mu\nabla^2 u_1 +(\nabla.\mathbf{u})\frac{\partial\lambda}{\partial x_1}+2\frac{\partial u_1}{\partial x_1}\left(\frac{\partial\mu}{\partial x_1}\right)=\rho\ddot{u}_1 \qquad (6.15.1)$$

$$(\lambda+\mu)\frac{\partial}{\partial x_2}(\nabla.\mathbf{u})+\mu\nabla^2 u_2 +\left(\frac{\partial u_1}{\partial x_2}+\frac{\partial u_2}{\partial x_1}\right)\left(\frac{\partial\mu}{\partial x_1}\right)=\rho\ddot{u}_2 \qquad (6.15.2)$$

Please note that $\dfrac{\partial\lambda}{\partial x_2}=\dfrac{\partial\mu}{\partial x_2}=0$ was used in the above equations as the material properties are the function of x_1 only. Next, the in-plane elastic wave displacement function was decomposed into two modes (longitudinal L and transverse T), satisfying the Helmholtz decomposition. The displacement functions were written as

$$\mathbf{u}_L =\left(u_{L_1}(x_1)\hat{\mathbf{e}}_1 + u_{L_3}(x_1)\hat{\mathbf{e}}_3\right)e^{-ikx_3}e^{i\omega t}\ \text{ satisfying } \nabla\times\mathbf{u}_L =0 \qquad (6.15.3)$$

$$\mathbf{u}_T =\left(u_{T_1}(x_1)\hat{\mathbf{e}}_1 + u_{T_3}(x_1)\hat{\mathbf{e}}_3\right)e^{-ikx_3}e^{i\omega t}\ \text{ satisfying } \nabla\cdot\mathbf{u}_T =0 \qquad (6.15.4)$$

Please note that total displacement is $\mathbf{u}=\nabla.\mathbf{u}_L +\nabla\times\mathbf{u}_T$ according to the Helmholtz decomposition. Substituting Eq. 6.15.3 and 6.15.4 in Eqs. 6.15.1 and 6.15.2, the longitudinal and transverse modes will separate and satisfy the following equations independently.

$$(\lambda+\mu)\underline{\nabla}(\nabla.\mathbf{u}_L)+\mu\nabla^2\mathbf{u}_L +\frac{\partial\lambda}{\partial x_1}(\nabla.\mathbf{u}_L)\hat{\mathbf{e}}_1 +2\frac{\partial\mu}{\partial x_1}(\hat{\mathbf{e}}_1\cdot\nabla)\mathbf{u}_L =-\rho\omega^2\mathbf{u}_L \qquad (6.15.5)$$

$$\mu\nabla^2\mathbf{u}_T +\frac{\partial\mu}{\partial x_1}\hat{\mathbf{e}}_1\times(\nabla\times\mathbf{u}_T)+2\frac{\partial\mu}{\partial x_1}(\hat{\mathbf{e}}_1\cdot\nabla)\mathbf{u}_T =-\rho\omega^2\mathbf{u}_T \qquad (6.15.6)$$

And further, the decoupled scalar equation for $u_{L_3}(x_1)$ and $u_{T_1}(x_1)$ would read

$$(\lambda+2\mu)\left\{\frac{d^2 u_{L_3}}{dx_1^2}-k^2 u_{L_3}\right\}+2\frac{d\mu}{dx_1}\frac{du_{L_3}}{dx_1}=-\rho\omega^2 u_{L_3} \qquad (6.15.7)$$

$$\mu\left\{\frac{d^2 u_{T_1}}{dx_1^2}-k^2 u_{T_1}\right\}+2\frac{d\mu}{dx_1}\frac{du_{T_1}}{dx_1}=-\rho\omega^2 u_{T_1} \qquad (6.15.8)$$

Further, as indicated in Ref [66], to remove the first-order derivative terms, a transformation is required and recommended as $u_{L_3}(x_1)=\psi^L(x_1)/e^{\int\frac{d\mu}{dx_1}\cdot\frac{1}{(\lambda+2\mu)}dx_1}$ and $u_{T_1}(x_1)=\psi^T(x_1)/\mu$. With this transformation, purely polarization-dependent two scalar governing equations will read as follows:

$$\left[-\frac{d^2}{dx_1^2}+U^L(x_1)\right]\psi^L=-k^2\psi^L \tag{6.15.9}$$

$$\left[-\frac{d^2}{dx_1^2}+U^T(x_1)\right]\psi^T=-k^2\psi^T \tag{6.15.10}$$

Here, comparing the above equations with the Schrödinger equation, the elastic Hamiltonian could be realized as $H^{L,T}=-\frac{d^2}{dx_1^2}+U^{L,T}(x_1)$ where the potential functions $U^{L,T}(x_1)$ could be explicitly written as follows:

$$U^L(x_1)=-\omega^2\frac{\rho}{(\lambda+2\mu)}-\frac{1}{(\lambda+2\mu)^2}\frac{d\mu}{dx_1}\frac{d(\lambda+\mu)}{dx_1}+\frac{1}{(\lambda+2\mu)}\frac{d^2\mu}{dx_1^2} \tag{6.15.11}$$

$$U^T(x_1)=-\omega^2\frac{\rho}{\mu}+\frac{1}{\mu}\frac{d^2\mu}{dx_1^2} \tag{6.15.12}$$

Similarly, the equations could be derived for the remaining components (e.g., $u_{L_1}(x_1)$ and $u_{T_3}(x_1)$).

6.7.1.2 Super Symmetry (SUSY) Ladder Operations

The objective of this section is to demonstrate and facilitate the separation of L and T modes of elastic waves in spatial domain. In a continuous elastic solid, the L and T modes are decoupled in time (having different velocity) but coupled in space. They both are entangled along the wave propagation direction. To achieve spatial separation of the longitudinal and shear modes while the material is being elastic has been difficult. It was realized that mode splitting is possible in a continuum material but requires special material model and deformation. For example, hyper elastic neo-Hookean material with defined shear deformation could also break the L and T modes in spatial domain. Without going through quantum understanding, employing continuum mechanics and energy models (refer to Chapter 2) for material constants could help achieve the spatial separation [64] of L and T modes. In this section, keeping the material within the elastic regime, creating metamaterial architecture, the spatial mode separation is achieved through quantum mechanical understanding of SUSY ladder operation.

First please refer to section 5.2.2 in Chapter 5 for Ladder operators and properties. It was shown in Eqs. 5.11.12 and 5.11.13 that a Hamiltonian could be expressed as a summation of Hermitian adjoint pair and energy where the Hermitian adjoint pair A^+A^- or A^-A^+ is composed of two non-Hermitian complex operators. It investigated what these operators do to the wave functions. It was found that the Hamiltonian of the newly operated wave functions ($A^+\Psi$ or $A^-\Psi$) transfer the system to eigenenergy states of the next level up or next level down, respectively like climbing up or down along the eigenladder. This very notion creates the utility of the ladder operation. Degeneracy due to supersymmetric (SUSY) ladders arises by simultaneously destroying/annihilating bosonic quantum state (spin-0) and creating fermionic

Waves in Periodic Media 311

quantum state (spin-1/2). Here in elastic waves, it was proposed that the use of supersymmetric (SUSY) transformation would factorize Hamiltonian by enabling a ladder construct of wave potentials and destroy its ground state. Factorization of Hamiltonian and its connection to SUSY transformation relies on the fact that the ground state wave function is known completely [67]. In quantum mechanics it is realized that factorizability of Hamiltonian and supersymmetric properties are inevitable for one dimensional potential. Now the question is how to judicially factorize the Hamiltonian and construct super-partner Hamiltonian that are related by the SUSY ladder. This process is demonstrated in the following paragraphs.

In all cases, the fundamental eigenmode of fundamental Hamiltonian $H_0 = -\dfrac{d^2}{dx_1^2} + U^0(x_1)$ would satisfy $H_0 \psi_n^0 = -k_n^2 \psi_n^0 = E_0^n \psi_n^0$ where n is the mode number. Applying the adjoint operators, the fundamental and original Hamiltonian would read.

$$H_0 = A^+{}_0 A^-{}_0 + E_0 = A^+{}_0 A^-{}_0 - k_0^2 \qquad (6.15.13)$$

Please note that $E_0 = -k_0^2$ and ψ_0^0 are the ground state eigenvalues and eigenfunction of Hamiltonian H_0. The objective of the SUSY ladder transformation is to create a super-partner Hamiltonian in such a way that it annihilates the ground state. This ground state annihilation with iso-spectrality (i.e., no change in fundamental modal eigenvalues) helps access modal filtering with a series of applications in the future. Mathematically, the non-Hermitian operators could be expressed with a differential operator and a super potential as follows.

$$A^+ = -\frac{d}{dx_1} + W(x_1) \text{ and } A^- = \frac{d}{dx_1} + W(x_1) \qquad (6.15.14)$$

where $W(x_1)$ was proposed [66] to be the solution of the Riccati equation (which is a first-order ordinary differential equation that has a quadratic factor of the unknown function) with arbitrary selection of the eigenstate E.

$$\frac{dW(x_1)}{dx_1} = -W(x_1)^2 + U(x_1) + E \qquad (6.15.15)$$

This allows the introduction of super-partner Hamiltonian.

$$H_S = -\frac{d^2}{dx_1^2} + U^S(x_1) = A^+{}_0 A^-{}_0 + E_0 \qquad (6.15.16)$$

with super potential as $U^S(x_1) = U^0(x_1) - 2\dfrac{dW(x_1)}{dx_1}$. With $E_0 = -k_0^2$ as a fundamental ground state eigenvalue, the solution to the Riccati equation in Eq. 6.15.15 would be $W(x_1) = \dfrac{d\left(\log \psi_0^0\right)}{dx_1}$. Now multiplying the non-Hermitian operator

$A^+ = -\dfrac{d}{dx_1} + W(x_1)$ to the ground state eigenfunction ψ_0^0 will result null. Thus, in this special case with special potential, it shows the super-partner ladder Hamiltonian removes the ground state E_0. But like traditional ladder operations, the energy state differs by 1 from the original Hamiltonian, which says $E_S^n = E_0^{n+1}$. This will lead to a ladder of super-partner Hamiltonian H_S^n. This iso-spectrality with removal of ground state helps sequential filtration of wave modes using SUSY transformation.

Here, what should be the potential has not been discussed so far. Next, it remains to express the potential functions under SUSY formalism and find an appropriate spatial variation of the material constant that could lead to this situation of modal separation.

Specific to the elastic waves, there are two Hamiltonians for L and T polarization of the wave modes as $H^{L,T} = -\dfrac{d^2}{dx_1^2} + U^{L,T}(x_1)$ respectively. Figure 6.39 shows a hypothetical example of how the SUSY ladder could be represented starting from L and T potential functions originated from specific spatial variations of material properties. A family of SUSY potentials U_n^L and U_m^T are independently described, and it is shown how lower modes n for L and m for T are annihilated through operations

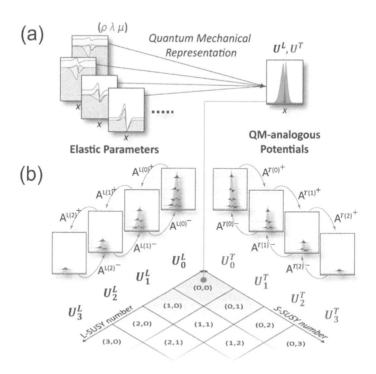

FIGURE 6.39 a) A schematic showing a hypothetical case for spatial variation of material parameters and their respective quantum analog potential functions for L and T polarization, respectively, b) schematic showing SUSY transformed elastic potentials and removal of respective ground state independently for L and T polarization potentials [66].

using A^{L+} and A^{T+}. Specific to any metamaterial design, a step-by-step process below describes how the SUSY ladder is constructed and how the SUSY ladder can contribute to the inhomogeneous material properties, and vice versa. Spatial decoupling of each polarization would require spatial distribution of L and T potentials to be spatially demultiplexed from the beginning. This could be achieved by designing a specific metamaterial. Exploiting the parametric diversity (this requires an offline parametric study of the effective material properties of the metamaterial to be used) in a metamaterial design and their judicial spatial arrangement could help create polarization-dependent SUSY ladders. Such metamaterial could help implement the mode splitter with proper selectivity, realizing continuous SUSY potentials. Please note that the features in the metamaterial should be of sub-wavelength scale, satisfying both the L and T modes. The steps are as follows:

Step 1: First, it is necessary to decide how the polarization would be demultiplexed or separated in a metamaterial. For example, in Ref [66] it was decided that the L-polarized wave will be propagated to the left, while the T-polarized wave will be propagated toward right when a mixed mode wave is injected into a metamaterial.

Step 2: Then, create an elastic metamaterial template with internal architecture that intuitively supports L-mode and T-mode individually. For example [66], a metallic block with a rectangular hole was proposed (shown in Fig. 6.40a). The hole had two dimensions, d_1 and d_2. The variation of these two parameters will change the effective material properties ρ, μ, λ of the unit cell and could be calculated using effective medium theory. For an arbitrary input waveguide with arbitrary parameters d_1 and d_2, specific effective properties ρ, μ, λ could be calculated. However, from the material property map as a function, d_1 and d_2 could help find the ground state potentials $U_0^L(x_1)$ and $U_0^T(x_1)$ using Eq. 6.15.11 and 6.15.12.

Step 3: By knowing how these parameter spaces cause change to the material constants, intuitively continuous SUSY ladder potentials could be designed separately

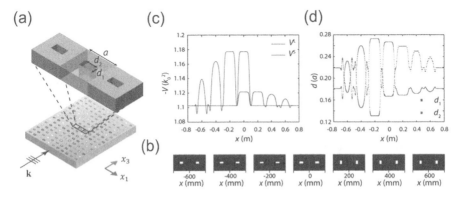

FIGURE 6.40 a) Elastic metamaterial design for separating L-polarized and T-polarized elastic wave mode, b) arrangement of the blocks along x_1 axis to create the separation between left (for L-mode) and right (for T-mode), c) analytically computed, intentionally designed SUSY ladder potentials for U^L and U^T, d) required metamaterial parameters d_1 and d_2 that will result the SUSY ladders in c) [66].

for L-mode and T-mode. Figure 6.40c shows the profile of the elastic potentials required for polarization separator SUSY ladders. The super partner families of SUSY potentials could be calculated using $U_{n+1}^{L,T}(x_1) = U_n^{L,T}(x_1) - 2\dfrac{dW(x_1)}{dx_1}$ where $W(x_1) = \dfrac{d\left(\log\psi_0^0\right)}{dx_1}$ for both L-mode and T-mode. In this specific case governed by Step 1, U^L was placed with left skewness and U^T was kept with right skewness long the x_1 axis about the center.

Step 4: After the SUSY potentials are arranged (Fig. 6.40c) for targeted separation of polarization, parameter space was investigated as a backward problem that causes such SUSY potential ladders. Figure 6.40d shows the variation of d_1 and d_2 along the x_1 axis that results in the spatial distribution of SUSY ladder along x_1 axis.

Step 5: Based on this finding, different but desired d_1 and d_2 values were used to create different metamaterial unit blocks. Modulated metamaterial templates governed as shown in Fig. 6.40d were then arranged side by side in a specific pattern presented in Fig. 6.40b. Here only seven blocks were created. Any higher odd number of unit cell blocks could be created in a wider metamaterial. If a similar linear arrangement of such seven (arbitrarily selected) metamaterial blocks is repeated along the x_3 axis, it is assured that the L-mode and T-mode will separate in the metamaterial along the x_3 axis.

Step 6: Perform a simulation or experiment to test the applicability of SUSY ladder operation for modal decoupling of elastic waves. Ref [66] shows an example with wave field plots that shows the separation of L and T elastic mode.

6.7.2 KLEIN-GORDON EQUATION AND DIRAC EQUATION

The spinor eigenstate, which is a signature behavior in quantum mechanics, is intended to be presented for elastic and acoustic waves equivalently. Thus, the quantum analogous application of elastic wave requires the depiction of the spin eigenstates. Traditionally, we know that the spin eigenstate does not exist in elastic and acoustic waves. Thus, it was difficult to perceive the elastic and acoustic wave equation with spin-equivalent eigenstates. Only recently has a unique spin behavior of acoustic wave been experimentally demonstrated [68] by creating a specific metamaterial. This propels us to wonder, What if, under special circumstances, a quantum equivalent spinor state exists for both elastic and acoustic waves? Although not generally manifested, is it possible to exploit or create a material or a method for spin-mediated device applications? A few such intrinsic cases are presented in this section. Please refer to section 5.4 in Chapter 5, where relativistic particles with spin are described. It is necessary to understand the philosophy of deriving the Klein-Gordon equation followed by the Dirac equation and Hamiltonian of spin ½ particles. In the previous sections, breaking of time-reversal symmetry was demonstrated using an active method; however, in quantum mechanics, spin-mediated topological effects sometime do not need an active field. Intrinsic breaking of symmetry requires the creation of new innovative metamaterials and their analysis, using different

Waves in Periodic Media   **315**

perspectives of the wave equation. Intrinsic symmetry breaking requires local resonance, breaking of structural features associated with symmetry, like chirality of a structural element. By comparing the wave equations from two different realms (quantum waves and elastic waves) would shed some light how quantum analog applications in acoustics and elastic wave could be achieved.

6.7.2.1 Elastic Klein-Gordon Equation

Let's start comparing the elastic wave equation with Klein-Gordon equation for spin-0 bosonic particles presented in section 5.4.1 in Chapter 5. Please note, that unlike the Schrödinger equation, where a first-order time derivate is present, the Klein-Gordon equation has a second-order time derivate resembling a wave equation. The Klein-Gordon equation derived from the square energy form (Eq. 5.28.1) is written in Eq. 5.28.5 and is reiterated herein.

$$\left[\left(\frac{\partial^2}{\partial x_1^2} + \frac{\partial^2}{\partial x_2^2} + \frac{\partial^2}{\partial x_3^2} \right) - \frac{m^2 c^2}{\hbar^2} \right] \Psi = \frac{1}{c^2} \frac{\partial^2}{\partial t^2} \Psi \tag{6.16.1}$$

Next, the elastic wave equation in an unbounded elastic homogeneous medium for P-wave or L-mode with longitudinal wave velocity c_p could be written as (refer to Eq. 3.46 in Chapter 3)

$$\left[\left(\frac{\partial^2}{\partial x_1^2} + \frac{\partial^2}{\partial x_2^2} + \frac{\partial^2}{\partial x_3^2} \right) - F^2 \right] \phi = \frac{1}{c_p^2} \frac{\partial^2}{\partial t^2} \phi \tag{6.16.2}$$

Similarly, for each shear wave component of S-waves or T-mode with shear wave velocity c_s, respective shear wave potentials satisfy a similar equation (refer to Eq. 3.47 in Chapter 3)

$$\left[\left(\frac{\partial^2}{\partial x_1^2} + \frac{\partial^2}{\partial x_2^2} + \frac{\partial^2}{\partial x_3^2} \right) - F^2 \right] \psi = \frac{1}{c_s^2} \frac{\partial^2}{\partial t^2} \psi \tag{6.16.3}$$

where F is the forcing function, generally ignored to solve eigenwave modes in elastic solids. To resemble the Klein-Gordon equation, the square of the forcing function is added.

6.7.2.2 Elastic Dirac Equation

As described in section 5.4.3 in Chapter 5, Dirac proposed to write the equation Eq. 6.16.1 in the following form:

$$\left[\sqrt{\left[\left(\frac{\partial^2}{\partial x_1^2} + \frac{\partial^2}{\partial x_2^2} + \frac{\partial^2}{\partial x_3^2} \right) - \frac{1}{c^2} \frac{\partial^2}{\partial t^2} \right]} \right]^2 \Psi = \frac{m^2 c^2}{\hbar^2} \Psi \tag{6.16.4}$$

316 Metamaterials in Topological Acoustics

Similarly, taking any elastic wave equation for a specific mode, but only one, the elastic Dirac form could be expressed as

$$\left[\sqrt{\left[\left(\frac{\partial^2}{\partial x_1^2}+\frac{\partial^2}{\partial x_2^2}+\frac{\partial^2}{\partial x_3^2}\right)-\frac{1}{c_p^2}\frac{\partial^2}{\partial t^2}\right]}\right]^2 \phi = F^2\phi \tag{6.16.5}$$

Referring to Eq. 5.28.13, specially formulated by Dirac as ingenious mathematical treatment, the square root of the operator could be written as

$$\sqrt{\left(\frac{\partial^2}{\partial x_1^2}+\frac{\partial^2}{\partial x_2^2}+\frac{\partial^2}{\partial x_3^2}\right)-\frac{1}{c_p^2}\frac{\partial^2}{\partial t^2}} = i\left(\left(\gamma^1\frac{\partial}{\partial x_1}+\gamma^2\frac{\partial}{\partial x_2}+\gamma^3\frac{\partial}{\partial x_3}\right)+\gamma^0\frac{\partial}{\partial t}\right) \tag{6.16.6}$$

Hence, the new first-order wave equation (the elastic Dirac equation) with unit velocity would read

$$i\left(\left(\gamma^1\frac{\partial}{\partial x_1}+\gamma^2\frac{\partial}{\partial x_2}+\gamma^3\frac{\partial}{\partial x_3}\right)+\gamma^0\frac{\partial}{\partial t}\right)\phi = F\phi \tag{6.16.7}$$

Following similar steps as in section 5.4.3 in Chapter 5, the Dirac equation and its conjugate was multiplied, and it is imposed that it should derive the elastic Klein-Gordon equation, because both must be true.

$$\phi^*\left(-i\gamma^0\frac{\partial}{\partial t}-i\gamma\cdot\nabla+F\right)\left(i\gamma^0\frac{\partial}{\partial t}+i\gamma\cdot\nabla-F\right)\phi = 0 \tag{6.16.8}$$

Ignoring the linear terms for momentum, and comparing the Klein-Gordon equation with the Laplace operator, the following conditions must be true.

$$\left(\gamma^0\right)^2 = 1,\ \left(\gamma^v\right)^2 = -1\ \gamma^\mu\gamma^\xi+\gamma^\xi\gamma^\mu = 0\ \text{for}\ \forall\ \mu\neq\xi \tag{6.16.9}$$

with $v = 1, 2, 3,\ \mu, \xi = 0, 1, 2, 3,$ which presents an anticommutation relation. Referring to Pauli's matrices in Eq. 5.24.16, the possible solutions that satisfy this relation are

$$\gamma^0 = \begin{pmatrix} I & 0 \\ 0 & -I \end{pmatrix}\ \gamma^v = \begin{pmatrix} 0 & \sigma^v \\ -\sigma^v & 0 \end{pmatrix} \tag{6.16.10}$$

Pauli's matrices (refer to section 5.2.5) are reiterated herein with required equivalent notation.

$$\sigma^1 = \sigma_{x_1} = \begin{pmatrix} 0 & 1 \\ 1 & 0 \end{pmatrix}\ \sigma^2 = \sigma_{x_2} = \begin{pmatrix} 0 & -i \\ i & 0 \end{pmatrix}\ \sigma^3 = \sigma_{x_3} = \begin{pmatrix} 1 & 0 \\ 0 & -1 \end{pmatrix} \tag{6.16.11}$$

Waves in Periodic Media

So, the gamma matrices could be read in full as follows:

$$\gamma^0 = \begin{pmatrix} 1 & 0 & 0 & 0 \\ 0 & 1 & 0 & 0 \\ 0 & 0 & -1 & 0 \\ 0 & 0 & 0 & -1 \end{pmatrix} \quad \gamma^1 = \begin{pmatrix} 0 & 0 & 0 & 1 \\ 0 & 0 & 1 & 0 \\ 0 & -1 & 0 & 0 \\ -1 & 0 & 0 & 0 \end{pmatrix}$$

$$\gamma^2 = \begin{pmatrix} 0 & 0 & 0 & -i \\ 0 & 0 & i & 0 \\ 0 & i & 0 & 0 \\ -i & 0 & 0 & 0 \end{pmatrix} \quad \gamma^3 = \begin{pmatrix} 0 & 0 & 1 & 0 \\ 0 & 0 & 0 & -1 \\ -1 & 0 & 0 & 0 \\ 0 & 1 & 0 & 0 \end{pmatrix}$$

(6.16.12)

Substituting Eq. 6.16.12 into Eq. 6.16.8, two sets of four-vector conjugate solutions would be achieved, and the Dirac equations could be written as

$$\Biggl(\Biggl(\Biggl(\begin{pmatrix} 0 & 0 & 0 & 1 \\ 0 & 0 & 1 & 0 \\ 0 & -1 & 0 & 0 \\ -1 & 0 & 0 & 0 \end{pmatrix} \frac{\partial}{\partial x_1} + \begin{pmatrix} 0 & 0 & 0 & -i \\ 0 & 0 & i & 0 \\ 0 & i & 0 & 0 \\ -i & 0 & 0 & 0 \end{pmatrix} \frac{\partial}{\partial x_2}$$

$$+ \begin{pmatrix} 0 & 0 & 1 & 0 \\ 0 & 0 & 0 & -1 \\ -1 & 0 & 0 & 0 \\ 0 & 1 & 0 & 0 \end{pmatrix} \frac{\partial}{\partial x_3} + \begin{pmatrix} 1 & 0 & 0 & 0 \\ 0 & 1 & 0 & 0 \\ 0 & 0 & -1 & 0 \\ 0 & 0 & 0 & -1 \end{pmatrix} \frac{\partial}{\partial t} - iF \Biggr) \phi = 0$$

(6.16.13)

And

$$\Biggl(\Biggl(\Biggl(\begin{pmatrix} 0 & 0 & 0 & 1 \\ 0 & 0 & 1 & 0 \\ 0 & -1 & 0 & 0 \\ -1 & 0 & 0 & 0 \end{pmatrix} \frac{\partial}{\partial x_1} + \begin{pmatrix} 0 & 0 & 0 & -i \\ 0 & 0 & i & 0 \\ 0 & i & 0 & 0 \\ -i & 0 & 0 & 0 \end{pmatrix} \frac{\partial}{\partial x_2}$$

$$+ \begin{pmatrix} 0 & 0 & 1 & 0 \\ 0 & 0 & 0 & -1 \\ -1 & 0 & 0 & 0 \\ 0 & 1 & 0 & 0 \end{pmatrix} \frac{\partial}{\partial x_3} + \begin{pmatrix} 1 & 0 & 0 & 0 \\ 0 & 1 & 0 & 0 \\ 0 & 0 & -1 & 0 \\ 0 & 0 & 0 & -1 \end{pmatrix} \frac{\partial}{\partial t} + iF \Biggr) \phi^* = 0$$

(6.16.14)

with solutions $\phi = \begin{Bmatrix} \varphi_1 \\ \varphi_2 \\ \varphi_3 \\ \varphi_4 \end{Bmatrix}$ and $\phi^* = \begin{Bmatrix} \overline{\varphi}_1 \\ \overline{\varphi}_2 \\ \overline{\varphi}_3 \\ \overline{\varphi}_4 \end{Bmatrix}$. Please note that for elastic potential solu-

tion ϕ, which was originally one potential function with one amplitude, is now split into four orthogonal components of a vector potential. This shows a mathematically possible situation. However, what it physically means is a bit tricky. The Dirac equation traditionally breaks a spin-0 state into two spin-1/2 states with the help of Pauli's matrix. An elastic wave traditionally has a spin-0 state, but in the above equations Pauli's matrix are used. So, are the above potential eigenstates representative of hidden spin state for elastic waves? To answer this question, a simpler problem in 1D with a spring-mass system is used to demonstrate the spin state in section 6.7.2.4. Equations 6.16.13 and 6.16.14 are complex and require further research and discovery to explain elastic waves from new perspectives.

6.7.2.3 Pseudospin State of Elastic Wave Modes

Here are some preliminary concepts of breaking the longitudinal mode (L) and transverse mode (T) in elastic waves, presented with their possible orthogonal pseudospin states.

Longitudinal wave mode (L): It is visualized that the longitudinal motion of the P-wave is assumed to be the manifestation of two pseudospin states acting inherently in the material and coupled together. They are inseparable, like two spin states in quantum mechanics. To illustrate, two circular discs (*+disc* and *−disc*), each carrying a vector that is rotating on the discs, are imagined (see Fig. 6.41). These two vectors are rotating in opposite directions with normal to the rotational plane opposite to each other but orthogonal to the direction of the longitudinal wave propagation. Here the pseudospin discs are situated on the $x_1 - x_2$ plane. Figure 6.41a shows the perspective from the pseudospin $-\frac{1}{2}$, or *−disc*. Here, at one instant, the rotating vector is making an angle θ^- on the *−disc* but rotating counterclockwise on the disc. From the same perspective of *−disc*, another vector with the same magnitude is rotating on the pseudospin $+\frac{1}{2}$ or on the *+disc* in the clockwise direction. At that same instant the vector is making an angle θ^+ on the *+disc*. Figure 6.41b shows a similar perspective from the pseudospin $+\frac{1}{2}$ or *+disc* with the vector rotating in the clockwise direction on the *+disc* but rotating counterclockwise in the *−disc*. As the magnitude of the vector is the same due to the identical radius of the discs, they produce a superposed longitudinal component of the displacement along the propagating axis but nullify the vertical components. This gives a sinusoidal wave propagation mode along the propagating axis with a longitudinal wave mode as shown in Fig. 6.41. It is hypothesized that if a metamaterial is designed to break this symmetric superposed situation of the pseudospin discs that essentially creates the longitudinal wave mode, the spin states could be extracted and a new quantum analogous application of elastic and acoustic waves could be devised.

Waves in Periodic Media

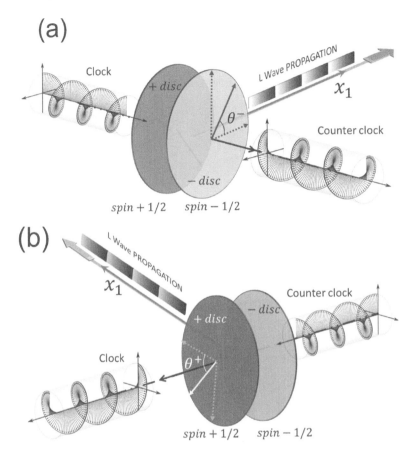

FIGURE 6.41 Schematics showing split spin state of longitudinal wave modes in elastic materials: a) axis aligned towards the observer (−disc) shows the required clockwise and counterclockwise spin discs for forward propagating longitudinal modes, b) axis aligned towards the observer (+disc) shows the required clockwise and counterclockwise spin discs for forward propagating longitudinal modes.

Transverse wave mode (T): Transverse or shear wave modes could be of two types as discussed in Chapter 3. They are vertically polarized SV or TV mode and horizontally polarized SH or TH mode. It is visualized that the transverse motion of the wave is the manifestation of two pseudospin states acting inherently in the material and coupled together. They are inseparable, like two spin states in quantum mechanics. To illustrate, two circular discs (+*disc* and −*disc*), each carrying a vector that is rotating on the discs are imagined (see Fig. 6.42a for TV and Fig. 6.42b for TH wave modes). Here two vectors are rotating on these transverse discs in opposite directions with normal to the rotational plane aligned to the direction of the wave propagation (i.e., x_1 axis). Here, the pseudospin discs are situated on the $x_2 - x_3$ plane. Figure 6.42a shows the pseudospin $-\frac{1}{2}$ and pseudospin $+\frac{1}{2}$ states with the −*disc* and +*disc* placed side by side. Here at an instant the rotating vector is making

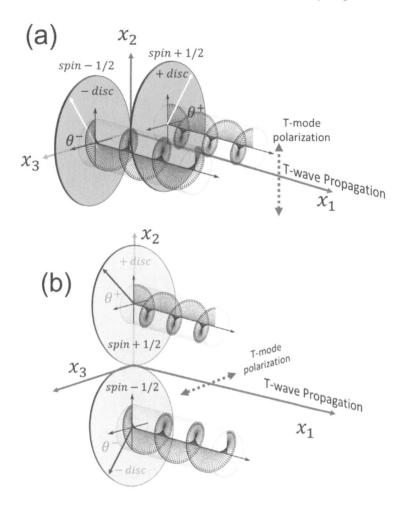

FIGURE 6.42 Schematics showing split spin states of transverse wave modes in elastic materials: a) SV or TV mode, b) SH or TH mode.

an angle θ^- on the $-disc$ but rotating counterclockwise on the disc. From the same perspective of $+disc$, another vector with the same magnitude is rotating on the pseudospin $+\frac{1}{2}$ or on the $+disc$ in the clockwise direction. At that same instant the vector is making an angle θ^+ on the $+disc$. Components of these two rotating vectors are always superposed to create the displacement along the x_2 direction, but the horizontal components cancel each other. Figure 6.42b shows a similar perspective for the horizontally polarized transverse wave modes. Here the pseudospin discs are placed on the $x_2 - x_3$ plane but on top of each other, assuming the wave propagation direction along the x_1 axis. Here again, the two vectors are rotating on these transverse discs in opposite directions with normal to the rotational plane aligned to the direction of the wave propagation (i.e., x_1 axis). In this case of horizontally polarized

Waves in Periodic Media

transverse wave modes, the components of the two rotating vectors on +*disc* and −*disc* are always superposed to create the displacement along the x_3 direction, but the vertical components cancel each other. It is hypothesized that if a metamaterial is designed to break this symmetric superposed situation of the pseudospin discs for the TV and TH modes that essentially creates the transverse wave modes with orthogonal polarities, the spin states could be extracted, and new quantum analogous application of elastic and acoustic waves could be devised.

6.7.2.4 Spring-mass System for Topological Elastic Waves

Here a coupled one-dimensional harmonic crystal (Fig. 6.43) is constructed to solve the wave propagation without external force in long wavelength limit, resembling an elastic wave. The stiffness constant of the springs along the length of the chains connecting two neighboring material points is assumed to be K_L. Similarly, the stiffness constant of the springs orthogonal to the length of the chains connecting upper and lower chains is assumed to be K_T. Mass of the material particles in the lower chain is M_l and the mass of the material particles in the upper chain is M_u. Displacement of the upper and lower material particles is u_u and u_l respectively. The motion of the *i*-th material particles could be written using two coupled differential equations as follows

$$M_u \ddot{u}_u^i + K_L \left(u_u^i - u_u^{i+1} \right) + K_L \left(u_u^i - u_u^{i-1} \right) - K_T \left(u_u^i - u_l^i \right) = 0 \tag{6.17.1}$$

$$M_l \ddot{u}_l^i + K_L \left(u_l^i - u_l^{i+1} \right) + K_L \left(u_l^i - u_l^{i-1} \right) + K_T \left(u_u^i - u_l^i \right) = 0 \tag{6.17.2}$$

These equations can be further modified to

$$M_u \ddot{u}_u^i = K_L \left(u_u^{i+1} - 2u_u^i + u_u^{i-1} \right) - K_T \left(u_u^i - u_l^i \right) = 0 \tag{6.17.3}$$

$$M_l \ddot{u}_l^i = K_L \left(u_u^{i+1} - 2u_u^i + u_u^{i-1} \right) + K_T \left(u_u^i - u_l^i \right) = 0 \tag{6.17.4}$$

For simplicity, let's assume $M_u = M_l$, and dividing the equations by M_l,

$$\frac{\partial^2 u_u^i}{\partial t^2} = \frac{K_L}{M_l} \left(u_u^{i+1} - 2u_u^i + u_u^{i-1} \right) - \frac{K_T}{M_l} \left(u_u^i - u_l^i \right) = 0 \tag{6.17.5}$$

$$\frac{\partial^2 u_l^i}{\partial t^2} = \frac{K_L}{M_l} \left(u_u^{i+1} - 2u_u^i + u_u^{i-1} \right) + \frac{K_T}{M_l} \left(u_u^i - u_l^i \right) = 0 \tag{6.17.6}$$

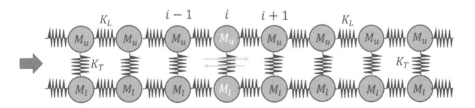

FIGURE 6.43 A schematic of a phononic metastructure with 1D crystals represented using spring-mass system to demonstrate equivalency of Dirac equation in elastic waves.

Metamaterials in Topological Acoustics

Assuming long wavelength limit, the above dynamic equations could be presented in a continuum form with a second-order derivative of space.

$$\frac{K_L}{M_l}\frac{\partial^2 u_u^i}{\partial x^2} - \frac{\partial^2 u_u^i}{\partial t^2} = \frac{K_T}{M_l}\left(u_u^i - u_l^i\right) \tag{6.17.7}$$

$$\frac{K_L}{M_l}\frac{\partial^2 u_l^i}{\partial x^2} - \frac{\partial^2 u_l^i}{\partial t^2} = -\frac{K_T}{M_l}\left(u_u^i - u_l^i\right) \tag{6.17.8}$$

Assuming $c_L = \sqrt{\dfrac{K_L}{M_l}}$ and $c_T = \sqrt{\dfrac{K_T}{M_l}}$, the above equations could be written in a matrix form as follows:

$$\begin{bmatrix} 1 & 0 \\ 0 & 1 \end{bmatrix} \begin{Bmatrix} \dfrac{\partial^2 u_u^i}{\partial x^2} - \dfrac{1}{c_L^2}\dfrac{\partial^2 u_u^i}{\partial t^2} \\ \dfrac{\partial^2 u_l^i}{\partial x^2} - \dfrac{1}{c_L^2}\dfrac{\partial^2 u_u^i}{\partial t^2} \end{Bmatrix} + c_T^2 \begin{bmatrix} -u_u^i & u_l^i \\ u_u^i & -u_l^i \end{bmatrix} = 0 \tag{6.17.9}$$

or

$$\left\{ \left(\frac{\partial^2}{\partial x^2} - \frac{1}{c_L^2}\frac{\partial^2}{\partial t^2}\right)\begin{bmatrix} 1 & 0 \\ 0 & 1 \end{bmatrix} + c_T^2 \begin{bmatrix} -1 & 1 \\ 1 & -1 \end{bmatrix} \right\} \begin{Bmatrix} u_u^i \\ u_l^i \end{Bmatrix} = 0 \tag{6.17.10}$$

It can be realized that the above equation can take the form of the Klein-Gordon equation (refer to eq. 6.16.2) as follows, where the wave function ϕ is equivalent to the vector $\{u_u^i \quad u_l^i\}^T$:

$$\left[\frac{\partial^2}{\partial x^2}\mathbf{I} - (c_T\mathbf{F})^2\right]\begin{Bmatrix} u_u^i \\ u_l^i \end{Bmatrix} = \frac{1}{c_L^2}\frac{\partial^2}{\partial t^2}\mathbf{I}\begin{Bmatrix} u_u^i \\ u_l^i \end{Bmatrix} \tag{6.17.11}$$

where $\mathbf{F}$ matrix is the square root of the matrix $\begin{bmatrix} -1 & 1 \\ 1 & -1 \end{bmatrix}$ and $\mathbf{I}$ is the identity matrix. The above Klein-Gordon equation now could be presented in Dirac form (refer to Eq. 6.16.5) as follows:

$$\left[\sqrt{\left[\frac{\partial^2}{\partial x^2} - \frac{1}{c_L^2}\frac{\partial^2}{\partial t^2}\right]}\mathbf{I}\right]^2 \phi = (c_T\mathbf{F})^2 \phi \tag{6.17.12}$$

following Eq. 6.16.8 the Dirac derivation presented in section 6.7.2.2 The Dirac equation could be written as

$$\phi^*\left(-i\gamma^0\frac{\partial}{\partial t} - i\gamma^1\frac{\partial}{\partial x} + c_T\mathbf{F}\right)\left(i\gamma^0\frac{\partial}{\partial t} + i\gamma^1\frac{\partial}{\partial x} - c_T\mathbf{F}\right)\phi = 0 \tag{6.17.13}$$

Waves in Periodic Media

Comparing the Klein-Gordon equation with Eq. 6.17.13, the following conditions must be true.

$$\left(\gamma^0\right)^2 = -1, \ \left(\gamma^1\right)^2 = 1 \ \gamma^0\gamma^1 + \gamma^1\gamma^0 = 0 \tag{6.17.14}$$

The Dirac equation for the elastic wave propagating in a 1D spring mass system shown in Fig. 6.43 could now create a solution with four elements in the vector of ϕ and ϕ^* with solutions $\phi = \begin{Bmatrix} \varphi_1 \\ \varphi_2 \\ \varphi_3 \\ \varphi_4 \end{Bmatrix}$ and $\phi^* = \begin{Bmatrix} \overline{\varphi}_1 \\ \overline{\varphi}_2 \\ \overline{\varphi}_3 \\ \overline{\varphi}_4 \end{Bmatrix}$

The equations could be written in matrix form as follows:

$$\left[c_L \begin{pmatrix} 0 & 0 & 0 & 1 \\ 0 & 0 & 1 & 0 \\ 0 & 1 & 0 & 0 \\ 1 & 0 & 0 & 0 \end{pmatrix} \frac{\partial}{\partial x} + \begin{pmatrix} 1 & 0 & 0 & 0 \\ 0 & 1 & 0 & 0 \\ 0 & 0 & -1 & 0 \\ 0 & 0 & 0 & -1 \end{pmatrix} \frac{\partial}{\partial t} - i\frac{c_T}{\sqrt{2}} \begin{pmatrix} 1 & -1 & 0 & 0 \\ -1 & 1 & 0 & 0 \\ 0 & 0 & 1 & -1 \\ 0 & 0 & -1 & 1 \end{pmatrix} \right] \begin{Bmatrix} \varphi_1 \\ \varphi_2 \\ \varphi_3 \\ \varphi_4 \end{Bmatrix} = 0 \ldots \tag{6.17.15}$$

and

$$\left[c_L \begin{pmatrix} 0 & 0 & 0 & 1 \\ 0 & 0 & 1 & 0 \\ 0 & 1 & 0 & 0 \\ 1 & 0 & 0 & 0 \end{pmatrix} \frac{\partial}{\partial x} + \begin{pmatrix} 1 & 0 & 0 & 0 \\ 0 & 1 & 0 & 0 \\ 0 & 0 & -1 & 0 \\ 0 & 0 & 0 & -1 \end{pmatrix} \frac{\partial}{\partial t} + i\frac{c_T}{\sqrt{2}} \begin{pmatrix} 1 & -1 & 0 & 0 \\ -1 & 1 & 0 & 0 \\ 0 & 0 & 1 & -1 \\ 0 & 0 & -1 & 1 \end{pmatrix} \right] \begin{Bmatrix} \overline{\varphi}_1 \\ \overline{\varphi}_2 \\ \overline{\varphi}_3 \\ \overline{\varphi}_4 \end{Bmatrix} = 0 \ldots \tag{6.17.16}$$

i.e.,

$$\left(\gamma^1 \frac{\partial}{\partial x} + c_L\gamma^0 \frac{\partial}{\partial t} - i\frac{c_T}{\sqrt{2}} \mathbf{D} \right)\phi = 0 \tag{6.17.17}$$

$$\left(\gamma^1 \frac{\partial}{\partial x} + c_L\gamma^0 \frac{\partial}{\partial t} + i\frac{c_T}{\sqrt{2}} \mathbf{D} \right)\phi^* = 0 \tag{6.17.18}$$

Here ϕ and ϕ^* are both the solutions of the system but they are not symmetric and do not change under certain transformation. They are called non-self-dual solutions. Here the Dirac equations will be changed when a certain inversion is done. If Eqs. 6.17.17 and 6.17.18 are considered individually, they do not satisfy the time-reversal symmetry ($T - symmetry$) or the space inversion symmetry ($P - symmetry$), if the operations ($t \rightarrow -t$ or $x \rightarrow -x$) are done independently. But when both the inversions are applied together, the equations demonstrate symmetry and are mutually recoverable. To understand $\mathcal{T}-symmetry$ and $\mathcal{P}-symmetry$, please refer to section 5.7 in Chapter 5.

Next, considering any one equation from the above and assuming a positive solution of the wave function as $\phi = A_j e^{i(kx+\omega t)}$, an eigenvalue problem could be obtained. Solving the eigenvalue problem, four solutions could be obtained. Two set of bands will be obtained, one will be the lower, symmetric wave mode, and another will be the upper antisymmetric wave mode, both with positive and negative frequency eigenvalues for all wave numbers. They have special spin-like character, which is discussed here, but the detailed trivial solution method of the above equations is omitted and can be found in Ref [6]. Fundamentally due to the two layers of chains in the frequency wave number space, two bands could be found. The lower band is symmetric. Wave propagation in both the positive (forward propagating wave) and negative (backward propagating wave) frequency axes can be presented using a planar wave solution with arbitrary amplitudes (A_j). This character is very much like bosons. Bosonic behavior could be tested by interchanging the chains with their respective amplitudes and the solution will remain unaltered. These are well known facts with the waves in a two-chain system but provide a quantum perspective to the solution.

The upper band, on the contrary, could provide some intriguing quantum insight. The associated wave function with the upper band shows spin-like behavior. It is understood that the two layers of the crystal system, although 1D, are coupled and related to each other. This system also breaks the chiral symmetry (refer to section 5.7 on symmetric operations in Chapter 5). That being said, if the amplitude of the forward-propagating wave is altered, then the backward propagating wave has to be modified for the second band. The basic amplitude constraints of the upper band impart a unique spin-mediated topological behavior to their eigenmode, which is not common for the lower band. A similar understanding could be found from Figs. 6.41 and 6.42. If one of the discs (+$disc$ or −$disc$) is rotated, it is impossible to hold back the other disc to have the wave propagation in the forward direction or in the backward direction. Forward and backward propagating waves are continuously coupled. They are antisymmetric wave modes. This is more like a fermion behavior. Following the previous discussion in Chapter 5 and in section 6.6.1 to support fermion-like behavior, there must be a manifold (Fig. 6.44) at zero wave number that can support both the solution of ϕ and ϕ^*. It is more like a parallel transport of the ϕ wave functions, which flips to demonstrate ϕ^* beyond a point as if the solution is passing through a manifold. Thus, they acquire the geometric phase. This ensured their orthogonality and symmetry, simultaneously.

This shows the differences in topology of the elastic pertaining to the two bands in this simple phononic structure.

FIGURE 6.44 A manifold supporting the solution ϕ and ϕ^* acquiring geometric phase.

Waves in Periodic Media 325

6.7.3 INTRINSIC SPIN STATES OF ELASTIC WAVE

Many of the topics discussed in this book will be brought together in this section and will be referred as needed. Beyond hypothetical pseudospin states that creates quantum valley Hall effect and QSHEs (see Chapter 7 for further details), elastic wave warrants discussions on its intrinsic spin states and spin angular momentum (SAM) density ($\boldsymbol{s}$). By now it is apparent that spin is one of the most important physical properties that are fundamental to the geometrical and topological behaviors of waves. Spin may create necessary geometrical phase for the topological behavior. Hence, spin properties of elastic waves that support both transverse and longitudinal waves are necessary to explore. Before going further in this section, it is highly recommended to read section 3.2.2.1.5 in Chapter 3 that provides the understanding of Helmholtz decomposition and the wave potentials necessary for P and S waves that are synonymous to the L and T waves in this section. It is important to review the sections 2.4 and 2.5 in Chapter 2 on material derivative in *Lagrangian and Eulerian* coordinate system and section 5.2.3 on angular momentum operators in Chapter 5.

6.7.3.1 Elastic Spin Angular Momentum

Elastic spin could be explored in a rotated elastic system. Total angular momentum (TAM) is a conserved quantity under rotation which could be derived from the first principles. Here, Noether's theorem is instrumental to describe the Noether current which happens to be the TAM under rotation. Emmy Noether, a prominent German mathematician, worked with David Hilbert during 1913–1918 at the University of Göttingen. Her works on mathematical invariants paved the way for natural symmetry driven understanding of conserved quantities [69].

Starting from the fundamentals of Hamilton principle described in Chapter 2 (refer Eqs. 2.5, 2.10, and 2.11), Euler-Lagrange equation can be instrumental to find the elastodynamic equation (refer section 2.13 in Chapter 2). Under spatial rotation of θ of the elastic system, the coordinate x_j changes to $x_j + \in_{jlm} \theta_l x_m$, where, $\in_{jlm}$ is the permutation symbol described in Chapter 2. Please note that here not only the coordinates are rotated, the elastic field parameters, e.g., displacements ($u_i(x_j)$) velocities ($v_i(x_j)$), and accelerations ($a_i(x_j)$), are also rotated. Hence, in addition to the variation of the coordinate system $\in_{jlm} \theta_l x_m$, the field varies due to its own rotation and the total variation of the displacement filed u_i, e.g., would read $\delta u_i = -\dfrac{\partial u_i}{\partial x_j}\left(x_j + \in_{jlm} \theta_l x_m\right) + \in_{kij} \theta_k u_j$. The new variation of the displacement field could be plugged in to the Lagrangin described in Chapter 2. Although not derived explicitly, one could easily show that the variation of the Lagrangian due to the variation of the field under rotation could be written as

$$\delta \mathcal{L} = \theta_i \frac{\partial J_{ij}}{\partial x_j} \tag{6.18.1}$$

where J_{i0} is the Noether current known as the TAM derived as

$$J_{i0} = \rho\left[-\in_{ijk} x_j \dot{u}_l \frac{\partial u_l}{\partial x_k} + \in_{ijk} u_j \dot{u}_k\right] \tag{6.18.2}$$

The TAM in index notation has two parts, namely orbital angular momentum (OAM) and the SAM. Please note that TAM, OAM, and SAM are previously discussed in section 5.2.3 in Chapter 5 in context of quantum mechanics. Let's discuss the similar expression alternatively derived in much simpler way using vector notations.

A material point $\mathbf{P}$ in Lagrangian coordinate system could be written as $\mathbf{P} = \mathbf{p} - \mathbf{u}$, where $\mathbf{p}$ is the position vector of the point in Eulerian coordinate system after displacement $\mathbf{u}$ (refer Fig. 2.4b). The $\mathbf{p} = x_j \hat{\mathbf{e}}_j$ here should not be confused with the linear momentum described in Chapter 5 in the context of quantum mechanics. One can find the TAM $\mathbf{J} = \mathbf{P} \times \left(\rho \dfrac{d\mathbf{P}}{dt} \right)$ or $\mathbf{J} = \mathbf{P} \times \left(-\rho \dfrac{d\mathbf{u}}{dt} \right)$, where $\dfrac{d\mathbf{u}}{dt}$ is the velocity of the particle in Eulerian coordinate system. Please note that in Eulerian system, the time derivative is a material derivative. As the Eulerian system can be dynamic, with superposition of translation and rotation motion, here velocity of the coordinate system itself should be considered (discussed in Chapter 2). Considering space-time variation of the velocity field and nonzero local vorticity due to rotating coordinate system, one can write the material derivative of displacement as

$$\frac{d\mathbf{u}}{dt} = \frac{\partial \mathbf{u}}{\partial t} + \left(\frac{\partial \mathbf{u}}{\partial t} \cdot \nabla \right) \mathbf{u} + \frac{\partial}{\partial t} (\nabla \times \mathbf{u}) \times \mathbf{u} \tag{6.18.3}$$

Using the above velocity field, now one can calculate the total time average angular momentum (TAM) $\mathbf{J}$ (refer Eq. 5.17.1 in Chapter 5) as follows, which is also happens to be the summation of OAM and SAM discussed in Chapter 5, $\mathbf{J} = \mathbf{L} + \mathbf{S}$:

$$\mathbf{J} = (\mathbf{p} - \mathbf{u}) \times \left(-\rho \frac{d\mathbf{u}}{dt} \right) \tag{6.18.4}$$

$$\mathbf{J} = \rho (\mathbf{p} - \mathbf{u}) \times \left[-\frac{\partial \mathbf{u}}{\partial t} - \left(\frac{\partial \mathbf{u}}{\partial t} \cdot \nabla \right) \mathbf{u} - \frac{\partial}{\partial t} (\nabla \times \mathbf{u}) \times \mathbf{u} \right] \tag{6.18.5}$$

$$\mathbf{J} = -\rho \left(\mathbf{p} \times \frac{\partial \mathbf{u}}{\partial t} \right) + \rho \left(\mathbf{u} \times \frac{\partial \mathbf{u}}{\partial t} \right) - \rho (\mathbf{p} - \mathbf{u}) \times \left(\frac{\partial \mathbf{u}}{\partial t} \cdot \nabla \right) \mathbf{u}$$

$$- \rho (\mathbf{p} - \mathbf{u}) \times \left(\frac{\partial}{\partial t} (\nabla \times \mathbf{u}) \times \mathbf{u} \right) \tag{6.18.6}$$

Let's explore Figs. 6.41 and 6.42 again. Here, pseudospin states of L and T modes with the polarization of the elastic waves are graphically presented. These spin states are not easily explored unless nontrivial states are intentionally created. This area of research is still at its infancy. However, to explore the intrinsic spin state, there are few nontrivial states that could be created discussed in this section. Let's divide the displacement field $\mathbf{u}$ into L-wave and T-wave fields discussed in section 6.7.1. Let's reiterate $\mathbf{u} = \nabla.u_L + \nabla \times \mathbf{u}_T$ according to the Helmholtz decomposition. Please note the u_L is equivalent to ϕ and $\mathbf{u}_T$ is equivalent to Ψ used in Chapter 3. Fundamental geometrical restrictions of the L and T waves govern $\nabla \times \mathbf{u}_L = 0$ and $\nabla.\mathbf{u}_T = 0$,

Waves in Periodic Media

respectively. Substituting $\mathbf{u} = \mathbf{U}^L + \mathbf{U}^T$, where $\mathbf{U}^L = \nabla.\mathbf{u}_L$ and $\mathbf{U}^T = \nabla \times \mathbf{u}_T$, in Eq. 6.18.6 could read

$$\mathbf{J} = \left[\rho \left(\left(\mathbf{U}^L + \mathbf{U}^T \right) \times \left(\frac{\partial}{\partial t} \mathbf{U}^L + \frac{\partial}{\partial t} \mathbf{U}^T \right) \right) \right]$$

$$-\rho \left(\mathbf{p} - \mathbf{U}^L - \mathbf{U}^T \right) \times \left(\frac{\partial}{\partial t} \left(\mathbf{U}^L + \mathbf{U}^T \right) \cdot \nabla \right) \left(\mathbf{U}^L + \mathbf{U}^T \right)$$

$$-\rho \left(\mathbf{p} - \mathbf{U}^L - \mathbf{U}^T \right) \times \left(\frac{\partial}{\partial t} \left(\nabla \times \mathbf{U}^T \right) \times \left(\mathbf{U}^L + \mathbf{U}^T \right) \right) \tag{6.18.7}$$

where for a monochromatic wave in isotropic homogeneous solid $\rho \left(\mathbf{p} \times \frac{\partial}{\partial t} \left(\mathbf{U}^L + \mathbf{U}^T \right) \right) = 0$ and $\nabla \times \mathbf{U}^L = 0$ and $\nabla.\mathbf{U}^T = 0$, conditions are utilized. A monochromatic wave displacement field could be written as $\mathbf{u} = \mathbf{A}^L(\mathbf{p})e^{-i\omega t} + \mathbf{C}^T(\mathbf{p})e^{-i\omega t}$ or in index notation $\mathbf{u} = u_i \hat{\mathbf{e}}_i = \left(A_i^L(x_j) + C_i^T(x_j) \right)e^{-i\omega t}\hat{\mathbf{e}}_i$. Please note that the amplitude definitions ($\mathbf{A}^L$ and $\mathbf{C}^T$) are consistent with the amplitudes used for L and T waves in section 3.2.2.1.5 in Chapter 3. With close observation, the second and the last term could be visualized as an extrinsic origin dependent quantity and synonymous to the OAM. However, the first term in Eq. 6.18.3 could be visualized as an origin independent SAM quantity. In wave propagation where, displacements are small, assuming $\mathbf{p} \gg |\mathbf{U}^L + \mathbf{U}_T|$, the OAM and SAM can be further simplified to OAM as $\mathbf{L}$ and SAM as $\mathbf{S}$, respectively.

$$\mathbf{L} = \frac{-i\rho\omega}{2} \mathbf{p} \times \left[Im\left[\mathbf{A}_L^*.(\nabla)\mathbf{A}_L \right] + Im\left[\mathbf{C}_T^*.(\nabla)\mathbf{C}_T \right] \right.$$
$$\left. + Im\left[\mathbf{A}_L^*.(\nabla)\mathbf{C}_T \right] + Im\left[\mathbf{C}_T^*.(\nabla)\mathbf{A}_L \right] \right] \tag{6.18.8}$$

$$\mathbf{S} = \frac{\rho\omega}{2} \left[Im\left[\left(\mathbf{A}_L^* \times \mathbf{A}_L \right) + \left(\mathbf{C}_T^* \times \mathbf{C}_T \right) + \left(\mathbf{A}_L^* \times \mathbf{C}_T \right) + \left(\mathbf{C}_T^* \times \mathbf{A}_L \right) \right] \right] \tag{6.18.9}$$

where, * indicated the complex conjugate of the respective amplitude functions. Alternatively referring Eq. 6.18.2, using $\mathbf{p} = x_j \hat{\mathbf{e}}_j$, the OAM and SAM could be written in index notation as follows:

$$\mathbf{L} = \hat{\mathbf{e}}_i \frac{\rho\omega}{2} \in_{ijk} x_j u_l^* \frac{\partial u_l}{\partial x_k} \; ; \mathbf{S} = \hat{\mathbf{e}}_i \frac{\rho\omega}{2} u_k^* \hat{S}_{kl}^i u_l \tag{6.18.10}$$

where $u_i = \left(A_i^L(x_j) + C_i^T(x_j) \right)$ and $\hat{S}_{kl}^i$ is the quantum spin operator which is a 3×3 matrix along the i direction.

Further denoting L-wave related SAM as $s^L = Im\left[\left(\mathbf{A}_L^* \times \mathbf{A}_L \right) \right]$, T-wave related SAM as $s^T = Im\left[\left(\mathbf{C}_T^* \times \mathbf{C}_T \right) \right]$, and hybrid state of superposition of L and T wave related SAM as $s^H = Im\left[\left(\mathbf{A}_L^* \times \mathbf{C}_T\right) + \left(\mathbf{C}_T^* \times \mathbf{A}_L\right)\right]$, the total SAM density $\mathbf{S}$ can further be written as

$$\mathbf{S} = \frac{\rho\omega}{2}\left(\mathbf{s}^L + \mathbf{s}^T + \mathbf{s}^H\right) \tag{6.18.11}$$

6.7.3.2 Elastic Spin Operators

It was previously discussed in Chapter 3 that the amplitude direction of a wave is identified as a polarization vector and they are momentum dependent. Here, further polarization is utilized in association with the understanding of TAM, OAM, and SAM. Without exploring nontrivial states, individually the spin states could not be explored in L and T wave modes. It is discussed in Chapter 5 that when the $\mathcal{T}$−symmetry is not broken, intrinsic spin Hall effect (SHE) can still occur due to spin-orbit interactions. In QSHE unidirectional wave transport happens due to the edge states with opposite spins (pseudospin in acoustics) propagating in opposite directions (Fig. 6.16). Such topological states with spin-momentum locking are responsible for topological behavior. Question we can ask is there any such spin-momentum locking exists in elastic waves?

To explore, let's use three superposition wave states. When these wave modes individually or together are superposed in wave number domain $(\mathbf{k})$, the possibility of existence of the spin states emerge. If two L waves propagating at two different directions on a $x_1 - x_2$ plane are superposed, and displacement polarization profile (real and imaginary) is plotted over space, a spin state could be recognized. Please refer Eq. 3.54.3 for the displacement function for an L wave. Two L waves with amplitudes $\mathcal{A}_1$ and $\mathcal{A}_2$ at equal frequency $\omega = 50$ kHz, but different wave numbers $(\mathbf{k}^{L1} = k_1^{L1}\hat{\mathbf{e}}_1$ and $\mathbf{k}^{L2} = k_1^{L2}\hat{\mathbf{e}}_1 + k_2^{L2}\hat{\mathbf{e}}_2)$ are considered propagating, a) one along the x_1 direction and the other b) at an angle of θ with the x_1 axis, respectively. Hence, $k_1^{L2} = \left|\mathbf{k}^{L2}\right|\cos\theta$ and $k_2^{L2} = \left|\mathbf{k}^{L2}\right|\sin\theta$. The displacement vectors without the time harmonic part for the waves $L1$ and $L2$ individually, and the total superposed potential could be written as, respectively,

$$\mathbf{U}^{L1} = \hat{\mathbf{e}}_1 i k_1^{L1} \mathcal{A}_1 e^{i\left(k_1^{L1} x_1\right)} \tag{6.18.12.1}$$

$$\mathbf{U}^{L2} = \left\{\left[i\mathcal{A}_2 \left|\mathbf{k}^{L2}\right|\cos\theta\right]\hat{\mathbf{e}}_1 + \left[i\mathcal{A}_2 \left|\mathbf{k}^{L2}\right|\cos\theta\right]\hat{\mathbf{e}}_2\right\} e^{i\left(\mathbf{k}^{L2} \cdot \mathbf{x}\right)} \tag{6.18.12.2}$$

$$\mathbf{U}^L = \mathbf{U}^{L1} + \mathbf{U}^{L2} \tag{6.18.12.3}$$

The polarization vector of the superposed wave over a unit length at an interval of 0.002 is plotted in Fig. 6.45a. A twisted helical nature of the displacement vectors on the $x_1 - x_2$ plane is apparent. To explore and quantify this helical nature, lets calculate the spin momentum density of the superposed longitudinal wave along the vertical axis along x_3. Spin momentum density along the x_3 direction could be calculated using the following equation [70] as derived in Eq. 6.18.5.

$$s^L = \left\langle \mathbf{A}_L \middle| \hat{\mathbf{S}}^3 \middle| \mathbf{A}_L \right\rangle = \mathbf{A}_L^* \cdot \left(\hat{\mathbf{S}}^3\right)\mathbf{A}_L = Im\left[\left(\mathbf{A}_L^* \times \mathbf{A}_L\right)\right] \tag{6.18.13}$$

Waves in Periodic Media

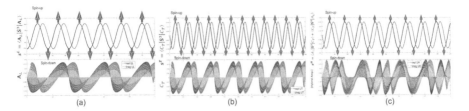

FIGURE 6.45 Spin states along x_3 direction for elastic waves when only the in-plane $(x_1 - x_2)$ propagation is consider a) superposition of two L-wave propagating at an angle $\theta = 25°$, b) superposition of two T-wave propagating at an angle $\theta = 25°$, and c) hybrid state with superposition of one L-wave mode and another T-wave mode propagating at an angle $\theta = 25°$ with respect to the x_1 axis.

where, $\hat{S}^i$ is the quantum spin operator for the elastic wave along the i-th direction introduced in Eq. 6.18.10. The spatial distribution of SAM is plotted in Fig. 6.45a, where spin-up $\left(+\dfrac{1}{2}\right)$ and spin-down $\left(-\dfrac{1}{2}\right)$ states are identified. When two L-waves are superposed and spin angular moment is calculated along x_1 and x_2 directions with quantum spin operators $\hat{S}^1$ and $\hat{S}^2$, respectively, it would result zero. In these equations, the Spin operator $\hat{S}^i$ could be explicitly written as [71].

$$\hat{S}^1 = -i\begin{pmatrix} 0 & 0 & 0 \\ 0 & 0 & 1 \\ 0 & -1 & 0 \end{pmatrix}; \hat{S}^2 = -i\begin{pmatrix} 0 & 0 & -1 \\ 0 & 0 & 0 \\ 1 & 0 & 0 \end{pmatrix}; \hat{S}^3 = -i\begin{pmatrix} 0 & 1 & 0 \\ -1 & 0 & 0 \\ 0 & 0 & 0 \end{pmatrix} \quad (6.18.14)$$

Please note that the amplitude of L and T wave modes, $\mathbf{A}_L$ and $\mathbf{C}_T$ are three-dimensional vector representing three amplitudes of the displacement along three different directions. Generalized spin momentum density for any arbitrary direction of propagation in three dimensions could be written as

$$s^L = \left(U_n^{L*}\hat{S}^i_{mn}U_n^L\right)\hat{\mathbf{e}}_i \quad (6.18.15.1)$$

$$s^T = \left(U_n^{T*}\hat{S}^i_{mn}U_n^T\right)\hat{\mathbf{e}}_i \quad (6.18.15.2)$$

$$s^H = \left(U_n^{L*}\hat{S}^i_{mn}U_n^T + U_n^{T*}\hat{S}^i_{mn}U_n^L\right)\hat{\mathbf{e}}_i \quad (6.18.15.3)$$

Spin states for the superposition of two T-waves (s^T) and hybrid spin state from superposition of one L and one T-wave (s^H) could also be explored. They are superposed and shown in Fig. 6.45b and 6.45c, respectively (self-explanatory). Please note that the s^T state is well known to demonstrate nontrivial spin density and may result QSHE as discussed in this chapter. However, the s^L and s^H state are hidden, and only nontrivial states due to special spin-orbit coupling could manifest such behavior as presented in this section. Under any situation, the conditions $\nabla \times \mathbf{U}_L = 0$ and $\nabla \cdot \mathbf{U}_T = 0$ are not violated. Mathematical equations for s^T and s^H spin momentum

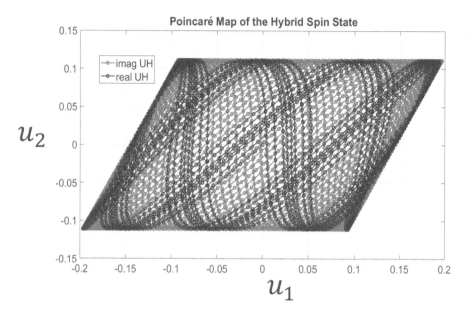

FIGURE 6.46 Poincaré map of hybrid displacement fields when superposition of one L-wave mode and another T-wave mode propagating at an angle $\theta = 25°$ with respect to the x_1 axis are studied.

density can be read from Eqs. 6.18.15.2 and 6.18.6 and in the respective figures. While exploring the spin state the real and imaginary part of the displacement amplitude could be plotted considering the state-space. Figure 6.46 shows the Poincaré map of the hybrid displacement filed obtained from Fig. 6.45c which shows the spin fabric and its spatial bounds.

Displacement polarization vectors and abnormal polarities are discussed in Chapter 3. It is to reiterate (Figs. 3.5 and 3.10) that the momentum dependent geometrical condition of elastic wave can show that the polarity of the $\mathbf{U}_L$ is normal to the wave number sphere and the polarity of the $\mathbf{U}_T$ is tangential to the wave number sphere (**k**). This geometrical boundary condition is essentially the underlying mechanism for the spin-orbit coupling in simplest form. Next, various abnormal polarity conditions in anisotropic solids could be explored that can cause spin-orbit coupling and spin momentum density. At the junction of isotropic and anisotropic solids, the abnormal polarity could open new doors to explore the elastic waves and would provide opportunity by new ways of controlling the waves using spin states for direction propagation that are topological in nature (a few such examples of abnormal polarities are presented in section 3.2.24 in Chapter 3).

6.7.3.3 Topological Behavior with Wave Vortex

Beyond normal polarization, if circularly polarized T-waves are considered (refer Chapter 5, section 5.2.3, Figs. 5.3 and 5.5 on wave vortex), one could calculate the Berry connections and Berry curvatures (Eq. 5.29.12 in Chapter 5) followed by topological Chern number, which is an integral of the Berry curvature over the

Waves in Periodic Media

longitudinal and transverse wave vector spheres ($\mathbf{k}^L$ and $\mathbf{k}^T$ sphere). A mandatory step discussed in Chapter 5 for justifying the topological behaviors. When the Chern number is calculated, the L-wave Chern number should be equal to zero. But the circularly polarized T-wave Chern number should be nonzero. Thus, the spin Chern number will be nonzero under nontrivial topological state of longitudinal wave.

6.7.3.4 Power Flow with Spin Angular Momentum

Along the Poynting vector $\mathbf{P}_E$, the energy flux density $-\dfrac{1}{2}Re\left[\sigma^*.\dfrac{\mathbf{du}}{\mathbf{dt}}\right]$ is derived in section 2.12.4 in Chapter 2. The magnitude of the real part is the average power flowing along the Poynting vector. The total energy density can further be divided into two terms: $|\mathbf{P}_E| = P_{OAM} + P_{SAM}$. Considering Eqs. 6.18.8 and 6.18.9, the total power can further be divided, when the L and T waves are superposed and propagating along any arbitrary directions as follows:

$$|\mathbf{P}_E| = P_{OAM}^L + P_{OAM}^T + P_{OAM}^H + P_{SAM}^L + P_{SAM}^T + P_{SAM}^H \qquad (6.18.16)$$

Utilizing the constitutive relation described in Eq. 3.98.2 in Chapter 3 for an isotropic homogeneous solid, the power terms can be calculated as

$$P_{OAM}^L + P_{OAM}^T = -\frac{\omega}{2}\left[c_p^{\,2}Re\left[i\mathbf{A}^{L*}.\left(\nabla\mathbf{A}^L\right)\right]+c_s^{\,2}Im\left[i\mathbf{C}^{T*}.\left(\nabla\mathbf{C}^T\right)\right]\right] \qquad (6.18.17.1)$$

$$P_{SAM}^L + P_{SAM}^T = \frac{\omega}{2}\left[\frac{\lambda}{2}\nabla\times Im\left[\mathbf{A}^{L*}\times\mathbf{A}^L\right]+\frac{\mu}{2}\nabla\times Im\left[\mathbf{C}^{T*}\times\mathbf{C}^T\right]\right] \qquad (6.18.17.2)$$

$$P_{OAM}^H + P_{SAM}^H = -\frac{\omega}{2}\left[c_p^{\,2}Re\left[i\mathbf{C}^{T*}.\left(\nabla\mathbf{A}^L\right)\right]+c_s^{\,2}Re\left[i\mathbf{A}^{L*}.\left(\nabla\mathbf{C}^T\right)\right]\right]$$

$$+\frac{\omega\rho^2 c_p^{\,2}}{4}\left[\nabla\times Im\left[\mathbf{A}^{L*}\times\mathbf{C}^T\right]+\nabla\times Im\left[\mathbf{C}^{T*}\times\mathbf{A}^L\right]\right] \qquad (6.18.17.3)$$

where, the L and T wave velocities are described in section 3.2.2.1.2 in Chapter 3 as $c_p = \sqrt{\dfrac{(\lambda+2\mu)}{\rho}}$ and $c_s = \sqrt{\dfrac{\mu}{\rho}}$. Hence, it is apparent that the spin state naturally carries power during the elastic and acoustic wave propagation.

6.7.3.5 Elastic and Acoustic Spin Mediated Skyrmion

Based on the foregoing discussion, it is necessary to explore nontrivial state for spin mediated topological behavior of elastic and acoustic waves. Skyrmions are homotopically (refer section 5.6.1 in Chapter 5) nontrivial state and hence could be useful to explore the spin states of elastic and acoustic waves. Usually, vortex like configuration that are statically stable in the bulk are called skyrmions. If a material or a structure is laid out in such a way that it exhibits nonzero spin behavior with a stable spiral nature, it could be characterized as a skyrmion. Thus, the description is topological in nature and hence, a new topological index should be used to describe

332 Metamaterials in Topological Acoustics

such state beyond the typical topological Chern number. Please note that the Chern number is described in Chapter 5 and was used throughout this chapter to explain the topological behavior. The newly introduced topological index is called the Skyrmion Number (Sr). If the displacement field or the velocity filed in known on the entire two-dimensional bulk media, then one could easily calculate an integral quantity and the Skyrmion Number defined as follows [72]:

$$Sr = \frac{1}{4\pi} \int\int \mathbf{u}.\left(\frac{\partial \mathbf{u}}{\partial x_1} \times \frac{\partial \mathbf{u}}{\partial x_2}\right) dx_1 dx_2 \qquad (6.18.18)$$

A nonzero skyrmion number resolves a nontrivial topology which in fact leads to the stability of the skyrmion field [72]. Trivial topological invariant is not immune to any defects in a bulk media. But a nontrivial state could be robust against defects. It is discussed in the previous section that the superposition of L and T elastic waves could result hybridization which results the hybrid SAM (s^H). Hence, it is possible that due to this strong spin-momentum locking, a nontrivial state could be designed where topological behavior would be immune to the defects, as shown in Ref [73], in a hexagonal perforated solid media. It was found that neither the skyrmion fields are dependent on the hexagonal architecture of the phononic crystals nor it must be actively mediated like it is proposed in Ref [73]. Here, 'active' means that it requires multiple sources simultaneously activated around all the boundaries of a bulk media. On the contrary using a passive mechanism (i.e., only one source at any location of the boundary), similar topological state in the bulk that is robust against defect can be generated exploring the Dirac-like cone behavior. This scenario is discussed in Chapter 7, where a phenomenon called topological black hole in a square lattice of phononic bulk media is demonstrated near the Dirac-like cone frequencies.

6.8 APPENDIX

6.8.1 Circular Phononic Crystal in a Host Matrix

A metamaterial is made of different constituent materials and arranged in a periodic fashion. Here a metamaterial made of two constituent materials (material I and II) is considered, where a circular phononic crystal inclusion (material I) is placed in a host matrix made of material II as shown in Fig. A.6.1a. The objective of this section is to present a generic framework and find the mathematical expressions for the Fourier coefficients of the periodic material properties to be used in section 6.2. As shown in Fig. A.6.1b, a fraction of the unit cell made of host material II is occupied by the inclusion made of material I. The integrated unit cell is repeated in a periodic fashion. Hence, it is necessary to find the filling fraction of the inclusion. The filling fraction of a circular inclusion of radius r_0 in a square unit cell of dimension a could be written as

$$f = \frac{\pi r_0^2}{a^2} \qquad (A.6.1)$$

Waves in Periodic Media

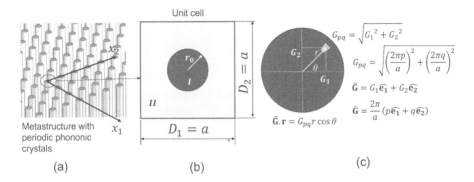

FIGURE A.6.1 a) Circular inclusion in a host matrix b) unit cell, c) Bloch wave vector in polar coordinate.

Referring to Eqs. 6.8.21 and 6.8.22, it is realized that, irrespective of different material constants, their respective expressions with Fourier coefficients in periodic media will look very similar. Hence, to perform the mathematical derivation for the Fourier coefficients of a material constant in a periodic medium the material constant is replaced by a generic notation α herein.

$$\alpha(\mathbf{x}) = \sum_{G_p}\sum_{G_q} \alpha_{(\bar{G}-G)} e^{i[\bar{G}\cdot\mathbf{x}]} = \sum_{G_p}\sum_{G_q} \alpha_{\bar{G}} e^{i[\bar{G}\cdot\mathbf{x}]} = \sum_{p}\sum_{q} \alpha_{pq} e^{i[\bar{G}\cdot\mathbf{x}]} \quad (A.6.2)$$

where $\alpha(\mathbf{x}) \equiv \mu \equiv \rho$. Please note that this equivalency is valid only for their respective mathematical expressions but is not valid for their unit or dimension. The Fourier coefficients of the properties are $\alpha_{(\bar{G}-G)} = \alpha_{\bar{G}}$ or α_{pq} as depicted in Eq. A.6.2 and appeared in the eigenvalue problem in Eq. 6.8.32. Applying the orthogonality condition through integral, the Fourier coefficients will take the form

$$\alpha_{pq} = \frac{1}{D_1}\frac{1}{D_2} \int_{-D_1}^{D_1}\int_{-D_2}^{D_2} \alpha(\mathbf{x}) e^{-i\bar{G}\cdot\mathbf{x}} dx_1 dx_2 \quad (A.6.3)$$

where D_1 and D_2 are the periodicity of the metasystem along the x_1 and x_2 directions, respectively. It can be realized that $D_1 \cdot D_2$ is the area of the unit cell. If the unit cell is a square of dimension a, then it means that it has the same periodicity in two orthogonal directions (i.e., $D_1 = D_2 = a$) and the area is a^2. For any generic media with two-dimensional periodicity having unit cell periodic area A_m, the Fourier coefficients would read

$$\alpha_{pq} = \frac{1}{A_m} \iint \alpha(\mathbf{x}) e^{-i\bar{G}\cdot\mathbf{x}} \cdot d\Gamma \quad (A.6.4)$$

where $d\Gamma$ is infinitesimal area on the surface of the periodic unit cell.

Now, considering a circular phononic crystal in a square matrix, the above integral could be explicitly written as follows:

$$\alpha_{pq} = \frac{1}{a^2} \int_{-a/2}^{a/2} \int_{-a/2}^{a/2} \alpha(\mathbf{x}) e^{-i\bar{G}.\mathbf{x}} dx_1 dx_2 \qquad (A.6.5)$$

Please note that the above expression is valid for both modulus and density. Thus, as a common derivation for both modulus and density Fourier coefficients, let's assume the $\alpha(\mathbf{x})$ for circular phononic crystal is α_I and $\alpha(\mathbf{x})$ for the host matrix is α_{II}. The integral for the unit cell must be broken into two integrals for two material types. As the inclusion is circular, let's transform the integral into the polar coordinate system and split the integral for inclusion and host material.

$$\alpha_{pq} = \frac{\alpha_I - \alpha_{II}}{a^2} \int_0^{r_0} \int_0^{2\pi} r.e^{-i\bar{G}(r).\mathbf{r}} dr d\theta + \frac{\alpha_{II}}{a^2} \int_0^{a} \int_0^{2\pi} r.e^{-i\bar{G}(r).\mathbf{r}} dr d\theta \qquad (A.6.6)$$

While performing the above integral, two scenarios are apparent. First, scenario is when the reciprocal Bloch wave vector $\bar{\mathbf{G}} = 0$ i.e. $\left(\tilde{\mathbf{G}} = \mathbf{G}\right)$ and the second scenario is when the Bloch wave vector $\bar{\mathbf{G}} \neq 0$ i.e., $\left(\tilde{\mathbf{G}} \neq \mathbf{G}\right)$. Please refer to the discussions preceding Eq. 6.8.27, where the meaning of $\tilde{\mathbf{G}}$ and $\mathbf{G}$ is explained. When $\bar{\mathbf{G}} = 0$, the integral in Eq. A.6.6 will read

$$\alpha_{pq} = 2\pi \frac{\alpha_I - \alpha_{II}}{a^2} \int_0^{r_0} r.dr + 2\pi \frac{\alpha_{II}}{a^2} \int_0^{a} r.dr = (\alpha_I - \alpha_{II}) \frac{\pi r_0^2}{a^2} + \alpha_{II} \qquad (A.6.7)$$

It can be self-explanatory that the Fourier coefficients would retrieve average property based on the filling fraction of the inclusion (refer to Eq. A.6.1).

$$\alpha_{pq} = \alpha_I f + (1 - f) \alpha_{II} \qquad (A.6.8)$$

But when $\bar{\mathbf{G}} \neq 0$, the above integral needs to be performed with the Bloch wave vector transformed into polar coordinate as shown below. Followed by Eq. A.6.6 for $\bar{\mathbf{G}} \neq 0$,

$$\alpha_{pq} = (\alpha_I - \alpha_{II}) F(\bar{\mathbf{G}}); \quad F(\bar{\mathbf{G}}) = \frac{1}{a^2} \int_0^{r_0} \int_0^{2\pi} r.e^{-i\bar{G}(r).\mathbf{r}} dr d\theta \qquad (A.6.9)$$

Please refer to Fig. A.6.1c, where the reciprocal wave vector in the polar coordinate is explained. Now it is required to find the integral in Eq. A.6.9. The dot product $\bar{\mathbf{G}}.\mathbf{r}$ could be written as $G_{pq} r \cos\theta$ (refer to Fig. A.6.1c).

$$F(\bar{\mathbf{G}}) = \frac{1}{a^2} \int_0^{r_0} \int_0^{2\pi} r\, e^{-i G_{pq} r \cos\theta} dr d\theta \qquad (A.6.10)$$

To exploit the substitution method and utilize a few identities for a complex integral, let's assume $-i G_{pq} r = m$. First, performing the integral over the angle, the integral will read

$$F(\bar{\mathbf{G}}) = \frac{1}{a^2} \int_0^{r_0} r \left[\int_0^{2\pi} e^{m.\cos\theta} d\theta \right] dr = \frac{1}{a^2} \int_0^{r_0} r.2\pi I_0(m) dr \qquad (A.6.11)$$

Waves in Periodic Media 335

Here, $\int_0^{2\pi} e^{m.\cos\theta} d\theta = 2\pi I_0(m)$ identity is used where $I_0(m)$ is the modified Bessel function of first find with 0-th order. An additional identity could be useful which is $I_p(m) = i^{-p} J_p(im)$, where $J_p(im)$ is the Bessel function of first kind of p-th order and $I_p(m)$ is the modified Bessel function of first kind of p-th order. Substituting the identities for 0-th order and $m = -iG_{pq}r$, the integral in Eq. A.6.11 would read

$$F(\bar{\mathbf{G}}) = \frac{2\pi}{a^2} \int_0^{r_0} r J_0(G_{pq}r) dr \qquad (A.6.12)$$

Further assuming $G_{pq}r = \xi$ and $G_{pq}dr = d\xi$ or $dr = \dfrac{d\xi}{G_{pq}}$, the integral will further modify to

$$F(\bar{\mathbf{G}}) = \frac{2\pi}{a^2} \cdot \frac{1}{G_{pq}^2} \int_0^{G_{pq}r_0} \xi J_0(\xi) d\xi \qquad (A.6.13)$$

Next, using the identity $\int \xi J_0(\xi) d\xi = \xi J_1(\xi)$ and performing the integral followed by replacing the integral limits, the final form of the structure factor $F(\bar{\mathbf{G}})$ will be

$$F(\bar{\mathbf{G}}) = \frac{2\pi}{a^2} \cdot \frac{1}{G_{pq}} r_0 J_1(G_{pq}r_0) = \frac{2\pi r_0^2}{a^2} \cdot \frac{1}{G_{pq}r_0} J_1(G_{pq}r_0) = \frac{2 f J_1(G_{pq}r_0)}{G_{pq}r_0} \qquad (A.6.14)$$

For $\bar{\mathbf{G}} \neq 0$ the Fourier coefficients for periodic material properties can be written as

$$\alpha_{pq} = (\alpha_I - \alpha_{II}) \frac{2 f J_1(G_{pq}r_0)}{G_{pq}r_0} \qquad (A.6.15)$$

This structure factor in Eq. A.6.14 is specific to the circular inclusion in a host matrix. In much of the literature, the above derivation is not presented but rather mentioned that it is easy to derive. However, as can be seen, the above derivation is not so easy, and it is explicitly derived here for the reader's understanding and follow-up derivations if needed.

6.8.2 Square Phononic Crystal in a Host Matrix

In this section, the mathematical derivation of the Fourier coefficients for a periodic metastructure/metamedia (with periodicity a_1 and a_2) is presented in a step-by-step process. For this derivation, please refer to Fig. A.6.2.

Applying the Heaviside step function, the $\alpha(\mathbf{x})$ function for the metamedia in Fig. A.6.2 is expressed as

$$\alpha(\mathbf{x}) = \alpha_{sq} H(\mathbf{x}), \qquad (A.6.16)$$

where $H(\mathbf{x}) = H(x_k) = H(x_1, x_2) = H(x_1). H(x_2)$

$$= \left[H\left(x_1 + \frac{b_1}{2}\right) - H\left(x_1 - \frac{b_1}{2}\right) \right] \left[H\left(x_2 + \frac{b_2}{2}\right) - H\left(x_2 - \frac{b_2}{2}\right) \right] \qquad (A.6.17)$$

and α_{sq} is the numerical value of the property of a square phononic crystal.

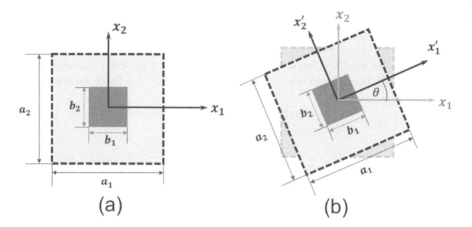

FIGURE A.6.2 a) A typical periodic metamedia with square PVC PnCs in an air matrix. b) Periodic media with rotated square PnCs.

Following Eq. A.6.2,

$$\alpha(\mathbf{x}) = \sum_{G_p}\sum_{G_q} \alpha_{pq} e^{i[\mathbf{G}\cdot\mathbf{x}]} \tag{A.6.18}$$

Alternatively, we can write

$$\alpha(\mathbf{x}) = \sum_{p}\sum_{q} \alpha_{pq} e^{-i\bar{\mathbf{G}}_{pq}\cdot\mathbf{x}} = \sum_{p}\sum_{q} \alpha_{pq} e^{-i(G_1 x_1 + G_2 x_2)} \tag{A.6.19}$$

where $\bar{\mathbf{G}}_{pq} = G_1 \hat{\mathbf{e}}_1 + G_2 \hat{\mathbf{e}}_2 = G_p \hat{\mathbf{e}}_1 + G_q \hat{\mathbf{e}}_2 = \dfrac{2\pi p}{a_1}\hat{\mathbf{e}}_1 + \dfrac{2\pi q}{a_2}\hat{\mathbf{e}}_2$

Using the Fourier transform, the Fourier coefficients α_{pq} are expressed as

$$\alpha_{pq} = \frac{1}{a_1 a_2} \int_{-\frac{a_1}{2}}^{\frac{a_1}{2}} \int_{-\frac{a_2}{2}}^{\frac{a_2}{2}} \alpha_{sq} \cdot \mathbf{H}(x_1, x_2) e^{i(G_1 x_1 + G_2 x_2)} dx_1 dx_2 \tag{A.6.20}$$

$$\alpha_{pq} = \frac{\alpha_{sq}}{a_1 a_2}\left[\int_{-\frac{a_2}{2}}^{\frac{a_2}{2}}\left[\int_{-\frac{a_1}{2}}^{\frac{a_1}{2}}\mathbf{H}(x_1)e^{i(G_1 x_1)} dx_1\right]\mathbf{H}(x_2)e^{i(G_2 x_2)}dx_2\right] \tag{A.6.21}$$

Before conducting the integration, a few basic identities are important to note, such as

$$\frac{dH(x)}{dx} = \delta(x),\ \int_{-\infty}^{\infty}\delta(x)\,dx = H(x),\ \frac{de^{i(G_j x_j)}}{dx_j} = i(G_j)e^{i(G_j x_j)},$$

and $\int_{-\infty}^{\infty} e^{i(G_1 x)}\delta(x-c)\,dx = e^{i(G_1 c)}$.

Waves in Periodic Media

Hence, Eq. A.6.21 changes to

$$\alpha_{pq} = \frac{\alpha_{sq}}{a_1 a_2}\left[\frac{1}{iG_2}\int_{-\frac{a_2}{2}}^{\frac{a_2}{2}}\left[\frac{1}{iG_1}\int_{-\frac{a_1}{2}}^{\frac{a_1}{2}}\mathbf{H}(x_1)d\left(e^{i(G_1 x_1)}\right)\right]\mathbf{H}(x_2)d\left(e^{i(G_2 x_2)}\right)\right] \tag{A.6.22}$$

Conducting the integral by parts separately for two orthogonal directions obtains

$$\int_{-\frac{a_1}{2}}^{\frac{a_1}{2}}\left[H\left(x_1 + \frac{b_1}{2}\right)d\left(e^{i(G_1 x_1)}\right)\right]dx_1$$

$$= \frac{1}{iG_1}\left[H\left(\frac{a_1 + b_1}{2}\right)e^{i\left(G_1\frac{a_1}{2}\right)} - H\left(-\frac{a_1}{2} + \frac{b_1}{2}\right)e^{-i\left(G_1\frac{a_1}{2}\right)}\right] - \int_{-\frac{a_1}{2}}^{\frac{a_1}{2}}e^{i(G_1 x_1)}dH\left(x_1 + \frac{b_1}{2}\right)$$

$$\tag{A.6.23}$$

Of note, $H\left(\dfrac{a_1 + b_1}{2}\right) = 1$, $H\left(-\dfrac{a_1}{2} + \dfrac{b_1}{2}\right) = 0$, and

$$dH\left(x_1 + \frac{b_1}{2}\right) = \delta\left(x_1 + \frac{b_1}{2}\right)dx_1; \int_{-\frac{a_1}{2}}^{\frac{a_1}{2}}e^{i(G_1 x_1)}\delta\left(x_1 + \frac{b_1}{2}\right)dx_1 = e^{-i\left(G_1\frac{b_1}{2}\right)}$$

Substituting these formulas into Eq. A.6.23 obtains

$$\int_{-\frac{a_1}{2}}^{\frac{a_1}{2}}\left[H\left(x_1 + \frac{b_1}{2}\right)d\left(e^{i(G_1 x_1)}\right)\right]dx_1 = \frac{1}{iG_1}\left[e^{i\left(G_1\frac{a_1}{2}\right)} - e^{-i\left(G_1\frac{b_1}{2}\right)}\right] \tag{A.6.24}$$

Similarly,

$$\int_{-\frac{a_1}{2}}^{\frac{a_1}{2}}\left[H\left(x_1 - \frac{b_1}{2}\right)d\left(e^{i(G_1 x_1)}\right)\right]dx_1 = \frac{1}{iG_1}\left[e^{i\left(G_1\frac{a_1}{2}\right)} - e^{i\left(G_1\frac{b_1}{2}\right)}\right] \tag{A.6.25}$$

Thus,

$$\left[\int_{-\frac{a_1}{2}}^{\frac{a_1}{2}}\mathbf{H}(x_1)e^{i(G_1 x_1)}dx_1\right] = \frac{1}{iG_1}\left[e^{i\left(G_1\frac{b_1}{2}\right)} - e^{-i\left(G_1\frac{b_1}{2}\right)}\right] \tag{A.6.26}$$

Following similar integration steps for the x_2 direction,

$$\left[\int_{-\frac{a_2}{2}}^{\frac{a_2}{2}}\mathbf{H}(x_2)e^{i(G_2 x_2)}dx_2\right] = \frac{1}{iG_2}\left[e^{i\left(G_2\frac{b_2}{2}\right)} - e^{-i\left(G_2\frac{b_2}{2}\right)}\right] \tag{A.6.27}$$

Substituting Eqs. A.6.26 and A.6.27 into Eq. A.6.22 obtains

$$\alpha_{pq} = -\frac{\alpha_{sq}}{a_1 a_2 G_1 G_2}\left[e^{i\left(G_1\frac{b_1}{2}\right)} - e^{-i\left(G_1\frac{b_1}{2}\right)}\right]\left[e^{i\left(G_2\frac{b_2}{2}\right)} - e^{-i\left(G_2\frac{b_2}{2}\right)}\right] \tag{A.6.28}$$

Using the identity called Euler's formula $e^{i\Theta} = \cos\Theta + i\sin\Theta$ obtains

$$\alpha_{pq} = \frac{4\alpha_{sq}}{a_1 a_2 G_1 G_2} \left[\sin\left(G_1 \frac{b_1}{2}\right) \right]\left[\sin\left(G_2 \frac{b_2}{2}\right) \right] \qquad \text{(A.6.29.1)}$$

Multiplying $b_1 b_2$ with the numerator and denominator, the equation will be

$$\alpha_{pq} = \frac{4f\alpha_{sq}}{b_1 b_2 G_1 G_2} \left[\sin\left(G_1 \frac{b_1}{2}\right) \right]\left[\sin\left(G_2 \frac{b_2}{2}\right) \right] \qquad \text{(A.6.29.2)}$$

or

$$\alpha_{pq} = f\alpha_{sq} \left[\frac{\sin\left(G_1 \dfrac{b_1}{2}\right)}{G_1\left(\dfrac{b_1}{2}\right)} \right]\left[\frac{\sin\left(G_2 \dfrac{b_2}{2}\right)}{G_1\left(\dfrac{b_2}{2}\right)} \right] = \alpha_{sq} F\left(\bar{\mathbf{G}}\right) \qquad \text{(A.6.29.3)}$$

or

$$\alpha_{pq} = f\alpha_{sq} \left[\frac{\sin\left(G_p \dfrac{b_1}{2}\right)}{G_p\left(\dfrac{b_1}{2}\right)} \right]\left[\frac{\sin\left(G_q \dfrac{b_2}{2}\right)}{G_q\left(\dfrac{b_2}{2}\right)} \right] = \left(\alpha_I - \alpha_{II}\right) F\left(\bar{\mathbf{G}}\right) \qquad \text{(A.6.29.4)}$$

where $f = \dfrac{b_1 b_2}{a_1 a_2}$ is the filling fraction of the unit cell.

When the square rods are rotated at an angle θ, as shown in Fig. A.6.2b, the reciprocal Bloch wave vector is transformed into

$$G_{1'} = \left(G_1 \cos\theta + G_2 \sin\theta\right)\ G_{2'} = \left(-G_1 \sin\theta + G_2 \cos\theta\right) \qquad \text{(A.6.30)}$$

or

$$G_{p'} = \left(G_p \cos\theta + G_q \sin\theta\right)\ G_{q'} = \left(-G_p \sin\theta + G_q \cos\theta\right) \qquad \text{(A.6.31)}$$

6.8.3 PARALLEL TRANSPORT OF A VECTOR

How do we know if a vector is parallelly transported on a surface or on a manifold? Answering this question is not easy. On a flat surface it is a lot easier to deal with than on a curved surface. A vector field on a flat surface is a constant vector field if the covariant derivative of the vector field is zero. A constant vector field means all the vectors on the surface have similar length and are facing the same direction. If covariant derivative of the vector field is zero, it means that two vectors at two different locations on a flat surface is interchnagable. It also means that if the vectors are transported parallelly, they will mutually result in a same vector. This may not be the case on a curved surface. Thus, we

Waves in Periodic Media

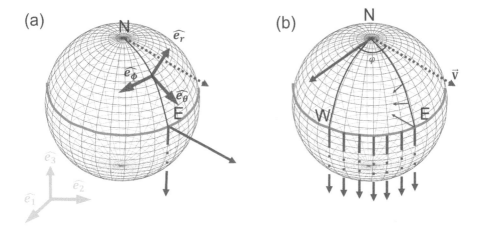

FIGURE A.6.3 a) A spherical curved surface with vector demonstrating parallel transport b) parallel transport in a closed loop on a curved surface acquires a phase angle called geometric phase. Also refer Fig. A.2.2 for the definition of the spherical coordinate system.

need an explanation of how the vectors change on a curved surface. For example, let's compare two vectors on a curved surface (i.e., sphere) in Fig. A.6.3a, which is reused from Chapter 2 Fig. A.2.2. with all its coordinate definition intact. There are two vectors, one dotted at point **N** and another solid at point **E**. It is not straightforward to just slide one over another to see if they are the same vectors. From the perspective of the elements on that surface of the sphere at **N** and **E**, the vectors are different. The dotted vector is pointing towards the horizon, making it tangential to the surface, whereas the solid vector is pointing upwards, making it normal to the surface. For this reason, to compare vectors on a curved surface the mathematical concept of parallel transport is required. In simple words, parallel transport can be defined as a way of moving vectors around on a surface so that the vector remains straight, carrying similar properties (e.g., tangential vector remains tangential or an orthogonal vector remain orthogonal to the surface). Hence, a parallelly transported vector at **E** must be tangential as shown in Fig. A.6.3a. Now if the dotted vector is parallelly transported, keeping the above requirement from point **N** to **E** to **W** and back to **N**, then unique characteristics emerge. The dotted vector and the solid vector recovered at point **N** are not the same vectors, and there is an angular difference between them. Please refer to Fig. A.6.3 b.

Next, let's represent this phenomenon mathematically. Parallel transport along a geodesic is interesting.

6.8.3.1 Christoffel Symbol and Geodesic

It is known that geodesics are the shortest path between two points on any surface. Hence, on a flat 2D surface, the geodesic is a straight line. But for any arbitrary surface the geodesic is not easy to find, but a typical process with an example of a spherical coordinate system is demonstrated here. Let's assume a position vector of a point on the surface is $\vec{P}(x_j)$ when the surface is defined by a parameter λ. The surface is now assumed to be parameterized using two in plane-unit vector and an out-of-plane normal vector. For a spherical surface these unit vectors are $\hat{\mathbf{e}}_r$, $\hat{\mathbf{e}}_\theta$, and $\hat{\mathbf{e}}_\phi$ as shown in Fig. A.6.3. $\hat{\mathbf{e}}_\theta$, and $\hat{\mathbf{e}}_\phi$ are the in-plane basis vectors. A spherical surface of unit radius depicted in Fig. A.2.2 can be written in terms of parameter θ and ϕ as follows

$$x_1 = \cos(\phi)\sin(\theta); \; x_2 = \sin(\phi)\sin(\theta); \; x_3 = \cos(\theta) \tag{A.6.32}$$

where $\lambda = \boldsymbol{e} = [\theta,\phi]$, and the position vector of a point is $\vec{P}(x_j)$ could also be written as $\vec{P}(e^j)$ or $\vec{P}(\theta,\phi) = (x_1(\theta,\phi),\, x_2(\theta,\phi),\, x_3(\theta,\phi))$. The Christoffel symbol for any curved surface that is traced by its own two tangential basis vectors ($\hat{\mathbf{e}}_\theta$, and $\hat{\mathbf{e}}_\phi$ for sphere) while the normal vector ($\hat{\mathbf{e}}_r$ for sphere) remains normal, can be written as

$$\Gamma_{ij}^k = \frac{\partial^2 \vec{P}(e^j)}{\partial e^i \partial e^j} \cdot \frac{\partial \vec{P}(e^j)}{\partial e^l}\, \mathfrak{g}^{lk} \tag{A.6.33}$$

where i, j, and k take values 1 and 2; $\lambda = \boldsymbol{e} = [\theta,\phi]$; $e^1 = \theta$, $e^2 = \phi$, and $\mathfrak{g}^{lk}$ is the inverse metric tensor described below when standard Einstein notations are implied. The Christoffel symbol essentially tracks how a basis vector changes from point to point on a surface; thus rate of change of the position vector with respect to parameters θ and ϕ, representing two tangential directions, is very important. Further these rates of changes are placed in a matrix form as follows and called a metric tensor.

$$g_{ij} = \begin{bmatrix} \dfrac{\partial \vec{P}}{\partial \theta} \cdot \dfrac{\partial \vec{P}}{\partial \theta} & \dfrac{\partial \vec{P}}{\partial \theta} \cdot \dfrac{\partial \vec{P}}{\partial \phi} \\[3ex] \dfrac{\partial \vec{P}}{\partial \phi} \cdot \dfrac{\partial \vec{P}}{\partial \theta} & \dfrac{\partial \vec{P}}{\partial \phi} \cdot \dfrac{\partial \vec{P}}{\partial \phi} \end{bmatrix} \tag{A.6.34}$$

where

$$\frac{\partial \vec{P}}{\partial \theta} = \cos(\phi)\cos(\theta)\frac{\partial \vec{P}}{\partial x_1} + \sin(\phi)\cos(\theta)\frac{\partial \vec{P}}{\partial x_2} - \sin(\theta)\frac{\partial \vec{P}}{\partial x_3} \tag{A.6.35}$$

$$\frac{\partial \vec{P}}{\partial \phi} = -\sin(\phi)\sin(\theta)\frac{\partial \vec{P}}{\partial x_1} + \cos(\phi)\sin(\theta)\frac{\partial \vec{P}}{\partial x_2} \tag{A.6.36}$$

Substituting the above values, the metric tensor and the inverse of the metric tensor for spherical surface would read

$$g_{ij} = \begin{bmatrix} 1 & 0 \\ 0 & \sin^2(\theta) \end{bmatrix}; \; \mathfrak{g}^{ij} = (g_{ij})^{-1} = \begin{bmatrix} 1 & 0 \\ 0 & \dfrac{1}{\sin^2(\theta)} \end{bmatrix} \tag{A.6.37}$$

Waves in Periodic Media

The geodesic between two points on any curved surface could be solved using the following two coupled differential equations:

$$\frac{d^2 e^k}{d\lambda^2} + \Gamma_{ij}^k \frac{de^i}{d\lambda} \frac{de^j}{d\lambda} = 0 \tag{A.6.38}$$

or explicitly,

$$\frac{d^2 \theta}{d^2 \lambda} + \Gamma_{22}^1 \frac{d\phi}{d\lambda} \frac{d\phi}{d\lambda} = 0 \tag{A.6.39}$$

$$\frac{d^2 \phi}{d^2 \lambda} + \Gamma_{12}^2 \frac{d\theta}{d\lambda} \frac{d\phi}{d\lambda} + \Gamma_{21}^2 \frac{d\phi}{d\lambda} \frac{d\theta}{d\lambda} = 0 \tag{A.6.40}$$

where λ is the parameter that defines the change in the parameter space, causing the motion along the surface. Specifically, if a circle of latitude is used as a curve on the spherical surface, then the solution of the geodesic curve would be the equator or any great circle on the sphere. If a solution of a geodesic is found on a curved surface, then the parallel transport of a vector field along that geodesic path would be required further to find if there is any phase accumulation due to the geometry. This phase was fined earlier as geometric phase or topological phase.

6.8.3.2 Parallel Transport along a Curved Path

When the vectors are parallelly transported (see Fig. A.6.3b), the change between two consecutive vectors along the path is a vector that is always normal to the surface, and here the normal vector points towards the center of the sphere. While parallelly transporting a vector, the rate of change of that vector is always normal to the surface. Mathematically, the rate of change of a vector is equal to the normal vector. If the normal vector $\hat{\mathbf{n}}$ to the surface is subtracted from the rate of change of the vector, the result should be equal to zero as follows

$$\frac{d\vec{\mathbf{V}}}{d\lambda} - \hat{\mathbf{n}} = 0 \tag{A.6.41}$$

It is impossible to define a constant vector field on a curved surface, but by parallel transporting a vector along a curve on that surface, the closest thing could be achieved. On a curved surface the covariant derivatives are used to find a constant vector field that is the result of a parallel transport. The covariant derivative is defined by equation A.6.41, where the rate of change of a vector field $\vec{\mathbf{V}}$ in a direction is changed by a parameter λ defined for the parameterized surface, with the normal component subtracted. When the covariant derivative is zero, it means that the vector is parallelly transported. In the case of a sphere, θ and ϕ define the curves along longitude and latitude, respectively. And the tangent basis vectors on the sphere are given as

$$\frac{\partial \vec{P}}{\partial \theta} = \hat{\mathbf{e}}_\theta; \; \frac{\partial \vec{P}}{\partial \phi} = \hat{\mathbf{e}}_\phi \tag{A.6.42}$$

$\hat{\mathbf{e}}_\theta$ basis vectors are the tangent vectors along the θ, curve and $\hat{\mathbf{e}}_\phi$ basis vectors are the tangent vectors along the ϕ curve. Expanding the $\vec{\mathbf{V}}$ vector as a linear combination of $\hat{\mathbf{e}}_\theta$ & $\hat{\mathbf{e}}_\phi$ basis vector, Eq. A.6.41 would read

$$\frac{d\vec{\mathbf{V}}}{d\lambda} - \vec{\mathbf{n}} = \frac{\partial}{\partial e^i}\left[V^\theta\hat{\mathbf{e}}_\theta + V^\phi \hat{\mathbf{e}}_\phi\right] - \vec{\mathbf{n}} = \frac{\partial}{\partial e^i}\left[V^j\hat{\mathbf{e}}_j\right] - \vec{\mathbf{n}} \qquad (A.6.43)$$

where $\lambda = e = [\theta,\phi]$; $e^1 = \theta$, $e^2 = \phi$ as is described in the previous section, and standard Einstein notations are implied. Next, taking the derivative and using the Christoffel symbol from Eq. A.6.33, the above equation can be rewritten as

$$\frac{d\vec{\mathbf{V}}}{d\lambda} - \vec{\mathbf{n}} = \left[\frac{\partial V^k}{\partial e^i} + V^j\Gamma^k_{ij}\right]\hat{\mathbf{e}}_k \qquad (A.6.44)$$

In this derivation, it can be found that the second fundamental form of the Christoffel symbol is the normal component with a unit normal vector with it. As, by definition, the normal vector is further subtracted from the expression, they cancel out and give the final form in Eq. A.6.44. Next, perform the following operations in a sequence to calculate the Christoffel symbol and necessary mathematical terms for a sphere (but equally applicable for any surface parameterized by two tangent vectors).

$$\frac{\partial\hat{\mathbf{e}}_j}{\partial e^i} = \Gamma^k_{ij}\hat{\mathbf{e}}_k + L_{ij}\hat{\mathbf{n}}; \;\; \frac{\partial\hat{\mathbf{e}}_j}{\partial e^i}\cdot\hat{\mathbf{e}}_l = \Gamma^k_{ij}\hat{\mathbf{e}}_k\cdot\hat{\mathbf{e}}_l = \Gamma^k_{ij}g_{kl}; \;\; \frac{\partial\vec{e_j}}{\partial u^i}\cdot\vec{e_l}g^{lm} = \Gamma^k_{ij}\delta^m_k \;\; (A.6.45)$$

Please note that the metric tensor and the inverse metric tensor are related (Eq. A.6.37) and hold $g_{kl}g^{lm} = \delta_{km}$ where δ_{km} is the Kronecker delta defined in this book multiple times. Thus the Christoffel symbol for a sphere reads

$$\Gamma^m_{ij} = \left(\frac{\partial\hat{\mathbf{e}}_j}{\partial e^i}\cdot\hat{\mathbf{e}}_l\right)g^{lm} \qquad (A.6.46)$$

$$\Gamma^1_{22} = -\frac{1}{2}\sin(2\theta); \;\; \Gamma^2_{12} = \Gamma^2_{21} = \cot(\theta) \qquad (A.6.47)$$

Plugging the Christoffel symbols into Eq. A.6.44, specific to the spherical surface the covariant derivatives will read

$$\nabla_\theta\vec{\mathbf{V}} = \left[\frac{\partial V^\phi}{\partial\theta} + V^\phi\cot(\theta)\right]\hat{\mathbf{e}}_\phi \qquad (A.6.48)$$

$$\nabla_\phi\vec{\mathbf{V}} = \left[\frac{\partial V^\theta}{\partial\phi} - V^\phi\frac{1}{2}\sin(2\theta)\right]\hat{\mathbf{e}}_\theta + \left[\frac{\partial V^\phi}{\partial\phi} + V^\theta\cot(\theta)\right]\hat{\mathbf{e}}_\phi \qquad (A.6.49)$$

6.8.3.3 Example: Parallel Transport

Let's take a curve that travels along an equator from **W** to **E** (i.e., $\phi = 0$ to $\phi = \pi/2$) on a spherical surface as shown in Fig. A.6.4. The formula for this curve is a function of the curve parameter λ where $\lambda \to \left(\theta = \dfrac{\pi}{2}, \phi = \lambda \right)$. Here θ is the equator's latitude and the vector travels along the curve as λ increases. Let's assign a vector field to the curve as

$$\vec{V} = \cos(\phi)\hat{e}_\theta + \sin(\phi)\hat{e}_\phi \tag{A.6.50}$$

Let's find the covariant derivative of the vector field along the curve parameter λ, which is the $\hat{e}_\phi$ basis vector. As $\theta = \dfrac{\pi}{2}$ and is constant, it is just required to calculate Eq. A.6.49 as follows:

$$\nabla_\phi \vec{V} = \left[\frac{\partial V^\theta}{\partial \phi} \right]\hat{e}_\theta + \left[\frac{\partial V^\phi}{\partial \phi} \right]\hat{e}_\phi = -\sin(\phi)\hat{e}_\theta + \cos(\phi)\hat{e}_\phi \neq 0 \tag{A.6.51}$$

Equation A.6.51 shows that the covariant derivative is nonzero. In fact, the vectors are rotating on the tangent plane. So the rate of change of the vector is not completely normal to the surface, but there always exists a tangential component. If the vector field is changed to $\hat{e}_\phi$, then it could be easily shown that the rate of change of the vector has only a normal component, which always points towards the center of the sphere.

Next let's take a different curve on the sphere, which is at $\theta = \dfrac{\pi}{4}$ and $\phi = 0$ to $\phi = \pi/2$. Let's take an initial vector $\vec{V}_0$ along the $\hat{e}_\theta$ basis vector at $\phi = 0$. Parameterizing the curve would be

$$\lambda \to \left(\theta = \frac{\pi}{4}, \phi \right) \text{ and } \vec{V}_0 = 1.\hat{e}_\theta + 0.\hat{e}_\phi \tag{A.6.52}$$

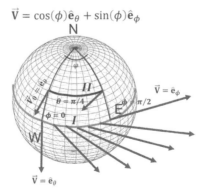

FIGURE A.6.4 Parallel transport of vectors along different curved path.

Parallelly transporting the vector $\vec{V}_0$ means taking the covariant derivative along curve II shown in Fig. A.6.3. However, how the parallelly transported vector would look on this curve at different points is not known. To solve for the parallelly transported vector at any point on curve is expressed as $\vec{V} = V^\theta \hat{\mathbf{e}}_\theta + V^\phi \hat{\mathbf{e}}_\phi$. Hence, the $\vec{V}_0$ vector would serve as an initial condition. By taking the covariant derivative of the vector $\vec{V} = V^\theta \hat{\mathbf{e}}_\theta + V^\phi \hat{\mathbf{e}}_\phi$ at $\theta = \dfrac{\pi}{4}$, which should be equal to zero, the eq. (A.6.53) or (A.6.54)

$$\nabla_\phi \vec{V} = \left[\frac{\partial V^\theta}{\partial \phi} - V^\phi \frac{1}{2}\sin\left(\frac{\pi}{2}\right)\right]\hat{\mathbf{e}}_\theta + \left[\frac{\partial V^\phi}{\partial \phi} + V^\theta \cot\left(\frac{\pi}{4}\right)\right]\hat{\mathbf{e}}_\phi = 0 \quad \text{(A.6.53)}$$

or

$$\left[\frac{\partial V^\theta}{\partial \phi} - \frac{V^\phi}{2}\right] = 0; \left[\frac{\partial V^\phi}{\partial \phi} + V^\theta\right] = 0 \quad \text{(A.6.54)}$$

Solving Eq. A.6.54, the vector components of vector $\vec{V}$ can be found as

$$V^\theta(\phi) = A\cos\left(\frac{\phi}{\sqrt{2}}\right) + B\sin\left(\frac{\phi}{\sqrt{2}}\right); \; V^\phi(\phi) = C\cos\left(\frac{\phi}{\sqrt{2}}\right) + D\sin\left(\frac{\phi}{\sqrt{2}}\right) \quad \text{(A.6.55)}$$

where A, B, C, and D are arbitrary coefficients to be solved using the initial conditions. Thus, after finding the coefficients, the parallelly transported vector along a curve II would be

$$\vec{V} = \cos\left(\frac{\phi}{\sqrt{2}}\right)\hat{\mathbf{e}}_\theta - \sqrt{2}\sin\left(\frac{\phi}{\sqrt{2}}\right)\hat{\mathbf{e}}_\phi \quad \text{(A.6.56)}$$

6.8.3.4 Notes on Parallel Transport for Wave Vectors

Finding a nonzero covariant derivative of a vector field on a curved surface is the key to knowing if the vector field will acquire any *geometric phase* when transported along the parameter space. A wave vector space is synonymous with a parameterized space. On a frequency–wave number topology map, the was a wave displacement vector field modifies due to the parallel transport when moved along a path (along wave number domain enclosing the Brillouin zone) may unlock many hidden features of wave propagation.

6.8.4 Computer Code to Explore Elastic Spin

```
%% THIS SAMPLE CODE EXPLORES THE INTRINSIC SPIN OF ELASTIC
   WAVES            %%
%%%%%%%%%%%%%%%%%%%%%%%%%%%%%%%%%%%%%%%%%%%%%%%%%%%%%%%%%%%%%%%%%%%

%% REFER CHAPTER 3 Eq. 3.61.1 and 3.62.2
```

Waves in Periodic Media

```matlab
% Code written by Prof. Sourav Banerjee Date 06/24/2022

%%%%%%%%%%%%%%%%%%%%%%%%%%%%%%%%%%%%%%%%%%%%%%%%%%%%%%%%%%%%%%%%%%%%%%%%%%%%

%% WAVE - AMPLITUDES
% Amplitude of P-waves
A1 = 0.10; % amplitude of L-Wave 1
A2 = 1.00; % amplitude of L-Wave 2
% Amplitude of S-waves
B1 = 0.10; % amplitude of T-Wave 1
B2 = 0.10; % amplitude of T-Wave 2
%% WAVE - VELOCITIES
% Velocities of P-wave
Cp1 = 3450; % m/sec wave velocity of P-wave 1
Cp2 = 3450; % m/sec wave velocity of P-wave 2
% Velocities of S-wave
Cs1 = Cp1/2; % m/sec wave velocity of S-wave 1
Cs2 = Cp2/2; % m/sec wave velocity of S-wave 2
%% FREQUENCY & DENSITY
w=50000;   % in Hz frequency
rho=2.3;      % in Kg/m^3
SF=rho*w/2; % a Spin Factor
%% WAVENUMBER
% Wavenumber of P-waves
kp1 = w/Cp1; % wavenumber of P-wave 1
kp2 = w/Cp2;  % wavenumber of P-wave 2
% Wavenumber of S-wave
ks1 = w/Cs1; % wavenumber of wave 1
ks2 = w/Cs2;  % wavenumber of wave 2
%% 1-D SPATIAL DISCREATIZATION
Sp_samprate=0.002;
X_limit=5;
%% WAVE PROPAGATION DIRETION OF WAVE 1 and WAVE 2 &
WAVEVECTOR
theta1 = 0 ; % wave propagation diraction of P-wave 1
theta2 = 25; % wave propagation diraction of P-wave 2
% P-wave Vecotors for wave 1 and wave 2
k1_1p=kp1*cosd(theta1);
k2_1p=kp1*sind(theta1);
k1_2p=kp2*cosd(theta2);
k2_2p=kp2*sind(theta2);
% S-wave Vecotors for wave 1 and wave 2
k1_1s=ks1*cosd(theta1);
k2_1s=ks1*sind(theta1);
k1_2s=ks2*cosd(theta2);
k2_2s=ks2*sind(theta2);
%% WAVE POLARIZATION
% Wave polarization direction of the P-wave
L_theta1 = 0 ; % wave polarization diraction of P-wave 1
L_theta2 = 25; % wave polarization diraction of P-wave 2
% Wave polarization direction of the S-wave
```

```
T_theta1=90-theta1; %wave polarization diraction of S-wave 1
T_theta2=90-theta2; %wave polarization diraction of S-wave 2
%% SPATIAL DISCRETIZATION
i=sqrt(-1);
x1=0:Sp_samprate:X_limit; % discretize the x1 axis
x2=0:Sp_samprate:X_limit; % discretize the x2 axis
x2_plot=(zeros(length(x1),1))';
x3=(zeros(length(x1),1))';
%% WAVE POTENTIALS
% P-wave potential
fi_1exp = A1.*exp(i*((k1_1p.*x1)+(k2_1p.*x2)+(kp1.*cosd(90)
.*x3)));
fi_2exp = A2.*exp(i*((k1_2p.*x1)+(k2_2p.*x2)+(kp2.*cosd(90)
.*x3)));
% S-wave potential
Zi_1exp = B1.*exp(i*((k1_1s.*x1)+(k2_1s.*x2)+(ks1.*cosd(90)
.*x3)));
Zi_2exp = B2.*exp(i*((k1_2s.*x1)+(k2_2s.*x2)+(ks2.*cosd(90)
.*x3)));
%% WAVE DISPLACEMENT AMPLITUDE
% P-wave
up_1_amp = A1*i*k1_1p+A1*i*k2_1p+A1*i*kp1*cosd(90);   %
Amplitude comes from gradient of the potential fi_1
up_2_amp = A2*i*k1_2s+A2*i*k2_2p+A2*i*kp2*cosd(90);   %
Amplitude comes from gradient of the potential fi_2
%S-wave
us_1_amp = B1*i*k2_1s;   % Amplitude comes from Cross Product
with Zi_1
us_2_amp = B2*i*k2_2s;   % Amplitude comes from Cross Product
with Zi_2

%% WAVE DISPLACEMENTS WITH POTENTIALS
%P-wave
up_1 = up_1_amp.*fi_1exp;
up_2 = up_2_amp.*fi_2exp;
%S-wave
us_1 = us_1_amp.*Zi_1exp;
us_2 = us_2_amp.*Zi_2exp;

%% SUPERPOSED WAVE DISPLACEMENT ALONG X1 and X2
   DIRECTIONS
%% 2-P-wave
up_x1=(up_1.*cosd(L_theta1)+up_2.*cosd(L_theta2));
up_x2=(up_1.*sind(L_theta1)+up_2.*sind(L_theta2));
%up_x3=ones(1,length(up_x1));
%% 2-S-wave
us_x1=(us_1.*cosd(T_theta1)+us_2.*cosd(T_theta2));
us_x2=(us_1.*sind(T_theta1)+us_2.*sind(T_theta2));
%up_x3=ones(1,length(up_x1));
%% 1P-1S-Hybrid wave P along theta1, S along theta2
uH_px1=(up_1.*cosd(L_theta1));
```

Waves in Periodic Media

```matlab
uH_sx1=(us_2.*cosd(T_theta2));
uH_px2=(up_1.*sind(L_theta1));
uH_sx2=(us_2.*sind(T_theta2));

uH_x1=uH_px1+uH_sx1;
uH_x2=uH_px2+uH_sx2;

%% EXPLORING THE INTRINSIC SPIN ONLY X1-X2 PLANE
%% 2-P-wave
UL_vec=[up_x1 ; up_x2];
UL_conj=[conj(up_x1) ; conj(up_x2)];
sample = size(UL_vec,2);               % At every position along
                                         X

for i = 1:sample
    ULcross = UL_conj(:,i)'*UL_vec(:,i);
    Spin_L_real(i)=real(ULcross);
    Spin_L_imag(i)=imag(ULcross);
    Spin_L_abs(i)=abs(ULcross);
end
%% 2-S-wave
UT_vec=[us_x1 ; us_x2];
UT_conj=[conj(us_x1) ; conj(us_x2)];
sample = size(UT_vec,2);               % At every position along
                                         X

for i = 1:sample
    UTcross = UT_conj(:,i)'*UT_vec(:,i);
    Spin_T_real(i)=real(UTcross);
    Spin_T_imag(i)=imag(UTcross);
    Spin_T_abs(i)=abs(UTcross);
end
%% 1P-1S-Hybrid wave P along theta1, S along theta2
UL_vec=[uH_px1 ; uH_px2];
UT_vec=[uH_sx1 ; uH_sx2];

UL_conj=[conj(uH_px1) ; conj(uH_px2)];
UT_conj=[conj(uH_sx1) ; conj(uH_sx2)];

sample = size(UT_vec,2);               % At every position along
                                         X

for i = 1:sample
    UHcross =
    (UL_conj(:,i)'*UT_vec(:,i))+(UT_conj(:,i)'*UL_vec(:,i));
    Spin_H_real(i)=real(UHcross);
    Spin_H_imag(i)=imag(UHcross);
    Spin_H_abs(i)=abs(UHcross);
end
%% VISUALIZATION OF SPIN
%% 2-P-wave
```

```
figure(1);quiver(x1,x2_plot,real(up_x1),real(up_x2),'k-');
hold on
figure(1);quiver(x1,x2_plot,imag(up_x1),imag(up_x2),'r-');
legend('real uL','imag uL')
figure(2);plot(up_x1,up_x2,'k','LineWidth',4); hold on;
        plot(imag(up_x1),imag(up_x2),'r.-','LineWidth',4);
figure(3);plot(Spin_L_real,'k','LineWidth',4); hold on
        plot(Spin_L_imag,'r.-','LineWidth',4);

%% 2-S-wave
figure(4);quiver(x1,x2_plot,real(us_x1),real(us_x2),'k-');
hold on
figure(4);quiver(x1,x2_plot,imag(us_x1),imag(us_x2),'r-');
legend('real uT','imag uT')
figure(5);plot(us_x1,us_x2,'k','LineWidth',4); hold on;
        plot(imag(us_x1),imag(us_x2),'r.-','LineWidth',4);
figure(6);plot(Spin_T_real,'k','LineWidth',4); hold on
        plot(Spin_T_imag,'r.-','LineWidth',4);
%% %% 1P-1S-Hybrid wave P along theta1, S along theta2
figure(7);quiver(x1,x2_plot,real(uH_x1),real(uH_x2),'k-');
hold on
figure(7);quiver(x1,x2_plot,imag(uH_x1),imag(uH_x2),'r-');
legend('real uH','imag uH')
figure(8);plot(imag(uH_x1),imag(uH_x2),'ro-',
'LineWidth',1.5);hold on;
        plot(uH_x1,uH_x2,'ko-','LineWidth',1.5);
        legend('imag UH','real UH');
        Title('Poincare map of Hybrid Spin State');
figure(9);plot(Spin_H_real,'k','LineWidth',4); hold on
        plot(Spin_H_imag,'r.-','LineWidth',4);

%% END
```

6.9 SUMMARY

Waves in periodic media are discussed in easy-to-follow simple mathematical steps. How to analyze the band diagrams in acoustic and elastic metamaterial is discussed through several examples. A sample computer code is also provided for first-time metamaterial researchers and students. Metamaterials to explore different novel wave phenomena such as wave tunneling, wave splitting, wave bending, orthogonal wave transport, zero index materials, and acoustic computing are discussed. Further, topological phenomena by symmetry breaking were mathematically explained and implemented. The primary objective of this chapter was to create a framework to understand the quantum analogous behaviors in acoustic and elastic waves such that advanced topological phenomena could understood and new phenomena could further be discovered. Ladder operations and the Klein-Gordon equation to Dirac equations for elastic waves were discussed through easy-to-follow simple mathematical steps for future quantum type application of acoustic and elastic waves.

REFERENCES

1. Ahmed, R.U., Banerjee, S., *Wave propagation in metamaterial using multiscale resonators by creating local anisotropy.* International Journal of Modern Engineering, 2013, **13**(2): p. 51.
2. Ahmed, R., Banerjee, S. *Introduction of novel split ring metamaterial for acoustic wave control* in *Society of Engineering Science, 50th Annual Technical Meeting and ASME-AMD Annual Summer Meeting.* 2013.
3. Ahmed, R., Banerjee, S. *Novel split ring metamaterial for multiple band gaps and vibration control.* in *Health Monitoring of Structural and Biological Systems 2013.* 2013. SPIE Smart Structure and NDE Conference, San Diego, CA, USA.
4. Ahmed, H., et al., *Hybrid Bessel beam and metamaterial lenses for deep laparoscopic nondestructive evaluation.* Journal of Applied Physics, 2021, **129**(16): p. 165107.
5. Brillouin, L., *Wave Propagation in Periodic Structures.* 1946, New York: Dover Publication.
6. Deymier, P., Runge, K., *Sound topology, duality coherence and wave-mixing: An introduction to the emerging new science of sound,* in *Solid-State Science.* 2017, Cham, Switzerland: Springer.
7. Li, Y., et al., *Unidirectional acoustic transmission through a prism with near-zero refractive index.* Applied Physics Letters, 2013. **103**(5): p. 053505.
8. Xie, L., Xia, B., Liu, J., Huang, G., Lei, J., *An improved fast plane wave expansion method for topology optimization of phononic crystals.* International Journal of Mechanical Sciences, 2017, **120**: p. 171–181.
9. Hegedűs, C.J., *Laurent expansion of an inverse of a function matrix.* Periodica Mathematica Hungarica, 1975, **6**: p. 75–86.
10. Hussein, M.I., Leamy, M.J., Ruzzene, M.J.A.M.R., Dynamics of phononic materials and structures: Historical origins, recent progress, and future outlook. Applied Mechanics Reviews, 2014, **66**(4): 040802 (38 pages).
11. Palermo, A., Marzani, A., *A reduced Bloch operator finite element method for fast calculation of elastic complex band structures.* International Journal of Solids and Structures, 2020, **191-192**: p. 601–613.
12. *www.comsol.com*, ed. C.M.v. 4.3. 2018, Stockholm, Sweden.
13. Goffaux, C., Vigneron, J., *Theoretical study of a tunable phononic band gap system.* Physical Review B, 2001, **64**(7): p. 075118.
14. Vasseur, J., et al., *Experimental and theoretical evidence for the existence of absolute acoustic band gaps in two-dimensional solid phononic crystals.* Physical Review Letters, 2001. **86**(14): p. 3012.
15. Laude, V., et al., *Full band gap for surface acoustic waves in a piezoelectric phononic crystal.* Physical Review E, 2005. **71**(3): p. 036607.
16. Benchabane, S., et al., *Evidence for complete surface wave band gap in a piezoelectric phononic crystal.* Physical Review E, 2006. **73**(6): p. 065601.
17. Indaleeb, M.M., Ahmed, H., Banerjee, S., *Acoustic computing: At tunable pseudospin-1 Hermitian Dirac-like cone.* The Journal of the Acoustical Society of America, 2022, **152**(3): p. 1449–1462.
18. Indaleeb, M.M., Banerjee, S., *Simultaneous Dirac-like cones at two energy states in tunable phononic crystals: An analytical and numerical study.* Crystals, 2021, **11**(12): p. 1528.
19. Li, F., Huang, X., Lu, J., Ma, J., Liu, Z., *Weyl points and Fermi arcs in a chiral phononic crystal.* Nature Physics Letters, 2017, **14**: p. 30–35.
20. Chu, H., Zhang, Y., Luo, J., Xu, C., Xiong, X., Peng, R., et al., *Band engineering method to create Dirac cones of accidental degeneracy in general photonic crystals without symmetry.* Opt Express, 2021, **29**(18070).

21. Chen, Y., Jia, M.F., Li, B., Huang, G. X., *Inverse design of photonic topological insulators with extra-wide bandgaps*. Physica Status Solidi RRL, 2019, **13**(1900175).
22. Li, W., Meng, F., Chen, Y., Li, Y.F., Huang, X., *Topology optimization of photonic and phononic crystals and metamaterials: A review*. Advanced Theory and Simulations, 2019. **2**(1900175).
23. Lin, Z., Christakis, L., Li, Y., Mazur, E., Rodriguez, A.W., Lončar, M., *Topology optimized dual-polarization Dirac cones*. Physical Review B, 2018. **97**(081408(R)).
24. Lin, Z., Pick, A., Lončar, M., Rodriguez, A.W., *Enhanced spontaneous emission at third-order Dirac exceptional points in inverse-designed photonic crystals*. Physical Review Letters, 2016, **117**(107402).
25. Luo, J., Lai, Y., *Hermitian and non-Hermitian Dirac-like cones in photonic and phononic structures*. Frontiers in Physics, 2022, **10**: p. 63.
26. Liu, F., et al., *Dirac cones at $\vec{k} = 0$ in phononic crystals*. Physical Review B, 2011, **84**(22): p. 224113.
27. Chen, Z.-G., et al., *Accidental degeneracy of double Dirac cones in a phononic crystal*. Scientific Reports, 2014, **4**: p. 4613.
28. Liberal, I., Engheta, N., *The rise of near-zero-index technologies*. Science, 2017, **358**(1540-1): pp. 1540–1541.
29. Luo, L., et al., *Non-Hermitian effective medium theory and complex Dirac-like cones*. Optics Express, 2021, **29**(10): p. 14345–14353.
30. Miri, M.A., ALÙ, A., *Exceptional points in optics and photonics*. Science, 2019, **363**(eaar7709).
31. Özdemir, S.K., Rotter, S., Nori, F., Yang, L., *Parity-time symmetry and exceptional points in photonics*. Nature Materials, 2019, **18**(18): p. 783–98.
32. Zhen, B., Hsu, C.W., Igarashi, Y., Liu, L., Kaminer, I., Pick, A., Chua, S.-L., Joannopoulos, J.D., Soljačić, M., *Spawning rings of exceptional points out of Dirac cones*. Nature Letters, 2015, **525**: p. 354–358.
33. Li, Z.-W., Liu, J.-J., Chen, Z.-G., Tang, W., Chen, A., Liang, B., Ma, G., Cheng, J.-C., *Experimental realization of Weyl exceptional rings in a synthetic three-dimensional non-Hermitian phononic crystal*. Physical Review Letters, 2022, **129**: p. 084301.
34. Zhang, Z., et al., *Topological creation of acoustic pseudospin multipoles in a flow-free symmetry-broken metamaterial lattice*. Physical Review Letters, 2017, **118**(8): p. 084303.
35. Zhang, Z., et al., *Experimental verification of acoustic pseudospin multipoles in a symmetry-broken snowflakelike topological insulator*. Physical Review B, 2017. **96**(24): p. 241306.
36. Deng, Y., Lu, M., Jing, Y., *A comparison study between acoustic topological states based on valley Hall and quantum spin Hall effects*. Journal of the Acoustical Society of America, 2019, **146**(1): p. 721–728.
37. Li, J., et al., *Pseudospins and topological edge states in elastic shear waves*. AIP Advances, 2017, **7**(12): p. 125030.
38. Wang, S., et al., *Spin-orbit interactions of transverse sound*. Nature Communications, 2021, **12**(1): p. 1–9.
39. Long, Y., Ren, J., Chen, H., *Intrinsic spin of elastic waves*. Proceedings of the National Academy of Sciences, 2018, **115**(40): p. 9951–9955.
40. Ahmed, H., et al., *Multifunction acoustic modulation by a multi-mode acoustic metamaterial architecture*. Journal of Physics Communications, 2018, **2**(11): p. 115001.
41. Chadi, D.J., Cohen, M.L., *Special points in the Brillouin zone*. Physical Review B, 1973, **8**(12): p. 5747–5753.
42. Khelif, A., et al., *Complete band gaps in two-dimensional phononic crystal slabs*. Phys Rev E, 2006, **74**(4 Pt 2): p. 046610.

Waves in Periodic Media

43. Wu, T.-T., Huang, Z.-G., Lin, S., *Surface and bulk acoustic waves in two-dimensional phononic crystal consisting of materials with general anisotropy.* Physical Review B, 2004, **69**(9): p. 094301.
44. Giurgiutiu, V., *Structural Health Monitoring.* 2nd ed. July 2014, New York, NY: Elsevier AP.
45. Banerjee, S., Ahmed, R.U., *Phonon confinement using spirally designed elastic resonators in discrete continuum.* International Journal of Materials Science and Applications, 2014, **3**(1): p. 6.
46. Shelke, A., et al., *Wave guiding and wave modulation using phononic crystal defects.* Journal of Intelligent Material Systems and Structures, 2014, **25**(13): p. 1541–1552.
47. Zhang, X.L., Zha, *Superlenses to overcome the diffraction limit.* Nature Materials, 2008, **7**: p. 435–441.
48. Indaleeb, M.M., et al., *Deaf band based engineered Dirac cone in a periodic acoustic metamaterial: A numerical and experimental study.* Physical Review B, 2019, **99**(2): p. 024311.
49. Indaleeb, M.M., et al., *Deaf band-based prediction of Dirac cone in acoustic metamaterials.* Journal of Applied Physics, 2020, **127**(6): p. 064903.
50. Boolchand, P., Georgiev, D.G., Goodman, B., *Discovery of the intermediate phase in chalcogenide glasses.* Journal of Optoelectronics and Advanced Materials, 2001, **3**(3): p. 703–720.
51. Gump, J., Finkler, I., Xia, H., Sooryakumar, R., Bresser, W.J., Boolchand, P., *Light induced giant softening of network glasses observed near the mean-field rigidity transition.* Physical Review Letters, 2004, **92**(245501).
52. Srinivasan, A.V., Mcfarland, D.M., *Smart Structures.* 2001, Cambridge, UK: Cambridge University Press.
53. Danas, K., Kankanala, S., Triantafyllidis, N., *Experiments and modeling of iron-particle-filled magnetorheological elastomers.* Journal of the Mechanics and Physics of Solids, 2012, **60**(1): p. 120–138.
54. Chen, Y., et al., *Enhanced flexural wave sensing by adaptive gradient-index metamaterials.* Scientific Reports, 2016, **6**(1): p. 1–11.
55. Casadei, F., et al., *Piezoelectric resonator arrays for tunable acoustic waveguides and metamaterials.* Journal of Applied Physics, 2012, **112**(6): p. 064902.
56. Li, F., et al., Wave transmission in time- and space-variant helicoidal phononic crystals. Physical Review E, 2014, **90**(5): p. 053201.
57. Trainiti, G., Ruzzene, M., *Non-reciprocal elastic wave propagation in spatiotemporal periodic structures.* New Journal of Physics, 2016, **18**(8): p. 083047.
58. Raghu, S., Haldane, F.D.M., *Analogs of quantum-Hall-effect edge States in photonic crystals.* Physical Review A, 2008, **78**(3): p. 033834.
59. Fang, A., et al., Anomalous Anderson localization behaviors in disordered pseudospin systems. PNAS, 2017, **114**(16): p. 4087–4092.
60. Sanchis-Alepuz, H., Kosevich, Y.A., Sánchez-Dehesa, J., *Acoustic analogue of electronic Bloch oscillations and resonant Zener tunneling in ultrasonic superlattices.* Physical Review Letters, 2007, **98**(134301).
61. Miri, M.-A., Heinrich, M., Christodoulides, D.N., *Supersymmetry-generated complex optical potentials with real spectra.* Physical Review A, 2013, **87**(043819).
62. Banerjee, S., Leckey, C.A.C., *Computational Nondestructive Evaluation Handbook.* 2020, Boca Raton, FL: CRC Press, Taylor & Francis Group. **560**.
63. Kweun, J.M., Lee, H.J., Oh, J.H., Seung, H.M., Kim, Y.Y., *Transmodal Fabry-Pérot resonance: Theory and realization with elastic metamaterials.* Physical Review Letters, 2017, **118**(205901).

64. Chang, Z., Guo, H.-Y., Li, B., Feng, X.-Q., *Disentangling longitudinal and shear elastic waves by neo-Hookean soft devices*. Applied Physics Letters, 2015, **106**(161903).

65. Liu, F., Liu, Z., *Elastic waves scattering without conversion in metamaterials with simultaneous zero indices for longitudinal and transverse waves*. Physical Review Letters, 2015, **115**(175502).

66. Cho, C., Yu, S., Park, N., *Elastic Hamiltonians for quantum analog applications*. Physical Review B, 2020, **101**(13): p. 134107.

67. Lahiri, A., Roy, P.K., Bagchi, B., *Supersymmetry and the ladder operator technique in quantum mechanics: The radial Schrödinger equation*. International Journal of Theoretical Physics, 1989, **28**(2): p. 183–189.

68. Shi, C., Zhao, R., Long, Y., Yang, S., Wang, Y., Chen, H., Ren, J., Zhang, X., *Observation of acoustic spin*. National Science Review, 2019, **6**(4): p. 707–712.

69. Noether, E. *Invariante Variationsprobleme*. Nachrichten von der Gesellschaft der Wissenschaften zu Göttingen. Mathematisch-Physikalische Klasse, 1918: p. 235–257.

70. Long, Y., Ren, J., Chen, H., *Intrinsic spin of elastic waves*. PNAS, 2018, **115**(40): p. 9951–9955.

71. Bliokh, K.Y., Smirnova, D., Nori, F., *Quantum spin Hall Effect of light*. Science, 2015, **348**: p. 1448–1451.

72. Ge, H., Xu, X-Y., Liu, L., Xu, R., Lin, Z-K, Yu, S-Y., Bao, M., Jiang, J-H., Lu, M-H., Chen, Y.-F., *Observation of acoustic skyrmions*. Physical Review Letters, 2021, **127**: p. 144502.

73. Cao, L., Wan, S., Zeng, Y., Zhu, Y., Assouar, B., Observation of Phononic Skyrmions based on hybrid spin of elastic waves. Science Advances, Vol. 9, Issue 7, 2023 online DOI: 10.1126/sciadv.adf3652.

7 Topological Acoustics in Metamaterials

Mustahseen M. Indaleeb and Sourav Banerjee

7.1 TOPOLOGY

Topological geometric phase is discussed in Chapter 5. Some of the discussion will be repeated here but is necessary to describe the topological acoustic phenomena. Mathematically, topology relates to the global properties of space, preserved under continuous deformation. In topology, distances and angles are apparently irrelevant. It is often described as a rubber-sheet geometry. Objects in traditional geometries, such as triangles, circles, planes, and polyhedrals, are considered rigid, with well-defined distances and angles between points and edges/surfaces respectively. But in topology, objects are allowed to be molded to change shape without ripping apart. That's why a torus and a sphere are said to be topologically distinct as they cannot be molded or shaped to one another due to a hole present in the middle, which is a unique characteristic of a torus (Fig. 7.1). Such a torus can, however, be molded into a coffee mug. That is why they can be said to be topologically equivalent. The number of holes in a complex surface can be characterized by its genus. The topological index remains constant even when the surface is deformed but without any inclusion, cut, or glue. That is how the torus is topologically distinct from the sphere but equivalent to a coffee mug since both have a handle-like feature in their bodies. This phenomenon can be described mathematically as

$$\oiint_M K \, d\Gamma = 2\pi \chi_M = 4\pi (1 - g) \tag{7.1}$$

where K is the Gauss curvature, χ_M is the topological index of an arbitrary manifold M, and g is called the genus of the surface. Keeping the topology unchanged, χ_M is the invariant of the manifold, when the manifold M is deformed adiabatically (refer to Chapter 5). This is called the Gauss-Bonnet theorem [1], which remained popular and well known in the mathematical community for a long period. After a while, physicists noticed a similarity of the theorem with certain explanations of the quantized Hall conductivity [1], where a 2D periodic Brillouin zone can be treated as a closed manifold in an abstract space. Here, the integer quantized Hall coefficient can be regarded as the topological index, χ_M of the complex geometry. The topological index (χ_M) of a physical system can be assumed to be a global parameter assigned from the collective behaviors of the particles or waves in the system. Irrespective of quantum or acoustic waves, the topological index, which can be quantized, could not change its value under the deformation of the system continuously. Hence, the topological index or the invariant is the conserved quantity under continuous deformation. That is why the integral of the local curvature of any surface can be called

DOI: 10.1201/9781003225751-7

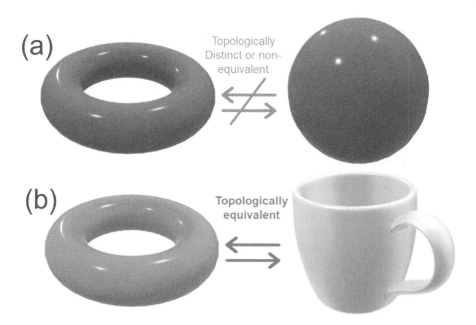

FIGURE 7.1 a) The torus and the sphere are not topologically equivalent, b) but the torus is topologically equivalent to the coffee mug.

the topological invariant. Exploiting the mathematical understanding of a topological invariant with different phase transitions of a material, physicists came up with a separate group of materials, known as topological insulators. Inspired by quantum Hall effects, certain metamaterials have properties with bulk-boundary distinction, where the bulk is insulated and the boundary is conductive, called topological insulators. In this chapter, another topological situation is explained in which the bulk is acoustically conductive and the boundary is insulated.

7.1.1 Topology in Crystals

In the early 1900s, Landau helped characterize stable phases of matter by a local order parameter that vanishes for the disordered phases (e.g., symmetry breaking) and becomes nonzero for the ordered phases [2]. This symmetry-breaking theory categorizes the phases with nonzero order parameters and the way they transform under symmetric operations. It remained accepted and unchallenged by the greater scientific community, believing that this theory can describe all the possible phases of matter as well as all the possible continuous phase transitions. But the discovery of the Berezinskii-Kosterlitz-Thouless (BKT) [3] transition transformed the map of quantum mechanics research. It was found that explaining the phase transition in a two-dimensional model cannot be explained by symmetry-breaking theory [4, 5]. With this, the concept of the integer quantum Hall effect emerged [3]. The quantum Hall effect (QHE) is observed in 2D electron systems with low temperatures and a strong magnetic field. (That means adding an external field as discussed in

Topological Acoustics in Metamaterials

section 5.8 in Chapter 5.) Here, the bulk state behaves like an insulator but cannot be characterized by any local order parameters. A robust edge state at the boundaries exists even under external deformation. Such behavior intuitively contradicts our understanding of Anderson localization [6], where the edge state should be affected by impurities and deformation. Thus, the existence of a truly distinct phase of matter is established by QHE. Here, the bulk is characterized by a topological invariant as the bulk order parameters fail to describe the new phase. The quantized Hall conductance is an invariant that does not change with the arbitrary phase appearing in the eigenfunctions.

An analytic derivation of the quantized Hall conductivity in periodic crystals was given by Thouless, Kohomoto, Nightingale, and den Nijs (TKNN) in 1982 [7]. They considered a 2D electronic gas under a uniform perpendicular magnetic field and an arbitrary lattice potential described by a single particle Hamiltonian. The conductivity was expressed utilizing the Kubo formula [8] of linear response as an integral involving valence band state.

$$\sigma_{xy} = C_n \frac{e^2}{\hbar} \tag{7.2.1}$$

where

$$C_n - \frac{i}{2\pi} \int_{BZ} \sum_{E_m < E_f} 2iIm \left\langle \frac{\partial}{\partial k_1} u_m(\mathbf{k}) \left| \frac{\partial}{\partial k_2} u_m(\mathbf{k}) \right\rangle dk_1 dk_2 = \frac{i}{2\pi} \int_{BZ} Tr(\mathcal{R}) \in \mathbb{Z} \tag{7.2.2}$$

Here, the summation is taken over the occupied periodic crystal, and the integral is over the magnetic Brillouin zone [9]; $\mathcal{R}$ is the Berry curvature [10] expressed in Eq. 5.29.12 in Chapter 5. The topology of the occupied band is characterized by the TKNN integer, v, also known as the first Chern number. E_m and E_f indicates all the branches below a particular band gap, describing the topological properties of that band gap. BZ means the integration over the first Brillouin zone. (Please refer to Chapter 5 for an understanding.) For a nonzero Chern number ($C_n \neq 0$), the band gap is called a topologically nontrivial band gap, and for $C_n = 0$, it is called a topologically trivial band gap [7]. A gap with a certain topological invariant cannot be smoothly transformed into another gap associated with a different topological invariant. The topological difference of the band gap distinguishes the quantum Hall insulators from conventional insulators and gives rise to the bulk-edge correspondence [11] or bulk boundary distinction. The topological invariant represents the number of the chiral edge states. Even in the presence of large impurities, it ensures conduction on the surface without dissipation or backscattering. Figure 7.2 illustrates the chiral edge state at the interface between a quantum Hall insulator and a conventional insulator [12].

However, such topological behaviors pose great challenges and are difficult to overcome in an electronic system. Hence, the researchers have extended this study to photonics and phononic systems. Fabrication and control over the system get easier in the case of photonics and phononics. Moreover, a multi-regional and wide frequency spectrum could be covered by photonic/phononics systems, without limiting only to

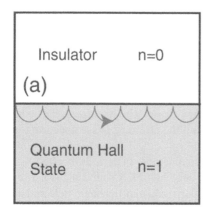 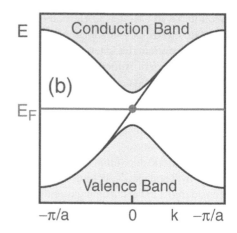

FIGURE 7.2 a) Chiral edge state at the interface of a quantum hall insulator and a conventional insulator, b) the band structure with A single edge state linking the valence band to a conduction band. The figures are adapted from Ref [11].

the Fermi levels. However, photons/phonons do not possess any half-integer spin to interact with the applied magnetic field. Breaking the time-reversal symmetry poses a greater challenge in such cases. Especially, phononic systems in acoustics lack the transverse polarization degree of freedom that can be used to construct pseudospins to mimic the spin up and spin down in the electronic system. The first analog QHE in photonics was proposed and demonstrated in 2008 by Haldane [13]. A gyromagnetic material subjected to a magnetic field was used to break the time-reversal symmetry for a 2D photonic crystal.

Haldane also introduced periodic magnetic flux, substituting an external applied field to concur the breaking of time-reversal symmetry [14] or $\mathcal{T}$-symmetry (refer to section 5.7.2 in Chapter 5). Such an effect, named the quantum anomalous Hall effect (QAHE) is similar to the QHE. They can both be expressed in terms of the same topological invariant. Later, similar QHE studies were extended to replace the periodic magnetic flux by a ferromagnetic insulator with strong spin-orbit coupling [15–17]. This discovery led to one of the most groundbreaking phenomena in the field of condensed matter physics – the topological insulator, which can also be called the quantum spin Hall effect (QSHE), shown in Fig. 7.3 [18–22].

In photonics, although the electrons propagate in both forward and backward directions at the same edge, a nonmagnetic impurity may lead to destructive interference of backscattering, which contributes to a perfect transmission of the edge state. When the time-reversal symmetry or the $\mathcal{T}$-symmetry is broken, the backscattering remains no longer destructive. If the $\mathcal{T}$-symmetry is preserved, the Chern number will be equal to zero and the total Hall conductance vanishes due to the opposite spins that counter-propagate and contribute to opposite quantized Hall conductivities. This state is called the quantum spin Hall effect (QSHE).

Kane and Mele first proposed the QSHE to exist in graphene with spin-orbit coupling, shown in Fig. 7.3 [12, 24]. But due to an extremely small gap, induced by the spin-orbit coupling, a new search began. Searching for a better alternative continued

Topological Acoustics in Metamaterials 357

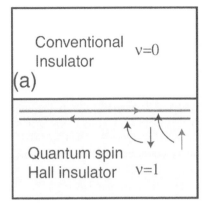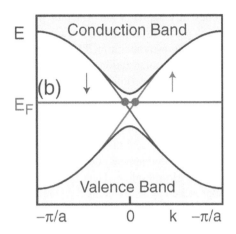

FIGURE 7.3 a) The helical edge states in the interface between a quantum spin Hall insulator and a conventional insulator, b) the band structure of the corresponding structures; pairs of helical edge state connect the valence band to the conduction band. The figures are adapted from Ref [11].

and ultimately was obtained in HgTe quantum wells in 2007, leading to the discovery of topological insulators [23]. Such HgTe-based quantum well heterostructure can host pair of counter-propagating edge states that are related to each other through $\mathcal{T}$-symmetry. From there, the name 'topological insulator' emerged, because the bulk of the material is insulated while exhibiting topologically protected metallic/conductive states at the boundary. The band structure of one topological insulator material is shown in Fig. 7.4 [23, 25, 26]. Here, a gapless surface state with a single Dirac cone with a linear E-$\mathbf{k}$ relation is observed. The surface electrons behave like massless Dirac fermions with ultra-high mobility. (Refer to the discussion on Dirac

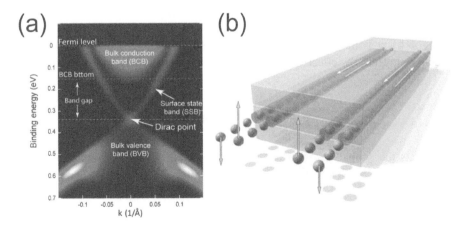

FIGURE 7.4 Topological insulators and the spin-momentum locking mechanism: a) energy band structure of Bi_2Se_3, b) illustration of spin-polarized helical edge channel in a TI material. The figures are adapted from ref (König et al. 2007) [23].

cones in Chapter 5 and Chapter 6.) The most essential feature that distinguishes topological insulators from traditional insulators is the mechanism of locking the surface spin-momentum. This mechanism dictates the spin-orbit coupling, resulting in the spin polarization of the surface electrons and inevitably being locked to its momentum direction. Hence, with applied external current, electrons with opposite spins separate along the edges, causing such topologically insulating phenomenon, (Fig. 7.4) [23, 25, 26]. As the $\mathcal{T}$-symmetry is protected, backscattering by the non-magnetic impurities is forbidden during the electron transport, leading to the QSHE. Hence, topological insulators are widely known as QSH insulators as well [27–31].

Chapter 5 discussed how local dispersion behavior could be achieved by a locally defined Hamiltonian. Hence, studying the behavior of relativistic particles can be approached by reproducing the Dirac equation under the frame of band theory [32, 33]. But topology, being a global property of the band in the Brillouin zone, cannot be confined to the continuous limit. An unconfined lattice model in the high symmetry point is more appropriate. However, a lattice model can indeed be mapped near the critical point of topological quantum phase transition. The topology of the lattice remains unchanged when the energy band gap persists without closing. Then the Hamiltonian of a 2D lattice model as an electron in an effective magnetic field would read

$$H = \boldsymbol{d}(\boldsymbol{k}) \cdot \sigma \tag{7.3.1}$$

And its dispersion relation would read

$$E_{k,\pm} = \pm |\boldsymbol{d}(\boldsymbol{k})| \tag{7.3.2}$$

For 2×2 modified Dirac models with time-reversal symmetry preserved, the following expression must hold with time-reversal operator, $\Theta = i\sigma_y K$, where K is the conjugate,

$$\Theta \mathbf{d}(\mathbf{k}) \cdot \sigma \Theta^{-1} = -\mathbf{d}(-\mathbf{k}) \cdot \sigma \tag{7.3.3}$$

where $\mathbf{d}(\mathbf{k}) \cdot \sigma$ is the spin-up ($\uparrow$) component and $-\mathbf{d}(-\mathbf{k}) \cdot \sigma$ is the spin-down ($\downarrow$) component. Thus, the effective Hamiltonian for a simple QSHE can be obtained as

$$H_{QSHE} = \begin{pmatrix} \mathbf{d}(\mathbf{k}) \cdot \sigma & 0 \\ 0 & -\mathbf{d}(-\mathbf{k}) \cdot \sigma \end{pmatrix} \tag{7.3.4}$$

This is the simplest version of the QSHE, where further spin-orbit couplings like Rashba spin-orbit coupling and Zeeman splitting could be added and could be expanded further [34].

A few key points are summarized below:

- Ordinary Hall effect is achieved by inducing a magnetic field (**H**) to a conductor, and Hall voltage is created. Hall voltage is generated without the accumulation of any spin. $\mathcal{T}$-symmetry is broken.

Topological Acoustics in Metamaterials

- Anomalous Hall effect is achieved through magnetization (**M**) of a conductor that carries spin polarization. Hall voltage is generated with simultaneous accumulation of spin. $\mathcal{T}$-symmetry is broken
- No magnetic field is necessary to break $\mathcal{T}$-symmetry, and pure spin Hall effect can be achieved where no Hall voltage is generated but spins are accumulated at the boundary.
- Quantum spin Hall effect can be visualized as two copies of QHE experiencing opposite magnetic field.
- In QSHE, helical edge state is developed where two states with opposite spin counters propagate (Fig. 7.3) to the edge, giving bulk-boundary distinction.

7.1.2 TOPOLOGY IN PHONONICS

Similar phenomena to those discussed above now warrant discussion for acoustics. Understanding the physics, acoustic metamaterial made of phononic crystals, or discrete or continuous interconnected periodic structures can be intentionally created, which may manifest similar behaviors. However, one fundamental key challenge in acoustics is that there is no inherent spin. The following section discusses how the topological quantum trio phenomena could be achieved in acoustics.

7.1.2.1 Breaking $\mathcal{T}$-symmetry: QHE

For QHE, $\mathcal{T}$-symmetry must be broken. Breaking time-reversal symmetry or simple $\mathcal{T}$-symmetry (refer to section 5.7.2 in Chapter 5) in acoustics is perhaps more challenging. However, it is shown that utilizing a dynamic fluid flow in a metamaterial made of phononic crystal can break the time-reversal symmetry. Artificially induced fluid-flow serves the purpose of the magnetic bias by splitting the degeneracy of two counter-propagating azimuthal resonant modes inside the phononic crystals [35–38]. The wave equation for the acoustic wave propagating in a circulating airflow with velocity **V** would read

$$\left[\left(\nabla - i\mathbf{A}_{eff}\right)^2 + \omega^2/c^2 + \left(\nabla\rho/2\rho\right) - \nabla^2\rho/\left(2\rho\right)\right]\phi = 0 \tag{7.4.1}$$

where ϕ is the velocity potential, ω is the angular frequency, c is the sound speed, and ρ is the mass density of air. For nonzero **V**, the term $\mathbf{A}_{eff} = -\omega\mathbf{V}/c^2$ creates an effective vector potential, generating an effective magnetic field,

$$\mathbf{B}_{eff} = \nabla \times \mathbf{A}_{eff} \tag{7.4.2}$$

To understand a magnetic field, please refer to Chapter 4. The $\mathbf{B}_{eff}$ helps breaking time-reversal symmetry in the phononic metamaterial system. Figure 7.5 shows the hexagonal lattice and honeycomb lattices, respectively, used in developing QHE analogues in acoustics. A pair of Dirac-like degeneracies emerge due to intrinsic lattice symmetry at the BZ boundary for **V** = 0. After introducing the fluid flow, the degeneracy is lifted due to the broken time-reversal symmetry by opening a bandgap. According to the bulk-edge correspondence principle, one-way edge states prevail with such signature topologically nontrivial systems. Here, acoustic wave

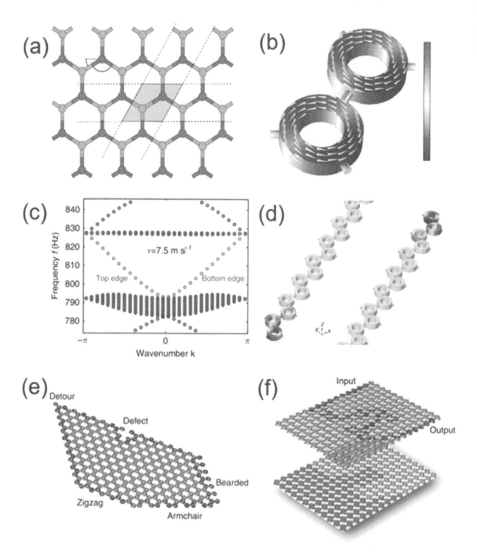

FIGURE 7.5 a) Diatomic lattice forming an acoustic analog of graphene. b) One unit cell of the lattice modeled in finite element, with acoustic pressure distribution shown for one of the Dirac modes of interest. The gray arrows indicate the direction of airflow in the resonators. Structure dimensions: inner and outer radius of the cavity are R_{in} = 5.08 cm and R_{out} = 9.21 cm, respectively; the height of the cavity is H = 4.45 cm. c) acoustic band structure for a supercell of 20 unit cells and a uniform rotational bias velocity v_{air} = 7.5 m s 1. Bulk modes are shown by blue and edge modes by black, green, and red colored markers. d) Acoustic pressure profiles of the one-way edge mode localized at the bottom and top of the supercell, respectively. e) A one-way (counterclockwise) edge mode propagates along different cuts of the acoustic graphene lattice and flows around a deliberately introduced defect (top zigzag cut) without backscattering and formation of standing-wave patterns. f) Excitation of a one-way (counterclockwise) edge mode and its propagation (top subplot) along an irregularly shaped domain wall created by the reversal of the Doppler bias (bottom subplot). The figures are adapted from Ref [39].

Topological Acoustics in Metamaterials

propagation exhibits unidirectional behaviors, which are topologically protected and robust against various defects and sharp bends.

A graphene-like resonator lattice is also reported to allow a one-way edge state with circular airflow around the resonator [39]. Unlike a nontopological guided edge mode, this system shows strong robustness against structural defects and irregularities such as sharp corners, defects, detours, armchairs, and many more. This edge state allows the idea of reflectionless routing along arbitrarily defined pathways. The wave is reconfigurable in real time by creating line boundaries within the lattice with opposite applied angular momenta on the two sides.

7.1.2.2 Without Breaking $\mathcal{T}$-symmetry: QSHE

It might be assumed that to achieve topological behavior, breaking of time-reversal symmetry or the $\mathcal{T}$-symmetry is required. However, breaking of time-reversal symmetry is not the only possible way to achieve the topological behaviors in acoustics. Preserving the $\mathcal{T}$-symmetry, topological states could still be present by breaking spatial inversion symmetry. This new state exhibits a robust edge state at the interface of the spatial features where the spatial symmetry is broken. Here, two states with opposite spin polarization counterpropagating at a given edge resemble two copies of the QHE simultaneously, as is discussed in section 7.1.1. The spins are correlated to the direction of propagation of the wave.

In acoustics, however, spins are not present. Apparently, spin-1/2 fermions are not manifested. Phonons may have full integer spin-like spin-1 but not spin-1/2. A fundamental difference between the spin-1/2 fermions and spin-1 photon/phonons are the Kramer's doublets [40] and is a mandatory condition for QSHE. This is also absent in acoustics if superficially seen. However, by understanding the physics with its prerequisites for QSHE, *it is realized that a metamaterial unit cell must have more degrees of freedom to simulate the spin.* Then Kramer's doublets could be artificially created in acoustics as a pair of pseudospin states like electrons. Hence, in acoustics, due to the longitudinal curl-free nature, the waveguide modes for QSHE are traditionally described with *spin-like modes (pseudospin)* rather than real spin angular momentum. The equivalence between acoustic pseudospin and real spin angular momentum is absent in this case. However, an intrinsic acoustic spin angular momentum of an acoustic field is obtainable by a velocity field with rotational polarized properties despite being curl-free. The spin angular momentum in acoustic wave can be expressed as

$$s = \frac{\rho}{2\omega} Im\left[v^* \times v\right] \tag{7.5.1}$$

where ρ is the mass density, ω is the frequency, v is the velocity of particle vibration, and v^* is the conjugate of v. Hence, by absorbing the acoustic spin angular momentum, a small particle will receive a torque proportional to s.

In acoustics, inducing the pseudospin is also challenging. Parametric modulation of the spatial properties of the unit cell is the only viable way to create two different states in a periodic metamaterial. A primary feature that must be present in the dispersion behavior of a metamaterial exhibiting QSHE is a presence of a state of double Dirac cone at a specific geometric configuration (i.e., geometric parameters). Double Dirac cones are discussed in section 6.4.9 in Chapter 6 and depicted in Fig. 6.16. As discussed, this two-fold degeneracy has different properties at different geometric parameters. By varying the filling fraction or geometric dimension of constituent phononic crystals or a metamaterial, the double twofold degeneracy could be modulated. Synonymous to the quantum orbital of p-orbital and d-orbital demonstrated in photonics a pair of dipolar resonance states accompanied by a pair of quadrupolar resonance states are called *p-type* and *d-type* bands. Interestingly, similar to p and d orbitals of electrons, even and odd symmetrical to the axes x_1/x_2 dipoles are even or odd symmetrical bands called p_x and p_y. When the quadrupole mode is odd symmetrical to the axes x_1 and x_2, the mode is called d_{xy}. When the mode is even symmetrical to the axes x_1 and x_2, the mode is called $d_{x^2-y^2}$. This part is also discussed in Chapter 6. Refer to Figs. 7.6 and 7.7 to follow the following discussions.

As a trivial case, the dipolar modes must be entitled to the degeneracies at a lower frequency, but the quadrupolar mode must be entitled to the degeneracies at a higher frequency. However, at a specific geometric configuration (i.e., geometric parameters), it is possible that the dipolar modes are degenerated at a higher frequency than the frequency where the quadrupolar modes are degenerated. This is a nontrivial state. Again, at another geometric configuration, the situation could be flipped. It signifies that there is a topological phase transition between the degeneracies (flipping of the dipolar and quadrupolar mode along the frequency scale) at a double Dirac frequency as the geometric parameter changes. If such two states of the metamaterials are placed side by side, creating an arbitrary interface of trivial and nontrivial states, the space inversion symmetry will be broken. Intentionally creating this transition, a bulk-boundary distinction is achieved, and the acoustic wave propagates only along the confined interface, which is immune to the backscattering and internal defects.

Figure 7.6a and b illustrates a phononic crystal unit cell with the dispersion relation of the lattice showing a double Dirac cone at the Γ point [41]. The mode of spin angular momentum is introduced by the circular polarization of the velocity field. The dispersion relations for both trivial and nontrivial cases are given in Fig. 7.6c. Topological phase transition occurs between the trivial case and nontrivial case.

p_x and p_y are the two dipole modes for trivial case that undergoes band inversion to a nontrivial state (Fig. 7.6d). Spin angular momentum distribution also affirms the occurrence of the band inversion. In band inversion, p-states get exchanged with the d-states by transitioning the topological phase. It is synonymous to the parallel transport of a vector along a mobius strip formed in the parameter space. To illustrate the mechanism of topological phase transition, **k.p** perturbation theory (refer to section 5.3.3.2 in Chapter 5) is used to derive the effective Hamiltonian and calculate Chern numbers. The acoustic eigenequation can be written as

$$-\nabla \cdot \left[\rho^{-1}(\mathbf{r}) \right] \nabla \phi_{n,\mathbf{k}}(\mathbf{r}) = \omega_{n,\mathbf{k}}^2 \left(\rho(\mathbf{r})c^2 \right)^{-1}(r)\phi_{n,\mathbf{k}}(\mathbf{r}) \qquad (7.5.2)$$

where $\phi_{n,\mathbf{k}}(\mathbf{r})$ is the wave function of the pressure field at a radial distance of $\mathbf{r}$ from the reference point. $\omega_{n,k}$ is the eigenfrequency of the acoustic Bloch state with wave vector $\mathbf{k}$ for the n-th band. $\rho(\mathbf{r})$ is the mass density and c is the acoustic velocity.

Topological Acoustics in Metamaterials

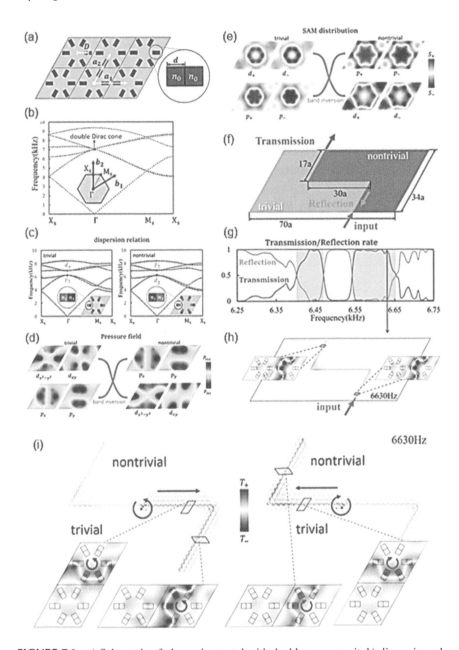

FIGURE 7.6 a) Schematic of phononic crystal with double square unit, b) dispersion relation of the lattice, c) dispersion relation of trivial and nontrivial unit cells, d) topological phase transition from trivial to nontrivial units depicting acoustic pressure field distribution with dipole and quadrupole modes, e) spin angular momentum distribution of p and d modes in trivial and nontrivial unit cells, f) schematic of test setup with Z-shape edge, g) transmission/reflection rate of Z-shape edge wave guide, h) spin angular momentum distribution on the Z-shape edge for 6.63 kHz, and i) topological one-way edge wave guide mode on Z-shape edge with CW and CCW spin denoting source mode. The figures are adapted from Ref [41].

The wave function around Γ point can be expressed as the summation of Bloch wave functions, so that the Hamiltonian can be expanded as a Taylor expansion of $\mathbf{k}$ with $\mathbf{k.p}$ perturbation form as follows

$$H_{nn'} = \omega_{n,\Gamma}^2 \delta_{nn'} + k \cdot P_{nn'} + k \cdot M_{nn'} + \cdots \tag{7.5.3}$$

In the second-order perturbation term, the contribution mainly derives from the diagonal elements. The major contribution here comes from the second-order perturbation term, $M_{nn'}$. The effective Hamiltonian (also refer to Eq. 6.12.9 in Chapter 6) near Γ point based on the four modes at the double Dirac cone for $\left(p_x, d_{xy}, p_y, d_{x^2-y^2} \right)^T$ is

$$H_{eff}(\mathbf{k}) = \begin{pmatrix} -\Omega - Bk^2 & Ak_+ & 0 & 0 \\ A^* k_+^* & \Omega + Bk^2 & 0 & 0 \\ 0 & 0 & -\Omega - Bk^2 & Ak_- \\ 0 & 0 & A^* k_-^* & \Omega + Bk^2 \end{pmatrix} \tag{7.5.4}$$

where $k_{\pm} = k_1 \pm ik_2$ and $\Omega = \dfrac{\omega_d - \omega_p}{2}$ is the frequency difference between the d-state and p-state, which is positive for the trivial state and negative for the nontrivial state. Due to the band inversion, the sign changes and could be captured through this formulation of the Hamiltonian. A is determined by the off-diagonal elements of the first-order perturbation term of effective Hamiltonian, and B is the diagonal elements of the second-order perturbation term of the Hamiltonian. k^2 for p and d modes are opposite, which are related to the curvature of their dispersion relation near the Γ point.

The spin-orbit momentum distribution in Fig. 7.6d shows a strong correlation with the pseudospin. This spin-orbit momentum is neither pure positive nor negative. Spin-orbit momentum is presented as a spatial distribution. The topological edge waveguide in a finite structure with a Z-shaped edge is shown in Fig. 7.6f. A plane wave is generated from the input point, and the transmission spectra are plotted in Fig. 7.6g. The gray region in Fig. 7.6g highlights the robust edge modes. The spin angular momentum distribution on the Z-edge at 6.63 kHz is given in Fig. 7.6h. The acoustic spin angular momentum when absorbed by an object to produce torque, it is represented as positive spin excitation, inducing a positive torque. This chiral spin excitation creates a topologically robust mode that will cross the sharp corner of the Z-shape edge without apparent backscattering (Fig. 7.6i). That is how an analog QSHE in acoustics is demonstrated, exploiting acoustic intrinsic spin angular momentum.

So far, to create the topological edge state, two metamaterial structures (A and B) with geometric parameters that give one trivial state and one nontrivial state should be created. These structures and their interface that create the boundary or a domain wall (DW) breaking spatial inversion symmetry are permanent and cannot be modulated in real time for device applications. However, an alternate method of creating such state is to rotate the elements of the phononic crystals as needed. This mechanism will help create a different geometric configuration of the DW between A and B structures, and the edge can be exploited in real time for diverse device applications. One such example of QSHE in acoustics is shown in Fig. 7.7a and b [42]. Pseudospin states are again transitioned by topological mode inversion method.

Topological Acoustics in Metamaterials 365

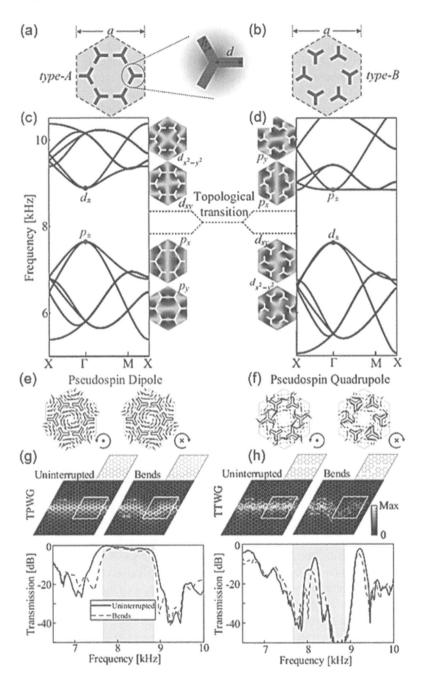

FIGURE 7.7 a) and b) Schematics of two snowflakes-like metamolecules, Type A and Type B; c) and d) corresponding dispersion of the topologically trivial and nontrivial structure; e) and f) sound intensity for pseudospin-down and pseudospin-up modes at Γ point for trivial and nontrivial regimes; and g) and h) simulated pressure field distribution along with transmission spectra showing the stronger robustness for topologically protected waveguide. The figures are adapted from Ref [42].

Here, a snowflake-like metamolecule of phononic triangular lattice is studied to achieve one-way edge state. Six three-legged elements are introduced inside a hexagonal unit cell of a phononic crystal to build the metamaterial. However, the three-legged elements have rotational degrees of freedom. By rotating the elements, trivial (no rotation, A) and nontrivial (with rotation, B) states are accessed. The topological transition between trivial and nontrivial structure is shown in Fig. 7.7c. Topological mode inversion between the pseudospin states is illustrated. Sound intensity I associated with pseudospin-down and pseudospin-up modes at the Γ point for both trivial and nontrivial regimes is given in Fig. 7.7e and f. Numerical simulations of absolute pressure fields at the frequency 8.2 kHz for both topologically protected waveguide (TPWG) and topologically trivial waveguide (TTWG) are shown in Fig. 7.7g and 7.7h. It can be immediately realized that a bigger metamaterial structure made of trivial state A could be transformed to a nontrivial state B, only partially leaving a boundary or a DW just by rotating the elements at selected and desired unit cells. One side would remain trivial and the other side of the DW would become nontrivial just by rotating the elements. Geometric patterns and various configurations of DWs could be created in real time as needed. Real-time and reconfigurable acoustic delay lines would be the best and most futuristic applications of QSHE.

7.1.2.3 Without Breaking $\mathcal{T}$-symmetry: QVHE

Now let's look at another phenomenon called quantum valley Hall effect (QVHE).

Fabrication of a metamaterial to achieve nonreciprocal edge states breaking $\mathcal{T}$-symmetry is extremely challenging for practical applications. Hence, obtaining unperturbed edge states, keeping the time-reversal symmetry protected, has been a quest for researchers for the past decade. Several studies have investigated the topological materials based on broken space-inversion symmetry, instead of time-reversal symmetry, as is discussed for QSHE. The effect of the asymmetry opens a topological bandgap at a Dirac near K and K' symmetry points, hence creating a condition for the existence of an edge state. These edge states cannot be explained by QHE or QSHE. The lattice may still have a trivial topology within the context of QHE, but the time-reversal symmetry is intact. It also has a trivial topology of QSHE as it lacks the spin degree of freedom. However, due to the large separation in k-space of the two valleys (K and K' symmetry points), valley-dependent topological invariants are defined and used to classify the topological states of the different lattices. This approach is called the quantum valley Hall effect (QVHE) [13, 43–46]. The QVHE exploits valley states instead of spin states, with the advantage that each lattice site needs to have only one degree of freedom. This concept helps obtain the configuration of reduced geometrical complexity. Valley degrees of freedom arise naturally in the systems with time-reversal symmetry and were predicted theoretically in graphene. In graphene, wave functions at opposite valleys demonstrate opposite polarizations and thus emulate spin-orbit interaction [44, 45, 47, 48]. For practical implantation, QVHE is less complicated than QHE and QSHE, but on the other hand, it results in a less robust (weak) topological effect.

Figure 7.8a and b portrays an example of QVHE in a photonic system, using a honeycomb lattice of coupled micropillars with a tunable energy detuning between A and B pillars [49]. Pillar diameters tune the gap size given in Fig. 7.8c. A perfect

Topological Acoustics in Metamaterials 367

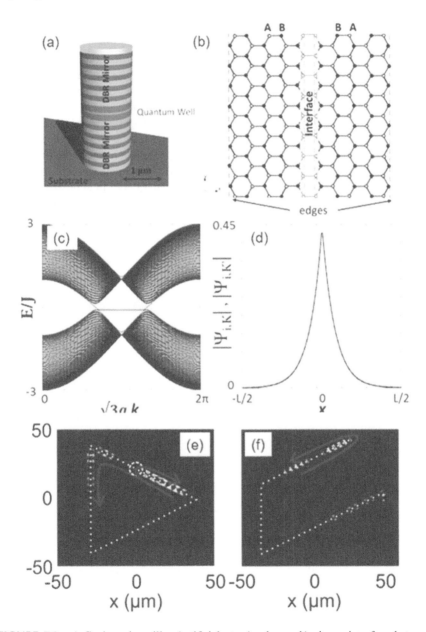

FIGURE 7.8 a) Cavity micropillar (artificial atom) scheme, b) zigzag interface between two lattices with opposite organization giving rise to zero lines modes and quantum valley hall effect with a proper choice of parameters, c) ribbon dispersion, d) corresponding in-gap interface wave-function absolute values projection on the transverse (x) direction for K and K'valleys, which are identical e) and f) QVHE edge state propagation with little backscattering perturbation on a localized defect and at the boundary respectively. The figures are adapted from Ref [49].

valley filter is shown, using a domain wall between QAH and QVH phases using mixed light-matter excitation-polariton quasiparticles. The QAH phase for polaritons is predicted to occur in different kinds of polariton lattices under Zeeman field. A zigzag interface between two state lattices with opposite organization is given in Fig. 7.8b. The dispersion and corresponding in-gap interface absolute values of wave-function at K and K' valleys are given in Figs. 7.8c and d. As mentioned, above, the QVHE implicates poor robustness while propagating along the edge state. Figure 7.8e and f shows two distinct situations, demonstrating localized defects leading to backscattering while maintaining the edge propagation somewhat intact.

Although the acoustic and elastic analog of QVHE is called acoustic valley Hall effect (AVHE), in this book the term QVHE is used. QVHE in acoustics does not exhibit any actual quantum mechanical behavior. Here, the edge state along the domain wall (DW) is shown in Fig. 7.9 [46]. As discussed under QSHE, DWs are

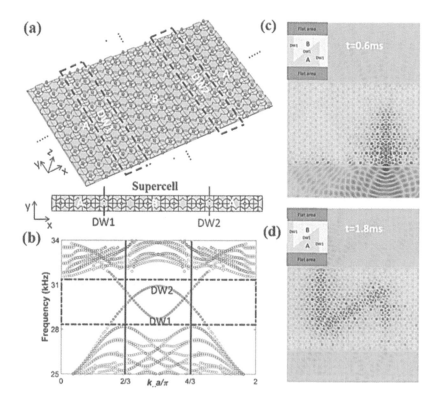

FIGURE 7.9 a) Schematics of an elastic waveguide having A- and B-state lattices connected along their zigzag edges. This configuration gives rise to two different domain walls, DW1 and DW2, as marked by the dashed black boxes. The primitive supercell of such a waveguide is also shown. b) The dispersion relations of the waveguide clearly show the existence of edge states in the topological band. For clarity, only the flexural modes are drawn in the dispersion plot. c) Plots of the eigenstates of the edge modes supported at the domain walls DW1 and DW2. The plots illustrate different symmetry patterns (either symmetric or antisymmetric) taking place with respect to the interface plane (marked by dashed lines). The figures are adapted from Ref [32].

Topological Acoustics in Metamaterials **369**

the interfaces that separate two adjacent phases of the topological states. A schematic view of the fundamental lattice structure is shown in Fig. 7.9. This analysis contains two different state lattices with a broken space inversion symmetry. Two different DWs are created in an elastic waveguide at the interface between the two states (DW1 and DW2). The supercell is also shown in Fig. 7.9, which helps to obtain the dispersion properties of the composite waveguide. In a frequency range around the topological bandgap, periodic boundary conditions are applied to all the sides of the supercell before solving the dispersion relations. The edge states at DW1 and DW2 are shown in Fig. 7.9, where the edge modes are between 28 and 31 kHz. These edge states are almost immune to backscattering due to a large separation in momentum space between forward and backward traveling modes. Full-field transient simulation demonstrates the propagation of wave bursts at two successive time instants. The energy can travel along the sharp bends with very minimal backscattering.

7.2 TOPOLOGICAL PHONONICS AND QUANTUM TRIO: A GATEWAY OF QUANTUM TRANSPORTATION

As discussed above, one of the groundbreaking discoveries in the field of condensed matter physics is finding topological quantum states of matter [12, 50]. Emerging physics of QAHE, QSHE, and QVHE are explained by topological electronic states. In addition, applications like topological quantum computation, spin-transfer torques, and thermoelectrics are realized by topological electronic states [51–53]. The most important highlights of topological quantum states are the existence of unusual boundary/edge transport in the presence of bulk-boundary correspondence. Such edge states remain unaffected against local disorder or scattering. Phonons are also considered as an important elementary particle of condensed matter systems besides electrons. Phonons are the primary carriers of heat (energy quanta) due to lattice vibration. The emergence of topological electronic states has opened a new branch in the field of phononics, called 'topological phononics' [54–56]. Topological phononics uses some primary concepts of topological quantum states (geometric phase, pseudospin, and topology) to manage phononic behavior in a controlled manner. Although topological phononics have emerged following the discovery of topological quantum physics, they differ significantly from each other in many different aspects as depicted in Table 7.1.

Like topological electronic states, phononic systems in the periodic lattice can be defined by Bloch wave functions, thus yielding topological concepts like the Berry phase, Berry connection, Berry curvature, and Chern Number for the phonons. These terminologies are discussed in Chapters 5 and 6. Lately, the topological phononic states have been characterized by two different topological invariants – phononic Berry/Zak's phase, and phononic Chern number. The phononic Berry phase, φ_g^N (refer to Chapter 5) is a function of the phononic Berry connection, $A_N(\mathbf{k}) = \langle u_N(\mathbf{k}) | \nabla_k u_N(\mathbf{k}) \rangle$. The phononic Chern number C_n, is a function of the phononic Berry curvature, $\mathcal{R}_N(\mathbf{k}) = \nabla_k \times A_N(\mathbf{k})$ [56, 57].

TABLE 7.1

Difference between Electrons and Phonons

Electrons	Phonons
Electrons in topological electronic states are assumed to be spin-1/2 fermions, obeying Fermi-Dirac postulates.	Phonons in topological phononics are spin-0 bosons, obeying Bose-Einstein postulates.
Electrons follow the quantum mechanical Schrödinger equation, possessing both positive and negative eigenenergies.	Phonons obey classical Newtonian physics, having only positive eigenenergies.
Electrons are known to have multiple atomic orbitals, such as $s, p, d,$ and $f.$	Phonons are lattice vibrations and can vibrate only along $x_1, x_2,$ and x_3 directions, resembling the $p_{x,y,z}$ orbitals.
Electrons are charged, having spin degrees of freedom, interacting with electric and magnetic fields.	Phonons have no charge (neutral) and no spin, which cannot directly interact with electric and magnetic fields.

$$\text{Berry/Zak's phase, } \varphi_g^{\ N} = \oint_C d\mathbf{k} \cdot A_N(\mathbf{k}) \tag{7.6.1}$$

$$\text{Chern number, } C_n = \frac{1}{2\pi} \oiint_s dk_1 \, dk_2 \cdot \mathcal{R}_N(\mathbf{k}) \tag{7.6.2}$$

Topological phononic states pertaining to nonzero phononic Berry/Zak's phase may result in different unusual dispersion phenomena. Starting from a 1D acoustic band inversion [58] to 2D phononic Dirac cones [59, 60] to 3D phononic nodal lines [61], the phononic Berry phase has contributed to different phononic topological states (Fig. 7.10a). The nonzero phononic Chern number in topological states breaks time-reversal symmetry for a 2D phononic Dirac state and a 3D phononic Weyl point (Fig. 7.10b). Please refer to Chapter 6 for a detailed understanding of Weyl point.

Interestingly, unusual transportation behavior is observed at the edge of two phononic systems with different topological invariants. As a result of bulk-boundary correspondence, one of the most studied phenomena is the topological phononic boundary state. Plenty of useful applications, such as phonon waveguide [57], acoustic delay lines, [62], and acoustic antennas [63], can be achieved by exploiting topological phononic boundary states. However, achieving phononic analogs of quantum (A for anomalous, S for spin, V for valley) Hall states is challenging because of the non-intractability of phonons with magnetic fields. Such difficulties pose a prime obstacle for breaking of $\mathcal{T}$-symmetry. Kramer's degeneracy [40] remains absent due to the spinless nature of the phonons as well, which makes it difficult to obtain QSHE phase of phonons. Despite being spinless, phonons possess various types of pseudospins, which help in breaking $\mathcal{T}$-symmetry. In general, pseudospin is characterized as a degree of freedom that helps to express an effective Hamiltonian by Pauli matrices. A-type trivial and B-type nontrivial sublattices with pseudospins are physically

Topological Acoustics in Metamaterials 371

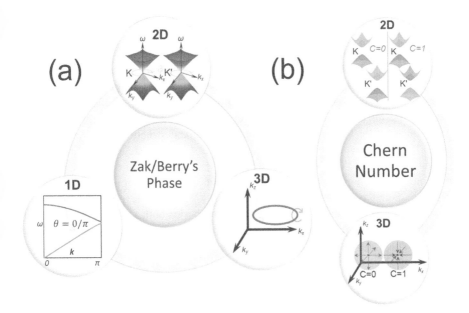

FIGURE 7.10 Schematics of topological phononic states in 1D, 2D, and 3D systems related to a) Berry/Zak's phase, b) Chern number.

trivial and difficult to control. However, pseudospins with controllable features may inject newer quantum degrees of freedom to control the phonons. It means that by increasing degrees of freedom transition from A to B state would be easier. In such cases, the physics of nonzero Berry curvature, nonzero Berry phase, and quantized pseudo-angular momentum are exploited to achieve topological behavior.

The topological properties of bulk bands are characterized by Berry/Zak's phase. A topological nontrivial system (B-type) possesses a quantized nonzero Berry/Zak's phase, whereas a topological trivial system (A-type) possesses a quantized zero Berry/Zak's phase [58, 64]. A nonzero phononic Chern number characterizes a 2D system with an analogous quantum anomalous Hall-like phononic state (QAHE). Such QAHE phononic states have also been reported in mechanical metamaterials and acoustic systems [1, 65–67]. $\mathcal{T}$-symmetry is necessary to be broken for QAHE. The simplest $\mathcal{T}$-symmetry breaking of phononic lattice can be accomplished by Coriolis force [68] or Raman spin-lattice interactions on lattice vibrations [66]. To host phononic QAHE, phonons are required to have a separate Lagrangian, $\tilde{\mathcal{L}} = \eta_{ij}\dot{u}_i u_j$, along with the typical lattice Lagrangian. (Refer to Chapter 2 for a Lagrangian description of motion.)

$$\mathcal{L}_0 = \frac{1}{2}(\dot{u}_k)^2 - \frac{1}{2}D_{ij}u_i u_j \tag{7.6.3}$$

where $u_i = \sqrt{m_i}\, x_i$. Here m_i and x_i mean the atomic mass and displacement of the i-th atom respectively; $D_{ij} = \dfrac{1}{\sqrt{m_i m_j}} \dfrac{\partial^2 \mathcal{U}}{\partial u_i\, \partial u_j}$ refers to the mass-weighted second-order force constant matrix element, where $\mathcal{U}$ refers to the potential energy. (Refer to Chapter 2 for details.) Hence, phononic state with lattice Lagrangian of $\mathcal{L} = \mathcal{L}_0 + \tilde{\mathcal{L}}$ can host QAHE states [57].

Quantum spin Hall-like states (QSHE) are also topological phononic states that do not require the breaking of $\mathcal{T}$-symmetry. Instead of breaking $\mathcal{T}$-symmetry, implementing some form of spin-analogue (as phonons being spinless) helps generate QSHE for phonons. Pseudospin degrees of freedom are utilized to acquire QSHE in acoustic systems [69], artificial mechanical structures, [70] and Kekule lattice [66]. However, like electrons, phonons are spinless excitations without possessing any intrinsic spin-orbit coupling [70, 71]. But, utilizing crystalline symmetry-protected phononic pseudospins, a form of spin-orbit coupling analog can be achieved, called 'pseudospin-orbit coupling' [70, 71]. Phononic topological states are able to showcase unusual phononic pseudospin physics both at the boundary and in the bulk, such as the pseudospin Hall effect [65, 66], the pseudospin sensitive optical selection rule [66], the pseudospin-momentum locked boundary states [72] and the pseudospin-induced phononic chiral transport [73].

7.2.1 BERRY PHASE AND BAND TOPOLOGY IN PHONONICS

The concept of the Berry phase was proposed for electrons and is discussed in Chapter 5 in detail. Berry phase and band topology make a large contribution in establishing concepts like electric polarization [64, 74, 75], orbital magnetization [76], anomalous Hall effect [77] and Spin Hall effect [78], topological insulators [79], and many more. Berry phase is a geometric phase, which is acquired by integrating the Berry connection along the closed loop inside Brillouin zone. Further, the Berry curvature is the curl of the Berry connection, playing the role as a magnetic field in the momentum space. The concept of Berry phase can be generalized by writing the Schrödinger equation in terms of an equation of phonons (refer to Chapter 5), as

$$\mathbf{H}_k \Psi_k = \omega_k \Psi_k \tag{7.6.4}$$

where $\mathbf{H}_k = \begin{pmatrix} 0 & iD_k^{1/2} \\ -iD_k^{1/2} & -2i\eta_k \end{pmatrix}$, $\Psi_k = \begin{pmatrix} D_k^{1/2} u_k \\ \dot{u}_k \end{pmatrix}$, D_k is the dynamic matrix, η_k is the term pertaining to $\mathcal{T}$-symmetry breaking, u_k, and $\dot{u}_k$ are the displacements and velocities respectively [56]. When $\mathcal{T}$-symmetry is intact, η_k becomes zero. Hence, η_k is nonzero only when the $\mathcal{T}$-symmetry is broken. Ψ_k is the wave function of phononic state expressed in displacement-velocity space [55, 80].

7.2.2 PHONONIC TOPOLOGICAL STATES

Topological phononic states can be described for 1D, 2D, and 3D metamaterials. For a 1D phononic state, a simple diatomic chain with an alternating spring system with spring force constants of $f \pm \delta$ can be studied. Topological band inversion between in-phase and out-of-phase modes can be acquired. By varying the unit cell geometry, such topological band inversion can be introduced artificially. When the band-inversion

Topological Acoustics in Metamaterials

occurs due to the bulk-boundary information exchange, simply called bulk-boundary correspondence, a consistent interface in-gap state appears. For a 2D phononic state, a honeycomb lattice is studied to acquire QAHE or QSHE. The intercept of both the higher and lower (identified as optical and acoustic branch, respectively) modes at the Brillouin zone boundaries (i.e., at K and K') yields a conical formation of the local bands called Dirac cone, which are discussed in Chapter 5 and 6.

In Fig. 7.11a, the phononic Dirac cones are described by the effective Hamiltonian described in Eq. 6.12.2 in Chapter 6. The equation is a phononic dispersion equation assumed to be gapless, protected by both the crystalline symmetry ($\mathcal{P}$-symmetry) and the $\mathcal{T}$-symmetry. When the $\mathcal{T}$-symmetry is intended to be broken, mathematically a Haldane-type mass term, $\mathbf{H}_T$ can be added to the effective Hamiltonian.

$$\mathbf{H}_T = m_T \xi \sigma_{x_3} \tag{7.6.5}$$

An effective Hamiltonian that mathematically describes the QAHE can be written as $\mathbf{H} + \mathbf{H}_T \cdot \mathbf{H}_T$ has opposite Dirac masses between the boundary points K and K', creating a nontrivial band topology. Such opposite Dirac masses demonstrate a

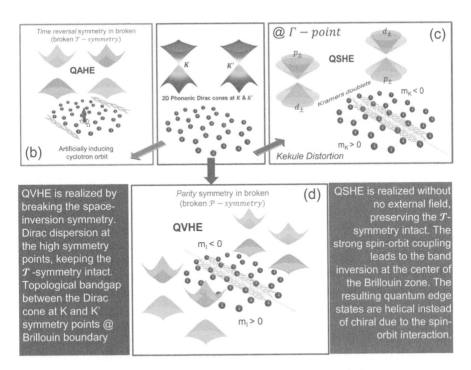

FIGURE 7.11 Schematic of QAH, QSH, and QVH effect by breaking symmetries of a 2D phononic Dirac cone system in a honeycomb lattice: a) 2D phononic Dirac cones at K and K' valleys in a honeycomb lattice, b) with introducing Kekule distortion, phononic Dirac cones become gapped into a QSH-like state with pseudospin-momentum locked state at the boundary, c) time-reversal symmetry is broken by an external magnetic field, Ω, allowing one-way edge states like QAH), d) inversion symmetry is broken, yielding QVH.

nontrivial band topology known as a QAHE state, shown in Fig. 7.11b. The Berry curvature is very small throughout the Brillouin zone for comparatively smaller m_T, except at the boundary points, K and K'. So, the valley yields a nonzero total Chern number by integrating the Berry curvature near K and K'. The nonzero total Chern number manifests the existence of one-way chiral edge state. In contrast to the QAHE, instead of a dipolar state, a quadrupolar state with opposite Chern numbers is known as QSHE. Spin-momentum locking is present for QSHE, having a helical boundary state. Possessing a certain pseudospin degree of freedom results in QSHE. Here, boundary points K and K' both fold towards the center of the Brillouin zone, resulting in strong valley-valley mixing (e.g., Kekulé Lattice distortion). Such valley mixing creates a couple of Kramer's-like doublets resembling $p_\pm$ (p_x, p_y) and $d_\pm$ (d_{xy}, $d_{x^2-y^2}$) atomic orbitals. Including a mass term with effective Hamiltonian gives the QSHE and can be mathematically described as $\mathbf{H} + \mathbf{H}_K$, where

$$\mathbf{H}_K = m_K \sigma_n \sigma_{x_1} \tag{7.6.6}$$

σ_n is the valley Pauli matrix parallel to $x_1 - x_2$ plane [66]. n signifies a normal direction at an angle θ made with the $\Gamma - K$ direction ($n = \cos\theta$, $\sin\theta$, 0). σ_n is off-diagonal in valley subspace; it interacts with boundary states from K and K'. The mixing of two boundary states makes it difficult to define the pseudospin. Hence, another form of pseudospin, $S = \sigma_{n'} \sigma_{x_2}$ can interact with the total effective Hamiltonian. When the mass m_K changes sign at the interface, a pair of pseudospin-momentum locked states emerge. This phenomenon is shown in Fig. 7.11c.

In the case of QVHE, phonons utilize the valley index at K and K' points in the Dirac phonon state. In a hexagonal honeycomb lattice, the QVHE can be mathematically described by including a specific mass-term, $\mathbf{H}_I = m_I \sigma_{x_3}$, and the effective Hamiltonian can be written as $\mathbf{H} + \mathbf{H}_I$ [56]. The total Chern number remains zero for both positive and negative m_I, as the $\mathcal{T}$-symmetry is intact or preserved. However, the valley-resolved Chern number changes with the change of sign of m_I, resulting in a pair of valley-polarized interface states with valley-momentum locking, shown in Fig. 7.11d. To summarize, the phononic analog of topological electronic states has been achieved through different techniques like QAHE, QSHE, and QVHE.

Key Points from Quantum Trio

As discussed above, QAHE, QSHE, and QVHE, commonly referred to as 'quantum trio', all belong to the quantized edge transport phenomena, though their intrinsic mechanisms are quite distinct. The major characteristics of these three phenomena are as follows:

- QAHE is realized in a Chern insulator, insulating the bulk because of discrete Landau levels, with a conduction channel existing only at the edges. The applied external magnetic field (in photonics) and artificially induced cyclotron orbit around nodes (in phononics) performs as a gauge field coupling with the momentum of the fermion state, resulting in cyclotron motions in real space. $\mathcal{T}$-symmetry is not preserved herein.
- QSHE is realized with no externally applied magnetic field (in photonics) and no artificially induced cyclotron field (in phononics), preserving the

Topological Acoustics in Metamaterials

$\mathcal{T}$-symmetry intact. The strong spin-orbit coupling leads to the band inversion at the center of the Brillouin zone. The resulting quantum edge states are helical instead of chiral due to the spin-orbit interaction.

- QVHE is realized by breaking the space-inversion symmetry in a lattice possessing Dirac dispersion at the high symmetry points, keeping the $\mathcal{T}$-symmetry intact. Opening a topological bandgap between the Dirac cone associated with the K and K' symmetry points (at the boundary of the Brillouin zone) results in valley-dependent edge state propagation.

7.2.2.1 Topological Insulators in Acoustics

Topological insulators (TIs), were originally proposed in quantum systems [81]. Due to simpler implementation TIs were designed [14, 82, 83] for photonic systems, and various extraordinary phenomena were observed. In acoustics, the analog quantum Hall effect was first conceived in 2015 [84]. The acoustic Chern insulator was virtually created by injecting external flow fields to break the time-reversal symmetry (Fig. 7.5). Later this was experimentally validated [85] in 2019. With further progress, and utilizing the pseudospin states, a feasible design of flow-free acoustic TIs [71, 86, 87] was proposed. By this means, robust one-way propagation based on topological edge states was demonstrated in several types of phononic crystals (Fig. 7.5). As an analog to the quantum spin Hall effect, this class of acoustic TIs has a double Dirac degeneracy in the band structure where topological phase inversion occurs. The pseudospin states at the double Dirac point form topological edge states, which propagate along the interface between ordinary phononic crystals and topological phononic crystals.

7.2.2.2 Acoustic Spinning and Topological Edge State in Acoustics

The valley Hall effect is introduced in optic, acoustic, and elastic wave systems. Single Dirac degeneracy in a graphene-like lattice is lifted when C6 symmetry is reduced. It forms a pair of separated valley vortex states. These valley vortex states have opposite chirality and can also realize topological edge states along the interface of two domains (A and B) with opposite orientations. Despite different requirements of excitation, valley Hall phononic crystals (VHPCs) also possess the ability to hold robust one-way propagation. Although varieties of designs for acoustic topological systems were proposed and investigated, to the best of our knowledge, a systematic comparative study between VHPCs and TIs has not been conducted yet. For example, while it has been demonstrated that TIs are robust against cavity defects [13, 88, 89], it is not clear whether this is also the case for VHPCs. Here, VHPCs are systematically compared with TIs for acoustic waves. Their fundamental differences and similarities highlight potential topological acoustic devices such as acoustic antennas [63] and delay line devices [62].

7.3 EMERGENCE OF TOPOLOGICAL BLACK HOLES

Edge state conduction, wave modulation, and innovative wave control in the fields of electromagnetism and acoustics have been widely studied in the past two decades. Due to the complexity of practical implementation and fabrication, advancement of topological research in the field of acoustics is minimal compared to photonics and electromagnetics. Hence, novel and innovative implementations of the topological physics that are known and the discovery of new physics with proper quantum explanations are absolutely needed for acoustics.

In this section, beyond quantum trio, a deaf band-based physics of a topological phenomenon is discussed for acoustics. A spin-mediated bulk conduction waveguide, keeping the edge state insulated, is proposed in this chapter. Such topological bulk containment of energy is reported in the literatures as acoustic Skyrmions (refer Chapter 6). To recapitulate herein, Skyrmions are homotopically (refer section 5.6.1 in Chapter 5) nontrivial state and hence could be useful to explore the spin states of elastic and acoustic waves. Usually, vortex-like configurations that are statically stable in the bulk are called Skyrmions. In most cases such state was achived in the bulk using active creation of incident wave field throguhout the boundaries of a hexagonal structure. However, in the field of acoustics, passive creation of bulk mode are not common in the literature. Periodic phononic crystals, called PnCs, are utilized in this acoustic design.

Encouraged by the deaf band-based energy modulation discussed in Chapter 6, topological bulk conduction like Skyrmion is achieved with negligible leakage. This phenomenon is named 'topological blackhole' (TBH). Please note that the topological black hole is not to be confused with the acoustic black hole, previously reported in the literature. Earlier, research was conducted on acoustic blackhole (ABH), where trapping of flexural waves based on local heterogeneities [88, 90] was proposed. ABH has mostly been exploited to achieve efficient passive control of structural vibration [91–95]. With the help of a wedge having decreasing thickness profile using a power law, the velocity of the incoming flexural wave was reduced to zero, hence resulting in a non-reflecting or a fully absorbing (termination of energy) metamaterial. This concept was confined to the wave reflectivity using ABH indentation, in the form of wedges, pits, spiral meta-structure, acoustic tubes, etc. as shown in Fig. 7.12. [90, 96, 97].

Dirac cone-based robust topological energy containment in the bulk is yet to be discussed in the chapter. A doubly degenerated state formed at the Γ point may manifest TBH under certain circumstances. If properly realized and the physics is understood, TBH could be utilized for a broader spectrum of physics applications, for example, quantum analog computing using acoustics where higher degree of information could be stored utilizing higher degrees of freedom. The proposed TBH possesses a unique bulk-boundary distinction, resulting in an acoustic energy sink inside the periodic meta-structures. This behavior is a counterpart of the topological acoustic insulator. The polarization of the two energy modes, degenerated at the Γ point needs to be orthogonal to each other to acquire the TBH. Such mismatched polarity results in an ambiguity within the energy propagating through the PnCs, resulting in alternate spinning to allow the energy to be trapped within. Changing the shape and size of the waveguide made of PnCs does not destroy the TBH, confirming the high robustness of the system. Thus, TBH is topologically protected. Regardless of any defects inside the meta structure made of the PnC array, the topological behavior persists when excited at the doubly degenerated frequency. The energy trapping is independent of the impurities within the metamaterial made of PnCs.

Topological Acoustics in Metamaterials

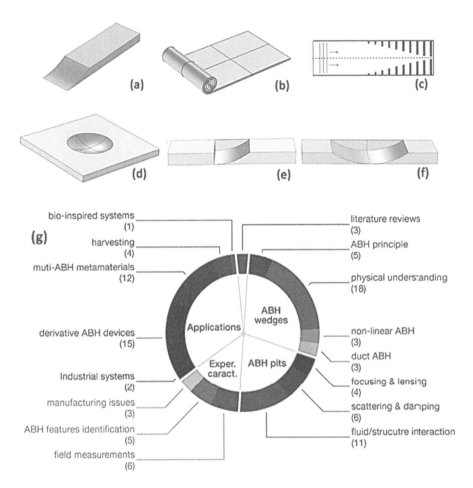

FIGURE 7.12 Examples of so-called retarding structures based on the concept of power-law taper: a) tapered wedge, b) spiral ABH, c) acoustic tube with axially varying impedance made with a collection of branch discs of increasing diameters, d) two-dimensional circular acoustic black hole, e) one-sided, and f) two-sided ABH slots, g) summary of the available literature related to the acoustic black hole effect. The figures are adapted from Ref [88, 90].

The topological insulator helps to conduct electrons along the boundary of the material, where the rest of the material acts as an insulator, prohibiting conduction inside the bulk media. Similarly, there might be a phenomenon where the bulk media act as a conductor, keeping the edge protected from the conduction. Here, the boundary exhibits insulating phenomena, keeping transportation away from the edges of the metamaterial. An acoustic model is created where this phenomenon was observed. As mentioned earlier, the counter phenomenon of topological insulators inspired to name it as a topological conductor also named as a topological blackhole (TBH). If a metamaterial made of an array of PnCs is excited from one end at a certain frequency, the meta-structure behaves like a TBH, keeping the edge state unperturbed.

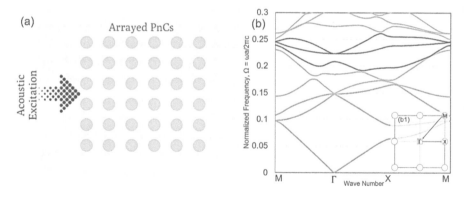

FIGURE 7.13 a) Experimental setup for frequency domain analysis, where the left-most edge/boundary has been excited with prescribed displacement, b) the band diagram for the cylindrical unit cell.

A sample schematic for desired TBH is given in Fig. 7.13a. Acoustic energy excited at a specific frequency generates the TBH. A plane wave excited from the left boundary, shown in Fig. 7.13a, impinges on the periodic PnCs array. The excitation frequency is swept for a wide range, where the normalized frequency, $\omega = 0.231 \cdot (2\pi c/a)$ generates the bulk propagation of the acoustic energy. The normalized frequency was calculated by

$$\Omega = \frac{\omega a}{2\pi c} \tag{7.7}$$

where Ω is the normalized frequency, ω is the real actuation frequency, a is the lattice constant and c is the wave speed in the traveling medium (in this case it is air). The frequency, which is called TBH frequency, is the frequency very close to a doubly degenerated state of two modes at the Γ point of the Brillouin zone, shown in Fig. 7.13b. It is discussed in Chapter 6 that the deaf band-based modulation can transform the double degeneracy into a triply degenerated Dirac-like point. However, in the current state, a directional bandgap (from $\omega = 0.208 \cdot (2\pi c/a)$ to $\omega = 0.231 \cdot (2\pi c/a)$) exists between the doubly degenerated point and a single mode below the bandgap. (Please refer to Dirac-like cone discussed in Chapter 6 to understand this statement.) As the directional bandgap is along the ΓX direction of the Brillouin zone, and the degenerated point is also at the Γ point, a plane wave excitation along the ΓX will result in TBH inside the bulk media.

7.3.1 Modal Analysis at Dirac Frequency

For the initial numerical study, an array of 41×41 PnCs are arranged in a square lattice formation. An excitor is placed at a distance of ≈ 115 mm from the first unit cell of PnCs. A normal pressure excitation is generated from the excitor along the ΓX direction towards the arrayed PnC matrix. A free host media is assumed beyond the PnC matrix to visualize wave transmission and scattering outside the meta

Topological Acoustics in Metamaterials 379

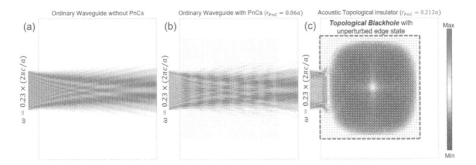

FIGURE 7.14 (Color) Formation of topological conduction phenomenon, using proposed PnC techniques: a) A plane wave generated from the left terminal of the model propagates linearly through the air media when no PnCs are present and with PnCs of radius r = 0.06a. b) When the parameter r is changed to 0.212a, the energy content tends to accumulate inside the PnC matrix, without interfering the edge states. c) This counterpart of topological insulation is being visualized as an 'acoustic topological blackhole'.

structure made of PnCs. To avoid unwanted reflective waves, the host boundaries are assumed to be non-reflective, and plane wave radiation boundary condition is implemented. The PnCs are assumed to have free boundary condition. The lattice constant, $a = 25.4$ mm, defines the Brillouin Zone, where only one PnC is placed at the center. The plane wave generated at TBH excitation frequency in an ordinary waveguide without PnCs propagates inside the medium with minimal to no interactions inside (Fig. 7.14a). The wave maintains an almost similar pattern, keeping the wavefronts intact even when 41×41 PnCs with radius, r = 0.06a are placed in the host matrix, where a is the lattice constant (Fig. 7.14b). However, based on the understanding from deaf band tuning (discussed in Chapter 6, section 6.5.2), when the radii of the PnCs are changed from r = 0.06a to r = 0.212a, the propagating energy gets trapped as a bulk inside the PnC array, with no or negligible leakage out of the array (Fig. 7.14c). Figure 7.14c portrays stagnant acoustic energy contained inside the PnC array, keeping the edge state unperturbed. Almost 99% of the energy excited by the excitor is confined inside the PnC array. Such unique phenomenon of bulk energy containment using periodic insertions in air is named 'topological blackhole' (TBH). This arrangement allows robust propagation and acts like a topologically protected system. Bulk containment of the wave energy is not affected by the edge deformation of the array. The arrangement of PnCs' size and shape was changed afterward to observe the topologically protected property of this topological conductor. It was found that, regardless of shape and size, the topological property of TBH remains unperturbed.

Figure 7.15 demonstrates two different cases that generate TBH, where in the first case (Fig. 7.15a), a square annular structure made of PnCs is created, which is like the PnCs being taken away from the matrix, creating a closed-loop waveguide. A proper formation of bulk containment of the wave energy prevailed inside the waveguide when excited at the TBH frequency. Offsetting the source and changing the shape of the annular waveguide does not interrupt the TBH properties, as shown in Fig. 7.15b. Hence, the robustness of this system is stronger than a usual valley-Hall

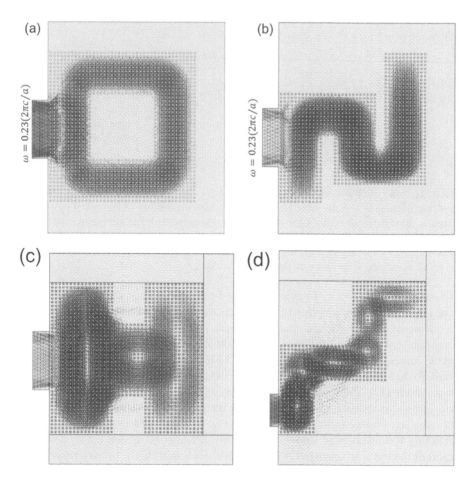

FIGURE 7.15 (a) Wave propagating inside O-shaped PnC array with one hole, topological charge 1, validating the robustness of the proposed TBH and conductor in the field of acoustics, (b) TBH through a S-shaped PnC array, (c) and (d) asymmetric H-joint and staircase structure with cylindrical PnCs matrices preserving the topological feature but not breaking edge state.

conduction. Although it is found that decreasing the width of the waveguide weakens the TBH propagation and containment (not shown here), TBH characteristics remain persevered. Tuning the TBH frequency apparently yields TBH harmonic modes with incremental change in frequency by 1Hz.

Figure 7.16 shows multiple higher-order modes of TBH that are generating inside the bulk meta structure made of PnC arrays, keeping the topological phenomena protected. Please note that the TBH mode shapes are discussed herein. They are not the mode shape of a specific band in the dispersion relation synonymously known as wave function in acoustics. As discussed above, the first bulk mode is achieved at the frequency, $\omega = 18.947\ kHz$. The wave contained within the bulk forms a single mode with a single central lobe at the center. The energy is distributed around the

Topological Acoustics in Metamaterials

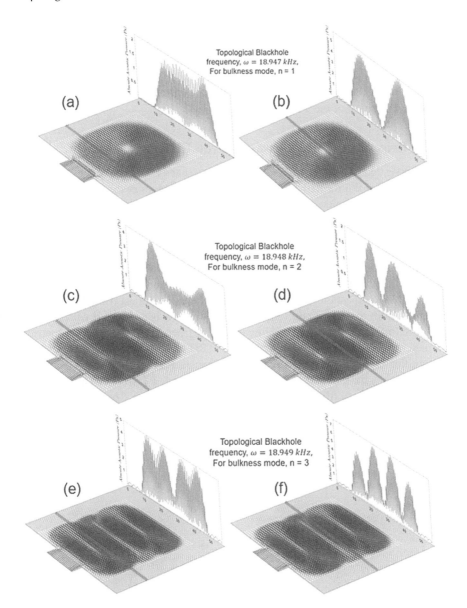

FIGURE 7.16 (Color) Absolute acoustic pressure distribution with pressure plots across 2D cut-lines through a side and the center of the PnC arrayed matrix, at frequencies a) and b) $\omega = 18.947\ kHz$, c) and d) $\omega = 18.948\ kHz$, and e) and f) $\omega = 18.949\ kHz$.

central lobe, similar to a whirling pattern, like the energy is sinking. The absolute acoustic pressure distribution plotted at the cut line shown in (Fig. 7.16a) and along the center (Fig. 7.16b) depicts the bulk-stagnation of the energy excited at the TBH frequency. The pressure along the side is almost uniform across the PnC arrays, while the pressure dips down to almost zero at the center of the meta structure. With increasing the frequency at $\omega = 18.948\ kHz$, a second harmonic of the TBH bulk mode is observed. The pressure plots across a side and at the center are given in Fig. 7.16c and d. Changing the frequency further yields more central lobes of TBHs inside the bulk. Figure 7.16e and f demonstrates the third harmonic mode of TBHs generated inside the meta structure. The pressure distribution across the cut lines shows the formation of the TBH.

7.3.2 Tuning of Geometric Parameters for TBHs

Further investigation by focusing on three unique bands is conducted to determine the robustness of the formed TBH. As discussed in Chapter 6, the bands below and around the TBH frequency are called the bottom band, deaf band, and top band respectively. By changing the physical parameters of the PnC inclusion (e.g., radius in the case of a circular PnC – 2D approximation for cylindrical PnCs), the frequency curves for the three bands are plotted in Fig. 7.17a and b shows the degenerated state of the three bands. From the figure, it is identified that at the cross hair with radius $r = 0.212a$ at frequency $\omega = 18.947\ kHz$, the current TBH state (Fig. 7.17c) is obtained. Now a question we can ask: Will this behavior persist if the PnC unit cell is altered? With this question in mind, an investigation is conducted with different geometric PnC parameters by changing the filling fraction. Altering the geometrical parameter of the PnCs (i.e., radius herein), the TBH phenomena was indeed achieved at a significantly wide range of actuation frequencies. Radius of the PnCs from $0.17a$ to $0.35a$ ensures the generation of the TBH. Hence, such a TBH is consistently present for a wide range of filling fractions, regardless of the lattice size of the cylindrical PnCs. But why is TBH identified only when r is in between $0.17a$ and $0.35a$? With

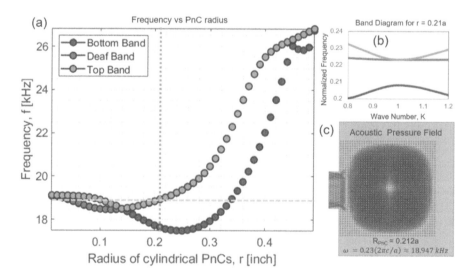

FIGURE 7.17 a) The band diagram for circular PnC, b) T band, deaf band, and B band frequencies for varying PnC radius are plotted. Here, the formation of Dirac-like cone is evident with r = 0.21a. c) TBH formation at 18.947 kHz.

Topological Acoustics in Metamaterials

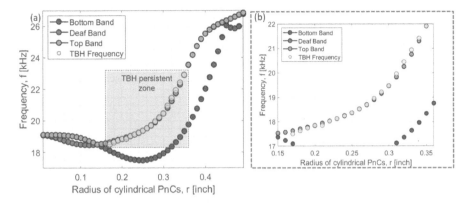

FIGURE 7.18 a) The frequency at which TBH exists for different radii of PnCs. It is to be noted that the trend of TBH formation merges and aligns with the deaf band frequencies band. Hence, impact of the deaf band on TBH formation is present. b) A closer look at the TBH frequency slope with respect to deaf band frequency slope for varied r.

close observation, it can be found that this segment of the degenerated (deaf-band and top band) bands starts after the triply degenerated state at a Dirac-like cone along the parameter space where the bottom band is lifted from the degeneracy and is identified as a nontrivial state. Investigating the parameter space for the Dirac-like cone, it can be seen that before the Dirac cone is formed, the bottom band and the deaf band are degenerated and the top band is lifted identified as a trivial state and no TBH is formed. It is also observed that this segment has a concave curvature (i.e., the second-order derivative of the frequency curve with respect to the parameter space is positive). Hence, the red square region in Fig. 7.18a shows the TBH zone. When the red zone is zoomed in, the TBH frequencies were found to be very close to the degenerated frequency of the deaf band and the top band, where the bottom band is lifted from the degeneracy forming a Dirca-like cone. The frequency–radius slope is almost equal for both the degenerated curve and the TBH frequency curve. This strengthens the proposal for the generation of such unique topological transportation and acquires higher robustness, preserving the edge state.

7.3.3 Real-time Tunable Metamaterial for TBHs

It is evident that to acquire a TBH state, different dimensions of the PnCs should be used and cannot be tuned in real time. On the other hand, rotation-driven QSHE is presented in section 7.1.2.2, where the concept of real-time formation of topological behavior is demonstrated. Circular cylinders remain symmetric under rotation and thus cannot be used for real-time tuning. Hence, a PnC element that breaks symmetry under rotation is conceptualized over the circular PnC elements. Instead of cylindrical or circular PnCs, square-shaped PnCs are investigated for the generation of TBHs. A square PVC rod of dimension $b = 0.342a$ and a = 1 inch is used as the lattice constant. Initially, the square PnCs are placed with no angular orientation with respect to the incident wave direction, shown in Fig. 7.19a. The frequency region of

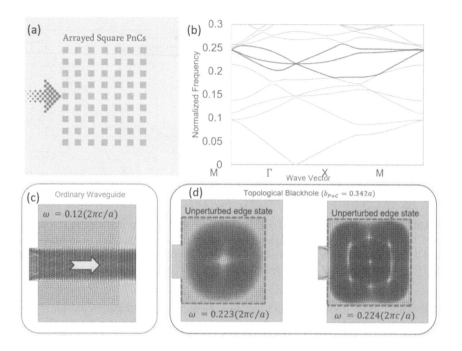

FIGURE 7.19 a) TBH and conduction for square resonator of b = 0.342a, b) the band diagram, c) at frequency 10 kHz plane wave propagates through the metamaterial (note the formation of single mode TBH lobe at frequency 18.48 kHz) and, d) multimode TBH at 18.56 kHz.

interest for obtaining possible TBHs is plotted in red in the band structure shown in Fig. 7.19b. For a normalized frequency of $\omega = 0.12(2\pi c/a)$, the plane wave propagates through the PnCs, keeping the planar wavefront unperturbed by the PnCs, as shown in Fig. 7.19c. However, the plane wave generated at a different frequency, $\omega = 0.223(2\pi c/a)$, a frequency almost equal to the deaf band frequency, administers the black-hole pattern inside the 41 × 41 PnC matrix. Figure 7.19d demonstrates the successful generation of bulk wave and its containment within the bulk, keeping the boundary insulated. Hence, like cylindrical PnCs, square-shaped PnCs also possess the same characteristics. The edge state is unperturbed, and almost 99% of the energy excited from the left excitor is conserved inside the bulk PnC array. Please note that Fig. 7.19d shows the first harmonic TBH mode and, by increasing the excitation frequency, reveals a higher-order mode inside the bulk, preserving the edge state. A sample higher harmonic mode is shown in Fig. 7.19e with an increase in excitation frequency of $\omega = 0.224(2\pi c/a)$. In this case, TBH frequency is almost like the degenerated frequency of the deaf band and the top band. A thorough investigation on the formation of the TBH with different modifications of PnCs is discussed further in this chapter.

In Chapter 6 tuning techniques are presented to explore the possibility of acoustic computing. A similar tuning method is implemented in this case. Rotating the square PnC rod clockwise and counterclockwise with angle θ (with respect to the direction of wave incident, along ΓX), the wave behavior inside the meta structure made of square PnCs is documented. Although such a tuning technique alters the dispersion

Topological Acoustics in Metamaterials

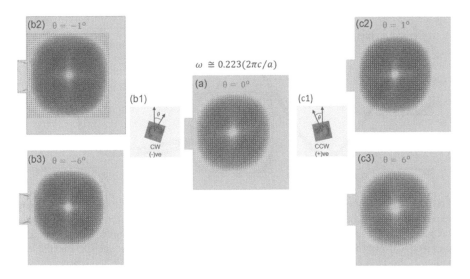

FIGURE 7.20 a) Persistence of TBH development for changing the incidence angle or the PnC angles with respect to the incident plane wave from left terminal. Almost identical TBH is formed for both b) $\theta = -1°, 0°$ and $1°$, and c) $\theta = -6°, 0°$ and $6°$.

modes in wavenumber space, surprisingly it does not hamper the robustness of the TBH formation. That means the robustness of the system remains preserved, with negligible leakage of energy from the edge, keeping the rest within the PnC array. Figure 7.20 shows the pressure field distribution with different PnC arrangements in terms of angular orientation. For square PnC orientation angle, $\theta = 0°$, the first harmonic bulk mode is shown in Fig. 7.20a. With counterclockwise rotation of $\theta = 1°$ and $\theta = 6°$, the TBH remains preserved, shown in Figs. 7.20c1 and c2, respectively. Similarly, rotating the PnCs in the opposite direction (i.e., in clockwise direction $\theta = -1°$ and $\theta = -6°$), TBH also persists as shown in Fig. 7.20b1 and b2 respectively. Such behavior persists till $\theta \approx 15°$, in either direction of the parameter space, which is rotational angle θ. The system starts losing the bulk TBH characteristics when the PnCs are rotated for more than $\theta \approx 15°$.

Figure 7.21 shows the frequency of each band (top, deaf, and bottom) with a parametric sweep. Band frequencies are identified at different angular orientation of the square PnCs at an increment of $1°$. In this case, the deaf band remains degenerated with the bottom band when $-6° < \theta < 6°$. Triple degeneracy, or the Dirac-like cone, forms at $\theta \cong 6.2°$. After that, the deaf band is merged with the top band, generating a directional bandgap between the degenerated bands and the bottom band. However, as the TBH phenomena exists between a range of $-15° < \theta < 15°$, the involvement of the deaf band is evident. For a certain range of parametric space, degenerated frequency of the top band and the deaf band is aligned with the frequency of TBH formation. Hence, the role of negative radii of curvature evidently has an impact on the TBH formation.

Apart from PnC rotation, next the dimension of the square PnCs was changed to study the TBH state. This changes the filling fraction (f, refer Chapter 6) of the unit cell by increasing or decreasing the dimension of the square PnCs and affects the TBH phenomenon. Only a small window of filling fraction, $0.13 < f < 0.109$, allows

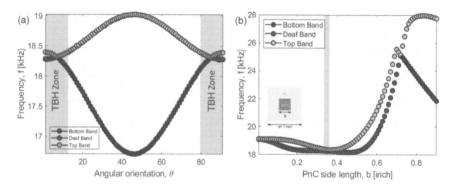

FIGURE 7.21 Bottom band, top band, and deaf band frequencies: a) frequencies plotted over a range of angle of rotation of the PnCs with respect to incidence wave, gray zone between 0° to 14° and from 76° to 90° indicates the TBH zone in rotational space, b) frequencies plotted over a range of square dimension of the PnCs, gray zone between 0.3 and 0.4 inch indicates the TBH zone (filling fraction, $0.13 < f < 0.109$) in PnC dimension space.

the TBH state. Figure 7.21b shows the change in frequency of the bands (top, deaf, and bottom bands) with the size of the PnCs. A small region where the Dirac-like cone is formed, allowing the conditions discussed above and to be discussed in this chapter, is identified where the TBH exists. Nevertheless, the deaf band frequency mode matches with the curvature of the TBH frequency mode as the orientation angle θ changes. Figure 7.22 shows the TBH frequencies for each of the PnC orientations, along with the dispersion relation of the top band, deaf band, and the bottom

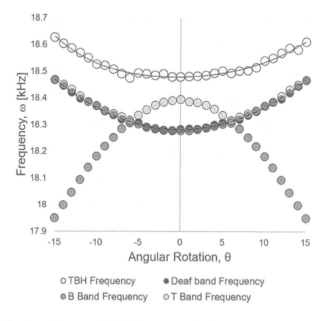

FIGURE 7.22 Coordinance of TBH frequency with the deaf band frequencies for PnC angles between −15° and 15°.

Topological Acoustics in Metamaterials

band. This proves that the generation of TBH is not accidental but rather controlled and tunable following the deaf band for varieties of PnCs with different orientations and filling fractions.

Absolute acoustic pressure in the air inside and outside the PnC array is calculated and plotted in Fig. 7.23. The absolute pressure was integrated over the 2D surface and plotted against a range of excitation frequency for both cylindrical and square PnCs. Figure 7.23a shows the pressure amplitude to be minimal till the frequency where TBH dominates, as the directional bandgap persists in that frequency region. At the frequency of degeneracy of the deaf band and the top band, the TBH originates. Maximum localization of the wave with a sudden spike in the pressure

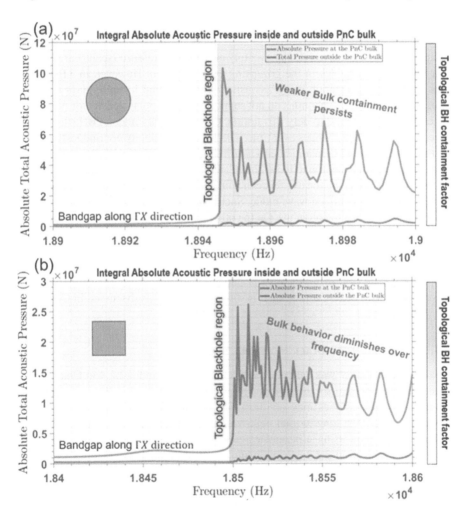

FIGURE 7.23 Surface integral of absolute acoustic pressure inside and outside of PnC arrayed matrix. The comparison shows the full containment of wave energy trapped inside the TBH, leaking very negligible energy out of the system for a) circular PnC and b) square PnC.

plot is observed (Fig. 7.23a) with circular PnCs. The total pressure plot outside the bulk PnCs, that is along the edges, is negligible at and after the TBH frequency. A similar pattern is observed for square PnCs (Fig. 7.23b). Please note that the TBH effect and its magnitude oscillate but start to weaken as the frequency increases shown in Fig. 7.23.

7.4 POLARIZATION ANOMALY: DEMONSTRATION OF ACOUSTIC SPIN

Anomalous polarity and its effect on wave propagation was also discussed in section 3.2.2.4 in Chapter 3. Wave polarization in acoustic waves is associated with the longitudinal wave modes only. Acoustics in fluids (liquid or gas) do not support shear waves, and thus the transverse wave mode does not exist. As discussed in Chapter 3, acoustic wave velocity (vector) or acoustic pressure (scalar) are the primary unknown variables in the governing wave equation. However, in TBH it was found that the polarization of the acoustic wave field, and its spin, plays a critical role. Although acoustic phonons do not support spin, the polarized vector field at the state of TBH was found to defy this notion. Global velocity under TBH state has an inherent polarity and changes over time. The polarized velocity continuously rotates, first in clockwise and then anticlockwise directions. Global spin of the acoustic field was calculated and was found to be counterintuitive.

One way of verifying the existence of spin is the spatial integration of the acoustic spin density for a localized wave mode. Such integration must result in zero to agree with spin-0 nature of the phonons. On the other hand, the spin-orbit interaction (SOI) generated by the coupling of spin and the orbital angular momentum (OAM) must be absent for longitudinal waves in acoustics. Hence, if the acoustic energy demonstrates any nonzero acoustic spin density, then spatially the acoustic energy must support the state of transverse wave or anomalous polarization of two longitudinal wave modes with different propagation directions (like it is discussed in Chapter 6, Fig. 6.45).

7.4.1 FORMATION OF BLACK HOLE-LIKE SINK

Polarization anomaly was described in sections 3.2.2.3 and 3.2.2.4 in Chapter 3. It was demonstrated that sometimes, due to effective anisotropy in the metamaterials, longitudinal wave modes may manifest vertical polarization and transverse wave modes may manifest longitudinal polarization. They were identified from the equifrequency contours using a parameter called polarity differential $\left[\theta_{pd} \right]$, or PoDi. Please refer to section 3.2.2.3.2 in Chapter 3 for more details. This may occur due to a specific degree of effective anisotropy of the material with specific conditions described in Chapter 3. However, they can be effectively demonstrated when a wave vector moves along a certain equifrequency contour and transitions from one wave mode to another wave mode present at the same frequency.

Sometimes, due to this anomaly, spin-dependent wave vortex may develop, causing the acoustic phonon to be trapped within a host structure. Like bosons, if certain conditions are met, transverse acoustic phonons are spin-1 in nature and may

Topological Acoustics in Metamaterials

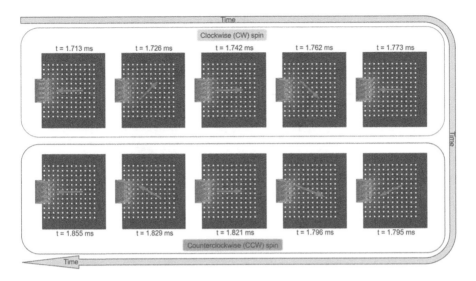

FIGURE 7.24 (Color) Instantaneous local acoustic velocity field with the calculated instantaneous field polarity. Starting from 1.713 msec, the polarity encounters a CW rotation of 2π at t = 1.773 msec, and then reverses the rotation in CCW by 2π at t = 1.855 msec.

manifest elastic wave properties. It can be an analog of chirality. In TBH, the existence of such a spin-dependent vortex is evident, as shown in Fig. 7.24. TBH frequency imposes that wave is contained within the bulk and certain conditions for anomalous polarity are met. The bulk wave was found to have local rotation/spin in TBH. With phase shifts, the energy in TBH spins clockwise and counterclockwise, alternately. Upon observing the time-domain analysis results with the square PnCs, ($b = 0.342a$) with $\theta = 9.7°$, a total wave field was integrated to find the resultant velocity vector at any instant. Over time, the directionality of the velocity vectors shows an interesting pattern and generation of the acoustic vortex. After entering the metamaterial made with the PnCs matrix, locally the micro-spin was initiated, which globally formed a wave sink, where the wave appeared to be trapped inside a blackhole without any escape. This phenomenon is topologically protected. Absorption of energy inside the PnC arrays is evident by forming constructive and destructive blackholes. Figure 7.24 shows a segment of the spin event between $2\pi - 4\pi$ clockwise and $4\pi - 6\pi$ anticlockwise. Figure 7.25 portrays the total segment of spin between $0 - 4\pi$ and $4\pi - 8\pi$ with the formation the acoustic TBH.

7.4.2 Counterinteractive Spin and OAM in TBH

Instead of analyzing local spin, taking integration over the instantaneous acoustic wave velocity field results in instantaneous polarity. This is very close to Skyrmion-like behavior discussed in Chapter 6. Such chiral property evolves with small time increments. Time-dependent study demonstrates the polarized energy spinning clockwise (CW) and counterclockwise (CCW) within the periodic media. Here wave propagation has no preferential direction and thus the wave does not propagate in

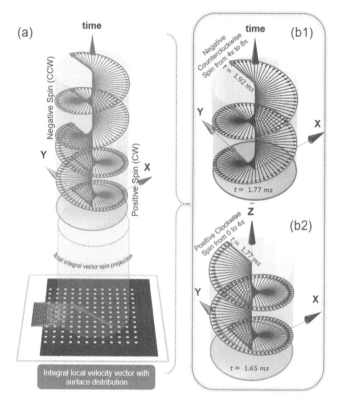

FIGURE 7.25 (Color) a) TBH yielding a polarity phase shift from 0 to 4π with CW spin, and 4π to 8π with CCW spin obtained from time-dependent study; b1) and b2) illustrate positive CW spin and negative CCW spin, respectively.

any specific direction but rather rotates. Global polarity of the velocity field first rotates with positive (CW) and then negative (CCW) spin. The counterpropagating spins result in the bulk phenomenon of the wave vortex as a TBH effect. Please note that wave vortex, discussed in Chapter 5, for electromagnetic wave has preferential direction of wave propagation along x_3. Irrespective of topological charge (Fig. 5.3 in Chapter 5), wave vortices have a preferred direction of propagation. However, in TBH the spins switch between CW and CCW continuously over time. To demonstrate this situation, a segment ($2\pi - 4\pi$ CW and $2\pi - 4\pi$ CCW) of the total spinning event $0 - 8\pi$ is shown in Fig. 7.24. Instantaneous polarities at different time steps are presented incrementally along the top row with CW spin and bottom row with CCW spin. At $t = 1.71\ 3\ msec$, the polarity is towards the direction where wave is generated. But with time, the CW spin shifts the polarity by π radian at $t = 1.742$. The polarity returns to the initial state after another rotation of π at $t = 1.773\ msec$. Please note that spin velocity is not constant. Spin velocity increases with increasing angle towards π and decreases while spinning towards 2π. Next a unique behavior is identified. Instead of continuing the CW spin, the polarity switched to CCW spin. With a 2π rotation along the CCW direction, the polarity comes back to the initial

Topological Acoustics in Metamaterials

state after $t = 1.855$ *msec*. This polarization is widely known as circular polarization but has its unique feature of switching spin between positive (CW) to negative (CCW), continuously. The spin due to circular polarization generates positive OAM. OAM results in shifting the geometric phase from 0 to 4π for each full CW rotation and similarly generates negative OAM in shifting the phase from 4π to 8π. To demonstrate this situation, a schematic is shown in Fig. 7.25. Figure 7.25 shows this SOI graphically. A time axis placed on top of the TBH system with velocity polarity helps understand the polarization over time. The circular polarization is projected to an extent for a better understanding polarity shift over the time in CW direction with a total phase shift of 4π. The initial time is 1.65 *msec*, and after the phase change by 4π, the time changes to 1.77 *msec*. Next, the phase change starts to shift backward, by rotating CCW from $t = 1.77$ *msec* with a total phase change of 4π again, reaching at $t = 1.92$ *msec*. It's like the resultant shift of polarity and its path takes 2 spiral stairs rotating CW $(0 - 4\pi)$ going up and two spiral stairs rotating CCW $(4\pi - 8\pi$ or $4\pi - 0)$ going down connected to each other at the phase of 4π, and continuously repeating itself. That is how the wave energy forms a localized energy sink/well, keeping the edge preserved. This unique SOI with zero resultant phase change results in the containment of the wave energy within the periodic bulk media. And that is how, topological blackhole phenomena emerged.

7.4.2.1 Mathematical Treatment to Find Polarity Anomaly

After numerically solving the wave field at TBH, the local instantaneous wave velocity components $v_1(t)$ and $v_2(t)$ were observed. The resultant local instantaneous velocity was calculated using this equation

$$\mathbf{v}(x_1, x_2, t) = \sqrt{v_1^2(t) + v_2^2(t)} \tag{7.8.1}$$

and phase angle $\theta_l(t) = tan^{-1}\left(\dfrac{v_2(t)}{v_1(t)}\right)$. Integrating the local velocity over the entire bulk media made of air inside the arrayed PnCs results in an integrated velocity vector with a specific polarity. Then the surface integral to achieve the integrated net velocity field would be

$$\mathbf{V}_g(t) = \iint_\Gamma \mathbf{v}(x_1, x_2, t) dx_1 \, dx_2 \tag{7.8.2}$$

It is shown in Fig. 7.25 that the polarity is changing with time obtained from this transient study. Here, Γ indicates the surface integral over the entire domain of the simulation. The global velocity field $\mathbf{V}_g(t)$ has its two components V_1 and V_2. The global phase of the wave field at an instant would be

$$\theta_g(t) = tan^{-1}\left(\frac{V_2}{V_1}\right) \tag{7.8.3}$$

$\theta_g(t)$ determines the polarity direction of the wave field at an instant t. At each time step, the global velocity vector, $\mathbf{V}_g(t)$, and the orientation polarity, $\theta_g(t)$,

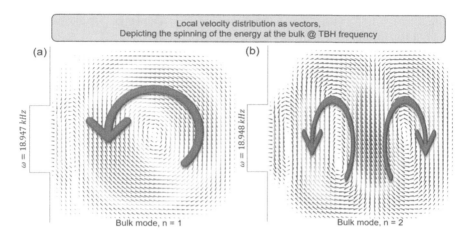

FIGURE 7.26 (Color) Local spatial distribution of velocity vectors obtained from frequency-dependent study. The spin-generated vortex is evident for both the harmonic bulk modes: a) for first bulk mode with frequency of 18.947 kHz with TBH topological charge 1, b) for the second bulk mode with a frequency of 18.948 kHz, TBH topological charge 2.

were calculated and were found to be changing. It was found that the $\theta_g(t)$ continually changes from -2π and $+2\pi$ (or $0-4\pi$) and then again from $+2\pi$ and -2π (or $4\pi - 8\pi$), continually oscillating as shown in Fig. 7.25. Figure 7.26 shows the local velocity distribution with small arrow vectors. Figure 7.26a shows the single spin mode with bulk topological mode =1 and could be assigned a topological charge = 1. Figure 7.26b shows the double spin mode with bulk topological mode =2 and could be assigned a topological charge = 2. It is concluded that the energy is trapped inside the Bulk media due to counter-interactive CW and CCW spins. The reason for this spin mediation is explained below using the modal polarity of the deaf band and the top band as follows.

As discussed earlier, the TBH frequency is very close to the degenerated frequency of the deaf band and the top band. Near the Γ point of the BZ, equifrequency surfaces (EFS) of both the bands (deaf and top) are shown in Fig. 7.27a. The normalized equifrequency contours (EFC) at TBH frequency are plotted in Fig. 7.27b. The red EFC belongs to the deaf band and the blue EFC belongs to the top band from the band structure. The EFCs of deaf band and the top band cannot intercept at the TBH frequency. However, considering the normalized EFCs of these bands, at multiple direction of the wave vectors, they intercept. Hence, the polarity for that specific intercepted point must align for a trivial case. The polarization characteristics of the PnCs are evaluated by θ_P, which denotes the angle of the polarization orientation. The polarization vectors are illustrated with arrows at each wave vector direction at an interval of $\theta_K = 1°$. The polarization vectors are presented on the EFS in Fig. 7.27c. The local polarity in each direction for both the bands could be similar or different. But two different polarities for a specific wavenumber is practically not

Topological Acoustics in Metamaterials 393

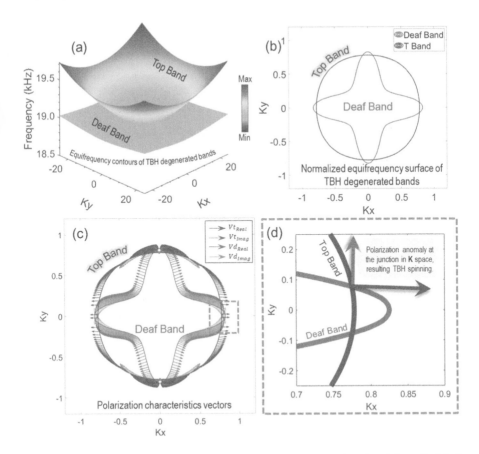

FIGURE 7.27 (Color) (a) Equifrequency contour of deaf band and top band for cylindrical PnCs, (b) normalized equifrequency surface for deaf and top band, (c) polarization characteristics vectors in wavenumber space, (d) polarization anomaly at the intercepted junction of two EFS modes, indicating the origin of the TBH bulk mode.

possible. They must produce a superposed unique polarity. However, the polarity at different direction of wave vectors rotates and creates a spin-like behavior. This is where the polarity anomaly occurs. Figure 7.27d portrays the polarity ambiguity for the deaf band and the top band, having orthogonal polarities at the interception. However, with close observation it can be found that the polarity of the deaf band and the top band are always orthogonal for all possible directions of wave vectors. Figure 7.28 illustrates this scenario where P-zone and S-zone intercept with no clear distinction between their polarities. Polarization orientation, θ_P merged for both the bands at $\theta_K = 45°$. Such merging of θ_P with θ_K validates the anomalous polarization. Thus, the generation of this unique TBH is materialized and can be utilized widely in the field of acoustics.

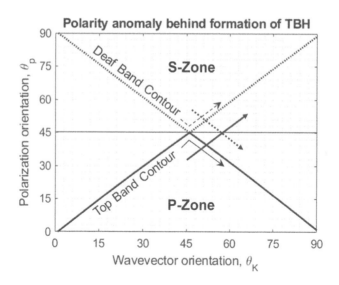

FIGURE 7.28 Anomalous polarity at the TBH appeared from polarity differential plots discussed in Chapter 3.

7.4.2.2 A Possible Link to the Acoustic Skyrmions

In Chapter 6, Skyrmion Number is introduced as a measure of nontrivial spin state for defect immune topological behavior. As demonstrated, the TBH has similar topological behavior in the bulk where the state is robust again all defects. An attempt has been made to calculate the Skyrmion Number (Sr) from the velocity filed presented in the previous section. It was found to be a non-zero number which it evident from the Fig. 7.25.

7.5 SUMMARY

Wave manipulation with different new patterns could be utilized in a wide spectrum of fields of acoustics and elastic waves. By altering the geometry and other material properties of the inclusion materials, the topological features could be activated by exploiting and understanding the band diagrams. By design, the frequency of application could be shifted to any frequency depending on the applications. The quantum trio effect in acoustics is presented with their respective physics. QHE in acoustics (i.e., QAHE) and its application were discussed where the active breaking of time-reversal symmetry was required. How the breaking space inversion symmetry by design could help activate QSHE and QVHE in acoustic metamaterials without breaking the time-reversal symmetry is discussed. Further, a unique acoustic phenomenon, named topological blackhole, is demonstrated. This phenomenon is topologically protected. The energy is contained in the bulk of the metamaterial without any leakage through the edges, which makes it a counter phenomenon that is present in topological insulators.

REFERENCES

1. Thouless, D.J., et al., *Quantized Hall conductance in a two-dimensional periodic potential. Physical Review Letters*, 1982, **49**(6): p. 405–408.
2. Stoof, H.T., Gubbels, K.B., Dickerscheid, D., *Ultracold Quantum Fields*, 2009: Springer.
3. Klitzing, K., Dorda, G., Pepper, M., New method for high-accuracy determination of the fine-structure constant based on quantized Hall resistance. *Physical Review Letters*, 1980. **45**(6): p. 494.
4. Berezinskii, V.L., *Destruction of long-range order in one-dimensional and two-dimensional systems having a continuous symmetry group I. Classical Systems*, 1971, **32**(3): p. 493–500.
5. Kosterlitz, J.M., Thouless, D.J., Ordering, metastability and phase transitions in two-dimensional systems. *Journal of Physics C: Solid State Physics*, 1973, **6**(7): p. 1181.
6. Fang, A., et al., Anomalous Anderson localization behaviors in disordered pseudospin systems. *Proceedings of the National Academy of Sciences of the United States of America* 2017. **114**(16): p. 4087–4092.
7. Thouless, D.J., et al., Quantized Hall conductance in a two-dimensional periodic potential. *Physical Review Letters*, 1982, **49**(6): p. 405.
8. Visscher, W.M., *Transport processes in solids and linear-response theory. Physical Review A*, 1974, **10**(6): p. 2461–2472.
9. Zak, J, *Magnetic Translation Group. Physical Review*, 1964, **134**(6A): p. A1602.
10. Berry, M.V., Sciences, P., Quantal phase factors accompanying adiabatic changes. *Proceedings of the Royal Society of London. Series A, Mathematical and Physical Sciences*, 1984, **392**(1802): p. 45–57.
11. Graf, G.M., Porta, M., Bulk-edge correspondence for two-dimensional topological insulators. Communications in Mathematical Physics, 2013, **324**(3): p. 851–895.
12. Hasan, M.Z., Kane, C.L., *Colloquium: Topological insulators. Reviews of Modern Physics*, 2010, **82**(4): p. 3045–3067.
13. Raghu, S., Haldane, F.D.M., *Analogs of quantum-hall-effect edge states in photonic crystals. Physical Review A*, 2008, **78**(3): p. 033834.
14. Haldane, F.D.M., Raghu, S., *Possible realization of directional optical waveguides in photonic crystals with broken time-reversal symmetry. Physical Review Letters*, 2008, **100**(1): p. 013904.
15. Onoda, M., Nagaosa, N, Quantized anomalous Hall effect in two-dimensional ferromagnets: Quantum Hall effect in metals. *Physical Review Letters*, 2003, **90**(20): p. 206601.
16. Liu, C.-X., et al., Quantum anomalous Hall effect in Hg 1– y Mn y Te quantum wells. Physical Review Letters 2008. **101**(14): p. 146802.
17. Qiao, Z., et al., Quantum anomalous Hall effect in graphene from Rashba and exchange effects. 2010, **82**(16): p. 161414.
18. Wu, C., Orbital analogue of the quantum anomalous Hall effect in p-band systems. *Physical Review Letters*, 2008, **101**(18): p. 186807.
19. Chang, C.-Z., et al., Experimental observation of the quantum anomalous Hall effect in a magnetic topological insulator. *SCIENCE*, 2013. **340**(6129): p. 167–170.
20. Jotzu, G., et al., Experimental realization of the topological Haldane model with ultra-cold fermions. *Nature*, 2014, **515**(7526): p. 237–240.
21. Kane, C.L., Mele, E.J., Z 2 Topological order and the quantum spin Hall effect. *Physical Review Letters*, 2005, **95**(14): p. 146802.
22. Bernevig, B.A., Zhang, S.-C, Quantum spin Hall effect. *Physical Review Letters*, 2006, **96**(10): p. 106802.

23. König, M., et al., Quantum spin Hall insulator state in HgTe quantum wells. *Science*, 2007, **318**(5851): p. 766–770.
24. Kane, C.L., Mele, E.J, Quantum *spin* Hall *effect in graphene*. *Physical Review Letters*, 2005, **95**(22): p. 226801.
25. Chen, Y.L., et al., Massive Dirac *fermion on the surface of a magnetically doped topological insulator*. *SCIENCE*, 2010, **329**(5992): p. 659–662.
26. Kou, X., et al., Review of *quantum* Hall *trio*. *Journal of Physics and Chemistry of Solids*, 2019, **128**: p. 2–23.
27. Zhang, H., et al., Topological *insulators* in Bi2Se3, Bi2Te3 and Sb2Te3 *with a single* Dirac *cone on the surface*. *Nature Physics*, 2009, **5**(6): p. 438–442.
28. Chen, Y., et al., Experimental *realization of a three-dimensional topological insulator*, Bi2Te3. *SCIENCE*, 2009, **325**(5937): p. 178–181.
29. Fu, L., Topological *crystalline insulators*. *Physical Review Letters*, 2011, **106**(10): p. 106802.
30. Xu, S.-Y., et al., Observation of a *topological crystalline insulator phase and topological phase transition in pb1– Xsnxte*. *Nature Communications*, 2012, **3**(1): p. 1–11.
31. Okada, Y., et al., Observation of Dirac *node formation and mass acquisition in a topological crystalline insulator*. *SCIENCE*, 2013, **341**(6153): p. 1496–1499.
32. Beenakker, C.W.J., Colloquium: Andreev *reflection* and Klein *tunneling in graphene*. *Reviews of Modern Physics*, 2008. **80**(4): p. 1337.
33. Neto, A.C., et al., The *electronic properties of graphene*. 2009, **81**(1): p. 109.
34. Imura, K.-I., et al., Zigzag *edge modes* in a Z 2 *topological insulator*: Reentrance and *completely flat spectrum*. *Physical Review B*, 2010, **82**(8): p. 085118.
35. Zhang, X., et al., Topological *sound*. 2018. **1**(1): p. 1–13.
36. Fleury, R., et al., Sound *isolation and giant linear nonreciprocity in a compact acoustic circulator*. 2014, **343**(6170): p. 516–519.
37. Ni, X., et al., Topologically *protected one-way edge mode in networks of acoustic resonators with circulating air flow*. *New Journal of Physics*, 2015, **17**(5): p. 053016.
38. Yang, Z., et al., *Topological Acoustics*. 2015. **114**(11): p. 114301.
39. Khanikaev, A.B., et al., Topologically *robust sound propagation in an angular-momentum-biased graphene-like resonator lattice*. *Nature Communications*, 2015, **6**(1): p. 1–7.
40. Xu, C., Moore, J.E., Stability of the quantum spin Hall effect: Effects of interactions, disorder, and Z 2 topology. *Physical Review B*, 2006, **73**(4): p. 045322.
41. Yang, C., et al., Real spin angular momentum and acoustic spin torque in a topological phononic crystal. *Journal of Applied Physics*, 2021, **129**(13): p. 135106.
42. Zhang, Z., et al., Experimental *verification of acoustic pseudospin multipoles in a symmetry-broken snowflakelike topological insul*ator. *Physical Review B*, 2017. **96**(24): p. 241306.
43. Xiao, D., Yao, W., Niu, Q., Valley-*contrasting physics in graphene: magnetic moment and topological transport*. *Physical Review Letters*, 2007, **99**(23): p. 236809.
44. Pal, R.K., Ruzzene, M., Edge *waves in plates with resonators: an elastic analogue of the quantum valley* Hall *effect*. *New Journal of Physics*, 2017. **19**(2): p. 025001.
45. Vila, J., Pal, R.K., Ruzzene, M., Observation of *topological valley modes in an elastic* hexagonal *lattice*. *Physical Review*, 2017, **96**(13): p. 134307.
46. Zhu, H., Liu, T.-W., Semperlotti, F., Design and experimental *observation of valley-Hall edge states in diatomic-graphene-like elastic waveguides*. *Physical Review*, 2018, **97**(17): p. 174301.
47. Liu, T.-W., Semperlotti, F., Tunable *acoustic valley-Hall edge states in reconfigurable phononic elastic waveguides*. *Physical Review*, 2018, **9**(1): p. 014001.
48. Lu, J., et al., Observation of *topological valley transport of sound in sonic crystals*. *Nature Physics*, 2017, **13**(4): p. 369–374.

Topological Acoustics in Metamaterials

49. Bleu, O., Solnyshkov, D., Malpuech, G., *Quantum valley Hall effect and perfect valley filter based on photonic analogs of transitional metal dichalcogenides. Physical Review*, 2017, **95**(23): p. 235431.
50. Qi, X.-L., Zhang, S.-C., *Topological insulators and superconductors. Reviews of Modern Physics*, 2011, **83**(4): p. 1057–1110.
51. Sarma, S.D., Freedman, M., Nayak, C., *Majorana zero modes and topological quantum computation. Npj Quantum Information*, 2015, **1**(1): p. 15001.
52. Mellnik, A.R., et al., *Spin-transfer torque generated by a topological insulator. Nature*, 2014, **511**(7510): p. 449–451.
53. Xu, Y., Gan, Z., Zhang, S.-C., *Enhanced thermoelectric performance and anomalous Seebeck effects in topological insulators. Physical Review Letters*, 2014, **112**(22): p. 226801.
54. Li, N., et al., *Colloquium: Phononics: Manipulating heat flow with electronic analogs and beyond. Reviews of Modern Physics*, 2012, **84**(3): p. 1045–1066.
55. Huber, S.D., *Topological mechanics. Nature Physics*, 2016, **12**(7): p. 621–623.
56. Liu, Y., Xu, Y., Duan, W., *Berry phase and topological effects of phonons. National Science Review*, 2018, **5**(3): p. 314–316.
57. Liu, Y., et al., *Model for topological phononics and phonon diode. Physical Review B*, 2017. **96**(6): p. 064106.
58. Zhang, W., et al., *Role of transparency of platinum–ferromagnet interfaces in determining the intrinsic magnitude of the spin Hall effect. Nature Physics*, 2015, **11**(6): p. 496–502.
59. Lu, J., et al., *Dirac cones in two-dimensional artificial crystals for classical waves. Physical Review B*, 2014, **89**(13): p. 134302.
60. Jin, Y., Wang, R., Xu, H., *Recipe for Dirac phonon States with a quantized valley berry phase in two-dimensional hexagonal lattices. Nano Letters*, 2018, **18**(12): p. 7755–7760.
61. Xiong, Z., et al., *Topological node lines in mechanical metacrystals. Physical Review B*, 2018, **97**(18): p. 180101.
62. Zhang, Z., et al., *Topological acoustic delay line. Physical Review Applied*, 2018, **9**(3): p. 034032.
63. Zhang, Z., et al., *Directional acoustic antennas based on valley-hall topological insulators. Advanced Materials*, 2018, **30**(36): p. 1803229.
64. Zak, J., *Berry's phase for energy bands in solids. Physical Review Letters*, 1989, **62**(23): p. 2747–2750.
65. Zhang, L., Niu, Q., *Chiral phonons at high-symmetry points in monolayer hexagonal lattices. Physical Review Letters*, 2015, **115**(11): p. 115502.
66. Liu, Y., et al., *Pseudospins and Topological Effects of Phonons in a Kekul'e Lattice. Physical Review Letters*, 2017, **119**(25): p. 255901.
67. Haldane, F.D., *Model for a quantum Hall effect without Landau levels: Condensed-matter realization of the "Parity Anomaly. Physical Review Letters*, 1988, **61**(18): p. 2015–2018.
68. Wang, Y.-T., Luan, P.-G., Zhang, S., *Coriolis force induced topological order for classical mechanical vibrations. New Journal of Physics*, 2015, **17**(7): p. 073031.
69. He, C., et al., *Acoustic topological insulator and robust one-way sound transport. Nature Physics*, 2016, **12**(12): p. 1124–1129.
70. Süsstrunk, R., Huber, S.D., *Observation of phononic helical edge states in a mechanical topological insulator. Science*, 2015, **349**(6243): p. 47–50.
71. Yang, Z., et al., *Topological Acoustics. Physical Review Letters*, 2015, **114**(11): p. 114301.
72. Mousavi, S.H., Khanikaev, A.B., Wang, Z., *Topologically protected elastic waves in phononic metamaterials. Nature Communications*, 2015, **6**(1): p. 8682.

73. Liu, Y., Xu, Y., Duan, W., *Three-dimensional topological states of phonons with tunable pseudospin physics. Research*, 2019, Article ID: 5173580. DOI: 10.34133/2019/5173580.
74. King-Smith, R.D., Vanderbilt, D., *Theory of polarization of crystalline solids. Physical Review B*, 1993, **47**(3): p. 1651–1654.
75. Resta, R., *Macroscopic polarization in crystalline dielectrics: the geometric phase approach. Reviews of Modern Physics*, 1994, **66**(3): p. 899–915.
76. Chang, M.-C., Niu, Q., *Berry phase, hyperorbits, and the Hofstadter spectrum: Semiclassical dynamics in magnetic Bloch bands. Physical Review B*, 1996, **53**(11): p. 7010–7023.
77. Nagaosa, N., et al., *Anomalous Hall effect. Reviews of Modern Physics*, 2010, **82**(2): p. 1539–1592.
78. Sinova, J., et al., *Spin Hall effects. Reviews of Modern Physics*, 2015, **87**(4): p. 1213–1260.
79. Bansil, A., Lin, H., Das, T., *Colloquium: Topological band theory. Reviews of Modern Physics*, 2016, **88**(2): p. 021004.
80. Süsstrunk, R., Huber, S.D., *Classification of topological phonons in linear mechanical metamaterials. Proceedings of the National Academy of Sciences*, 2016, **113**(33): p. E4767–E4775.
81. Kane, C.L., Mele, E.J., *Quantum spin Hall effect in graphene. Physical Review Letters*, 2005, **95**(22): p. 226801.
82. Wang, Z., et al., *Reflection-free one-way edge modes in a gyromagnetic photonic crystal. Physical Review Letters*, 2008, **100**(1): p. 013905.
83. Wang, Z., et al., *Observation of unidirectional backscattering-immune topological electromagnetic states. Nature*, 2009, **461**(7265): p. 772–775.
84. Skirlo, S.A., et al., *Experimental observation of large Chern numbers in photonic crystals. Physical Review Letters*, 2015. **115**(25): p. 253901.
85. Ding, Y., et al., *Experimental demonstration of acoustic Chern insulators. Physical Review Letters*, 2019. **122**(1): p. 014302.
86. Fleury, R., et al., *Sound isolation and giant linear nonreciprocity in a compact acoustic circulator. Science*, 2014, **343**(6170): p. 516–519.
87. Khanikaev, A.B., et al., *Topologically robust sound propagation in an angular-momentum-biased graphene-like resonator lattice. Nature Communications*, 2015, **6**(1): p. 8260.
88. Ganti, S.S., Liu, T.-W., Semperlotti, F., *Topological edge states in phononic plates with embedded acoustic black holes. Journal of Sound and Vibration*, 2020, **466**: p. 115060.
89. Liu, T.-W., Semperlotti, F., *Tunable acoustic valley-Hall edge states in reconfigurable phononic elastic waveguides. Physical Review Applied*, 2018, **9**(1): p. 014001.
90. Lee, J.Y., Jeon, W., Vibration damping using a spiral acoustic black hole. *Journal of Acoustical Society of America*, 2017, **141**(3): p. 1437–1445.
91. Romano, P.Q., Conlon, S.C., Smith, E.C., Investigation of contact acoustic nonlinearities on metal and composite airframe structures via intensity based health monitoring. *Journal of Acoustical Society of America*, 2013, **133**(1): p. 186–200. DOI: 10.1121/1.4770237.
92. Krylov, V.V., Acoustic'black holes' for flexural waves and their potential applications. *Procedia Engineering*, 2002, **199**(2017): p. 56–61.
93. Krylov, V.V., New type of vibration dampers utilising the effect of acoustic'black holes'. *Department of Aeronautical & Automotive Engineering, Loughborough*, 2004, **90**(5): p. 830–837.
94. Krylov, V.V., Winward, E., Experimental investigation of the acoustic black hole effect for flexural waves in tapered plates. *Journal of Sound and Vibration*, 2007, **300**(1–2): p. 43–49.

Topological Acoustics in Metamaterials

95. Bowyer, E., Krylov, V.V., Experimental *investigation* of *damping* *f*lexural *v*ibrations in *g*lass *f*ibre *c*omposite *p*lates *c*ontaining *o*ne-and *t*wo-*d*imensional *a*coustic *b*lack *h*oles. *Composite Structures*, 2014. **107**: p. 406–415.
96. Guasch, O., Arnela, M., Sánchez-Martín, P., *Transfer matrices to characterize linear and quadratic acoustic black holes in duct terminations*. *Journal of Sound and Vibration*, 2017, **395**: p. 65–79.
97. Pelat, A., et al., *The acoustic black hole: A review of theory and applications*. *Journal of Sound and Vibration*, 2020, **476**: p. 115316.

Index

A

abnormal polarity, 12, 95, 98, 100, 103–5, 107, 109, 114, 125, 330
abnormal polarity conditions, 94, 103–4, 330
abnormal polarization, 95, 102, 104, 106–7
accidental degeneracy, 274–75, 277, 281, 293, 349
Acoustical Society of America, 13, 125, 226, 349–50, 398
acoustic and elastic wave propagation, 42, 179
acoustic and elastic waves, xv–xvi, 117, 157, 165, 186, 225, 231, 236, 308, 315, 348
acoustic computing, xvi, 104, 226, 297, 299–300, 348–49, 384
acoustic energy, 288, 378, 388
acoustic metamaterials, xv, xvii, 8, 10, 11–13, 142, 272, 281, 284, 351, 359
acoustic pressure, absolute, 295, 387
acoustic pressure field, 43, 295–96
acoustic pseudospin multipoles, 350, 396
acoustics, xv–xvii, 8–10, 68, 142–43, 155, 157, 204–6, 223–25, 229, 270, 281–82, 359, 361, 364, 368, 375–77, 380, 388, 393–94, 398–99
acoustics and elastic waves, 117, 231, 238, 394
acoustic spin, 352, 388
acoustic spin angular momentum, 361, 364
acoustic systems, 371–72
acoustic topological insulators, xvi, 397
acoustic transparency, 284, 286–88
acoustic tubes, 376–77
acoustic valley Hall effect (AVHE), 368
acoustic wave equations, 138, 235, 283, 314
acoustic waves, 139–40, 173, 203, 206, 233, 235, 276, 298, 307–8, 314, 318, 331, 359, 361–62, 375–76
active breaking of time-reversal symmetry, 223, 300, 394
adaptive gradient-index metamaterials, 351
Ampere's Law, 132, 135, 138, 217
angular momenta, 173
angular momentum, 30, 32, 48, 129, 157–58, 160–64, 166, 171, 173, 220
    real spin, 361
    respective square, 173–74
angular momentum operators, 158, 160, 164, 325
anisotropic, 61, 74, 82–83, 86–87, 139
anisotropic materials, 68–69, 74, 82–84, 86, 109
anisotropic media, 82–83, 85–87, 89–90, 94, 100, 114–15, 125, 308
anomalous companions, 100–101, 103

anomalous polarity, 89–90, 125, 388–89, 394
anomalous polarization of elastic waves, 125
AQN. *See* azimuthal quantum number
axis, primary, 53, 218–19
axis in periodic media, 233
azimuthal quantum number (AQN), 167–68, 173, 176

B

backward propagating waves, 324
band degeneracies, 203, 273–74, 276, 280–81
band gap for surface acoustic waves, 349
band gaps, xvii, 273, 275–76, 285, 294, 304, 306, 349, 355
band inversion, 227, 283, 362, 364, 375
bands, 191–94, 202–4, 214–15, 272, 274, 276–78, 281, 283–84, 293–96, 306, 324, 380, 382–83, 385–87, 392–93
band structures, 12, 202, 211, 213–14, 292, 295–96, 305, 356–57, 375, 384, 392
band topology, 372
    nontrivial, 373–74
basis vectors, 21, 35, 56–57, 261, 342–43
beam and metamaterial lenses, 349
Berezinskii-Kosterlitz-Thouless (BKT), 224, 354
Berry, 207, 210, 215, 227, 395
Berry connection, 211, 213–14, 272, 330, 369, 372
Berry curvature, 211–15, 272, 276, 284, 330, 355, 369, 372, 374
Berry phase, 207, 210–17, 272, 276–77, 284, 300, 307, 369, 372, 397
Berry's phase for energy bands in solids, 227, 397
Berry/Zak's phase, 370–71
Bessel function, modified, 335
BKT (Berezinskii-Kosterlitz-Thouless), 224, 354
Bloch eigenfunction, 193, 195
Bloch eigenwave functions, 212–13
Bloch modes, 236, 238, 303
Bloch solution, 179–80, 188, 211, 215, 217, 224, 231, 235–36
Bloch solution for acoustic waves, 235
Bloch wave functions, 211, 215–16, 272, 364, 369
Bloch wave solutions in metamaterials, 261
Bloch wave vector, 211, 215, 240–41, 247, 252, 259, 264, 270, 302, 334
body force, 35, 39–41, 43–44, 67–68, 70, 86–87, 114, 236
bottom bands, 274, 276, 282–83, 293, 382–83, 385–86

401

402   Index

boundary conditions, 18, 48, 65, 113, 150, 152, 180–82, 212–13, 216, 231
boundary points, 373–74
breaking time-reversal symmetry, 359
breaking T-symmetry, 359, 361
Brillouin boundary, 12, 187, 203–4, 212, 214, 274–75, 284
Brillouin zone (BZ), 185–87, 200, 203–4, 209, 212, 214–15, 231–33, 275, 278, 284, 306–7, 355, 372, 374–75, 378–79
broken time-reversal symmetry, 359, 395
bulk acoustic waves, 351
bulk elastic waves, 226
BZ. *See* Brillouin zone

C

Cartesian coordinate system, 17, 20, 36, 51, 54, 57–58, 60, 66, 79, 130, 166
CCW, 294, 389–91
CCW spin, 363, 390, 392
Chern number, 214, 272, 277, 284, 307, 331–32, 356, 362, 369, 371
circular phononic crystals, 242, 248, 259, 292–93, 295, 332, 334
classical mechanics, 15–61, 127, 129, 146–47, 157, 160, 164, 207–8, 220
classical waves, 148, 397
closed path, 210–11, 213–15, 284
C_n symmetry, 218–19
coefficients, 52, 54, 78, 100–101, 181, 188, 192, 239, 241, 243, 247–48
   independent, 50–52, 91–92
commutation relations, 171, 173
complex operators, 159, 174–75, 177, 310
components, 33–34, 36, 69–71, 74, 77–78, 80, 137, 140–41, 164, 166, 171, 173–74, 178, 320–21, 358
conditions, plane-stress, 116
cones, 171–73, 202–4, 214, 226, 277, 279–80, 284, 292–95, 297, 349–50, 378, 382–83, 385–86
configurations, geometric, 12, 281, 286–87, 362, 364
conservation, 31–32, 38, 40, 106–7
conservation of mechanical energy, 40–41
constitutive equation, 52–53, 64–65, 67, 135, 139, 235
constitutive matrix, 49–50, 52–53, 91–92, 95–96, 100, 103
continuous body, 23, 25, 29, 31–32
coordinate system, 20–24, 27, 35, 61, 78, 84, 325–26
   generalized, 18
   new, 22–23
   transformed, 56, 59
covariant, 208, 338, 341, 343–44

crystals, 178–79, 186, 188, 191, 203, 207, 212–14, 217, 226, 349, 354
curl, 58, 60, 70–72, 122–23, 130, 132, 134–35, 142, 156, 213, 372
CW spin, 390
cylindrical coordinate system, 56–58

D

deaf band, xvii, 274–77, 293–95, 382–87, 392–93
   flat, 280
deaf band frequencies, 294–95, 384, 386
decomposition, 71, 81
defects, 332, 360–61, 376, 394
deformable body, 17–18, 20, 22, 24–25, 27, 29–31, 33, 35–36, 40–43, 61
deformation, 20–27, 31–33, 35, 38, 49, 70, 127, 130, 205, 353, 355
deformation tensor, 27, 30
degeneracies, 204, 214, 273–79, 281–82, 310, 359, 362, 383, 387
degeneracy points, 214, 279
density, spin momentum, 328, 330
differential equation, 150, 184
   ordinary, 150
Dirac, 197–98, 200, 203, 206, 226, 315–16, 357, 366, 376, 396
Dirac cone behaviors, 203, 297
Dirac cone parameter, 282
Dirac cones, 202–4, 206, 214, 274–79, 281, 284, 294, 298, 350–51, 373, 375
   non-Hermitian, 279
   phononic, 370, 373
Dirac equation, 195–96, 198–200, 202–4, 225, 308, 314, 316–18, 321–23, 348, 358
Dirac equation and Hamiltonian of spin, 314
Dirac frequency, 277, 280, 295–97, 378
Dirac points, 203–4, 276, 278–79
Dirac solution, 200
direction
   arbitrary, 64, 79, 167–68, 329, 331
   clockwise, 163, 318, 320, 385
   forward, 83, 324
   negative, 175, 231
   opposite, 36, 219, 222, 318–20, 328, 385
   preferential, 167–68, 389–90
   radial, 56, 130
directional bandgaps, 303, 378, 385, 387
direction cosines, 34, 83, 88
dispersion behavior, 187, 203, 231, 259, 269–70, 272–73, 286, 303, 361
dispersion curves, 258, 261, 269–71, 275, 284, 288–90, 303
dispersion diagrams, 295, 303–4
dispersion relation, 243, 248, 258, 271–72, 285–86, 293, 358, 362–64, 368–69, 380, 386

Index

403

displacement amplitudes, 292, 330
displacement field, 59, 70–71, 77, 87, 118–19, 122–23, 139, 165, 302, 326, 332
    solid, 295
displacement gradients, 18, 41, 43, 46, 48, 244
displacements, 17–18, 26–27, 39, 43–46, 48–49, 68–71, 76–78, 80–81, 86–87, 115–16, 132–34, 152–53, 239, 243, 290, 301–3, 320–21, 325–27, 371–72
    electric, 132–33, 220
displacement vectors, 25, 27, 30, 328
displacement wave field, 70–71, 75, 78, 87
divergence, 58, 60, 64–65, 71, 76, 118–19, 131–32, 136, 141
domain wall, 360, 364, 368
dot product, 66, 73, 75, 77, 87, 124, 137, 140–41, 189, 211, 264
double Dirac cone, 281–83, 350, 362, 364

E

edge modes, 360, 368–69, 396
edge states, 218, 224, 227, 328, 355–56, 361, 366, 368–69, 376–77, 379, 383–84
    chiral, 355–56
    double Dirac point form topological, 375
    helical, 357, 359
    one-way, 359, 361, 366, 373
effective Hamiltonian, 276–77, 279–80, 283, 358, 362, 364, 370, 373–74
eigenfunctions, 148, 152–55, 159–60, 168, 180, 182, 196, 199, 210, 213–14, 280, 284
eigenstates, 157, 159–60, 163, 168, 171, 173–76, 178, 198, 200, 210, 213
eigenvalue problem, 149, 156, 183, 186, 231, 237–39, 247–48, 261, 264, 266–67, 270, 324
eigenvalues, 100–102, 148–50, 152–53, 156–57, 159–60, 168, 177, 183, 210–11, 238–39, 255–57, 276–77, 279–81, 303, 305
    complex, 155, 281
    multiple, 148
    original, 160
    real, 155, 280
eigenvalues and eigenfunction, 153, 159, 210
eigenvalue solution, 150–53, 178, 238, 261, 269, 305
eigenvectors, 88, 100, 157, 202, 283
elastic Dirac equation, 315–16
elastic materials, 9, 319–20
elastic metamaterial, 14, 301, 348, 351
elastic metamaterial design, 313
elastic modulus, xv, 10, 250–51, 301
elastic waveguides, 368–69, 396
elastic wave modes, 310, 318

elastic waves, xv–xvii, 104, 117, 139–40, 155, 157, 186–87, 205–6, 225–27, 229, 231, 306–8, 310–12, 314–15, 318, 321, 325–26, 328–30, 348, 352
    guided, 10
    polarized, 103
elastic waves in metamaterial, 300
elastodynamics, xv, 8, 10, 16, 142, 173
electric field, 127–30, 132, 134, 136–37, 139, 141, 156, 161, 207, 213, 218, 220, 223
electric field equation, 135–38
electrodynamics, xv, xvii, 8, 10, 13, 142, 173, 217, 225
electromagnetic field, 13, 129, 142, 156, 161
electromagnetic wave equations, 135–38, 142
electromagnetic waves, xv, 103–4, 138–40, 142–43, 161–63, 206–7, 307–8, 390
    polarized, 103
electromagnetic waves and quantum waves, 308
electromagnetism, 103, 127, 129, 141–42, 375
electron density, 217
electrons, 144–45, 149–50, 152–53, 160, 164–65, 187, 196, 198, 200–202, 272, 281, 358, 361–62, 369–70, 372
energy, 32–33, 38–41, 140, 143–46, 148, 151, 153, 158, 196–97, 201–2, 204, 210–11, 217, 222–24, 376, 379–81, 384–85, 389
    equivalent, 144
    global, 38
    total, 32, 147, 187
energy bands, 186–87, 190, 193, 211, 213–15, 227, 397
energy eigenvalues, 185, 190, 194–95
energy equation, 195–96
energy levels, 203–4, 214
energy operator, 148, 158, 181, 196
energy propagation, 83–85, 204
energy states, 148–49, 186, 190, 200, 202–3, 214, 222, 226, 293, 312, 349
energy values, 203
equations, 19–20, 23, 37–38, 40–42, 45–46, 48–49, 53–54, 63–65, 69–72, 86–88, 113–15, 147–49, 184–85, 189–92, 194–98, 234, 236–38, 302–3, 321–24, 341–43
equifrequency contour map, 272, 274–75
equifrequency contours, 86, 106, 272–73, 290–91, 388
equivalency of Dirac equation, 321
Eulerian, 28–30, 32, 61, 325
Eulerian coordinate system, 23–26, 144, 326
Eulerian system, 26–27, 32, 144, 326
Euler-Lagrange equation, 20, 325
exceptional points, 204, 279–80, 350
excitation frequency, 297, 378, 384, 387

F

Faraday, 13, 127, 129, 134, 142
Faraday's Law, 135, 138, 156
fermions, 163–66, 168, 173, 199, 202–3, 214, 220, 226, 281, 284, 361
field
  constant vector, 338, 341
  following vector, 60
  scalar, 57, 59, 63–64, 70–72, 130
  skyrmion, 332
field theory, 127, 129–30
filling fraction, 242, 248–49, 275, 332, 334, 338, 362, 382, 385–87
first Brillouin zone, 185–86, 209, 232–33, 239, 278, 286, 288, 355
focal points, 289–92
Fourier coefficients, 180, 183–84, 237, 240–42, 245, 248, 254, 256, 262–64, 266, 332–36
frequency
  angular, 63, 87, 207, 271, 359
  higher, 362
  normalized, 378, 384
  ultrasonic, 285, 288
frequency eigenvalues, 277, 280
frequency range, 276, 288–91, 369
  audible, 287–88
frequency solutions, 231, 270
frequency space, 224, 305–6
frequency topology, 274–75
frequency topology map, 272–73, 290
frequency topology map of mode, 272
frequency values, 239
function of k-space, 186

G

Gauss's law, 130–33, 135–36, 217
generalized wave potentials, 72, 75–76
geodesic, 208, 339, 341
geometric dimensions, 276, 285–86, 362
geometric phase, 204–5, 207–8, 210, 212–13, 222–25, 227, 304, 306–7, 324, 339, 341, 344, 369, 372
geometric phase for elastic waves, 227
governing differential equation, 37–38, 43, 63, 68, 78, 109–10, 118–19, 150–53, 233, 302
Governing Differential Equations of Motion, 54, 58, 60, 68
governing equations, 18, 41, 113–14, 122, 231, 235–37, 239, 241, 261, 264, 309
gradient, 41, 44, 57, 59, 63–64, 67, 70–72, 119, 122–24, 130, 346
gradient operator, 44, 57, 59, 64, 117, 119, 164
graphene, 203, 213–14, 228, 230, 356, 360, 366, 395–96, 398

Green's deformation tensors, 27–28, 30, 42
ground state, 311–12
group velocity, 10, 84–85, 270–71, 276, 281, 288–89
ΓX, 286, 289, 292, 295, 303, 306, 378, 384
ΓX and MΓ direction, 288, 295–96
ΓX direction, 288–89, 296, 304, 378

H

Haldane, 225, 228, 351, 356, 395, 397
Hall effects, quantum valley, 117, 224, 275–76, 325, 366–67
Hall voltage, 224, 358–59
Hamiltonian, 44, 46–48, 147–50, 160, 184, 187, 190, 199, 204, 210, 212–14, 216–17, 221, 276–77, 310–12, 364
  evolving, 210
  original, 210, 311–12
  perturbed, 190, 193, 195
  reduced, 149, 190
  super-partner, 311
Hamiltonian matrix, 148
Hamiltonian operator, 157–59, 169, 190, 196, 210
heat energy, 33, 38–39, 42, 272
Helmholtz decomposition, 70–71, 75, 78, 87, 139, 308–9, 325–26
Hermitian, 157, 159, 177, 203–4, 277, 279–80, 350
Hermitian operators, 157
hexagonal metamaterial, 230
high-symmetry points, 397
hole, acoustic black, 398
homogeneous, 68–69, 83, 109–10, 139, 150
homogeneous materials, 69
host matrix, 9, 229, 234, 242, 248, 262–63, 269, 332–35, 379
hybrid spin of elastic waves, 352

I

in-plane, 79–80, 234, 309, 329
in-plane polarity, 243
insulators, quantum Hall, 355–56
internal energy, 32–33, 38, 40–42, 61
  total, 32–33, 41
internal energy density, 33, 39, 42
intrinsic spin of elastic waves, 350, 352
inversion symmetry, 215, 219, 373
isotropic materials, 52–54, 56, 68–69, 74, 82–85, 90–91, 94, 96

K

Kagomé lattice, 259–62
Kelvin-Christoffel matrix, 98, 100–103

Index

405

kinetic energy, 17, 31, 38, 40–41, 43–44, 147, 184, 187
Klein-Gordon equation, 196, 198, 225, 308, 314–16, 322–23, 348
Klein-Gordon equation for spin-0 bosonic particles, 315
Klein-Gordon equation to Dirac equations for elastic waves, 348
k-space, 186, 202–3, 209, 366

L

Ladder Operation for Elastic Waves, 308
Lagrangian, 17–18, 28–30, 32, 43–48, 325, 371
Lagrangian density, 18
Lagrangian strain tensor, 29–30
Laplace operator, 121–22, 157, 197–98, 316
lattice basis vectors, 259
longitudinal modes, 89, 318–19
longitudinal wave modes, 80, 89, 95, 105, 318–19, 388
longitudinal waves, 76, 83, 95, 302–3, 325, 331, 388

M

magnetic field, 127–38, 161, 213, 218, 220, 224–25, 301, 356, 358–59, 370, 372
  induced, 131, 134
magnetic field equation, 135–38
magnetic flux density, 131–32, 134, 220
magnetic quantum number (MQN), 167, 172–75
manifold, 214, 224, 284, 306–7, 324, 338, 353
material body, 18, 20–23, 27, 32, 35, 37–39
material coefficients, 61, 84, 100–103, 108, 114, 122, 236, 244–45, 301, 303
material constants, 88, 100, 110, 122–24, 245, 262, 264, 310, 312–13, 333
material coordinate system, 21–22
material derivatives, 24–25, 37, 39
material particles, 20–23, 25, 30–31, 34, 321
material points, 16, 20–21, 23, 26–27, 326
material properties, 9–10, 68, 84, 89, 103, 109, 114, 116–17, 128, 229, 240
  effective, 108, 313
materials, xv, 8–10, 21–22, 38, 50–53, 84–86, 100, 104, 109–10, 131–34, 202–3, 229–30, 250–51, 258–59, 261–62, 301–3, 309–10, 325–26, 331–32, 354
  generic, 235
  heterogeneous, 49, 109
material types, 9, 61, 204, 229, 334
mathematics, 15–16, 61, 148, 169, 197–98, 205, 228, 306
Maxwell-Faraday Equations, 134–35
mechanical energy, 40–41, 83
metamaterial design, 90, 109, 313

metamaterial geometry, 269
metamaterials
  designed, xv, 105
  photonic, 133, 203
metamaterial structures, 108, 230, 293, 295, 364, 366
metamaterial unit cell, 281, 361
metamaterial wedge, 105, 107
MΓ direction, 288–90, 295–96
Mobius strip, 206, 306–7, 362
modal wave function, 153
modern physics, 125, 127, 226, 395–98
modes
  bulk topological, 392
  qFS wave, 95, 97–98, 105–6
  qSS, 89, 92, 94–96
  quadrupolar, 295, 362
mode shape of point, 289
mode shapes, 153, 157, 288–90, 294–96, 380
momentum, 27, 29–30, 32, 144–46, 151, 158–60, 167, 169–71, 208, 210, 217, 220, 222, 224, 328
  spin-orbit, 364
momentum space, 209, 213–14, 224, 369, 372
monoclinic Materials, 51, 96–97
monopoles, 127, 214, 284, 307
MQN. *See* magnetic quantum number
multi-mode acoustic metamaterial architecture, 227, 350

N

Navier's equation of motion, 68, 70–71
non-Hermitian, 204, 277, 279–80, 310, 350
nontrivial state, 326, 329, 331–32, 362, 364, 366, 376, 383

O

OAM (orbital angular momentum), 160–64, 166–68, 171–73, 175, 326–28, 331, 388–89
OAM, internal, 161–62
observer, 21, 143–44, 319
operators, xv, 146–47, 155–59, 164–69, 171, 173–74, 176–78, 184, 221–22, 310, 316
orbital angular momentum. *See* OAM
orthogonal directions, 69, 76, 80, 86, 104, 107, 138, 224, 297, 302, 333, 337
orthotropic materials, 52, 91–93, 98, 115
out-of-plane polarity, 239

P

pair of Dirac cones, 278–79
parallel transport, 208, 223, 300, 306–7, 324, 338–39, 341, 343–44, 362

parameter space, 208, 210–16, 218, 281–82, 284, 314, 341, 344, 362, 383, 385
particle displacements, 75, 77, 79, 290
particle motions, 79–80, 83, 86–89
particles
    charged, 156, 224
    local, 40, 128
    neighboring, 33, 36
particle waves, 143
path in energy, 223
path lines, 24–25
periodic crystal, 9, 187, 207, 209, 212, 215, 355
periodic functions, 237
periodicity, 10, 180, 186, 188, 190–91, 229–31, 235, 239, 243, 333, 335
periodic metamaterials, xvii, 12, 63, 225, 230, 300, 306, 361
periodic system, xvii, 185–87, 218, 232, 275
    three-dimensional, 278
phase, dynamic, 204–5, 207, 210
phase difference, 74, 161, 163, 207
phase factor, 212, 222
phase velocities, 83–85, 88, 270–71
phononic Berry phase, 369–70
phononic crystals, 9–10, 125, 234, 241, 243, 248–51, 293, 298, 300, 349–50, 359, 363–64, 366
phononic topological states, 370, 372
phonons, 143, 173, 272, 281, 361, 369–72, 374, 388, 397–98
photonics, 142, 205, 222, 224, 281, 350, 355–56, 362, 374, 376
photons, 142–44, 163–64, 166, 173, 226, 272
plane
    reference, 89, 143–44
    rotational, 160, 318–20
plane of swing, 208
planes of symmetry, 51–52, 61, 102
plane waves, 65, 114, 181, 199, 239, 243, 295–96, 308, 364, 378–79, 384
    incident, 289, 385
    polarized, 234, 259
point, 21–23, 31, 34, 36, 45–46, 68–69, 74, 83–87, 127–28, 186–87, 191, 202–4, 217–20, 274–81, 288–93, 306–8, 339–41, 343–44, 364–66, 378
    degenerated, 292, 378
    extreme, 187
    unique, 203
pointing, 79, 155, 339
points in space, 21, 130
polarity, 88–100, 104–5, 107, 114, 137, 161, 165, 306–7, 330, 388, 390–94
    differential, 91, 98
    longitudinal, 105
    respective, 90–93, 95, 97–98
polarity components, 92, 94

polarization, 80, 83, 86, 88–89, 105–7, 139, 206–8, 307–8, 312–14, 326, 328, 388, 391
    circular, 161, 166, 362, 391
    prior, 133
polarization anomaly, 388, 393
polarization vectors, 87–88, 207–8, 328, 392
polarized waves, 89, 103–4, 108, 115–16, 133, 149, 161, 258–59
polarizers, 207
position vector, 21, 31, 40, 49, 73–74, 87, 146, 160, 164, 219–20, 339–40
potential energy, 17, 40–41, 43–44, 147, 149, 184, 187, 372
potential functions, 40–41, 43, 66–67, 87, 145, 178, 187, 312, 318
Poynting vector, 42–43, 331
Poynting vector and energy, 140
pressure, 63, 67, 128, 139–40, 381, 387
pressure fields, 63–64, 68, 70, 362
probability density function, 165
pseudospin, 281, 283, 318–19, 328, 356, 361, 364, 369–72, 374
pseudospin discs, 318–21
pseudospin states, 282, 318–19, 326, 364, 366, 375

Q

QAHE (quantum anomalous Hall effect), 225, 228, 356, 369, 371, 373–74, 394
qFS modes, 92, 94–95, 97–98, 107
QHE (quantum Hall effect), 117, 203, 206–7, 224–25, 228, 354–56, 359, 361, 366, 394–95, 397
QSHE (quantum spin Hall effect), 117, 224–25, 228, 281, 356, 358–59, 361, 364, 366, 368–69, 372–75, 394, 396
quantum, xv, xvii, 143–44, 225, 229, 307–8, 310, 314, 353, 356, 368, 370–71, 395–96
quantum anomalous Hall effect. *See* QAHE
quantum fields, 129
quantum Hall effect. *See* QHE
quantum mechanics, xv–xvii, 16, 117, 129, 142–43, 149–54, 156–61, 163–69, 204, 206–7, 220, 223–26, 281–82, 314, 318–19
quantum numbers, 160–62, 165, 172–76
    azimuthal, 167–68
quantum operators, 156–57
quantum spin Hall effect. *See* QSHE
quantum spin operators, 327, 329
quantum trio, 224, 369, 374, 376
quantum valley Hall effect. *See* QVHE
quantum wave equations, 155, 307
quantum wave function, 165, 231
quantum waves, 308, 315

Index

quantum waves and elastic waves, 315
QVHE (quantum valley Hall effect), 117, 224, 275–76, 325, 366–69, 374–75, 394, 397

R

real-time tunable metamaterial, 383
reciprocal space, 186, 202, 209, 211, 215, 231–33, 278
reduced wave vector, 258
reflection symmetry, 217–18, 220, 229
relativistic particles, 143–44, 195–96, 198–200, 314, 358
  massless, 145, 203
rotation, 161, 208, 218, 274, 294, 297–99, 307, 325, 366, 383, 389–90
rotational symmetry, 217–18

S

SAM (spin angular momentum), 160, 163, 165–66, 168, 172–73, 175–78, 325–29, 331, 361–62
Schrödinger equation, 143, 147, 155–56, 158, 178, 180–81, 183, 186–87, 189–91, 211, 215–16, 307–8, 310
  dependent, 147, 187
shear stresses, 33–34, 63, 65, 70–71, 105, 244
shear waves, 70, 72, 75, 77, 80, 91, 139
SH wave, 80–81, 89
SIS. *See* space inversion symmetry
Skyrmion Number, 332, 394
slowness profile EFC, 93–94, 96–97, 99
slowness surfaces, 83–86, 89, 102
sound topology, 227, 349
space, free, 130–36
space-inversion symmetry, 223, 375
space inversion symmetry (SIS), 213–14, 218, 220, 223, 274, 276, 284, 300, 323, 362
spatial coordinate system, 21–22, 24, 31
spawning, 204, 279–80
spectral energy density (SED), 305
spherical coordinate system, 59–61, 89, 130, 164, 167, 339
spin, 89, 160, 163, 165–66, 173, 176, 198–200, 208, 281, 314, 325, 347–48, 356, 358–59, 361, 370, 388–91, 395–96
  hybrid, 352
  pseudo, 277
spin angular momentum. *See* SAM
spin angular momentum distribution, 362–64
spin eigenstates, 166, 176, 283, 314
spin-momentum locking, 328, 374
spin-orbit coupling, 330, 356, 358
spinors, 166, 173, 176–78, 199–201, 226, 281, 283
  additional, 200

spinor states, 178, 200, 283
spin states, 12, 165, 176, 200–201, 318–19, 321, 326, 328–31, 366, 376
  hidden, 318
  showing split, 319–20
spin unit vectors, 166, 175, 283
spin-up, 202, 329, 358
spin velocity, 390
spring-mass system, 318, 321
square angular momentum, 171, 173, 176
square momentum operator, 196
square phononic crystals, 292–93, 297, 335
states
  electronic, 369–70, 374
  phononic, 371–73
  plane-stress, 114–15
  spin-0, 318
  stress, 33–35
  topological phononic, 369–72
state variables, 17, 20
strain-displacement relation, 30, 55, 58, 60, 236, 239
strain energy, 38, 41–43
strain matrix, 50, 119, 121, 139
strains, 20, 25, 27, 41, 44, 48–50, 54, 68–69, 86, 104–5, 128
stress and strains, 20, 49
stresses, 20, 22, 27, 33–34, 36, 47–51, 53–54, 68, 86, 113, 115, 128, 244–45
  normal, 33–34, 63, 70–71, 235–36
stress matrix, 50, 110, 122, 139
summation of kinetic energy, 40–41, 147
surface, 33, 35–37, 39, 42–43, 83–86, 130–31, 134, 140–42, 206, 208, 338–43, 353, 387, 391, 396
  spherical, 140–41, 340–43
surface electrons, 357–58
surface traction, 35–36, 41
SUSY, 308, 310–11, 314
SUSY ladder, 311–14
SV wave, 80–81
S-wave, 315, 345–46
symmetric operation, 218, 324, 354
symmetry, 50–53, 103, 203–4, 209, 214–15, 217–25, 229–31, 304–7, 314–15, 323–24, 348–49, 356–59, 361, 366, 370–75
  breaking, 223, 373
  chiral, 219, 223, 324
  helical, 217
  mirror, 51, 303–4
  parity, 219–20, 223
  particle-hole, 305–6
  scale, 217
  translational, 217
symmetry elements, 219–20
symmetry points, 366, 375
  high, 214, 358, 375

Index

system, 16–21, 32–33, 127, 147–49, 157–59, 182–87, 190, 210–13, 220, 222–23, 231, 238–39, 279, 323–24, 353, 371, 385
  electronic, 212, 355–56
  non-Hermitian, 279–80
  phononic, 355–56, 369–70

T

TAM (total angular momentum), 30–31, 160, 166–67, 173, 175, 325–26, 328
tangent, 25, 85, 90, 142, 271
tangential, 224, 330, 339
TBH (topological black hole), 12, 332, 375–80, 382–91, 394
TBH frequency, 378–84, 386, 388–89, 392
TBH state, 383, 385–86, 388
terms
  diagonal, 100–101, 241
  second-order perturbation, 283, 364
time-dependent Schrödinger equation, 154, 156
time-independent Schrödinger equation, 149, 188–89
time-reversal symmetry, 214, 220, 222–23, 225, 300, 305, 356, 358–59, 361, 366, 373, 375, 394
top bands, 274, 276–77, 282–83, 293–94, 382–87, 392–93
topological, 12, 109, 300, 304, 330–32, 353, 369–71, 374, 376, 395–96
topological bandgap, 366, 369, 375
topological band inversion, 372
topological behaviors, xv–xvi, 117, 125, 325, 328, 330–32, 355, 361, 371, 376, 383
  unique, 279, 281
  unique spin-mediated, 324
topological black hole. *See* TBH
topological blackhole, 376–77, 379
topological charge, 161–63, 272, 277, 279, 284, 380, 390, 392
topological Chern number, 330, 332
topological conductor, 377, 379
topological edge states, 350, 364, 375
topological effects, 82, 301, 366, 397
topological elastic waves, 321
topological features, 300, 380, 394
topological index, 332, 353
topological in nature, 330–31
topological insulators, xvii, 225, 354, 356–58, 372, 375, 377, 394–97
topological invariants, 205, 217–18, 224–25, 272, 279, 284, 354–56, 369–70
topological materials, 366
topological mode inversions, 282–83, 366
topological phases, 73, 117, 207–8, 227, 341, 362
topological phase transition, 362–63, 396

topological phenomena, xv, 117, 225, 300, 304, 348, 376, 380
topological phononic crystals, 375, 396
topological phononics, 369–70, 372, 397
topological physics, xvi, 376
topological properties, 205, 355, 371, 379
topological quantum computation, 369, 397
topological quantum states, 369
topological space, 209, 217, 220
topological states, 223, 328, 332, 361, 366, 369
topological wave behavior, 83, 95, 98, 109
topological waves, xvii, 68, 83, 89, 230, 275, 307
topology, 73, 204–9, 217, 220, 223, 261–62, 307, 324, 353–55, 358–59, 369
  non-conventional, 218, 222–23
  trivial, 366
topology and geometric phase, 223
topology optimization, 269, 349
total angular momentum. *See* TAM
transformation, 56, 206, 217, 309, 311, 323
transverse wave modes, 105, 308, 319–21, 388
trivial, 82, 362–63, 365–66, 371
T-waves, 327, 329
two-dimensional systems, 186, 228, 395

U

ultrasonic metamaterials, 13
ultrasonic waves, 13, 68, 70, 285
uncertainty principle, 166–69, 173–74, 220
unit cell, 9, 179, 229–31, 259, 262–64, 269, 276, 278, 285–86, 292–93, 332–34, 338, 360–61
unit vectors, 24, 59, 83–84, 88, 98, 137, 193, 259, 340

V

valley Hall phononic crystals (VHPCs), 375
vector, 20–24, 26, 70–71, 74–75, 77–78, 83–85, 88, 127–28, 183–84, 208, 211, 216–17, 240–41, 252–53, 306–8, 318–20, 322–23, 338–39, 341–44
  contravariant, 23, 47
  dotted, 339
  normal, 84, 130, 340–42
  parallelly transported, 339, 344
  reciprocal lattice, 180, 182, 188, 237–38, 241, 247
  rotating, 318–21
  tangent, 85, 342
vector field, 58, 64, 69–72, 127–28, 130, 140–42, 156, 208–9, 338, 341, 343–44
vector field identities, 71, 135
vector pointing, 20, 66
vector wave, 70

Index

velocity field, 25, 66, 107–8, 140, 165, 326, 361–62, 390
velocity profile, phase wave, 90–91
velocity profile EFC, 93–94, 96–97, 99
velocity surfaces, 84–85, 89
VHPCs (valley Hall phononic crystals), 375
vibration, 157, 205, 398–99

W

wave dispersion, 83, 209, 223, 225, 229, 249, 270–71
wave energy, 18, 83–85, 88, 108, 144, 296, 379, 387, 391
wave energy propagation, 83–84
wave equation for relativistic particle, 196
wave equations, 38, 43–44, 72, 109, 113, 115–17, 135, 142, 147, 155, 196, 198, 315
    elastic, 138, 157, 308, 315–16
    polarized, 114
wave field, 48, 73, 75–76, 84, 136, 139, 161, 163, 232, 287, 391
wave fronts, 66, 82–84, 161
wave function, 145–47, 152, 154–55, 157–58, 165–66, 170, 179–84, 187–88, 191–92, 195, 199, 207–12, 221–22, 224–25, 306–7, 324
    spatial, 169–70
    total, 193
wave function in periodic media, 182–83
wavelengths, 9–10, 143–44, 161, 163, 231, 234, 271, 291–92
wave modes, 85–86, 88–93, 95, 97–98, 104, 107, 238–39, 270–73, 275, 308, 312, 328–29, 388
wave number domain, 89, 91–99, 183, 185, 211, 231, 303, 328, 344
wave number planes, 86, 91
wave numbers, 65–66, 74–75, 86–87, 105–7, 137–38, 145–46, 151, 180, 182–85, 188, 190–91, 203–4, 209, 211–12, 231–32, 236, 238–39, 255–56, 270–72, 305–7
    characteristic, 182
    function of, 187, 212, 238
    normalized, 252–53
    possible, 185, 231
    shear, 72–73, 81
wave number space, 224, 280, 304–5, 307
wave number sphere, 330

wave number value, 107, 183, 190, 200, 238–39, 274
wave number vector, 189, 212
wave particle duality, 144–46, 155–56, 169, 199, 225
wave polarities, 83, 89–92, 95, 97, 100, 125
wave polarization, 83, 86, 89–90, 109, 345
wave potentials, 80–81, 113, 139, 155, 157, 311, 325, 346
wave propagation, 14–19, 38–39, 61, 63–70, 75, 77–85, 88–89, 93–94, 96–99, 104–5, 125, 137–38, 161–62, 233, 287, 307–8, 319–21, 324, 349, 388–90
wave propagation angle, 89–90
wave propagation direction, 77, 79–80, 83, 88, 93–97, 99, 104–5, 137, 288, 290, 292
wave propagation equations, 63–64, 233
waves, xv–xvii, 10, 65–67, 72–89, 91, 93, 103–4, 106, 116–17, 125, 143–45, 150, 154–55, 204–5, 207–9, 226, 229–351, 361, 378–80, 389–90
    1P-1S-Hybrid, 346–48
    acoustic pressure, 8
    incident, 9, 84, 105, 287
    in-plane, 309
    monochromatic, 86, 237, 327
    plane harmonic, 72
    refracted, 207, 291
    solid magnetic field, 139
    transmitted, 289–90
    transverse, 77, 104, 303, 352, 388
wave vector, 72, 74–75, 77, 79–80, 83–89, 212–13, 216, 252–53, 271–72, 275, 277–78, 388, 392–93
wave vector directions, 79, 88, 161, 233, 274, 278, 392–93
    possible, 100–101
wave vector space, 212–13, 284, 344
wave velocities, 63, 82, 88–89, 92, 98, 137–39, 143, 146, 149, 271, 331
wave velocity surfaces, 84–85, 89
wave vortex, 139, 161, 163, 330, 390
Weyl points, 275, 278–80, 370

X

XM directions, 286

Z

Zak Phase, 215, 217, 300